Biological Process Engineering

BIOLOGICAL PROCESS ENGINEERING

An Analogical Approach to Fluid Flow, Heat Transfer, and Mass Transfer Applied to Biological Systems

Arthur T. Johnson, Ph.D., P.E.

University of Maryland

A Wiley-Interscience Publication

JOHN WILEY & SONS, INC.

New York / Chichester / Weinheim / Brisbane / Singapore / Toronto

Copyright © 1999 by John Wiley & Sons, Inc. All rights reserved.

Published simultaneously in Canada.

Library of Congress Cataloging-in-Publication Data

Johnson, Arthur T. (Arthur Thomas)
    Biological process engineering: an analogical approach to fluid
  flow, heat transfer, and mass transfer applied to biological systems / Arthur T. Johnson.
      p. cm.
    "A Wiley-Interscience publication."
    Includes bibliographical references and index.
    ISBN 0-471-24547-X (cloth : alk. paper)
    1. Biochemical engineering.   I. Title.
TP248.3.J64 1998
  660.6'3–dc21   98-3744

This book is dedicated
To all the little children in my life
Who made me smile
Even on the grayest of days.

CONTENTS

PREFACE

"... criticism of scientific research into dinosaurs left me shaking my head in disbelief. He can't understand of what earthly benefit dinosaur hunting is. One might as well ask of what benefit music is. Or poetry. Poor fellow. He's forgotten that dinosaurs are the stuff dreams are made of. Creatures of immense size and power, they are alien to everything in our experience. They rescue us from the humdrum existence of school and the playground. *Tyranosaurus rex* would eat the local bully and whisk us away on his mighty back. Superman and Tinker Bell might rescue us also, but dinosaurs are different because they are really and truly real! The amazing fact that dinosaurs could actually have walked the earth, albeit zillions of years ago, was what introduced a lot of boys and girls to science. If it weren't for dinosaurs, many scientists might have become bankers, street cleaners, or shoe salesmen. Each new dinosaur discovered puts a twinkle in the eye of some future Pasteur or Einstein, or for that matter, a Steven Spielberg. That is a treasure beyond price."

Eric Lurio

Bob Cooke (Cornell University, Ithaca, NY) calls it "analogic thinking," the ability to transfer knowledge from a familiar applications area to another totally new. The field of transport processes has always been based on elements of analogic thinking because the traditional transport processes of fluid flow, heat transfer, and mass transfer have in common many of the fundamental governing equations. Nevertheless, there is a wide range of methods employed for presentation of transport processes, with this being an additional one.

Walter W. Frey, writing in the December 1990 issue of the *IEEE Spectrum*, compared left-brain- with right-brain-dominant individuals. Left-brain-dominant individuals think by forming textual or logical concepts, while right-brain-dominant individuals specialize in visual imagery and artistic/spatial/

intuitive approaches. He states that creativity, innovation, and flexibility are considered an engineer's greatest assets, but that these are largely right-brain functions. Schooling, he asserts, discriminates against right-brain functions in favor of left-brain functions.

Perhaps you can see what I am getting to. Analogic thinking should be done best by right-brain-dominant individuals, but transport processes are often taught in a very abstract, mathematically oriented manner. Thus the people who should best be able to understand transport process applications are the people who must struggle to learn them in the abstract.

I have made it a goal to try to present transport processes as much from a concrete viewpoint as I can. In addition, I try to make use of graphs and figures to make the material understandable. Certainly, equations and computer programs would present the material much more compactly. But the real breakthroughs in transport processes are those that come from imagination and understanding, not computation. Thus my method of approach has been defined to foster analogic thinking.

This approach, I suppose, follows from the educational philosophy instilled in me by my agricultural engineering background. That is, I believe there is a place for engineers trained in a general enough fashion that they can work between fields, acting as bridges between other disciplines and bringing a cross-fertilization to disparate applications. Therefore, I believe in a general, liberal, undergraduate engineering education, and specialization, if it occurs at all, should be postponed until after the bachelor's degree. This philosophy the reader will see reflected in this material: Examples and problems are drawn from a wide range of biological systems applications. It is important, I think, that engineering students have their horizons defined broadly, so that they can see the breadth of choice that lies ahead of them, before they are forced to follow inevitably narrower career paths.

The objective of this text, then, is to introduce biological engineering students to the concepts of transport processes. Should they wish to know more about the subject, students can move on later to the more theoretical mathematical approach used by other authors. An additional objective is to prepare biological engineering students to design products and processes using, or to be used with, biological systems. Thus I have included tables of parameter values, design charts, and empirical equations that allow a biological engineer to make a very close first approximation to the final design. Each biological example in the text is intended to introduce the reader to a specialized vocabulary as well as to give an appreciation for some of the specialized biological knowledge required before application of engineering methods. Although not usually a prerequisite for a transport processes book, some previous courses or, at least, prior knowledge of biological sciences will definitely help here.

Finally, when I came in 1975 to the University of Maryland and was assigned to teach the courses entitled "Biological Process Engineering" and "Engineering Dynamics of Biological Systems," Andy Cowan handed me a large number of folders containing the course material and wished me good luck. At that time my familiarity with transport processes was sketchy, but my knowledge of instrumentation systems and circuit design were strong, and my interest in biological systems was keen. Thus, as I have developed this material over the years, I have approached it from a different direction than usual. And I think it is more understandable as a result.

Arthur T. Johnson, Ph.D., P.E.
University of Maryland

References

Frey, W. W., 1990, Schools Miss Out on Dyslexic Engineers, *IEEE Spectrum* **27**(12): 6 (December).

Hook, S., 1988, The Closing of the American Mind: An Intellectual Best-Seller Revisited, *American Scholar* **58**: 123–135.

Johnson, A. T., and W. M. Phillips, 1995, Philosophical Foundations of Biological Engineering, *J. Engr. Educ.* **84**: 311–318.

ACKNOWLEDGMENTS

My thanks are extended for the hard work of many helpful individuals. At the top of the list is Thelma de Cheubel for her long efforts in typing and retyping the many drafts of this text. She cannot be thanked enough. Next, Lovant Hicks produced the many figures you see inside. Some of these figures almost gave Lovant a permanent squint, but he persevered and did a fine job. My students are to be thanked for using this text in its unfinished form and for helping to find places that needed improvement. Finally, there is no one person who knows enough about the details of all the biological systems and instruments appearing as examples in this text. So, I had to call on my friends. Following is a list of people who helped by giving me specialized information that they knew best. I thank you all.

Lou Albright, Agricultural and Biological Engineering, Cornell University
Joe Andrade, Bioengineering, University of Utah
Andy Baldwin, Biological Resources Engineering, University of Maryland
Erik Biermann, Biological Resources Engineering, University of Maryland
Paul Bottino, Plant Biology, University of Maryland
Kim Brown, Biological Resources Engineering, University of Maryland
Lew Carr, Biological Resources Engineering, University of Maryland
Karen Coyne, Biological Resources Engineering, University of Maryland
Ashim Datta, Agricultural and Biological Engineering, Cornell University
Jerry Deitzer, Horticulture, University of Maryland
Keith Dionne, Alza Corporation, Palo Alto, CA
Kimberly Edwards, Biological Resources Engineering, University of
    Maryland
Gary Felton, Biological Resources Engineering, University of Maryland
Tim Foutz, Biological and Agricultural Engineering, University of Georgia
Gene Giacomelli, Bioresources Engineering, Rutgers University
Paul Heineman, Biological and Agricultural Engineering, Penn State
    University
Dennis Heldman, Biological Engineering, University of Missouri

Peter Hillman, Agricultural and Biological Engineering, Cornell University

Ben Hurley, Kinesiology, University of Maryland

Jeffrey Janik, Water Resources Management, State of California, Sacramento

Sammy Joseph, Microbiology, University of Maryland

Pat Kangas, Biological Resources Engineering, University of Maryland

John Kleckler, Advanced Biosciences Laboratories, Rockville, MD

Bob Langer, Chemical Engineering, Massachusetts Institute of Technology

Chris Lausted, Biological Resources Engineering, University of Maryland

Charles Maulé, Agricultural and Bioresource Engineering, University of
Saskatchewan

Ian McFarlane, EA Engineering, Hunt Valley, MD

John McMillan, Biological Resources Engineering, University of Maryland

Irv Miller, College of Engineering, University of Akron

Skip Pierce, Zoology, University of Maryland

Marius Pruessner, Chemical Engineering, University of Maryland

D. Raj Raman, Agricultural and Biosystems Engineering, University of
Tennessee, Knoxville

Jack Reynolds, Private Practice Dentistry, Edgewood, Maryland

Marc Rogers, Kinesiology, University of Maryland

Dave Ross, Biological Resources Engineering, University of Maryland

John Sager, National Aeronautics and Space Administration, Kennedy Space
Center, FL

Manjit Sahota, Biological Resources Engineering, University of Maryland

Joseph Sampugna, Biochemistry, University of Maryland

Don Schlimme, Food Science and Nutrition, University of Maryland

Paul Schreuders, Biological Resources Engineering, University of Maryland

Neal Sereboff, Agricultural Engineering, University of Maryland

John Sheppard, Biosystems Engineering, McGill University

Adel Shirmohammadi, Biological Resources Engineering, University of
Maryland

Simon Smith, Racal Filter Technology, Brockville, Ontario, Canada

Talbot Smith, Biological Resources Engineering, University of Maryland

Theo Solomos, Horticulture, University of Maryland

Bernie Tao, Agricultural and Biological Engineering, Purdue University

Gene Taylor, Modern Mushroom Farms, Avondale, PA

Fred Wheaton, Biological Resources Engineering, University of Maryland

Tom Wickham, GenVec, Inc., Rockville, MD

Biological Process Engineering

1

SYSTEMS CONCEPTS FOR TRANSPORT PROCESSES

"A model can be likened to a caricature. A caricature picks on certain features (a nose, a smile, or a lock of hair) and concentrates on them at the expense of other features. A good caricature is one where these special features have been chosen purposefully and effectively. In the same way a model concentrates on certain features of the real world."

A. M. Starfield, K. A. Smith, and A. L. Bleloch

1.1 INTRODUCTION

The term *transport processes* is usually applied to the study of heat, mass, and momentum transfer (momentum transfer in flowing fluids). These three systems are linked because of analogies between them. Indeed, the list can be expanded to include other analogous entities such as electricity and mechanics.

heat transfer, mass transfer, and fluid flow are similar

What commonalties are there between them? Each has two classes of variables, and the relationships between them convey the same information no matter what the system. Understanding of these general relationships will allow comprehension of one system to be transferred to other systems. Thus, if you conceptualize best in fluid flow systems, all the rest of the systems can be conceptualized in terms of pumps, pipes, valves, and tanks. If you think best in terms of electrical systems, all the rest of the systems can be conceptualized in terms of batteries, wires, resistors, capacitors, and inductors. And so on.

basic units

Therefore, we begin by establishing the types of variables and the relationships between them. For the discussion to follow, a consistent set of metric units has been used (see Table 1.1.1). These are newtons, meters,

1

seconds, degrees Celsius, and coulombs, which reflect the basic dimensions of force, length, time, temperature, and electrical charge. There are, of course, other dimensions [for example, mass—expressible by Newton's second law $(F = ma)$ in terms of force, length, and time] and units (for example, volts $= \mathrm{N\,m/coulomb}$ and amperes $=$ coulomb/sec), but for consistency we stick with just those listed in Table 1.1.1.

sizes of units   How big are these units? Because the units chosen here are not part of the everyday heritage, a few words will be spent describing these units in familiar terms. First of all, consider newtons: a small-sized apple weighs about 1 N (a

Table 1.1.1   Some Conversion Factors

1 acre $= 4047\,\mathrm{m}^2$
1 atmosphere $= 1.01325 \times 10^5\,\mathrm{N/m}^2$
1 BTU $= 1055.1\,\mathrm{N\,m}$
1 BTU/h $= 0.29307\,\mathrm{N\,m/sec}$
1 BTU/(h ft °F) $= 1.7307\,\mathrm{N\,m/(m\,sec\,°C)}$
1 BTU/(h ft^2 °F) $= 5.6783\,\mathrm{N\,m/(m^2\,sec\,°C)}$
1 BTU/(lbm °F) $= 4186.8\,\mathrm{N\,m/(kg\,°C)}$
1 centimeter of water $= 98.062\,\mathrm{N/m}^2$
1 centipoise $= 10^{-3}\,\mathrm{N\,sec/m}^2$
1 clo $= 6.45\,\mathrm{m\,°C\,sec/N}$
1 cubic foot $= 0.028317\,\mathrm{m}^3$
1 darcy $= 10^{-12}\,\mathrm{m}^2$
1 dyne $= 1 \times 10^{-5}\,\mathrm{N}$
°F $= \frac{9}{5}$°C $+ 32$
1 foot $= 0.3048\,\mathrm{m}$
1 ft lb $= 1.3558\,\mathrm{N\,m}$
1 foot of water $= 2988.9\,\mathrm{N/m}^2$
1 gravitational acceleration $= 9.8066\,\mathrm{m/sec}^2$
1 hectare $= 10^4\,\mathrm{m}^2$
1 horsepower $= 745.70\,\mathrm{N\,m/sec}$
1 inch $= 0.02540\,\mathrm{m}$
1 inch of water $= 249.08\,\mathrm{N/m}^2$
1 joule $= 1\,\mathrm{N\,m}$
K $= 273.15 +$ °C
1 kilocalorie $= 4184.0\,\mathrm{N\,m}$
1 kilopond m/min $= 0.1635\,\mathrm{N\,m/sec}$
1 lbm $= 0.45359\,\mathrm{kg}$
1 liter $= 0.001\,\mathrm{m}^3$
1 mile/hour $= 0.4470\,\mathrm{m/sec}$
1 mm Hg $= 133.32\,\mathrm{N/m}^2$
1 newton $= 1\,\mathrm{kg\,m/sec}^2$
1 pascal $= 1\,\mathrm{N/m}^2$
1 psi $= 6894.7\,\mathrm{N/m}^2$
1 U.S. gallon $= 0.0037854\,\mathrm{m}^3$
1 volt $= 1\,\mathrm{N\,m/coulcomb}$
1 watt $= 1\,\mathrm{N\,m/sec}$

medium-sized apple weights about 1.5 N), a small (5 lb) bag of sugar weighs about 22 N, and an average adult male weighs about 680 N. Next, consider meters: a single pace (the distance between two consecutive times the same foot lands when walking) is about $1\frac{1}{2}$ m, an average female is about 1.7 m tall, a doorway is about $2\frac{1}{2}$ m high, and it would take about $2\frac{1}{2}$ min to run 1000 m. Consider degrees Celsius: 0°C is the temperature at which water freezes, 20°C is the temperature where most of us feel comfortable, 37°C is our body temperature, and 100°C is where water boils. Last, consider coulombs: The charge on an electron is 1.6×10^{-19} coul, 1 L of gaseous hydrogen ions would have a charge of about 5000 coulomb, and one coulomb would be enough charge to stick about 25,000,000,000 inflated balloons to the ceiling.

1.2 EFFORT VARIABLES

effort =
cause

Effort varaibles are those variables that cause action to happen. An example of an effort variable is temperature: A temperature difference causes heat to flow. Another example is voltage. Voltage causes an electrical current to flow. A list of common effort variables is found in Table 1.2.1.

Effort variables can be thought of as potential, or field, variables. A potential field defines the force on a susceptible particle to cause the particle to react, usually by moving.

1.3 FLOW VARIABLES

flow =
effect

These are the variables that move. Heat is the flow variable in thermal systems; current is the flow variable in electricity; and fluid flow is the flow variable in fluid systems. Thus each of these systems has common characteristics, although the terminology and symbology are different.

1.4 RELATIONSHIPS BETWEEN FLOW AND EFFORT VARIABLES

There are several important relationships between these two variables.

1.4.1 Power

power equals
effort times
flow

First of all, the product of effort and flow variables is usually defined as power:

$$P = ei \qquad (1.4.1)$$

where P is power (N m/sec); e, voltage (N m/coulomb or volts); and i, current (coulombs/sec or amperes). Power is, of course, the time rate of energy

Table 1.2.1 Analogous Transport Systems

System	Effort Variable	Flow Variable	Power	Resistance	Capacity	Inertia
Electricity	Voltage (e)	Current (i)	ei	e/i	$\dfrac{1}{e}\int i\,dt$	$e/(di/dt)$
Fluid	Pressure (p)	Fluid flow ($\dot{V}$)	$p\dot{V}$	$p/\dot{V}$	$\dfrac{1}{p}\int \dot{V}\,dt$	$p/(d\dot{V}/dt)$
Heat	Temperature (θ)	Heat flow ($\dot{q}$)	$\dot{q}$	$\theta/\dot{q}$	$\dfrac{1}{\theta}\int \dot{q}\,dt$	—
Mass	Concentration (c)	Mass flow ($\dot{m}$)	—	$c/\dot{m}$	$\dfrac{1}{c}\int \dot{m}\,dt$	$c/(d\dot{m}/dt)$
Mechanics	Force (F)	Velocity ($\dot{x}$)	$F\dot{x}$	$F/\dot{x}$	$\dfrac{1}{F}\int \dot{x}\,dt$	$F/(d\dot{x}/dt)$
Optics	Photon pressure (ϕ)	Photon flow (υ)	—	ϕ/υ	$\dfrac{1}{\phi}\int \upsilon\,dt$	—
Magnetics	Magnetomotive force (mmf)	Flux Φ	—	mmf/Φ	$\dfrac{1}{\mathrm{mmf}}\int \Phi\,dt$	—

production or use. Thus electrical power can be consumed as long as neither current nor voltage is zero.

Power has dimensions of force times length divided by time:

$$P \triangleq FL/T \tag{1.4.2}$$

where F is dimensional force; L, dimensional length; T, dimensional time; and $\triangleq$ denotes dimensional equivalence. F, L, and T can assume any consistent set of units.

Because of the dimensions associated with P, the analogies between systems are not quite complete. For instance, in fluid flow systems the product of pressure and volume rate of flow, $p\dot{V}$, equals power:

$$p\dot{V} \triangleq \left(\frac{F}{L^2}\right)\left(\frac{L^3}{T}\right) = \frac{FL}{T} \tag{1.4.3}$$

where p is pressure, in force/(length squared); and $\dot{V}$ is the volume rate of flow, in volume/time, but in thermal systems, the rate of heat flow ($\dot{q}$) by itself has units of power. Thus heat flow times temperature is not power:

$$\dot{q}\theta \triangleq \left(\frac{FL}{T}\right)(°) \neq P \tag{1.4.4}$$

where $\dot{q}$ is the rate of flow of heat (energy/time); θ, the temperature (degrees); and °, degrees. Nor is the product of effort and flow variables equivalent to power in mass transfer.

1.4.2 Resistance

resistance equals effort divided by flow

A basic relationship between effort and flow variables is given in electrical terms by Ohm's law:

$$R = \frac{e}{i} \tag{1.4.5}$$

where R is resistance (N m sec/coul^2 or ohms); e, the voltage difference from one end of the resistor to the other end (N m/coul or volts); and i, the current through the resistor (coul/sec or amps). Resistance impedes the flow variable. It dissipates power (usually as heat) and thus removes power from the system.

resistance dissipates power

Thermal resistance is, by analogy, the ratio of temperature to heat flow ($\theta/\dot{q}$, °C sec/N m) fluid resistance is the ratio of pressure to fluid volume flow rate ($p/\dot{V}$, N sec/m^5), and mass resistance is the ratio of concentration to mass flow rate ($c/\dot{m}$, sec/m^3). In each of these cases resistance can be thought of as limiting the flow variable for any given effort variable.

EXAMPLE 1.4.2-1. Effort Variable across a Resistance

Determine the pressure difference across a resistance of $35\,\text{N}\,\text{sec}/\text{m}^5$ if the flow is given by $(70 \sin 25t)\,\text{m}^3/\text{sec}$.

Solution

We know that $R = p/\dot{V}$; so

$$p = R\dot{V}$$

$$= \left(\frac{35\,\text{N}\,\text{sec}}{\text{m}^5}\right)(70 \sin 25t)\frac{\text{m}^3}{\text{sec}}$$

$$= (2450 \sin 25t)\,\text{N}/\text{m}^2$$

Remark

Pressure and flow are both given by sine waves and are said to be "in phase."

1.4.3 Capacity

In electrical systems, capacitance is given as:

$$C = \frac{Q}{e} = \frac{1}{e}\int_0^t i\, dt \qquad (1.4.6)$$

where C is the capacitance (coulomb^2/N m); Q, the electrical charge (coulombs); e, the voltage difference from one end of the capacitor to the other end (N m/coul); i, the current (coul/sec); and t, the time (sec). Capacitance is an element that stores energy. A voltage is required to store
capacity stores flow
electrical charge on a capacitor, but when the charge is stored, a voltage is available from the capacitor to drive current through an electrical circuit. The capacitor may serve as both an electrical source or electrical sink depending on capacitor voltage in relation to other circuit voltages.

It is conceivable that the effort variable and flow variable may both undergo simultaneous changes. In this case, both voltage and current could change, as with alternating current (ac). Capacity is defined here, however, while the effort variable remains constant and the flow variable accumulates. This is not to say that once the capacity value has been determined that effort and flow cannot vary simultaneously. They can. But for definition purposes, effort is assumed to remain constant.

Thermal capacitance is given by:

$$C_{\text{th}} = \frac{1}{\theta} \int_0^t \dot{q}\, dt \tag{1.4.7}$$

where C_{th} is the thermal capacitance ($\text{N m}/{}^{\circ}\text{C}$).

Thermal capacitance expresses the amount of heat that can be stored per degree of temperature. This heat is available (by the First Law of Thermodynamics) to perform work. Thus thermal capacitance can serve either as a thermal source or sink depending on the thermal capacitance temperature with respect to the surrounding temperatures.

Fluid capacitance, also called compliance, is given by:

$$C_{\text{f}} = \frac{1}{p} \int_0^t \dot{V}\, dt \tag{1.4.8}$$

where C_{f} is the fluid compliance (m^5/N). Compliance can be visualized as a balloon. Blow up the balloon, and a volume of air is stored at some pressure. Release the balloon, and the escaping air causes the balloon to fly around the room. Stored energy is translated into motion energy. Thus the compliance stores energy (is blown up) as long as the source pressure is greater than that inside the compliance; when source pressure is reduced, the compliance can deliver energy.

energy can be stored in capacity for later use

Mass capacitance is given by:

$$C_{\text{m}} = \frac{1}{c} \int_0^t \dot{m}\, dt \tag{1.4.9}$$

where C_{m} is the mass capacitance (m^3).

There can be a nonzero capacity term only if the substance (mass, fluid, heat, or electrical charge) can accumulate within a fixed volume due to a change in effort variable (concentration, pressure, temperature, or voltage). To illustrate this point, consider, as examples, water in two closed pots and in a balloon (Figure 1.4.1). To the first closed pot, we add heat, and the temperature of the water rises. Since heat accumulated as temperature changed, the first pot of water has a thermal capacitance. To the second closed pot, we add pressure. Since the walls of the pot are rigid and the water is practically incompressible, there is no additional water that will fit inside the pot as pressure is increased. The second pot of water has no fluid capacitance (or compliance). To the balloon we find that as water is added the pressure inside the balloon increases. The walls of the balloon, being elastic instead of rigid, give the balloon system fluid capacitance. Therefore, capacitance depends greatly on the physical properties of the substance and the boundaries of the problem.

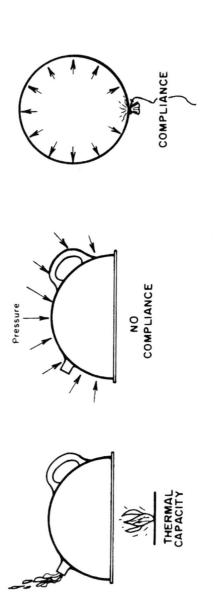

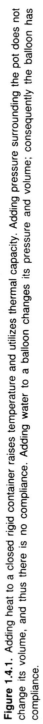

Figure 1.4.1. Adding heat to a closed rigid container raises temperature and utilizes thermal capacity. Adding pressure surrounding the pot does not change its volume, and thus there is no compliance. Adding water to a balloon changes its pressure and volume; consequently the balloon has compliance.

Incidentally, when the water in the pot reaches boiling temperature and additional heat causes the formation of steam, but no additional temperature rise, the incremental (or additional) thermal capacity of the pot of water becomes zero.

EXAMPLE 1.4.3-1. Effort Variable across a Capacity

Determine the temperature difference across a thermal capacity of $60 \, \mathrm{N\,m/^\circ C}$ if the flow is given by $(20 \sin 25t) \, \mathrm{N\,m/sec}$.

Solution

From Eq. 1.4.7, we know that $C_{th} = \frac{1}{\theta} \int_0^t \dot{q} \, dt$; so

$$\theta = \frac{1}{C_{th}} \int_0^t \dot{q} \, dt$$

$$\theta = \left(\frac{1}{60} \frac{^\circ C}{\mathrm{N\,m}}\right)\left(\frac{\mathrm{N\,m}}{\mathrm{sec}}\right)(\mathrm{sec}) \int_0^t (20 \sin 25t) \, dt$$

The units of seconds comes from the variable time in the integral.

$$= \left(\frac{1}{60}\right)\left(\frac{1}{25}\right)(-20 \cos 25t)_0^t (^\circ C)$$

$$= \frac{1}{75}(1 - \cos 25t)^\circ C$$

Remark

Temperature varies as a negative cosine, and flow varies as a positive sine wave. Thus, there is a $-\pi/2$ radian phase shift between effort and flow variables, with flow leading effort.

1.4.4 Inertia

Electrical inertia is usually called inductance, and is defined by

$$L = e \Big/ \left(\frac{di}{dt}\right) \tag{1.4.10}$$

inertia
maintains flow

where L is the inductance ($\mathrm{N\,m\,sec^2/coul^2}$). Inductance stores energy, as does capacitance, but inductance stores kinetic, not potential, energy. That is, inductance stores energy as current, and tends to maintain that current even when the voltage drops to zero. Thus an inductor can be an energy source or sink depending on the sign of (di/dt).

Likewise, thermal inertia could be defined as

$$I_{th} = \theta \Big/ \left(\frac{d\dot{q}}{dt}\right) \tag{1.4.11}$$

where I_{th} is the thermal inertia ($^\circ$C sec^2/N m), and would resist changes in the rate of heat flow in a thermal system. Thermal inertia, however, is only present when there is associated mass movement, as in convection processes.

Fluid inertia, also called inertance, is

$$I_f = p \Big/ \left(\frac{d\dot{V}}{dt}\right) \tag{1.4.12}$$

where I_f is the inertance (N sec^2/m^5).

Mass inertia is

$$I_m = c \Big/ \left(\frac{d\dot{m}}{dt}\right) \tag{1.4.13}$$

where I_m is the mass inertia (sec^2/m^3).

Notice that although the dimensions and units of each of these inertias vary, they all express a certain relationship between effort and flow variables, completely homologously between them. Each gives the rate of change of flow variable for a given amount of effort variable.

not all transport processes have inertia

As hinted above, inertia is not always an appropriate parameter for various transport processes. In mechanical systems, inertia has meaning because mass has inertia: A mass at rest tends to remain at rest, and a mass in motion tends to remain in motion (Newton's first principle). Inertia is appropriate for fluid systems because flowing fluid has mass, and mass has inertia. In electrical systems, the magnetic field surrounding flow of current in a conductor resists changes in current flow, and an equivalent of inertia is thus present. In heat transfer, inertia is not present unless an associated flow of mass is also present. The phrase in common use to describe hot buildings after the evening air has cooled following a hot day is "thermal inertia." However, this really is not a sign of thermal inertia, but thermal capacity. The flow of heat is not maintained if the temperature difference is changed, and so inertia is not present. Likewise, unless viewed on a very microscopic scale, diffusion mass transfer inertia does not exist. Only convective mass transfer can be said to possess inertia.

EXAMPLE 1.4.4-1. Effort Variable across Inertia

Determine the voltage difference across an inductance of 2.1×10^{-3} N m sec^2/coul^2 if the current is given by $(70 \sin 25t)$ coul/sec.

Solution

From Eq. 1.4.10, we know that

$$L = e/(di/dt), \text{ so}$$
$$e = L(di/dt)$$
$$= (2.1 \times 10^{-3} \text{N m sec}^2/\text{coul}^2)(1750 \cos 25t \text{ coul/sec}^2)$$

The units of coul/sec^2 are the units of (di/dt).

$$e = 3.675 \cos 25t \text{ N m/coul}$$

Remark

Thus there is a $+\pi/2$ radian phase shift between effort and flow variables, with effort leading flow.

1.4.5 Nonlinearities

If biological materials are characterized by anything, it is their nonconstant, nonideal behavior. Thus the elements of resistance, capacity, and, perhaps, inertia introduced in prior sections can only be considered to be constant in biological systems over relatively small regions.

It is hard to imagine why inertia should be different in biological systems from that in nonbiological systems. Inertia is present whenever a mass requires acceleration, and biological mass is the same as physical mass.

biological nonlinearities

For other elements, however, biological systems may act much differently from nonbiological systems. Steel, for instance, exhibits a linear stress–strain relationship over a large range of axial stress and obeys Hooke's law, but most biological materials tend to become stiffer as they stretch (Figure 1.4.2). This type of behavior is found in the pressure–volume behavior of the lung and chest wall, the pressure–volume behavior of the blood capillaries, and the stress–strain behavior of tendons, ligaments, and muscles, to name a few.

For modeling purposes, the curve in Figure 1.4.2 can be mathematically described as:

$$1/\text{Def} = a + e^{b(c-F)} \tag{1.4.14}$$

where Def is the deformation (appropriate units); F, the force (N); and a, b, c are constants (appropriate units), and the curve can be linearized by subtracting a from both sides and then taking logarithms. The pressure–volume curve for the lung has been found to fit the equation (Johnson, 1984):

$$\text{VC/V} = 1.00 + e^{(1.0-1.8\times10^{-3}\text{p})} \tag{1.4.15}$$

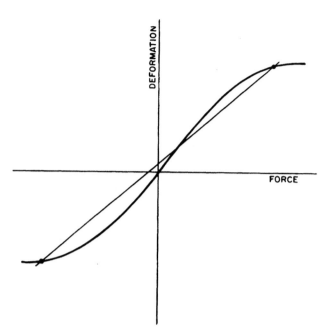

Figure 1.4.2. A nonlinear force–deformation relationship characteristic of biological systems. To deal with this relationship, a chord is often drawn between two pertinent points. The chord may not always pass through the origin.

where V is the lung volume (m^3); VC, the lung vital capacity, (m^3); and p, the pressure (N/m^2).

From its definition, the slope of the line in Figure 1.4.2 is equal to capacity. Thus capacity varies from some maximum value near the origin to zero at the extremes. Which value should be used? Sometimes the value near zero is of most interest, and that slope is used. Sometimes, when the material is at one extreme or the other, a local value close to zero should be used. Most often, a chord is drawn between two points that define the limits of interest on the curve, and the slope of the chord becomes the capacity value used.

There is no reason to expect the curve to be symmetrical about the origin. Indeed, real values often are not. Nor is there any reason to expect that the curve would be exactly retraced each time the material is loaded or unloaded, thus leading to hysteresis (Figure 1.4.3).

hysteresis   The presence of hysteresis indicates that energy used to deform the material is not completely recovered when the material returns to its undeformed shape. Arrows in Figure 1.4.3 indicate the directions of loading and unloading, and the area enclosed within the loop is the work ($\int F \, dx$) not recovered during unloading. Even the position of the curves depends somewhat on the starting point for the process, thus confounding the situations even further. Fortunately, the degree of hysteresis is often relatively small, and can be ignored. If not,

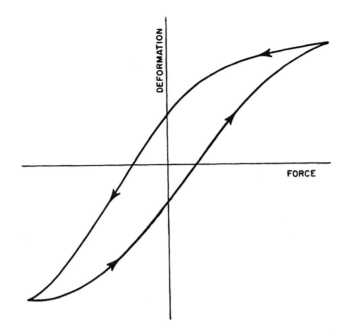

Figure 1.4.3. Hysteresis is often present in biological materials. Engineers often ignore differences between loading and unloading curves.

then hysteresis presents a severe challenge to analysis and design involving these materials.

nonconstant resistance

Resistances, too, are often nonideal when dealing with biological systems. The presence of valves in the circulatory system (heart and veins) causes resistance to be finite in one directional of flow and infinite in the other direction. Different flow regimes and distensibilities of airways cause the flow resistance or the respiratory airways to be nonconstant (Johnson, 1984):

$$R_{aw} = K_1 + K_2 \dot{V} + \frac{K_3}{V} \qquad (1.4.16)$$

where R_{aw} is the respiratory airways resistance ($N \sec/m^5$); K_1, K_2, K_3 are modified Rohrer coefficients ($N \sec/m^5$, $N \sec^2/m^8$, $N \sec/m^2$, respectively); V, the lung volume (m^3); and $\dot{V}$, the volume flow rate ($m^3/\sec$).

Because of active control, resistance values can change. Vascular resistance to blood flow is determined by the firing rates of baroreceptors and by local carbon dioxide levels, among other things. Even more amazing effects are present in cell membranes. Because sodium and potassium are pumped against their natural concentration gradients in animal neurons and muscle cells (it appears that calcium and chloride are pumped in vegetative cells), apparent resistances to mass movement in these cells become negative. Hence, mass flow and concentration gradients have opposite signs. The only means by which this situation can be maintained is through the expenditure of energy, but it is maintained.

Living organisms can perform otherwise highly improbable feats by expending energy to accomplish otherwise irreversible processes. A reversible process is one that meets two criteria:

1. The process may be caused to occur in exactly reverse order and the system and all surrounding systems return to their exact initial states.

2. All energy used during the process is returned upon reversal in form, location, and amount.

Nonreversible processes are those that do not meet these two criteria. The difference between the two can be illustrated by means of the following story:

"We have a skunk in the basement," shouted the caller to the police dispatcher. "How can we get it out?"

"Take some bread crumbs," said the dispatcher, "and put down a trail from the basement out to the backyard. Then leave the cellar door open."

Later the resident called back. "Did you get rid of it?" asked the dispatcher.

"No," replied the caller. "Now I have two skunks in there!"

The dispatcher expected an irreversible process, but the caller got a reversible process instead. But, that's life!

Engineers designing processes dealing with or applied to biological systems can use the concepts of resistance, capacity, and inertia to help conceptualize their designs and guide their decisions. They must realize, however, that working with biological systems is a bit more involved than working with inert materials.

EXAMPLE 1.4.5-1. Capacity in Nonlinear System

Using Eq. 1.4.15 expressing the relationship between pressure and volume, determine lung compliances at lung volumes of 50% and 85% of lung vital capacity. Compare these to the lung compliance obtained from the slope of a line connecting the two lung volumes. Lung vital capacity for an average healthy male is $4.8 \times 10^{-3} \, m^3$.

Solution

Compliance is given as the slope of the line relating pressure to lung volume. Compliances at the two points of 50% and 85% of vital capacity can be calculated from:

$$C_0 = \frac{dV}{dp}$$

From Eq. 1.4.15,

$$V = VC[1.00 + \exp(1.0 - 1.8 \times 10^{-3}p)]^{-1}$$

$$\frac{dV}{dp} = C_0 = VC[1.00 + \exp(1.0 - 1.8 \times 10^{-3}p)]^{-2}$$

$$\times (-1)(1.8 \times 10^{-3}) \exp(1.0 - 1.8 \times 10^{-3}p)$$

Inserting the values for vital capacity, and multiplying:

$$C_0 = \frac{(8.64 \times 10^{-6}) \exp(1.0 - 1.8 \times 10^{-3}p)}{[1.00 + \exp(1.0 - 1.8 \times 10^{-3}p)]^2}$$

We now must calculate pressure at the point $V/VC = 0.50$ using Eq. 1.4.15:

$$\frac{V}{VC} = 0.50 = [1.00 + \exp(1.0 - 1.8 \times 10^{-3}p)]^{-1}$$

$$p = \frac{1 - \ln(1/0.50 - 1)}{1.8 \times 10^{-3}} = 556 \frac{N}{m^2}$$

The pressure at $V/VC = 0.85 = 1519 \, N/m^2$.

Inserting these values for pressure into the equation for C_0 above, we obtain $C_{0.50} = 2.16 \times 10^{-6} \, m^5/N$ and $C_{0.85} = 1.10 \times 10^{-6} \, m^5/N$. Comparing these two values, we see that the slope of the line becomes flatter as lung volume increases, as illustrated by the upper right quadrant of the curve in Figure 1.4.2.

The second part of the example requires determination of the slope of the line connecting $V/VC = 0.50$ and $V/VC = 0.85$.

$$C = \frac{\Delta V}{\Delta p} = \frac{(4.8 \times 10^{-3})(0.85 - 0.50)}{(1519 - 556) \, N/m^2}$$

$$= 1.74 \times 10^{-6} \, m^5/N$$

Remark

This value is between the values of $C_{0.50}$ and $C_{0.85}$ determined earlier. It is not the same as the average of $C_{0.50}$ and $C_{0.85}$.

EXAMPLE 1.4.5-2. Resistance in a Nonlinear System

Determine the respiratory airways resistance at flows of $0 \, m^3/sec$ and $3 \times 10^{-3} \, m^3/sec$ at a lung volume of $2.4 \times 10^{-3} \, m^3$. Values for Rohrer's coefficients can be taken as

$$K_1 = 7.29 \times 10^4 \text{ N sec/m}^5$$
$$K_2 = 4.2 \times 10^7 \text{ N sec}^2/\text{m}^8$$
$$K_3 = 66.6 \text{ N sec/m}^2$$

Solution

From Eq. 1.4.16,

$$R_{aw} = K_1 + K_2\dot{V} + \frac{K_3}{V}$$

At a flow of $0 \text{ m}^3/\text{sec}$,

$$R_{aw} = 7.29 \times 10^4 \frac{\text{N sec}}{\text{m}^5} + \frac{66.6 \text{ N sec/m}^2}{2.4 \times 10^{-3} \text{ m}^3}$$
$$= 1.01 \times 10^5 \text{ N sec/m}^5$$

At a flow of $3 \times 10^{-3} \text{ m}^3/\text{sec}$,

$$R_{aw} = 7.29 \times 10^4 \frac{\text{N sec}}{\text{m}^5} + 4.2 \times 10^7 \frac{\text{N sec}^2}{\text{m}^8} 3 \times 10^{-3} \frac{\text{m}^3}{\text{sec}}$$
$$+ \frac{66.6 \text{ N sec/m}^2}{2.4 \times 10^{-3} \text{ m}^3}$$
$$= 2.27 \times 10^5 \text{ N sec/m}^5$$

1.4.6 Biological Variation

Measurements made on samples of biological materials, organisms, or systems usually do not exactly coincide. There can even be a great deal of variability in the measurements, not only from sample to sample, but also from the same sample over time. Most of the measurements of effort, flow, resistance, capacity, or inertia will likely be clustered around some average value, but some can be quite different. If measurements are made on a large number of samples, we could group the measurements into magnitudes that fall into certain ranges, and then plot the number of measurements falling within the group against the average magnitude within the group range. The result would be a plot similar to Figure 1.4.4.

As the number of measurements increases, the magnitude ranges of the groups can decrease and the number of groups can increase. In the limit, as the number of measurements tends to infinity, the data in Figure 1.4.4 would

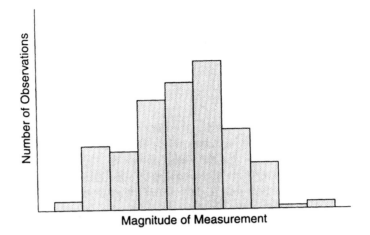

Figure 1.4.4. If a large number of measurements is grouped into ranges, and the number of measurements for each group is plotted against measurement magnitude representing those groups, the result would be a plot that shows central tendency.

normal
distribution
become more like the curves in Figure 1.4.5. This is the Gaussian, or normal, distribution of data.

This curve is called a *density function*, which means that the area under the curve must always equal 1.0

$$\int_{-\infty}^{\infty} f(x) \, dx = 1 \tag{1.4.17}$$

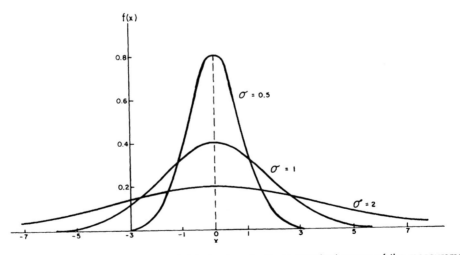

Figure 1.4.5. The normal distribution of data is given by these standard curves of the measurement probability against normalized magnitude.

where $f(x)$ is the ordinate value for the curve (dimensionless); and x is the abscissa value for the curve, with units of the measurement.

The equation for the curve is:

$$f(x) = \frac{1}{\sigma\sqrt{2\pi}}e^{-(x-\mu)^2/2\sigma^2} \tag{1.4.18}$$

where μ is the average, or mean, of the data (units of the measurements); and σ, the standard deviation of the data (units of the measurement).

distribution spread

The reason for the three curves in Figure 1.4.5 is to show that some measurements are spread out more than others; they are dispersed more; they do not cluster very tightly around the average. The parameter that expresses this spread is the standard deviation. The greater the standard deviation, the smaller is the peak due to the fact that the area underneath the curve must equal 1.0.

These curves can also shift left or right depending on the mean of the data. Shifting the curve does not change its shape or the area under the curve.

The probability of the occurrence of all possible measurements is 1.0, which is equal to the area under the curve. The probability of the occurrence of a measurement within a certain range x_1 to x_2 is less than 1.0, and is equal to the area under the curve from x_1 to x_2 (Figure 1.4.6).

$$P(x_2 < x < x_2) = \int_{x_1}^{x_2} f(x)\, dx \tag{1.4.19}$$

If the mean of x values is zero, and $x = \pm\sigma$, then $P(-\sigma < x < \sigma) = 0.683$, or 68.3% of the values of the measurement are expected to be within a range of plus or minus one standard deviation from the mean. Also, 95.4% of the values are in the range of plus or minus two standard deviations and 99.7% of the values are within the range of plus or minus three standard deviations from

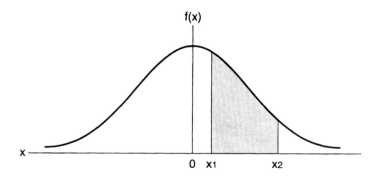

Figure 1.4.6. The probability of a measurement falling in the range $x_1 < x < x_2$ is just the area beneath the curve between x_1 and x_2.

the mean. Thus the portion of the curve three standard deviations below the mean to three standard deviations above the mean encompass nearly the entire area beneath the curve.

Enough measurements of parameters of biological systems are not usually made to define a normal distribution completely, as in Figure 1.4.5. However, when multiple measurements are made, they are often expressed by two numbers: the mean and the standard deviation. The latter is related to the variance except that it is calculated for relatively small numbers of measurements. The mean is calculated from

mean, or average

$$\bar{x} = \sum x_i / \sum i = \sum x_i / N \qquad (1.4.20)$$

where $\bar{x}$ is the mean of the measurements (units of the measurement); x_i, each individual measurement (units of the measurement); i, the counter for each measurement (dimensionless); and N, the total number of measurements (dimensionless).

standard deviation

The standard deviation is estimated from

$$\sigma = \left[\left(\sum (x_i - \bar{x})^2 \right) / (N - 1) \right]^{1/2}$$
$$= \left\{ \left[\sum x_i^2 - \left(\sum x_i \right)^2 / N \right] / (N - 1) \right\}^{1/2} \qquad (1.4.21)$$

where σ is the standard deviation (units of the measurement).

other data distributions and measurements

Before going further, it must be noted that not all data from biological systems are distributed with a normal distribution. Nor is the mean always the best measure of central tendency: One may use the mode (the measurement with the greatest frequency of occurrence), or the median (the measurement with 50% of the observations greater and 50% of the observations less). The mean, median, and mode are all the same value for the normal distribution. Other measures of dispersion that may be used are ranges and percentiles (percentile expresses the percentage of measurements that are less than the measurement in question). There are cases where these are appropriate.

addition of absolute errors

Measurements of biological parameters are often further manipulated to yield useful results. For example, measurements of airways resistance, lung tissue resistance, and chest wall resistance may be added to form total respiratory system resistance. Whenever parameters that have values with some degrees of uncertainty are combined, the overall uncertainty of the result is larger than any of the uncertainties of the original components. When two elements are added or subtracted, for example, the values of the uncertainties add together:

$$(A \pm \Delta A) + (B \pm \Delta B) = (A + B) \pm (\Delta A + \Delta B) \qquad (1.4.22)$$

where A and B are the original parameters, and ΔA and ΔB are the uncertainties in A and B. Thus

$$\Delta(A + B) = \Delta A + \Delta B \qquad (1.4.23)$$

addition of
relative errors When two elements are multiplied, we find using the chain rule:

$$\Delta(AB) = A\,\Delta B + B\,\Delta A \qquad (1.4.24a)$$

$$\frac{\Delta(AB)}{AB} = \frac{\Delta B}{B} + \frac{\Delta A}{A} \qquad (1.4.24b)$$

Thus, for multiplication or division of the two elements, the overall relative (or percent) error is the sum of the component relative (or percent) errors. If we consider $\Delta(AB^2)$, then the overall relative error is $\Delta A/A + 2\,\Delta B/B$. More complex manipulations result in more complicated expressions of uncertainty.

For the biological engineer using measurement data from biological systems to formulate a design, biological variation must be considered. In most of the mathematical manipulations in this and similar texts, it is often not clear that variability exists. One reason for this is that simple and practical means to incorporate uncertainties in parameter values into mathematical expressions do not exist. The biological engineer must be aware that average values are often used, but these may or may not be sufficient to formulate a workable design. If it is critical that a design be effective for nearly the entire population rather than just the average of the population, then parameter values applicable to a 98th or 99th percentile must be used in place of the average (50th percentile).

conservative
engineering
choices Conservative engineering choices should also be made. Choosing a pump size somewhat larger than calculated or thermal insulation thickness somewhat more than calculated are two such choices. There is always a penalty to be paid for such choices (somewhat higher initial or operating costs, for instance); so the choices cannot be made without a basis in reality, but payment of some penalty is often preferable to an ineffective design. First, and foremost, the biological engineering design must serve its intended purpose.

very precise
calculations
unnecessary If there is an inevitable variation in parameter values characterizing biological systems, then ultraprecise calculations are unwarranted. Many digits to the right of the decimal point available on your calculator or your computer does not ascribe to them any meaning. There is definitely a limitation to the precision of any calculation if the value of an essential parameter could vary by ±50%.

The technical study of chaos shows that the endpoint of a process subject to an uncertain starting condition is often not predictable (Glass and Mackey, 1988). Indeed, even the intermediate states for the process may be so variable that the specific mathematical means or equations used to predict results may not matter. This is not to imply that all the equations and concepts in this text could be discarded without consequence. However, the biological engineer should always keep in mind that the very nature of biological systems, including variability and nonlinearity, adds a special challenge to producing workable analyses and designs.

EXAMPLE 1.4.6-1. Lethal Doses of Paraquat

Data for the lethal doses for 50% mortality of test rats (LD50) injected with the herbicide Paraquat are given below. What are the means and standard deviations of the data?

LD50	(mg/kg)
	236
	240
	235
	231
	227
	238
	236
	234
	239
	241

Solution

We calculate from the data:

$$\sum x_i = 2357$$

$$\sum x_i^2 = 555709$$

Using Eq. 1.4.20 to find the mean:

$$\bar{x} = 2357/10 = 235.7 \text{ mg/kg}$$

Using Eq. 1.4.21 to find the standard deviation:

$$\sigma = \{[555709 - (2357)^2/10]/9\}^{1/2} = 4.27 \text{ mg/kg}$$

EXAMPLE 1.4.6-2. Growth of Lettuce

Growth of head lettuce in a greenhouse is being monitored by measuring the diameters ·of the heads with a caliper and ruler. If the uncertainty of the diameter measurement is $\pm 10\%$, what are the uncertainties of the volumes of the heads?

Solution

The volume of a sphere is $\pi d^3 / 6$. Thus, using differential calculus,

$$\Delta V = \left(\frac{\pi}{6}\right)(3d^2)\, \Delta d$$

$$\frac{\Delta V}{V} = \frac{\left(\frac{\pi}{6}\right)(3d^2)\, \Delta d}{\left(\frac{\pi}{6}\right)(d^3)} = 3\,\frac{\Delta d}{d}$$

The uncertainties in the volumes of the heads of lettuce are therefore (3) (10%) = $\pm 30\%$. This shows that measurement errors can easily multiply to unacceptable values.

1.5 SOURCES

ideal effort sources Ideal effort variable sources can deliver as much of the corresponding flow variable as required by the external environment (Figure 1.5.1). That means that these sources have no internal resistance, possess an infinite amount of capacity, and have no inertia. For example, an ideal voltage source (battery) of 12 V maintains its voltage no matter what current is drawn. Conversely, it can deliver huge amounts of current if the resistance connected to it is low enough. This is what happens to your automobile battery when it drives the starter. Typical starters have 0.05 ohm resistance, drawing peak currents of about 240 amps from their batteries. Ideal pressure sources are not as easily found as ideal voltage sources. In order to maintain steady pressure despite varying amounts of fluid flow, a large capacity is required.

examples of effort sources Ideal temperature sources are usually made from change-of-state substances. A block of ice, for instance, maintains its temperature at 0°C as long as there is sufficient ice present to absorb all heat delivered to it. Eutectic mixtures can be used as similar constant-temperature sources.

Certain hygroscopic materials are used to maintain specific relative humidities within closed spaces, and thus act as concentration sources. Various sulfuric acid concentrations are used for this purpose. When humidity falls, the sulfuric acid releases water to the air; when humidity rises, the sulfuric acid absorbs water. A sulfuric acid solution with a density of 1200 kg/m^3 (28% H_2SO_4) maintains a relative humidity of 80.5% (water vapor pressure of 1866 N/m^2 at a temperature of 20°C). Other solutions can be used to maintain other concentrations.

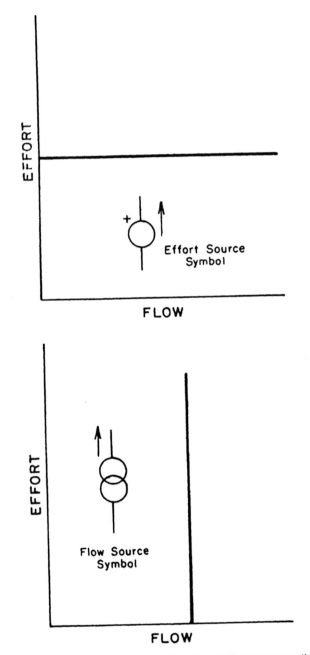

Figure 1.5.1. Diagrams of ideal sources, with an ideal effort source on the top and an ideal flow source on the bottom. The amount of flow is unlimited, and the effort is constant for the effort source. The effort is unlimited, but the flow is constant for the flow source.

Because magnitudes of the flow variable corresponding to the effort variable must sometimes change rapidly in response to external demands, inertia association with the ideal effort source must be nil. If this were not the case, the source would not maintain constant effort.

ideal flow sources

Ideal flow variable sources maintain flow at the expense of variable effort (Figure 1.5.1). They can thus produce as much effort as required to deliver a constant flow. That means that they possess no capacitance, have infinite resistance, and have infinite inertia.

Ideal electrical current sources are difficult to find. Some ionic currents across neural membranes can act as ideal current sources.

Ideal fluid flow sources are easier to find. Any positive displacement pump (such as a piston pump or gear pump) acts as a flow source. Because a piston pump must deliver a certain volume of liquid for each stroke of the piston, pressures high enough to cause pump damage can be produced if pipe resistance increases sufficiently.

Heat flow sources usually take the form of fuel burning at a constant rate. This heat can raise temperatures to extremely high levels if enough thermal resistance (insulation) is present.

examples of flow sources

Mass flow sources occur in various forms. The solar wind is a mass flow source, and so are the various ionic pumps in cellular membranes. Sufficient resistance to flow of these materials can raise concentrations to very high levels.

Ideal flow sources must be able to deliver rapid changes in their corresponding effort variables. Thus capacity must be nil because capacity resists effort variable change. Inertia must be high because inertia resists flow variable change. Resistance must be high because the flow must be maintained at the same level regardless of external resistance changes.

nonideal sources

Nonideal sources do not maintain constant effort variable values. Instead of the diagrams appearing in Figure 1.5.1, effort and flow variables for real sources appear as in Figure 1.5.2. An embodiment of these sources is diagrammed in Figure 1.5.3. Connecting either source in Figure 1.5.3 to a short circuit to ground results in a flow of $\dot{\omega}_0$. Leaving either source open circuited results in an effort of ϕ_0 (as long as both resistors R are the same value and as long as $\phi_0 = R\dot{\omega}_0$). There is thus no measurement of effort and flow at the output terminals that distinguishes between the two real sources. (There are, however, measurements not based on effort and flow at the output terminals that can separate one from the other.) Because they both have the same effort and flow characteristics, the two are said to be "duals" of one another.

Thevenin and Norton equivalents

The source on the left, with an ideal effort source in series with a resistor R, is called a Thevenin equivalent source. The rightmost source, with an ideal flow source in parallel with a resistor R, is called a Norton equivalent source. Both are equivalent representations of the same nonideal source.

If, instead of a short circuit or open circuit connected to the output terminals, a resistor is connected between the output terminal and ground,

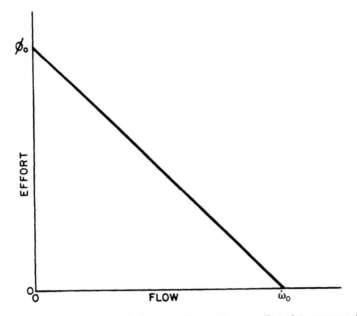

Figure 1.5.2. Diagram of a nonideal source. At zero flow, an effort of ϕ_0 appears. At zero effort, a flow of $\dot{\omega}_0$ happens. The locus of effort and flow values from the source is found on the straight line connecting ϕ_0 to $\dot{\omega}_0$. The slope of the line $(-\phi/\dot{\omega})$ has units of resistance and is the negative of the resistance of the source.

there would result flow and effort values that lie somewhere on the line connecting ϕ_0 and $\dot{\omega}_0$ in the diagram of Figure 1.5.2. Thus the line connecting ϕ_0 and $\dot{\omega}_0$ is the locus of effort and flow values obtained as the resistor connected to the output terminals is varied from zero to infinity.

The slope of the line has the value $\Delta\phi/\Delta\dot{\omega} = -\phi/\dot{\omega}_0$, which can be seen to be the negative of the value of resistor R.

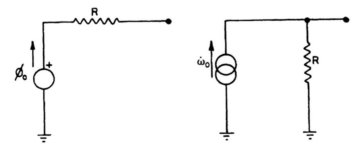

Figure 1.5.3. Two nonideal sources with identical characteristics. As long as the two resistors have the same value, and if $\phi_0 = \dot{\omega}_0 R$, then both sources will have an open-circuit voltage of ϕ_0 and a short-circuit current of $\dot{\omega}_0$. The left-hand source is based on an effort source, and the right-hand source is based on a flow source. The two sources are called "duals" of each other.

The distinguishing feature of an ideal source is not that it maintains either a constant flow or constant effort, because the source can be variable, but that its value is unaffected by external features. Thus an ideal flow source is such because the flow rate is totally determined by the source and not resistances, capacities, or inertia elements to which it is connected. The flow rate may vary, but not because of external elements. Likewise, an ideal effort source maintains its effort level no matter what it is connected to. The ac power outlet in the wall is a good example of an ideal effort source that varies in magnitude but is almost immune to resistance levels of devices plugged into it.

Nonideal effort and flow sources have less than ideal values of capacity, resistance, and inertia. Thus they do not maintain flow or effort constant in the face of varying external challenges. Nevertheless, the notion of ideal sources facilitates transfer of knowledge developed for one set of transport processes to another set.

1.6 COMBINATION OF ELEMENTS

The various elements involving effort and flow variables can be combined in ways to result in overall solutions to problems. There are some combinations that will result, not in solutions, but in causing additional problems.

series and parallel combinations The two combinations possible when connecting elements together are series and parallel (Figure 1.6.1). In a series connection, one terminal of one element is connected to one terminal of the other element. The other terminal of each element is left free or attached to other elements. When two elements are connected in series, the effort variables of both add together, and the flow through one element is the same as the flow through the other element.

In a parallel connection, both terminals of both elements are connected together. Connection to other elements is made with the combined terminals at both ends. When two elements are connected in parallel, the flow variables of both add together in the combination, and the effort across one element is equal to the effort across the other element.

Elements are connected in combinations that reflect reality. Representational elements are connected in series when the objects they represent are connected serially; representational elements are connected in parallel when the objects they represent are parallel to one another.

1.6.1 Sources

parallel effort sources cause trouble Effort variable sources must be combined in series, never in parallel. The reason for this is that effort sources in parallel attempt to impose their levels of effort on each other. Remember that effort sources expend whatever flow is necessary to maintain their levels of effort. The conflict between sources of different effort levels in parallel can thus lead to huge flow levels and explosive situations. Imagine connecting 6 V and 12 V automobile storage

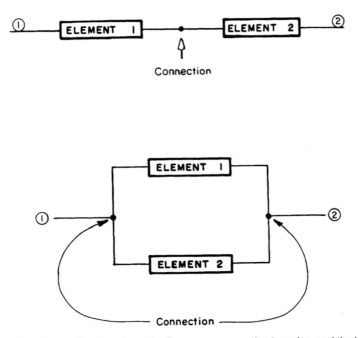

Figure 1.6.1. Connecting two elements: The upper connection is series, and the bottom connection is parallel.

batteries in parallel. The 6 V battery attempts to bring the 12 V battery to 6 V, and the 12 V battery attempts to bolster the 6 V battery. Since neither battery has much internal resistance, immense currents flow. Other effort sources connected in parallel may not perform as dramatically as the voltage sources just cited, but the effect is similar.

Effort sources connected in series produce effort as the sum of the effort levels of the sources. Thus a 12 V automobile storage battery is really six 2 V cells in series. A two-stage centrifugal pump is really two separate pumps in series to generate higher pressure than either pump could generate singly. In both of these examples the total flow of the combination proceeds through both sources.

series flow sources cause trouble

Flow sources should always be connected in parallel and never in series. Flow sources in series battle for supremacy in the same manner as effort sources in parallel. Each flow source in the series combination attempts to determine the level of flow through both of them, with disastrously high levels of effort produced as a result. Thus two positive displacement liquid pumps in series can lead to disaster, especially if the larger-capacity pump feeds into the smaller capacity pump.

Flow sources in parallel work as a team to produce a level of flow equal to the sum of the two flows. Thus a duplex piston pump can produce a more even flow than a simplex piston pump because one pump section supplies flow while the other is in the suction phase. Two lasers trained on a target can together penetrate a target that neither alone could supply enough energy to penetrate. Two parallel ion beams can supply more material than either separately can supply.

1.6.2 Resistances

Resistances in *series* add their values to result in the final value

$$R_{tot} = R_1 + R_2 + \cdots \tag{1.6.1}$$

and *parallel* resistances add inversely

$$1/R_{tot} = 1/R_1 + 1/R_2 + \cdots \tag{1.6.2}$$

where R_{tot} is the total resistance (N sec/m^5); and R_i, the component resistance (N sec/m^5).

For two resistances in parallel, total resistance is the product divided by the sum of the component resistances:

$$R_{tot} = (R_1 R_2)/(R_1 + R_2) \tag{1.6.3}$$

resistances
combined

Resistances in series will sum to a value larger than any of the component resistances. Resistances in parallel will combine to form a value smaller than any of the components. Thus series resistance combinations require higher effort values to maintain the flow variable at the levels that exist for each component taken singly. Parallel resistance combinations produce higher flow levels for the same effort values.

thermal
analog

A brick-faced concrete wall with a window in it (Figure 1.6.2) can be considered to have three thermal resistances: One resistance represents the concrete portion of the wall, another resistance represents the brick facing, and a third resistance represents the window. The concrete resistance and brick resistance are placed in series, because the same heat that flows through the concrete also flows through the brick. The interface between the brick and concrete has a temperature intermediate between the inside and outside temperatures.

The window resistance is placed in parallel with the other two resistances because heat that passes through the window does not pass through either the concrete or brick. The inside temperature of the window is the same as the inside temperature of the concrete, and the outside temperature of the window is the same as the outside temperature of the brick. Thus these resistances are not connected conceptually capriciously, but are connected to reflect rational reality.

EXAMPLE 1.6.2-1. Thermal Resistance Combinations

Determine the overall thermal resistance of the wall depicted in Figure 1.6.2 if the wall is 5 × 5 m and the window covers 4.5 m^2 of the area. The resistance of

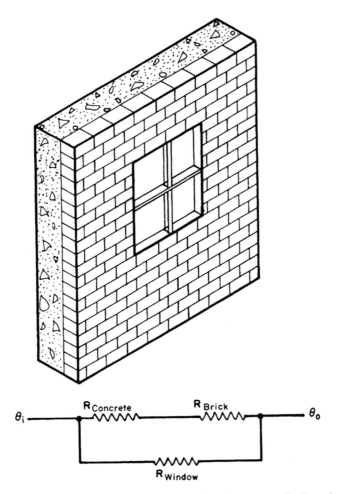

Figure 1.6.2. From a thermal standpoint, the brick-faced concrete wall with a window can be considered to be a series–parallel combination of three resistances.

the brick facing is $1.40 \times 10^{-2}\,^{\circ}\text{C}\,\text{sec}/(\text{N m})$, the resistance of the concrete is $2.54 \times 10^{-2}\,^{\circ}\text{C}\,\text{sec}/(\text{N m})$, and the resistance of the window is $9.98 \times 10^{-4}\,^{\circ}\text{C}\,\text{sec}/(\text{N m})$.

Solution

From the diagram in Figure 1.6.2, the solution is the series combination of R_{concrete} and R_{brick} in parallel with R_{window}.

$$R_s = R_{\text{concrete}} + R_{\text{brick}} = 1.40 \times 10^{-2} + 2.54 \times 10^{-2}$$

$$= 3.94 \times 10^{-2}\,^{\circ}\text{C}\,\text{sec}/(\text{N m})$$

Total wall resistance is

$$R_{tot} = \frac{R_s R_{window}}{R_s + R_{window}} = \frac{(3.94 \times 10^{-2})(9.98 \times 10^{-4})}{(3.94 \times 10^{-2} + 9.98 \times 10^{-4})}$$

$$= 9.73 \times 10^{-4} {}^{\circ}C \; sec/(N \; m)$$

Remark

Notice how much influence the window has in determining the overall thermal resistance of the wall. Most heat passing through the wall passes by way of the window.

EXAMPLE 1.6.2-2. Soil Penetration of Effluent

Soil beneath an earthen manure storage has three layers: (1) a manure seal at the top with a penetration resistance of $10^6 \, N \, sec/m^5$, (2) a mechanically compacted layer with a penetration resistance of $10^8 \, N \, sec/m^5$, and (3) a thick undisturbed subsoil layer with penetration resistance of $2 \times 10^{10} \, N \, sec/m^5$. What is the total resistance of the soil layer to effluent movement into the groundwater?

Solution

These layers lie one on top of the other, with the same flow through all of them. They thus appear in series.

$$R_{tot} = R_1 + R_2 + R_3$$

$$= 10^6 + 10^8 + 2 \times 10^{10} \approx 2 \times 10^{10} \, N \, sec/m^5$$

If the area of the storage were doubled, the additional resistance would appear in parallel with the original resistance (total flow being equal to the sum of the flows through both columns), and the total resistance would thus be:

$$R_{tot} = \frac{R_1 R_2}{R_1 + R_2} = \frac{(2 \times 10^{10})(2 \times 10^{10})}{2(2 \times 10^{10})} = 1 \times 10^{10} \, N \, sec/m^5$$

Remarks

The resistances just calculated are the ratios of pressure drop ($\rho g h$) to volume flow rate ($\dot{V}$) through the soil. Hydrologists normally speak of hydraulic conductivity (k in m/sec), the ratio of hydraulic gradient (in m/m) to flow velocity (m/sec). Hydraulic conductivity and flow resistance are related by $R = \rho g h / A k$, where ρg is the water specific weight $= 9.8 \, \text{N/m}^3$; h, the layer thickness (m); and A, the surface area perpendicular to flow (m^2).

1.6.3 Capacity

capacity
elements
combined

Capacity elements connected in *parallel* are added to yield the overall equivalent value:

$$C_{tot} = C_1 + C_2 + \cdots \tag{1.6.4}$$

and capacity elements in *series* add reciprocally:

$$1/C_{tot} = 1/C_1 + 1/C_2 + \cdots \tag{1.6.5}$$

Thus capacity elements in parallel result in a higher overall capacity, but capacity elements in series result in a lower overall capacity value.

balloon analog

The capacity (compliance) for two inflated balloons (Figure 1.6.3) is calculated as the ratio of volume to internal pressure for each. If both balloons are inflated to the same internal pressure, then the overall compliance is just the sum of the two volumes divided by the common internal pressure. This is equivalent to one balloon having a volume equal to the sum of the

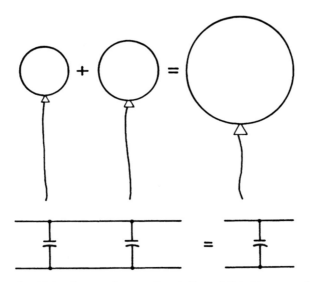

Figure 1.6.3. Combining the compliances of two balloons inflated to an equal pressure is similar to combining two electrical capacitors.

volumes of the other two balloons at the same pressure that inflates the other two.

EXAMPLE 1.6.3-1. Capacity Combinations

Determine the equivalent capacitance of three electrical capacitors in series. The individual capacitor values are 1.5×10^{-6}, 3.8×10^{-7}, and 5.7×10^{-6} coulomb^2/N m.

Solution

Capacitors in series add according to Eq. 1.6.5:

$$\frac{1}{C_{\text{tot}}} = \frac{1}{C_1} + \frac{1}{C_2} + \frac{1}{C_3}$$

$$= \frac{1}{1.5 \times 10^{-6}} + \frac{1}{3.8 \times 10^{-7}} + \frac{1}{5.7 \times 10^{-6}}$$

$$= 3.47 \times 10^6$$

$$C_{\text{tot}} = 2.88 \times 10^7 \text{ coulomb}^2/\text{N m}$$

Remark

Note that the total capacity of a series capacity combination, like the total resistance of a parallel resistance combination, is always smaller than the smallest individual contributor.

1.6.4 Inertia

inertia elements combined

Inertia elements in *series* add in the same manner as series resistances

$$I_{\text{tot}} = I_1 + I_2 + \cdots \tag{1.6.6}$$

and *parallel* inertia elements add reciprocally:

$$1/I_{\text{tot}} = 1/I_1 + 1/I_2 + 1/I_3$$

Sections of pipe can be considered to have inertia (inertance) in each one. Lengthening a pipe adds inertance by series combination, and makes the pipe more susceptible to water hammering, that loud clanking noise in pipes connected to the faucet (Figure 1.6.4).

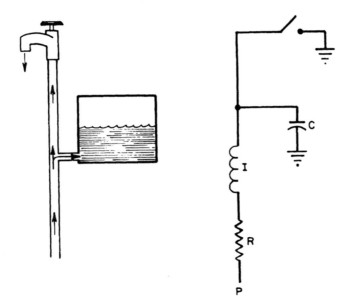

Figure 1.6.4. Water hammer is caused by inertia of the column of water in the pipe. One solution is to add compliance represented by the tank.

water hammer

Suddenly shutting a faucet can cause water hammering because the pressure produced by an inertance is directly related to the rate of change of flow. Shutting the faucet more slowly reduces the time rate of change of the flow and can decrease water hammer noises. Reducing the inertia of the flowing water column by decreasing either the length or diameter of the pipes can be used to reduce water hammering.

Water hammer is, incidentally, usually treated by adding a compliance element to the piping system. When the faucet is suddenly closed, the inertial flow of water pressurizes the pipe to a much smaller degree than it would have pressurized the pipe without the compliance. Hence, noise is reduced.

1.6.5 Combinations Involving Time

Combinations of different kinds of elements yield information about time responses of systems. The product of resistance and capacity has units of time, and is usually given the name time constant:

$$\tau = RC \tag{1.6.7}$$

where τ is the time constant (sec).

time constant

The time constant is a scaling parameter for systems with an exponential time response (Figure 1.6.5). The larger the value of the time constant, the

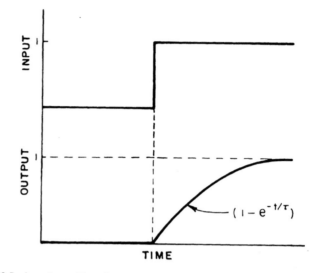

Figure 1.6.5. A system with resistance and capacity responds exponentially to a sudden change in input.

slower the system responds to a change. Very rapid time response is foretold by a very small time constant.

exponential time response   If a sudden change is made in the effort variable at one point in the system, then the effort variable at some other point responds exponentially as long as the system is dominated by resistance and capacity:

$$\phi = \phi_i(1 - e^{-t/\tau}) \qquad (1.6.8)$$

where ϕ is the generalized effort variable output; ϕ_i, the magnitude of the effort variable input; t, the time; τ, the time constant; and e, the base of natural logarithms.

When the time after the input change becomes equal to one time constant, then $(t/\tau) = 1$ and $\phi/\phi_i = (1 - e^{-1}) = 0.63$. When time equals three time constants, $\phi/\phi_i = (1 - e^{-3}) = 0.95$. The response is 99% complete once time becomes equal to 5 time constants.

Knowledge of the time constant is important to proper system design. Long time constant values translate into slowly responding systems, and these systems tend to maintain average values (Figure 1.6.6). These are sometimes called "low-pass filters," because they only respond to very slow changes in input values. Short time constant values are possessed by systems that respond rapidly. These systems respond to changes in input values without severely attenuating them.

pressure fluctuations dampened   In the case of a fluid flow system where the fluid is propelled by a piston pump, a pulsatile pressure wave due to intermittent pumping is propagated downstream. To smooth this wave so that pressure fluctuations effectively

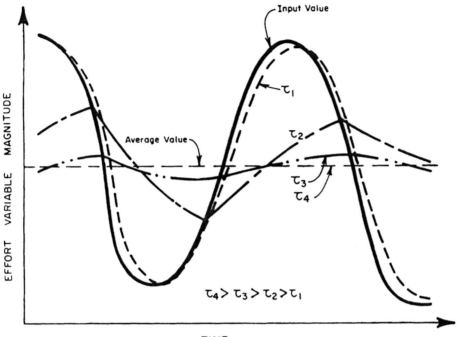

Figure 1.6.6. Output behavior of a system with exponential response to an oscillatory input. A very small time constant system (τ_1) follows the input very closely. As the time constant increases, the system output is smoothed, tending toward the average value. For a very long time constant (τ_4), the output is almost indistinguishable from the average input.

disappear, the system should be designed with a time constant large with respect to the on–off cycles of the pump. The system cannot respond rapidly to the pulsations, and they are not propagated. To achieve this long time constant requires a large resistance and compliance in the piping network. This is exactly the way pressure fluctuations are dampened in the animal cardiovascular system.

For the illustration of water hammer given in Figure 1.6.4, the addition of a large tank compliance that completely overwhelms fluid inertia allows the fluid system to be analyzed as a system with a long time constant. The system can no longer respond rapidly to pressure oscillations when they appear at the source, and average pressure is maintained at the faucet.

At the other extreme, small time constants allow rapid responses. Thermistors are temperature-sensing elements made from small beads of semiconductor material. The investigator usually needs to know the temperature of the medium surrounding the thermistor, rather than the temperature of the thermistor itself. Thus the thermistor is required to equilibrate its temperature rapidly with the temperature of its surroundings. Thermistors are thus designed with low resistance and small thermal capacity.

Since the amount of thermal capacity depends on the mass of the thermistor, smaller thermistors generally respond faster and give more accurate measurements of temperature transients.

The quotient of resistance and inertia also has units of time, and is also called a time constant.

$$\tau = I/R \tag{1.6.8}$$

where τ is the time constant (sec); R, the resistance; and I, the inertial element.

Since inertia relates to the rate of change of flow, this time constant usually refers to the time response of a flow system. The time response of this system to a sudden input change also follows an exponential curve.

Inertia is usually designed around, rather than designed with. Therefore, inertia is not usually an element that is adjusted to form a particular time constant. An exception to this is the design of bass-reflex speaker enclosures, where inertia is added by forming tubes for airflow. In most cases, however, it is much easier to add capacity elements.

The time responses of some systems are dominated by inertia and capacity, and little resistance is present. These systems show an oscillatory time response (Figure 1.6.7) with a natural frequency of

$$\omega_n = 1/\sqrt{IC} \tag{1.6.9}$$

where ω_n is the natural frequency (rad/sec); I, the inertia; and C, the capacity.

If the resistance is small, but not negligible,

$$\omega_n = \sqrt{\frac{1}{IC}\left(1 - \frac{R^2}{4I/C}\right)} \tag{1.6.10}$$

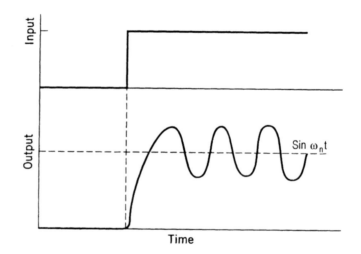

Figure 1.6.7. Systems containing both inertia and capacity can respond to sudden changes of input with oscillations.

and the system response appears to be an oscillation the magnitude of which decreases with time. Such a system is said to be *damped*.

When system response oscillates in response to a system input, it is usually considered to be out of control. As a consequence, systems are usually not designed to oscillate, and most biological systems are not found to oscillate either.

The response of a system dominated by inertia and capacity is shown in Figure 1.6.8. If the oscillatory input (for example, a sine wave) occurs at a frequency not close to the natural frequency of the system, the output magnitude of the system appears to be nearly the same as the input magnitude. If the input frequency is the same as ω_n, the output magnitude can be much greater than the input magnitude, and the system could possibly vibrate so much that it disintegrates.

mechanical vibrations
Much information about internal structures can be obtained from mechanical vibrations of objects. Thus ultrasound is used to image the interfaces between different tissues in humans and animals; seismic waves can be used to locate geological strata in the earth; damaged internal tissues can be ascertained by analyzing vibrations in fruits.

human viscera
The natural frequency of the human viscera occurs in the range of 38–50 radians per second. Vibrating the human body in this range causes extremely uncomfortable feelings to be generated because the internal organs vibrate with a magnitude that can be greater than the magnitude of the input

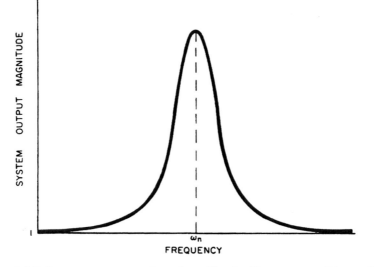

Figure 1.6.8. Frequency response of a system with natural frequency ω_n. At low and high oscillation frequencies, the system output magnitude is the same as the input magnitude. When the input frequency exactly matches ω_n, the system output magnitude is extremely high. The frequency range of the response around ω_n and the magnitude of the response at ω_n are inversely related, and are due to the relative amount of resistance present.

vibrations. Because damping of the human viscera is high (lots of resistance present), the human is usually in no danger of flying apart if vibrated at the natural frequency. However, from a comfort standpoint, vehicles and roadways must be designed to avoid vibrations in this range. Packaging systems designed to transport fresh fruits and vegetables face similar constraints to avoid damaging their flesh.

EXAMPLE 1.6.5-1. Soda Lime

A container of soda lime is used to absorb carbon dioxide from air flowing at $5\,L/sec$. The mass capacitance of the soda lime is $5.0 \times 10^{-4}\,m^3$. The convection resistance for mass transfer into the soda lime is $8.3 \times 10^{-5}\,sec/m^3$. If the concentration of carbon dioxide in the air is suddenly increased from 0 to 20% of the volume of the air, how much carbon dioxide will the soda lime absorb?

Solution

The time constant for this system is

$$\tau = RC = (8.3 \times 10^{-5}\ sec/m^3)(5.0 \times 10^{-4}\ m^3)$$

$$= 415\,sec$$

The absorption of carbon dioxide by the soda lime will be 99% complete in 5 time constants, or 2075 sec.

We assume that the soda lime removes all carbon dioxide from the air until its capacity is reached. In 2075 sec, $(5\,L/sec)\ (2075\,sec) = 10375\,L$ (or $10.375\,m^3$) of air flow through the absorber. Twenty percent of this (or $2.075\,m^3$) is carbon dioxide.

The density of carbon dioxide can be obtained from the ideal gas law:

$$\frac{m}{V} = \frac{p}{RT/M}$$

where M is the molecular weight of carbon dioxide, $44\,kg/kg$ mole. If the air is at 1 atmosphere pressure and 20°C,

$$\frac{m}{V} = \frac{101,325\ N/m^2 \times 44\,kg/kg\ mole}{8314.23\ \dfrac{N\ m}{kg\ mol\ K}\ 293\ K}$$

$$= 1.83\,kg/m^3$$

The maximum amount of carbon dioxide that can be absorbed is about $2.075 \, \mathrm{m}^3$, or $(2.075 \, \mathrm{m}^3) \, (1.83 \, \mathrm{kg/m}^3) = 3.80 \, \mathrm{kg} \, CO_2$.

EXAMPLE 1.6.5-2. Ultrasound Resonant Frequency

Ultrasound is used as a diagnostic tool on humans and animals. Mechanical vibrations are often produced from electrical oscillations by piezoelectric quartz crystals cut to a thickness needed to produce frequencies in the range of $6.28 – 12.6 \times 10^6 \, \mathrm{rad/sec}$. Determine the mass capacity of a quartz crystal $0.29 \times 1 \times 1 \, \mathrm{cm}$ vibrating at $6.28 \times 10^6 \, \mathrm{rad/sec}$.

Solution

We know from Eq. 1.6.9 that

$$C = 1/I\omega^2$$

From Table 1.2.1, $I = F/(\dot{x}/dt) = F/a = m$, from Newton's second law. Thus inertance is just the mass of the crystal. Since the density of quartz is $2650 \, \mathrm{kg/m}^3$,

$$m = I = (0.0029 \, \mathrm{m})(0.01 \, \mathrm{m})(0.01 \, \mathrm{m})(2650 \, \mathrm{kg/m}^3)$$
$$= 0.000768 \, \mathrm{kg}$$

and

$$C = 1/0.000768 \, \mathrm{kg})(6.28 \times 10^6 \, \mathrm{rad/sec})^2 = 3.30 \times 10^{-11} \mathrm{sec/kg}$$

Remark

Note that the natural frequency of the crystal depends only on the thickness of the quartz and not on its total mass. That indicates that mass capacity decreases as inertance increases when the length and width of the crystal change.

1.6.6 Alternative Representations

There are different ways to diagram combinations of resistance, capacity, and inertia. Two of the ways most often seen are electrical and mechanical analogs. Electrical analogs are most often used in this book, but it is instructive to compare the two different types of representations.

mechanical and electrical analogs of respiratory system

In Figure 1.6.9 are shown two representations of the pulmonary system impedance of dogs (Bates et al., 1988). One of these is electrical, with resistors and capacitors in series and parallel combinations. With the electrical representation, two elements, such as the resistor and capacitor combination, are in series if the same current flows through both and the voltage splits between them. Contrarily, elements are in parallel if the current splits and the voltage is the same across them. The series combination has one terminal of each element joined to the other; the parallel combination has two cojoined terminals.

The mechanical analog appears quite different. With dashpots analogous to resistors and springs analogous to capacitors, one might think that simply replacing resistor elements with dashpots and capacitors with springs would correctly represent the same system in mechanical terms. Such is not the case.

connections represent different things

The difference lies in the meaning of connections. In the electrical analog, connections that represent wires signify that current splits and voltage is the same. In the mechanical analog, connections that represent bolts or welds signify that the force splits and the velocity is the same. In general terms, then, the connection in the electrical representation requires that the effort variable is the same and the flow variable splits; the connection in the mechanical representation requires just the opposite.

Hence, the mechanical representation of series elements appears like the electrical representation of parallel elements, and the mechanical representa-

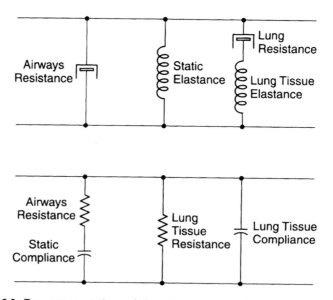

Figure 1.6.9. Two representations of the pulmonary impedances for dogs. The top representation is the mechanical analog proposed by Bates et al. (1988); the bottom representation is the corresponding electrical diagram. Although the two are analogs of each other, they appear to have different connections as well as basic elements.

tion of parallel elements appears like the electrical representation of series elements. That is why an electrical diagram cannot be converted into a mechanical diagram without changing the connections, and that is why some people who visualize resistance, capacity, and inertia elements in electrical terms have difficulty representing the same system in mechanical terms.

While there may be other means to diagram systems, there are none as common as the two above. Therefore, when considering thermal, fluid, mass, or other systems involving sources, resistances, capacity, or inertia, we do not have much choice except to diagram them with their electrical or mechanical analogs, and the electrical representation is by far the more popular. It is still good to be reminded, however, that a diagram of an electrical resistor, for instance, represents a length of pipe, or a block of material, or a membrane, or whatever is appropriate for the system being considered.

1.7 BALANCES

an equation represents a balance

Basic to nearly all areas of quantitative mathematics, science, and engineering is the concept of balance, or equality. Balance is the foundation of an equation, where the equal sign separates two different descriptions of quantitatively equal expressions. For example, the error function is defined by

$$\mathrm{erf}(\theta) = \frac{2}{\sqrt{\pi}} \int_0^\theta e^{-z^2}\, dZ \qquad (1.7.1)$$

Notice that any change in the value of θ is accompanied by a change in the value of $\mathrm{erf}(\theta)$ and the equality is maintained.

1.7.1 Chemical Balances

oxidation of glucose

Likewise, balance is the fundamental concept behind chemical equations. For example, aerobic oxidation of glucose (hexose) is given by

$$(CH_2O)_6 + 6\,O_2 \rightarrow 6\,CO_2 + 6\,H_2O \qquad (1.7.2)$$

Notice that each of the elements C, H, and O appear in equal numbers on both sides of the balance. The reason that there is an arrow rather than an equal sign on this balance is that, although both sides represent equal masses, this is not a complete equation. The formation of glucose and oxygen from water and carbon dioxide does not occur spontaneously. To convert this into a true balance requires the addition of an energy term on the right-hand side:

$$(CH_2O)_6 + 6\,O_2 = 6\,CO_2 + 6\,H_2O$$
$$+\, 2.816 \times 10^6\, \mathrm{N\ m/mol\ glucose} \qquad (1.7.3)$$

Now the equation describes a reversible process, as discussed in Section 1.4.5.

1.7.2 Force Balances

The engineering science of statics is based upon balances of forces and torques. In order for an object to remain motionless, the sum of the forces in all directions must be zero. In Cartesian coordinates, this becomes

$$\sum F_X + \sum F_Y + \sum F_Z = 0 \tag{1.7.4}$$

In this balance, one direction of each coordinate is arbitrarily assigned to be positive and positive forces must then equal negative forces. Since the three coordinates are orthogonal, each coordinate can have an independent balance:

$$\sum F_{+X} = \sum F_{-X} \tag{1.7.5}$$

Newton's
Second Law
    Newton's Second Law relates force imbalance to time rate of change of velocity (or acceleration):

$$\sum F_{+X} - \sum F_{-X} = m\,\frac{dv}{dt} \tag{1.7.6}$$

The above balances illustrate some aspects of all effort variable balances. Equations 1.7.4 and 1.7.5 are steady-state balances where the flow variable is constant. Equation 1.7.6 is a non-steady-state balance where the flow variable is increasing or decreasing. In the general case, force may be replaced by any effort variable and velocity may be replaced by any corresponding flow variable.

1.7.2.1 Law of Laplace   The law of Laplace is

$$p = \frac{2\tau\,\Delta r}{r} \tag{1.7.7}$$

where p is the internal pressure (relative to outside pressure, N/m^2); τ, the wall tensile stress (N/m^2); r, the sphere radius (m); and Δr, the sphere wall thickness (m).

relationship
among
pressure, wall
stress, and
size
    This law is important for interpreting the behaviour of hollow shapes. It indicates that for any given wall tensile stress and thickness (properties of the wall material), the pressure required to inflate a bubble is larger for smaller bubbles.

alveolar
application
    The gas exchange portions of the lung are composed of many small gas-filled pockets called alveoli. These alveoli are connected to each other through air passages that allow gas to flow relatively freely from one alveolus to another. The law of Laplace indicates that larger alveoli should contain lower

pressures than smaller alveoli. Since gas flows from higher to lower pressures, the gas in the smaller alveoli would be expected to empty into the larger ones; eventually one extremely large alveolus would remain open with the rest collapsed.

This does not happen. A complex organic macromolecular material called surfactant lines the inside of the lung tissue and modfies the wall tensile stress to equalize pressures required to maintain alveolar inflation.

capillaries

The law of Laplace for cylindrical shapes differs from Eq. 1.7.7 by omitting the factor 2. It indicates that smaller blood vessels can resist high blood pressures even with extremely thin walls. This is fortunate, since diffusion of blood gases and metabolites across capillary walls would be greatly reduced if the walls were required to be thick to resist blood pressure.

bacteria

Even the shapes of bacterial cell walls can be explained with the use of the law of Laplace. Peptidoglycan strands in the walls of Gram-negative (thin-walled) bacteria tend to orient themselves to resist the greatest wall stresses (in the direction across the largest radius). Gram-positive, rod-shaped bacteria (thick-walled) replace wall material more slowly in the end caps, where maximum stress is low, than in the side walls, where maximum stress is high. Other twisted shapes and transitional shapes during cell division can be explained as well.

The law of Laplace can be obtained from a force balance on a sphere that has been cut in half (Figure 1.7.1). As long as the wall is thin, the exposed area is $2\pi r \, \Delta r$, and wall force is $2\pi r \, \Delta r \, \tau$. Opposing the wall force is the force produced by the pressure inside the sphere. Only the component of the pressure in the vertical direction opposes the wall force.

$$\sum F_y = 0 \tag{1.7.8}$$

$$2\pi r \, \Delta r \, \tau = \int_{\text{sphere}} dF_y = \int_A p \cos \phi \, dA \tag{1.7.9}$$

where F_y is the vertical force component (N); A, the hemisphere surface area (m^2); and ϕ, the vertical spherical angle (rad).

For a sphere,

$$dA = r^2 \sin \phi \, d\phi \, d\theta \tag{1.7.10}$$

and thus

$$2\pi r \, \Delta r \, \tau = \int_0^{2\pi} \int_0^{\pi/2} pr^2 \sin \phi \cos \phi \, d\phi \, d\theta \tag{1.7.11}$$

where θ is the horizontal spherical angle (rad).

$$\int_0^{2\pi} \int_0^{\pi/2} pr^2 \sin \phi \cos \phi \, d\phi \, d\theta = 2\pi r^2 p \int_0^{\pi/2} \sin \phi \cos \phi \, d\phi \tag{1.7.12}$$

$$= \pi r^2 p [\sin^2 \phi]_0^{\pi/2} = \pi r^2 p$$

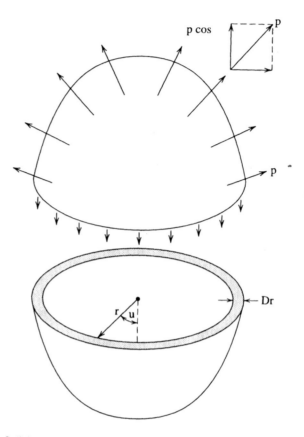

Figure 1.7.1. Splitting a sphere in half results in a pressure balance to determine the relationship between internal pressure and wall pressure. The result is the law of Laplace.

Returning to the original force balance,

$$2\pi r \, \Delta \tau = \pi r^2 p \tag{1.7.13}$$

and thus

$$p = \frac{2\tau \, \Delta r}{r} \tag{1.7.7}$$

For an object of arbitrary shape, the law of Laplace becomes

$$p = \tau \, \Delta r \left(\frac{1}{r_1} + \frac{1}{r_2} \right) \tag{1.7.14}$$

where r_1 and r_2 are values of radii in orthogonal directions (m).

EXAMPLE 1.7.2.1-1. Alveolar Walls

An individual alveolus $100 \, \mu m$ in diameter has a wall thickness of $0.5 \, \mu m$. When the pressure in the alveolus is $101.3 \, kN/m^2$, what is the tensile stress in the wall?

Solution

From the Law of Laplace,

$$\tau = \frac{pr}{2 \, \Delta r}$$

$$= \frac{(101,300 \, N/m^2)(50 \times 10^{-6} \, m)}{2(5.0 \times 10^{-7} \, m)}$$

$$= 5.065 \times 10^6 \, N/m^2$$

Remark

This is a very large tensile stress completely beyond the capability of alveolar wall tissue. The reason this value is wrong is that it assumes the pressure outside the alveolus to be a perfect vacuum. It is not. Tissue or gas surrounding each alveolus is at nearly the same pressure as the gas inside the alveolus. The correct value for p in the above expression is thus nearly zero, which makes the value for τ very small.

1.7.3 General Flow Balances

basic flow
balance

Flow balances used in the processes in the remainder of this book will have a general form similar to

$$
\begin{aligned}
&(\text{Rate of substance in}) - (\text{rate of substance out}) \\
&\quad + (\text{rate of substance generated}) \\
&\quad = (\text{rate of substance stored})
\end{aligned}
\tag{1.7.15}
$$

Comparing this general form with previous balances shows that there is no substance (force) generated in Eq. 1.7.6, and all previous balances were steady state, where there was no change in substance storage.

The flow balance in Eq. 1.7.15 is illustrated by two means in Figure 1.7.2. At the top is a systems diagram relating the general flow balance to the previous effort and flow sources and to the elements of resistance and capacity. At the bottom is a fluid flow representation. Both representations are equivalent in their essentials.

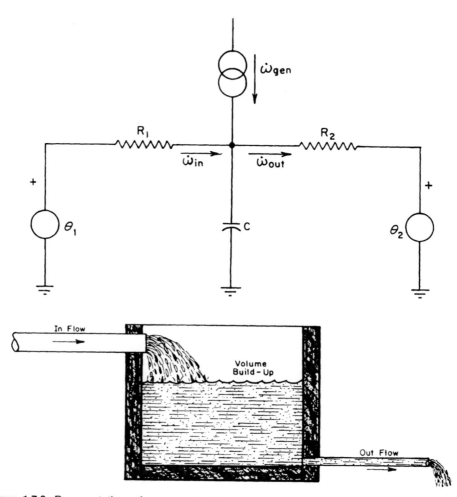

Figure 1.7.2. Representations of an unsteady balance. At the top is a general systems representation with flow of a substance in, flow of the substance out, and generation of the substance contributing to the accumulation of the substance in a capacity element. Below is a fluid flow representation, where there is no fluid generation term. If the outflow is less than the inflow, the volume of the stored fluid will increase.

Note, especially, that the flow balance specifically uses the words "rate of ..." One might be tempted to drop these words and give the balance in terms of amounts rather than rates. This would likely be wrong for the substance storage term. The only time it would not be wrong is if the storage term were reckoned from an absolute (not relative) zero reference. For example, a heat balance would describe the rate of heat in, rate of heat out, rate of heat generation, and rate of heat storage. Even if rates of heat in, heat out, and heat generation are zero, the *amount* of heat storage is *not* zero (unless the mass is at absolute zero temperature), although the *rate* of heat storage *is* zero.

Therefore, do not forget the "rate of ..." words unless you know where the storage reference level is.

In all our balances, there must be some substance that maintains its identity as it goes through the process. This substance cannot change form from one side of the balance to another. For example, an energy balance usually can be constructed for most processes; a potential energy balance usually is not constructed because potential energy changes form into kinetic energy (and vice versa).

materials
balance

Likewise, air can be considered to be an identifiable substance, despite the fact that it is not elemental, as long as it does not change form in the process. Thus air can be used as a substance in a psychrometric process material balance (which involves physical additions and separations), but not in an oxidation process chemical balance (which involves the consumption of a constituent of the air).

The general flow balance in Eq. 1.7.15 can thus apply to many kinds of flows, not just those previously classified as flow variables. To form a basis for topics to be covered later, the general flow balance equation will be expanded in three different directions to produce a mass or material balance, a field equation, and an energy balance.

EXAMPLE 1.7.3-1. Oxygen Balance in Exercise

Running on level ground at 4.47 m/sec requires an oxygen consumption of 6.70×10^{-5} m^3 O$_2$/sec. Maximum oxygen consumption for a young man is about 4.1×10^{-5} m^3 O$_2$/sec. What is the rate of oxygen debt incurred in a person exercising at the rate given?

Solution

From Eq. 1.7.15, the oxygen balance gives:

$$\text{rate of O}_2 \text{ in} - \text{rate of O}_2 \text{ out} + \text{rate of O}_2 \text{ generated}$$
$$= \text{rate of O}_2 \text{ stored}$$

$$\text{rate of O}_2 \text{ in} = 4.1 \times 10^{-5} \text{ m}^3 \text{ O}_2/\text{sec}$$
$$\text{rate of O}_2 \text{ out} = 6.7 \times 10^{-5} \text{ m}^3 \text{ O}_2/\text{sec}$$
$$\text{rate of O}_2 \text{ generated} = 0$$

Thus the rate of O$_2$ stored is negative (oxygen debt) at $(4.1 \times 10^{-5}) - (6.7 \times 10^{-5}) = -2.5 \times 10^{-5}$ m^3 O$_2$/sec.

1.7.3.1 *Mass and Materials Balance* The conservation of mass requires that mass be neither created nor destroyed. Thus Eq. 1.7.15 can apply as long as the rate of substance generated becomes zero.

In any arbitrary control volume, then, all paths where the mass can enter or leave must be accounted for. If mass is to be stored within the control volume, then the control volume itself must expand without changing its essential boundary characteristics (no change in ports for the mass to enter or leave) or the mass must be compressible (less volume per unit mass), or both. Equation 1.7.15 can hence be used for mass balance as long as the rate of substance generated term is omitted.

mass balance   If mass is given by density times volume, then the mass balance becomes:

$$\rho_{in} \dot{V}_{in} - \rho_{out} \dot{V}_{out} = \rho_{inside} \frac{dV}{dt} \qquad (1.7.16)$$

where ρ is the density (kg/m^3 or N sec^2/m^4); $\dot{V}$, the volume flow rate (m^3/sec); and V, the control volume (m^3). The subscripts represent: in, entering control volume; out, leaving control volume; and inside, inside control volume.

steady state   If steady-state conditions apply, no mass storage occurs, and

$$\rho_{in} \dot{V}_{in} = \rho_{out} \dot{V}_{out} \qquad (1.7.17)$$

A material balance differs from the mass balance in that a certain type of identifiable material may be generated within the control volume. For example, water can form from the elements hydrogen and oxygen. If a fuel is burning within a control volume, then water is being generated. This water must be added to the water entering the boundaries of the control volume, and adds to the water removal or storage. Since a material balance is concerned with a specific type of material, not just the total mass of material, the term for the rate of substance generated cannot be omitted automatically.

EXAMPLE 1.7.3.1-1. Materials Balance in the Lung

The volume of air that enters the respiratory system but that does not take part in gas exchange is called the respiratory dead volume. Determine the dead volume.

Solution

Dead volume can be determined by forming a balance on exhaled carbon dioxide:

Total exhaled $CO_2 = CO_2$ from dead volume $+ CO_2$ from aveoli

$$V_e F_{eCO_2} = V_{Ae} F_{AeCO_2} + V_{De} F_{DeCO_2}$$

where V_e is the exhaled volume (m^3); V_{Ae}, the exhaled volume from alveolar space (m^3); V_{De}, the exhaled volume from dead space (m^3); F_{eCO_2}, the average mixed volume fractional concentration of CO_2 from exhaled air $(m^3\,CO_2/m^3$ air); F_{ACO_2}, the fractional CO_2 concentration from alveolar space $(m^3\,CO_2/m^3$ air); and F_{DeCO_2}, the fractional CO_2 concentration from dead volume $(m^3\,CO_2/m^3$ air).

Because no gas exchange occurs in the dead volume, F_{DeCO_2} is the fractional CO_2 concentration in the inhaled air, usually assumed to be zero. Also, F_{AeCO_2} is assumed to be the concentration of CO_2 at the end of exhalation (end-tidal air). Thus

$$V_e F_{eCO_2} = (V_e - V_{De}) F_{AeCO_2}$$

$$V_{De} = V_e[(F_{AeCO_2} - F_{eCO_2})|F_{AeCO_2}]$$

This is called the Bohr equation. A typical resting value for exhaled volume $(V_e) = 500 \times 10^{-6}\,m^3$; the average mixed CO_2 concentration $(F_{eCO_2}) = 1.6\%$; the end-tidal CO_2 concentration $(F_{AeCO_2}) = 2.5\%$;

$$V_{De} = (500 \times 10^{-6} m^3)[(0.025 - 0.016)/0.025]$$

$$= 180 \times 10^{-6} m^3$$

Remark

This is a normal adult male value. Much higher dead volumes might indicate conditions of emphysema or chronic obstructive pulmonary disease (COPD).

EXAMPLE 1.7.3.1-2. Materials Balance in Food

Vitamin C (ascorbic acid) is destroyed during cooking. Assume the destruction occurs exponentially with a time constant of 67 min at a temperature of 116°C. How many milligrams of vitamin C must be added before processing to a food initially containing (50 mg vitamin C)/(100 g food) if, after processing for 30 min, a minimum of (40 mg vitamin C)/(100 g food) of vitamin C should remain?

Solution

Referring to the general materials balance, Eq. 1.7.15, we can write a balance on vitamin C as:

$$\text{Vit C in} - \text{Vit C out} + \text{Vit C gen} = \text{Vit C stored}$$

For each 100 g food,

$$(50\,\text{mg} + \text{Vit C added}) - (50\,\text{mg} + \text{Vit C added})(1 - e^{-30/67}) + 0 = 40\,\text{mg}$$

$$\text{Vit C added} = \frac{40\,\text{mg}}{e^{-30/67}} - 50\,\text{mg} = 12.6\,\text{mg}$$

1.7.3.2 *Field Equation*   The general flow balance equation can produce a field equation. A field equation describes the spatial and temporal distribution of effort variables. The field equation can be developed in general terms, and it will be seen later that the same equation, with more complete specification of parameters, can apply equally well to fluid, heat, and mass transfer systems. The equation will be developed first for a differential element of volume $dx\,dy\,dz$ in the Cartesian coordinate system. Six faces of the cubical element need to be considered (Figure 1.7.3).

A relationship between the general substance flow rate $\dot{\omega}$ and the general basic equation   effort variable ϕ needs to be assumed. For all the systems considered in this book, a linear relationship is usually accepted:

$$\dot{\omega} = -kA\,\frac{d\phi}{dx} \qquad (1.7.18)$$

where $\dot{\omega}$ is the rate of flow per unit time; k, a constant of proportionality; A, the cross-sectional area; ϕ, the effort variable; and x, the direction of flow of $\dot{\omega}$.

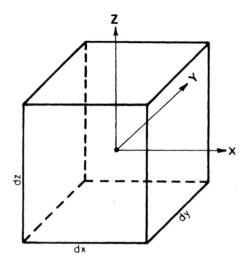

Figure 1.7.3. This differential volume in Cartesian coordinates is the basis for a flow balance given in the text.

The negative sign on the right-hand side of this equation is used to denote that flow is positive in the direction of negative effort variable gradient (flow occurs "downhill"). The parameter k is a property of the medium.

rate of material in and out

The rate of substance flow $\dot{\omega}$ into the surface perpendicular to the x axis is:

$$\dot{\omega}_x'' = -k_x \frac{\partial \phi}{\partial x} \tag{1.7.19}$$

where $\dot{\omega}_x''$ is the rate of flow per unit area (each "prime" mark indicates a length derivative); k_x, a constant of proportionality in the x direction; and ϕ, the effort variable.

Similar equations are written for flow into the other two surfaces:

$$\dot{\omega}_y'' = -k_y \frac{\partial \phi}{\partial y} \tag{1.7.20}$$

$$\dot{\omega}_z'' = -K_y \frac{\partial \phi}{\partial z} \tag{1.7.21}$$

The rate of change of flow in the x direction is $(\partial/\partial x)(\dot{\omega}_x'')$, and the difference in flow between the two parallel faces is:

$$-\frac{\partial}{\partial x}(\dot{\omega}_x'')\,dx = -\frac{\partial}{\partial x}\left(-k_x \frac{\partial \phi}{\partial x}\right)dx$$

$$= \frac{\partial}{\partial x}\left(k_x \frac{\partial \phi}{\partial x}\right)dx \tag{1.7.22}$$

The net rate of substance $\dot{\omega}$ due to flow in the x direction is:

$$\frac{\partial}{\partial x}\left(k_x \frac{\partial \phi}{\partial x}\right)dx\,A_x = \frac{\partial}{\partial x}\left(k_x \frac{\partial \phi}{\partial x}\right)dx\,dy\,dz \tag{1.7.23}$$

where A_x is the area of surface perpendicular to the x axis. Similar relations can be written for the other two directions.

rate of material generated

The substance generated in the differential volume will be considered to be generated diffusely throughout the volume at a rate of $\dot{\omega}'''$. Thus the total rate of substance generated is:

$$\dot{\omega}_{\text{gen}} = \dot{\omega}'''V = \dot{\omega}'''\,dx\,dy\,dz \tag{1.7.24}$$

The time rate of storage of substance $\dot{\omega}$ will be assumed to depend on the rate of change of the effort variable, and some property of the storage medium:

$$\dot{\omega}_{\text{stored}} = \xi V \frac{\partial \phi}{\partial t} = \xi \frac{\partial \phi}{\partial t}\,dx\,dy\,dz \tag{1.7.25}$$

general
balance
equation

It can be recognized that the elemental volume ($dx\,dy\,dz$) is common to all terms in the general balance, and can be cancelled from the equation. The general balance then becomes:

$$\frac{\partial}{\partial x}\left(k_x\,\frac{\partial\phi}{\partial x}\right) + \frac{\partial}{\partial y}\left(k_y\,\frac{\partial\phi}{\partial y}\right) + \frac{\partial}{\partial z}\left(k_z\,\frac{\partial\phi}{\partial z}\right) + \dot{\omega}''' = \xi\,\frac{\partial\phi}{\partial t} \qquad (1.7.26)$$

This general balance applies for fluid flow in porous media, electrical conduction in a diffuse medium, heat transfer by conduction, and mass transfer by diffusion, with appropriate definition of parameters and variables.

isotropic
medium

Consider some simplifications of Eq. 1.7.26. First, if the medium is isotropic, such that $k = k_x = k_y = k_z$, then:

$$\frac{\partial^2\phi}{\partial x^2} + \frac{\partial^2\phi}{\partial y^2} + \frac{\partial^2\phi}{\partial z^2} + \frac{\dot{\omega}'''}{k} = \frac{1}{\kappa}\,\frac{\partial\phi}{\partial t} \qquad (1.7.27)$$

where $\kappa = k/\xi$.

Fourier
equation

If, in addition, there is no distributed source of $\dot{\omega}$ within the volume,

$$\frac{\partial^2\phi}{\partial x^2} + \frac{\partial^2\phi}{\partial y^2} + \frac{\partial^2\phi}{\partial z^2} = \frac{1}{\kappa}\,\frac{\partial\phi}{\partial t} \qquad (1.7.28)$$

This is called the Fourier equation.

Laplace
equation

In the steady state, $\partial\phi/\partial t = 0$, and:

$$\frac{\partial^2\phi}{\partial x^2} + \frac{\partial^2\phi}{\partial y^2} + \frac{\partial^2\phi}{\partial z^2} = 0 \qquad (1.7.29)$$

which is called the Laplace equation. For one-dimensional flow, this becomes:

$$\frac{\partial^2\phi}{\partial x^2} = 0 \qquad (1.7.30)$$

Poisson
equation

Another circumstance is steady-state flow in an isotropic medium with a distributed source:

$$\frac{\partial^2\phi}{\partial x^2} + \frac{\partial^2\phi}{\partial y^2} + \frac{\partial^2\phi}{\partial z^2} + \frac{\dot{\omega}'''}{k} = 0 \qquad (1.7.31)$$

which is called the Poisson equation.

steady state,
flat plate

Solutions of the General Field Equation   Consider now steady-state one-dimensional flow with no flow source, as given in Eq. 1.7.30. The medium will be considered to be shaped as a flat plate with thickness L and with effort variable ϕ_1 on one surface and ϕ_2 on the other surface (Figure 1.7.4). The

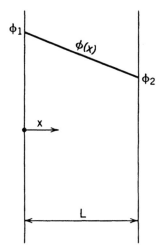

Figure 1.7.4. The solution to the general flow balance equation for a solid bounded by two flat planes predicts a constant slope for effort variable change with distance inside the solid.

mathematical description of the problem is:

$$\frac{\partial^2 \phi}{\partial x^2} = 0, \qquad \phi = \begin{cases} \phi_1 & \text{at } x = 0 \\ \phi_2 & \text{at } x = L \end{cases}$$

The equation can be integrated twice to obtain:

$$\phi = C_1 x + C_2 \qquad (1.7.32)$$

and the constants of integration, C_1 and C_2, are found from the boundary conditions to be:

$$\begin{aligned} C_2 &= \phi_1 \\ C_1 &= (\phi_2 - \phi_1)/L \end{aligned} \qquad (1.7.33)$$

effort
distribution

Hence

$$\phi = \phi_1 + (\phi_2 - \phi_1)x/L \qquad (1.7.34)$$

Differentiating this equation gives:

$$\frac{d\phi}{dx} = \frac{(\phi_2 - \phi_1)}{L} = \frac{(\phi_2 - \phi_1)}{(x_2 - x_1)} \qquad (1.7.35)$$

$$\dot{\omega} = -kA\,\frac{d\phi}{dx} = \frac{kA}{L}(\phi_2 - \phi_1) \qquad (1.7.36)$$

resistance   Since resistance $R = |\phi/\dot{\omega}|$,

$$R = \frac{L}{kA} \tag{1.7.37}$$

Thus the resistance to flow in a flat plate is given as the width divided by a property of the material times the surface area. This is the same resistance that we have studied before, and can be manipulated as previously explained. Once resistance is known, then the level of flow can be determined for any given effort variable level.

Notice also that only effort variable *differences* appear in Eq. 1.7.36. Thus the reference level for the effort variable is completely arbitrary and can be taken at any convenient level. For example, pressure in a flow system can be referred to absolute pressure, to atmospheric pressure, or to any other meaningful pressure level.

capacity   Capacity has been specifically defined in Eqs. 1.4.6–1.4.9. In the generic terms of Eqs. 1.7.18–1.7.26,

$$C = \frac{1}{\phi} \int \dot{\omega} \, dt = \xi V \tag{1.7.38}$$

time constant   Therefore, capacity is given as the term ξV in Eq. 1.7.25. Thus the time constant, the product of resistance and capacity, for rectangular coordinates is:

$$\tau = RC = \left(\frac{L}{kA}\right)(\xi V) = \frac{\xi L^2}{k} \tag{1.7.39}$$

Other Coordinate Systems   The general flow balance equation just developed is appropriate only for those shapes that are characterized by flat surfaces and 90° angle corners. There are other coordinate systems that lend themselves to other geometrical shapes, and make the solution of problems much easier than if Cartesian coordinates are used. Even if the geometrical shape of the outline of the problem does not fit one of these alternative coordinate systems, sometimes the problem itself lends itself to another system. For example, heat flow from a metal rod stuck in the ground is probably better analyzed using cylindrical coordinates than with Cartesian coordinates, because the heat flow can be considered to be one-dimensional radial flow in cylindrical coordinates rather than more complex two-dimensional x–y flow in Cartesian coordinates. Cylindrical coordinates simplify the analysis despite the fact that the outer boundary of the ground, if it is known at all, probably does not look like a cylinder. More obviously, pipes are best considered in cylindrical coordinates, and grapefruits in spherical coordinates.

There are other coordinate systems besides cylindrical and spherical that lend themselves to good representation of problem geometry. However, the only two systems considered here are cylindrical and spherical.

cylindrical
coordinates

Equation 1.7.27 expressed in cylindrical coordinates becomes (Figure 1.7.5)

$$\frac{\partial^2 \phi}{\partial r^2} + \frac{1}{r} \frac{\partial \phi}{\partial r} + \frac{1}{r^2} \frac{\partial^2 \phi}{\partial \theta} + \frac{\partial^2 \phi}{\partial z^2} + \frac{\dot{\omega}'''}{k} = \frac{1}{\kappa} \frac{\partial \theta}{\partial t} \tag{1.7.40}$$

where r, θ, and z are the coordinate axes and in spherical coordinates it becomes (Figure 1.7.6):

$$\frac{\partial^2 \phi}{\partial r^2} + \frac{2}{r} \frac{\partial \phi}{\partial r} + \frac{1}{r^2 \sin \theta} \frac{\partial}{\partial \theta} \left(\sin \theta \, \frac{\partial \phi}{\partial \theta} \right)$$

$$+ \frac{1}{r^2 \sin 2\psi} \frac{\partial^2 \phi}{\partial \psi^2} + \frac{\dot{\omega}'''}{k} = \frac{1}{\kappa} \frac{\partial \phi}{\partial t} \tag{1.7.41}$$

where r, θ, and ψ are the spherical coordinate axes.

resistance

Steady-state one-dimensional radial flow in a hollow cylinder with no flow source and of length L with inside (r_i) effort level ϕ_i and outside (r_o) effort level ϕ_o gives:

$$\phi = \phi_i - (\phi_i - \phi_o) \frac{\ln(r/r_i)}{\ln(r_o/r_i)} \tag{1.7.42}$$

$$\dot{\omega} = \frac{-2\pi kL(\phi_i - \phi_o)}{\ln(r_o/r_i)} \tag{1.7.43}$$

$$R = \frac{\ln(r_o/r_i)}{2\pi kL} \tag{1.7.44}$$

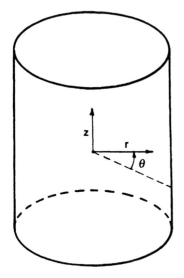

Figure 1.7.5. Cylindrical coordinate system used to develop the general flow balance equation.

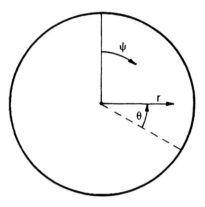

Figure 1.7.6. Spherical coordinate system for developing the general flow balance equation. The angles θ and ψ are orthogonal to one another.

spherical coordinates

Steady-state one-dimensional radial flow in a hollow sphere with no flow source and with inside (r_i) effort level ϕ_i and outside (r_o) effort level ϕ_o gives:

$$\phi = \frac{1}{1/r_i - i/r_o}\left(\frac{\phi_i - \phi_o}{r} + \frac{\phi_o}{r_i} - \frac{\phi_i}{r_o}\right) \tag{1.7.45}$$

$$\dot{\omega} = \frac{-4\pi k(\phi_i - \phi_o)}{1/r_i - 1/r_o} \tag{1.7.46}$$

$$R = \frac{1/r_i - 1/r_o}{4\pi k} \tag{1.7.47}$$

resistance

These resistance forms should be used when appropriate. The above two coordinate system resistances are valid only for hollow cylinder and hollow sphere. If inside radius is taken to be zero, infinite resistances result.

time constants

The time constants for these coordinate systems differ from that given in Eq. 1.7.39 because of the different forms for resistance:

$$\tau = \left[\frac{\ln(r_o/r_i)}{2\pi kL}\right](\xi V)$$

$$= \frac{\xi(r_o^2 - r_i^2)\ln(r_o/r_i)}{2k} \tag{1.7.48}$$

for cylindrical coodinates, and

$$\tau = \left[\frac{(1/r_i - 1/r_o)}{4\pi k}\right](\xi V)$$

$$= \frac{\xi(r_o^3 - r_i^3)(1/r_i - 1/r_o)}{3k} \tag{1.7.49}$$

in spherical coordinates.

One-Dimensional Flow with Distributed Constant $\dot{\omega}$ Source   The one-dimensional form of Eq. 1.7.31:

$$\frac{\partial^2 \phi}{\partial x^2} = \frac{-\dot{\omega}'''}{k} \qquad (1.7.50)$$

is integtrated twice to give:

$$\phi = -\frac{\dot{\omega}''' x^2}{2k} + C_1 x + C_2 \qquad (1.7.51)$$

If $\phi = \phi_1$ at $x = 0$ and $\phi = \phi_2$ at $x = L$, then

$$C_2 = \phi_1 \qquad (1.7.52a)$$

$$C_1 = \frac{(\phi_2 - \phi_1)}{L} + \dot{\omega}''' \frac{L}{k} \qquad (1.7.52b)$$

$$\phi = \frac{\dot{\omega}''' x (L - x)}{2k} + (\phi_2 - \phi_1)\frac{x}{L} + \phi_1 \qquad (1.7.53)$$

This solution can be seen to be the superposition of a flow generation term, a passive conduction term, and a reference term (Figure 1.7.7). The generation term by itself causes a parabolic relationship between ϕ and x and results in a maximum value at $x = L/2$. The second term is the same as that which arises from conduction without flow generation (Eq. 1.7.34).

distributed source in three coordinate systems
For one-dimensional flow in hollow cylindrical coordinates subject to an effort ϕ_i at $r = r_i$ and ϕ_o at $r = r_o$,

$$\frac{\phi - \phi_o}{\phi_i - \phi_o} = \frac{\dot{\omega}''' r_o^2}{4k(\phi_i - \phi_o)}$$

$$\times \left[\left(1 - \frac{r_i^2}{r_o^2}\right)\frac{\ln(r/r_o)}{\ln(r_o/r_i)} - \left(1 - \frac{r^2}{r_o^2}\right)\right] - \frac{\ln(r/r_o)}{\ln(r_o/r_i)} \qquad (1.7.54)$$

For a hollow sphere with $\phi = \phi_i$ at $r = r_i$ and $\phi = \phi_o$ at $r = r_o$,

$$\frac{\phi - \phi_i}{\phi_o - \phi_i} = \frac{\dot{\omega}'''}{(\phi_o - \phi_i)6k}\left[(r_i^2 - r^2) + (r_i^2 - r_o^2)\left(\frac{1}{r_i} - \frac{1}{r_o}\right)\left(\frac{1}{r} - \frac{1}{r_i}\right)\right]$$

$$+ \left(\frac{1}{r_i} - \frac{1}{r_o}\right)\left(\frac{1}{r_i} - \frac{1}{r}\right) \qquad (1.7.55)$$

point source in sphere
One-Dimensional Flow with Central $\dot{\omega}$ Source   A special case involving solid spheres, cylinders, and slabs can be useful at times. Ingersoll et al. (1954)

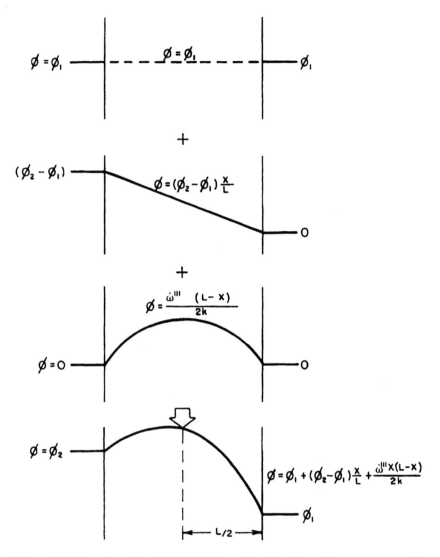

Figure 1.7.7. One-dimensional conduction in a slab with a diffuse $\dot{\omega}$ source can result in a maximum value appearing within the slab boundaries. If $\dot{\omega}'''$ is small enough, a maximum is not formed.

give the effort variable distribution for spherical steady-state flow where the $\dot{\omega}$ source is located at the center of an infinite solid (Figure 1.7.8). This case can arise when fluid, heat, mass, electrical current, or other flow is injected at a concentrated spot inside a large solid mass. Examples could include the release of drug from the tip of a hypodermic needle deep in muscle tissue, heat flow from an underground deposit of uranium, or the spread of pollution from

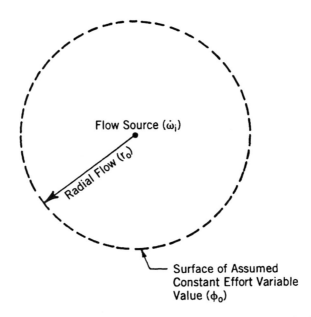

Flow Source ($\dot{\omega}_i$)

Radial Flow (r_o)

Surface of Assumed
Constant Effort Variable
Value (ϕ_o)

Figure 1.7.8. Geometry for assumed spherical flow when there is a point flow source located within an infinite solid.

a point source into the environment. The outer boundary of the solid does not have to be spherical, but it must make sense to assume that flow from the point source happens radially. Under these conditions

$$\phi = \phi_o + \frac{\dot{\omega}}{4\pi k r_o} \qquad (1.7.56)$$

where ϕ_o is the constant effort variable value at some radius r_o; $\dot{\omega}_i$, the value of flow source at center of spherical volume; r_o, the radial distance from point source to assumed surface of constant ϕ_o value; and k, the appropriate parameter.

resistance   Resistance is given by:

$$R = \frac{1}{4\pi k r_o} \qquad (1.7.57)$$

This value is different from that for the hollow sphere given in Eq. 1.7.47 because this resistance value is for a solid sphere with different boundary conditions.

line source in
cylinder

There may be times when a flow source appears in the shape of a line within an infinite solid. There would be every reason to expect that flow would occur radially from the line and that the surface of assumed constant ϕ_o value would be cylindrical rather than spherical. Flow of oxygen from a capillary into surrounding tissue, or drainage of excess water into a buried tile pipe, or heat flow from a nichrome wire inserted in a food material could be examples of this. Under these conditions,

$$\phi = \phi_o + \frac{\dot{\omega}_i}{2\pi Lk} \tag{1.7.58}$$

where $\dot{\omega}_i$ is the value of line flow source at center of cylindrical volume; L, the length of the flow source; and ϕ_o, a constant effort variable value at some radius r_o.

resistance   The value of L should be large relative to the value of r_o.

$$R = \frac{1}{2\pi Lk} \tag{1.7.59}$$

Solutions to other, more complicated geometries and boundary conditions can be found in *Conduction of Heat in Solids* (Carslaw and Jaeger, 1959).

One-Dimensional Non-Steady-State Solution The Fourier equation 1.7.28 can be simplified for one-dimensional flow to:

$$\frac{\partial^2 \phi}{\partial x^2} = \frac{1}{\kappa} \frac{\partial \phi}{\partial t} \tag{1.7.60}$$

where resistance occurs matters   It is often necessary to know how the effort variable ϕ varies with time in order to design systems to perform functions such as drying and cooling. The preceding equation is especially important if the resistance to flow inside the object of interest is non-negligible. If internal resistance is unimportant relative to resistance external to the object, then the value for the effort variable is uniform everywhere inside the object, and transient effects can be analyzed in the simplified manner given in Section 1.6.5.

If internal resistance is high enough to limit the flow variable, then solutions to equations such as Eq. 1.7.60 are required to determine changes in effort and flow with time.

Before proceeding to solve Eq. 1.7.60, we need to relate the equation to real transport systems. There are three of them that are commonly related: those for heat, momentum, and mass (Table 1.7.1). Momentum is important in fluid flow.

diffusivity   With suitable choices of the effort and flow variables, the constant parameter κ assumes units of m^2/sec in all three systems, and is called the "diffusivity." Diffusivity is important in both steady-state and transient flow systems, as will be discussed later.

Effort variables for heat, momentum, and mass are each expressed as concentrations per unit volume. Thus the effort variable for heat flow, important for this analysis, is heat per unit volume ($\rho c_p \theta$) instead of

Table 1.7.1 Analogous Values for Diffusivity

System	Effort Variable	Flow Variable	Steady-State Equation	Non-steady-state Equation	Diffusivity
Heat	$\rho c_p \theta$ (heat concentration)	$\dot{q}/A$ (heat flux)	$\dfrac{\dot{q}}{A} = -\alpha \dfrac{d(\rho c_p \theta)}{dx}$	$\dfrac{\partial \theta}{\partial t} = \alpha \dfrac{\partial^2 \theta}{\partial x^2}$	$\alpha = \dfrac{k}{\rho c_p}$
Units	N/m^2	N/(m sec)			m^2/sec
Momentum	ρv (momentum concentration)	τ (momentum flux)	$\tau = -v \dfrac{d(\rho v)}{dx}$	$\dfrac{\partial v}{\partial t} = v \dfrac{\partial^2 v}{\partial x^2}$	$v = \dfrac{\mu}{\rho}$
Units	kg/(m^2 sec)	kg/(m sec^2)			m^2/sec
Mass	c (mass concentration)	$\dot{m}/A$ (mass flux)	$\dfrac{\dot{m}}{A} = -D \dfrac{dc}{dx}$	$\dfrac{\partial c}{\partial t} = D \dfrac{\partial^2 c}{\partial x^2}$	D
Units	kg/m^3	kg/(sec m^2)			m^2/sec

temperature, as used elsewhere. Momentum concentration (ρv) and mass concentration (c) are also used.

variable forms   Flow variables are expressed in terms of fluxes, that is, rate of flow per unit area. Thus the flow variable for the heat system is the rate of heat flow per unit area ($\dot{q}/A$) instead of rate of heat flow only, as used elsewhere. Momentum flux (τ) and mass flux ($\dot{m}/A$) are also used.

Note that these are not the most convenient effort and flow variables to use for most purposes. They are important to understand the context for diffusivity, and other boundary layer analyses not covered in this text.

basic equation   Diffusivities are used in both steady-state equations of the form:

$$\dot{\omega} = \kappa \frac{d\phi}{dx} \tag{1.7.61}$$

and in transient equations of the form of Eq. 1.7.60.

In order to solve Eq. 1.7.60, we postulate a solution of the form:

$$\phi(x, t) = X(x)T(t) \tag{1.7.62}$$

From this,

$$\frac{\partial^2 \phi}{\partial x^2} = T \frac{\partial^2 X}{\partial x^2} \tag{1.7.63a}$$

$$\frac{\partial \phi}{\partial t} = X \frac{\partial T}{\partial t} \tag{1.7.63b}$$

separation of variables   Putting these into equation 1.7.60 gives

$$T \frac{\partial^2 X}{\partial x^2} = \frac{\partial^2 \phi}{\partial x^2} = \frac{1}{\kappa} \frac{\partial \phi}{\partial t} = \frac{1}{\kappa} X \frac{\partial T}{\partial t} \tag{1.7.64}$$

The two variables x and t can now be separated, or isolated on each side of the equation:

$$\frac{1}{X} \frac{\partial^2 X}{\partial x^2} = \frac{1}{\kappa T} \frac{\partial T}{\partial t} \tag{1.7.65}$$

Now the left side of this equation is only a function of the variable x, while the right side of the equation is only a function of the variable t. Since these two variables are independent, the only way for the equality to be true is if both sides are equal to the same constant. We call this constant $-\lambda^2$. Thus

$$\frac{1}{X} \frac{d^2 X}{dx^2} = -\lambda^2 \tag{1.7.66a}$$

$$\frac{1}{\kappa T} \frac{dT}{dt} = -\lambda^2 \tag{1.7.66b}$$

Both equations have been expressed as ordinary linear differential equations because each only involves one independent variable.

The solution of the first-order Eq. 1.7.66b is an exponential function:

$$T(t) = c_1 e^{-\lambda^2 \kappa t} \tag{1.7.67}$$

because the first derivative of an exponential function is also an exponential function, and this must be true to satisfy the original Eq. 1.7.66b.

At this point, several values for the constant $-\lambda^2$ may be eliminated. If λ^2 were negative, then the exponential function $e^{-\lambda^2 \kappa t}$ would increase without bound as time increased. There are not many remaining systems for which this is true; we thus reject negative values for λ^2 as being unrealistic. We can also reject $\lambda^2 = 0$, because at $\lambda^2 = 0$, the time response becomes static, and there are few, if any systems like this. Thus λ^2 must be positive.

The other differential equation 1.7.66a must have a solution composed of sines and cosines, because these functions, when differentiated twice, also yield sines and cosines. Thus

$$X(x) = C_2 \cos \lambda x + C_3 \sin \lambda x \tag{1.7.68}$$

Because sines and cosines are periodic functions, there is an infinite number of solutions to this equation. Because the sum of all solutions must also solve the equation, the final solution is:

$$\phi(x, t) = X(x)T(t) = \sum_{i=1}^{\infty} C_1 e^{-\lambda^2 \kappa t}(C_2 \cos \lambda x + C_3 \sin \lambda x) \tag{1.7.69}$$

The constants C_1, C_2, and C_3, and λ must be found from boundary conditions for the particular problem to be solved, and often the values for λ must be obtained from the solution of transcendental equations. For instance, the solution for the set of boundary conditions:

$$\phi = \phi_0, \qquad t = 0, x = x$$

$$\phi = \phi_1, \qquad t = t, x = 0$$

$$\phi = \phi_1, \qquad t = t, x = 2L$$

is

$$\frac{\phi_1 - \phi}{\phi_1 - \phi_0} = \frac{4}{\pi} \sum_{i=1,3,5,\ldots}^{\infty} \left[\frac{1}{i} \exp\left(\frac{-i^2 \pi^2 \kappa t}{4L^2} \right) \right] \left(\sin \frac{i\pi x}{2L} \right) \tag{1.7.70}$$

where exp denotes the exponentiation of the argument. Notice that the effort variable was transformed into a dimensionless and normalized (ranging from 0 to 1) difference ratio.

There are several instances in transport systems where solutions such as Eq. 1.7.70 will be necessary to use. Fortunately, the most common of these has been reduced to graphical form, and this will be the preferred method of solution herein.

relating diffusivity and resistance

As one last point, diffusivity can be related to previous discussions about resistance and capacity. From the definition of resistance (Section 1.4.2) and Eq. 1.7.61, the resistance becomes:

$$R = \frac{\Delta\phi}{\dot{\omega}} = \frac{L}{\kappa} \tag{1.7.71}$$

where R is the resistance (sec/m); κ, the diffusivity (m^2/sec); and L, the length (m).

Notice that using the effort and flow variables defined in Table 1.7.1 results in a resistance with units of sec/m for all three transport processes. This is not the case for resistances given in Section 1.4.2, with units that depend on the particular transport system.

capacity is unrelated to diffusivity

From the definition of the time constant (Section 1.6.5) and Eq. 1.7.67,

$$\tau = RC = 1/\kappa\lambda^2 \tag{1.7.72}$$

where τ is the time constant (sec); C, the capacity (m); and λ, a parameter (m^{-1}). Then

$$C = 1/\lambda^2 L \tag{1.7.73}$$

Therefore, diffusivity does not contribute to capacity with this system of effort variables.

1.7.3.3 Energy Balance The general flow balance Eq. 1.7.15 can also be used to form an energy balance. To be most useful, this balance must include both kinetic and potential energy. Distinctions can be drawn between rotational and translational energy, but this usually has no utility for the types of systems we are considering.

Energy can pass across the boundaries of a control volume, be stored inside the control volume, and be created, in effect, from chemical, mechanical, or other sources within the control volume. The energy balance thus becomes similar to a materials balance.

To a limited degree, energy can be considered to be a flow variable. Since energy flows spontaneously from high to low states, the energy state can be thought to be the effort variable. Sometimes this concept can be useful in visualizing the meaning of an energy balance.

1.7.4 Differences between Effort and Flow Balances

efforts cannot accumulate

Effort variables cannot accumulate: They cannot be delivered to any certain point and stored. In general, then, effort variable balances are steady state and immediate — there are no time derivatives or integrals of effort variables that appear in balances (excluding certain definitional equations). Thus the sum of forces makes sense; the integral of a force usually does not (this may not be true for some control systems).

Flow variables can accumulate, and they can change rate of flow. Thus derivatives and integrals of flow variables are commonly found in balances.

flow balances contain derivatives and integrals The general form for a material balance, Eq. 1.7.15, contains rates of change and storage. This form must, therefore, be a flow variable balance. Equation 1.7.5 is an effort variable balance that does not contain derivatives or integrals. Equation 1.7.6 is an effort variable balance that contains a derivative, but the derivative is of a flow variable.

Effort and flow variable balances assume different forms because of the different natures of the variables.

1.7.5 Kirchhoff's Laws

When depicted as networks of resistances, capacities, inertias, and sources, transport processes can be analysed using generalized Kirchhoff's laws, which are used for electrical networks (Figure 1.7.9). In general form these are:

node equation 1. The algebraic sum of the flows coming to any junction in a network of elements is always zero:

$$\sum \dot{\omega} = 0 \text{ at every point} \tag{1.7.74}$$

loop equation 2. The algebraic sum of the potential (effort) drops around any closed loop in a network of elements is always zero:

$$\sum \phi = 0 \text{ for every closed loop} \tag{1.7.75}$$

Although Kirchhoff's laws are valid whether the medium is diffuse or confined, they are much easier to apply when transport occurs through conduits. Hence, fluid in pipes, heat in ducts, mass transport in confined spaces, or electrical current in wires are the situations where Kirchhoff's laws are applied to advantage. These rules can be used to help determine flows or levels of effort that might be difficult to determine in other ways.

general node equation If flow is delivered to the junction from a number of origins containing flow sources, resistances, capacities, and inertias, then the junction equation must be

$$\sum_i \dot{\omega}_i + \frac{\sum_j \phi_j}{R_j} + \sum_k C_k \frac{d\phi_k}{dt} + \sum_n \frac{1}{I_n}\int \phi_n \, dt = 0 \tag{1.7.76}$$

where $\dot{\omega}_i$ are the flow rate source values; ϕ_j, ϕ_k, ϕ_1, effort variable values; R_j, resistance values; C_k, capacity values; and I_n, inertia values.

Notice that the terms in Eq. 1.7.76 are all flow terms and are the inverse forms for definitions of resistance, capacity, and inertance given earlier. These

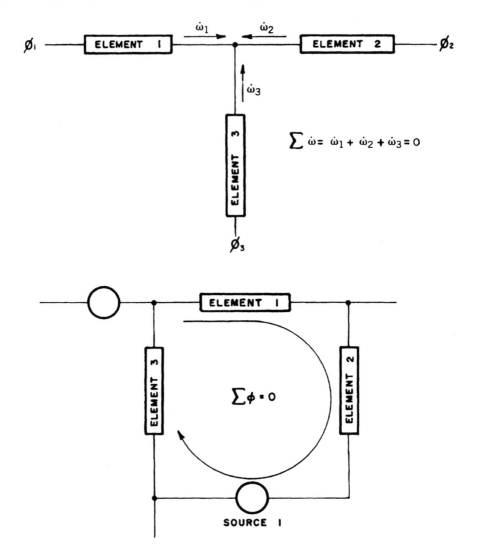

Figure 1.7.9. Kirchhoff's laws state that the sum of flows at any junction are zero (above) and the sum of efforts around any loop are zero (below).

forms may actually be practically more meaningful than the originals. For instance, general capacity was given $C = 1/\phi \int \dot{\omega} \, dt$. Usually, however, effort is not held constant while capacity is measured. Instead, flow is held constant, and the rate of rise of effort is measured. For example, thermal capacity is usually determined by adding heat at a constant rate and the rate of rise of temperature is measured. In this case, $C = \dot{\omega}/(d\phi/dt)$.

relation to balance equation

Careful comparison of Kirchhoff's junction equation (Eq. 1.7.76) with the general flow balance equation (Eq. 1.7.15) shows that the two are very closely related. The first two terms in Eq. 1.7.76 correspond to the (rate of substance in) − (rate of substance out) + (rate of substance generated) in Eq. 1.7.15, and the third term in Eq. 1.7.76 corresponds to the (rate of substance stored) term in Eq. 1.7.15. There is no inertia term in Eq. 1.7.15, which should be evident from Figure 1.7.2. Therefore, Kirchhoff's junction equation is a more general form for the general flow balance equation.

The terms ϕ_j, ϕ_k, and ϕ_n in Eq. 1.7.76 denote the fact that the effort variables in question are differences in effort variable values between their origins and the common junction. There may be any number of these effort variable differences depending on the particular geometry of the example under consideration.

general loop equation

When applying Kirchhoff's loop equation, all terms in the expanded equation must be given in terms of effort variables:

$$\sum_i \phi - \sum_j R_j \dot{\omega}_j - \sum_k \frac{1}{C_k} \int \dot{\omega}_k \, dt - \sum_n I_n \frac{d\dot{\omega}_n}{dt} = 0 \qquad (1.7.77)$$

where $\dot{\omega}_j$, $\dot{\omega}_k$, $\dot{\omega}_n$ are flow rates in different parts of the loop.

Since different flow rates may exist in different parts of the loop, care must be taken to consider only the flows that are appropriate to those parts of the loop.

EXAMPLE 1.7.5-1. Kirchhoff's Laws Applied to Boiling Water

Consider the simple case of heat being supplied to a pot of water from a flame produced when burning natural gas with the optimum amount of air (Figure 1.7.10). The flame temperature of the optimal natural gas–air mixture is 2300°C. A reasonable value for thermal resistance is 0.02°C sec/(N m) and, if the pot has a volume of 0.013 m³, the value for thermal capacity is 54,000 N m/°C. What is the rate of rise of temperature in the pot of water?

Solution

At the interface, or junction, between pot and flame, there are two components of heat flow: heat to the junction from the flame through the resistance, and heat from the junction to the thermal capacitor. To simplify this problem we assume no heat loss to the environment. The node (or junction) equation becomes:

$$\frac{\theta}{R} - C\frac{d\theta}{dt} = 0$$

$$\frac{2300°C}{0.02 \text{ sec}/(N \text{ m})} - 54,000 \text{ N m/sec}\left(\frac{d\theta}{dt}\right) = 0$$

$$\frac{d\theta}{dt} = 2.13°C/\text{sec}$$

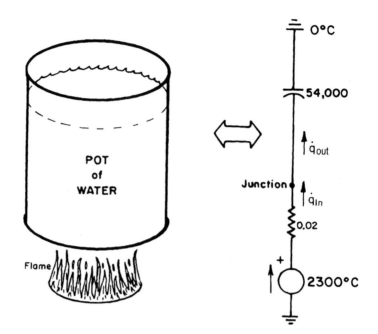

Figure 1.7.10. Thermal and electrical analogs of flow to a junction between resistance and capacity.

Remark

Notice in the figure that the thermal capacitance is assumed to terminate at a temperature of 0°C. When this temperature rises significantly above 0°C, the assumption of no heat loss to the environment can no longer be supported.

EXAMPLE 1.7.5-2. Ventilation Duct

Consider a ventilation duct with two air delivery points (Figure 1.7.11). A flow rate of $9.43 \times 10^{-4} \, \text{m}^3/\text{sec}$ is desired at one outlet, and a flow rate of $4.72 \times 10^{-3} \, \text{m}^3/\text{sec}$ is desired at the other outlet. The ventilation fan produces a pressure of $60 \, \text{N/m}^2$ at a total air delivery rate of $5.66 \times 10^{-3} \, \text{m}^3/\text{sec}$. Dampers in each duct section need to be adjusted to give the proper airflow rate. If the damper in the common duct is set to give a resistance of $8000 \, \text{N sec/m}^5$, what are the other two resistances?

Solution

There are a total of three loops in this example; any two of them give sufficient information to solve the problem, and the additional loop adds no new

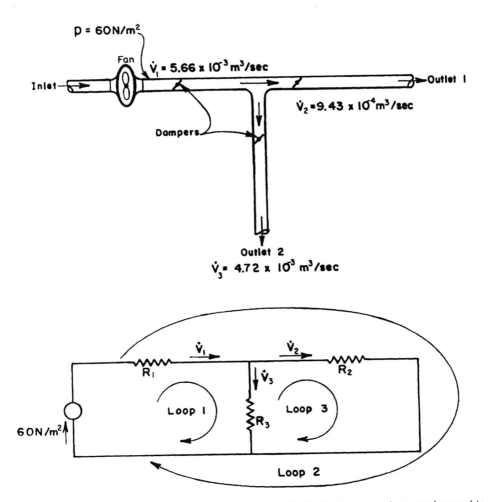

Figure 1.7.11. A ventilation duct and its electrical analog. Kirchhoff's loop equations can be used to solve for flow rates, resistances, or pressures. It should not be hard to see how the same technique can be applied to blood circulation or to caribou migration in a restricted passage.

information. Kirchhoff's equation for loop 1 is

$$60 \text{ N/m}^2 - (5.66 \times 10^{-3} \text{ m}^3/\text{sec})(8000 \text{ N sec/m}^5)$$
$$- (4.72 \times 10^{-3} \text{ m}^3/\text{sec})R_3 = 0$$

and for loop 3 is

$$- (9.43 \times 10^{-4} \text{ m}^3/\text{sec})R_2 - (-4.72 \times 10^{-3} \text{ m}^3/\text{sec})R_3 = 0$$

Solving for R_2 and R_3 values gives

$$R_2 = 15,610 \, \text{N sec}/\text{m}^2$$
$$R_3 = 3119 \, \text{N sec}/\text{m}^5$$

Remarks

There are two things to note: First, the sum of the flows at the junction between all three duct branches equals zero, as predicted from Kirchhoff's node equation, and second, flow rates going in the clockwise direction around the loop are considered positive, and counterclockwise flows are considered to be negative.

EXAMPLE 1.7.5-3. Kirchhoff's Equations for Respiratory Analog

Determine the relevant Kirchhoff's equations for the electrical analog of respiratory mechanics given in the bottom of Figure 1.6.9.

Solution

There are three junctions and three loops in the respiratory mechanics analog. One junction and one loop will be found to be redundant.

For the top junction, where the airways resistance (R_{aw}), lung tissue resistance (R_{lt}), and lung tissue compliance (C_{lt}) connect: from Eq. 1.7.76, $\sum \dot{\omega} = 0$; so

$$\dot{\omega}_{R_{aw}} + \dot{\omega}_{R_{lt}} + \dot{\omega}_{C_{lt}} = 0$$

or, in terms of volume flow rates and pressures,

$$\frac{p_{aw}}{R_{aw}} + \frac{p_{lt}}{R_{lt}} + C_{lt}\frac{dp_{pl}}{dt} = 0$$

where p_{aw} is the pressure across R_{aw}; and p_{pl}, the pressure across both R_{lt} and C_{lt}.

For the bottom junction, where static compliance (C_{st}), R_{lt}, and C_{lt} connect:

$$\dot{\omega}_{C_{st}} + \dot{\omega}_{R_{lt}} + \dot{\omega}_{C_{lt}} = 0$$

or

$$C_{st}\frac{d(p_{pl} - p_{aw})}{dt} + \frac{p_{pl}}{R_{lt}} + C_{lt}\frac{dp_{pl}}{dt} = 0$$

For the junction between R_{aw} and C_{st},

$$\dot{\omega}_{R_{aw}} + \dot{\omega}_{C_{st}} = 0$$

$$\frac{p_{aw}}{R_{aw}} + C_{st}\frac{d(p_{pl} - p_{aw})}{dt} = 0$$

For the leftmost loop containing C_{st}, R_{aw}, and R_{lt}, Eq. 1.7.77 indicates that $\sum \phi = 0$; so

$$\phi_{C_{st}} + \phi_{R_{aw}} + \phi_{R_{lt}} = 0$$

or

$$\frac{1}{C_{st}}\int \dot{V}_{aw}\,dt + R_{aw}\dot{V}_{aw} + R_{lt}\dot{V}_{lt} = 0$$

where $\dot{V}_{aw}$ is the volume flow rate through R_{aw}; and $\dot{V}_{lt}$, the volume flow rate through R_{lt}.

For the rightmost loop containing R_{lt} and C_{lt},

$$\phi_{R_{lt}} + \phi_{C_{lt}} = 0$$

or

$$R_{lt}\dot{V}_{lt} + \frac{1}{C_{lt}}\int \dot{V}_{ct}\,dt = 0$$

where $\dot{V}_{ct}$ is the volume flow rate through C_{lt}.

For the outer loop containing C_{st}, R_{aw}, and C_{lt},

$$\phi_{C_{st}} + \phi_{R_{aw}} + \phi_{C_{lt}} = 0$$

or,

$$\frac{1}{C_{st}}\int \dot{V}_{aw}\,dt + R_{aw}\dot{V}_{aw} + \frac{1}{C_{lt}}\int \dot{V}_{ct}\,dt = 0$$

Four of these six equations can be used to solve for all flows and pressures.

Remark

You may wonder how Kirchhoff's laws can be applied to the mechanical analog at the top of Figure 1.6.9. Actually, they cannot without modification. In the mechanical analog, loop equations would be of the form $\sum \dot{\omega} = 0$, and junction equations would be of the form $\sum \phi \neq 0$; so there is quite a bit of difference.

1.7.6 Visualizing Boundary Conditions

An equation of the form

$$\frac{\partial^2 \phi}{\partial x^2} + \frac{\partial^2 \phi}{\partial y^2} + \frac{\partial^2 \phi}{\partial z^2} + \frac{\omega'''}{k} = \frac{1}{\kappa} \frac{\partial \phi}{\partial t} \qquad (1.7.27)$$

is a partial differential equation because there are more than one independent variable (x, y, z, and t in this case). The visualization process to be described does not apply easily to partial differential equations.

When time is, or can be assumed to be, the only independent variable, then the resulting ordinary differential equation can be conceptualized as composed of resistance, capacity, and inertia elements, and flow and effort sources. The general form for Kirchhoff's loop equation (Eq. 1.7.77) is an example of this process.

Spatial dependencies of resistances, capacities, and inertial elements cannot normally be found except through mathematical techniques, but once expressioins for these elements have been found, those who understand better using conceptual visualization rather than mathematics can be helped by the analogies to be described next.

equation
solution
depends on
boundary
conditions
So much of a solution to a differential equation depends on the boundary conditions. It is these that determine the exact form of the solution. There are three types of boundary conditions to be considered here: (1) constant effort variable, (2) constant flux or flow variable, and (3) radiation or convection. These are all diagrammed in Figure 1.7.12.

The differential equation chosen to illustrate boundary condition effects is the generic, first-order differential equation with a constant coefficient, first introduced in Section 1.7.3.2:

$$\dot{\omega} = -kA \, \frac{d\phi}{dx} \qquad (1.7.18)$$

constant effort
condition
This equation can be diagrammed as a simple resistance element. The constant effort variable boundary condition, with effort variable value of ϕ_1 on one side and of ϕ_2 on the other side of the resistor leads to the solution:

$$\dot{\omega} \int dx = -kA \int d\phi$$

$$\dot{\omega} x = -kA\phi + C \qquad (1.7.78)$$

with boundary conditions:

$$\phi = \phi_1 \quad \text{at } x = 0$$

$$\phi = \phi_2 \quad \text{at } x = L$$

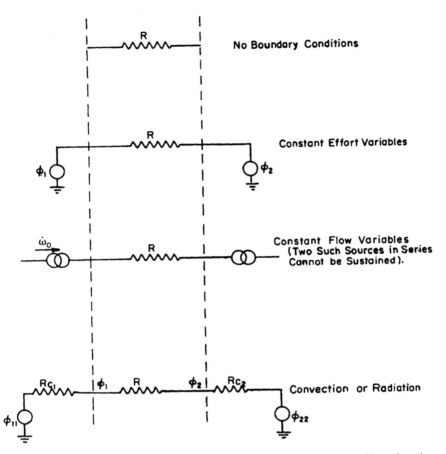

Figure 1.7.12. Visualization of the mathematical expression $\dot{\omega} = -kA\, d\phi/dx$ with various boundary conditions. The form of the answer to the problem depends very heavily on the boundary conditions.

Since there is only one constant of integration, only one boundary condition is required to determine its value. The simplest is the first, because the value for x becomes zero. Substituting the first boundary condition gives:

$$\dot{\omega}x = 0 = -kA\phi_1 + C$$
$$C = kA\phi_1 \tag{1.7.79}$$

and

$$\dot{\omega}x = -kA\phi + kA\phi_1 = kA(\phi_1 - \phi)$$

or

$$\dot{\omega} = \frac{kA}{x}(\phi_1 - \phi) \tag{1.7.80}$$

The second boundary condition eliminates the dependence on the variable x:

$$\dot{\omega} = \frac{kA}{L}(\phi_1 - \phi_2) \tag{1.7.81}$$

It was previously shown that the term (L/kA) is equivalent to a resistance R. Thus

$$\dot{\omega} = (\phi_1 - \phi_2)/R \tag{1.7.82}$$

The same solution would have been obtained if the second boundary condition had been used to determine the value of the constant of integration.

Therefore, the simple differential equation 1.7.18 is equivalent to a resistance element, and the constant effort variable boundary conditions determine the value of the flow through the resistor.

flow source conditions
Flux is the term normally given to a flow per unit area. Constant flow boundary conditions are often given in terms of flux. These are diagrammed, also in Figure 1.7.12, as flow sources. It has been previously explained in Section 1.5 that two flow sources should never be connected in series because each tries to dictate to the other the value of the flow. Thus the illustration in Figure 1.7.12 is meant to be conceptual rather than practical.

Equation 1.7.18 becomes

$$\dot{\omega} = -kA \frac{d\phi}{dx} = \text{const} = \dot{\omega}_0 \tag{1.7.18a}$$

which does not require integration to determine the value of the flow. However, to determine the effort variable difference across the resistance, integration gives

$$-ka\phi = \dot{\omega}_0 x + C$$
$$C = -kA\phi - \dot{\omega}_0 x \tag{1.7.83}$$

Although the effort variable values at each end of the resistor are unknown, they are defined as ϕ_1 (at $x = 0$) and ϕ_2 (at $x = L$). Thus

$$C = -kA\phi_1 - \dot{\omega}_0(0)(\phi_1 - \phi_2) = \dot{\omega}_0 R \tag{1.7.84}$$

in a similar result to the constant effort boundary condition.

convection boundary condition
Sometimes there is a surface process that also contributes to the resistance to flow through an object. These processes in heat and mass transfer are called convection, although some classical treatments of heat transfer equations call them radiation. These processes are diagrammed in Figure 1.7.12 as additional resistances connected to constant effort variables.

By now, it should start to become clear that the visualization through diagrams of the simple, but useful, Eq. 1.7.12 leads to equation solutions in a simpler manner than through mathematical solution. Indeed, it is easy to write down the expression for the flow variable value based on the effort variable difference and the combination of resistances in series

$$\dot{\omega} = \frac{(\phi_{11} - \phi_{22})}{R_{c_1} + R + R_{c_2}} \tag{1.7.85}$$

It can also be easily seen that the effort variable values at each surface are somewhere between, but not equal to, ϕ_{11} and ϕ_{22}. To determine what the surface effort variable values are, just use the fact that the flow is the same through resistances in series:

$$\dot{\omega} = \frac{(\phi_{11} - \phi_1)}{R_{c_1}} = \frac{(\phi_1 - \phi_2)}{R} = \frac{(\phi_2 - \phi_{22})}{R_{c_2}} \tag{1.7.86}$$

and any effort variable value can be obtained. The value for ϕ_1, for example, is

$$\phi_1 = \phi_{11} - \frac{R_{c_1}}{(R_{c_1} + R + R_{c_2})}(\phi_{11} - \phi_{22}) \tag{1.7.87}$$

Knowing the relationship between a mathematical expression and its schematic diagram often leads to a quick understanding of the essentials of a problem.

1.8 SYSTEM APPLICATIONS

the importance of conceptualization

In the previous sections the discussion was centered around general effort and flow variables, their interrelationships, means of combining various elements, and the formation of balances that can be used to determine unknown parameter values once the system is defined. It is good for the student to appreciate the general nature of transport processes because answers that are not always obvious in one type of process may be more easily obtained in another. That is, those students who conceptualize systems in electrical terms usually find it easier to deal with fluid systems as equivalent electrical circuits. Other students are more used to conceptualizing in fluid systems, or mechanical components, or heat transfer elements. Very few students view the world as equivalent mass-transfer systems.

transforming from one system to another

If you can visualize each of these systems in its own terms, that is excellent. To be able to visualize all types of systems in terms of one type is nearly as good. The general discussions preceding this section are meant to foster transferring information from one type of transport process into another.

However, general discussions alone are not usually sufficient to form the mental images required for complete understanding. For this we need applications to specific types of processes, and, in addition, we need examples. The following sections are intended to fill this need.

1.8.1 Flow through Porous Media

Fluid flowing through pipes is channelled and one dimensional. There are special relationships that have been developed for pipe flow. We will consider them in Chapter 2.

Fluid flowing through porous media may or may not be one dimensional. Furthermore, flow through porous media can be analyzed by the means previously given. Thus we will discuss this aspect of flowing fluids here.

There are many fluid/media examples to be found: water through soil, solvents through packed beds, air through charcoal, natural gas and oxygen through porous ceramic, and water through biological filters are some of these. In each case, the medium through which the fluid is flowing will be considered from a large-enough dimensional scale that it will be assumed not composed of finite particles with channels, but as a uniform medium that impedes fluid flow.

Darcy's law   Darcy's law is the equation usually used to describe one-dimensional flow through porous media:

$$\dot{V} = \frac{-kA}{\mu}\left(\frac{\partial p}{\partial L} + \rho g \sin \alpha\right) \qquad (1.8.1)$$

where $\dot{V}$ is the volumetric flow rate (m^3/sec); k, the permeability (m^2); μ, the fluid viscosity (N sec/m^2); A, the cross-sectional area (m^2); $\partial p/\partial L$, the pressure gradient along the flow path (N/m^3); ρ, the fluid density (kg/m^3); g, the acceleration due to gravity (m/sec^2 or N/kg); and α, the angle of inclination of the flow path with respect to the horizontal (rad).

The second term of Darcy's law takes into account the difference in potential energy of the fluid as it either rises or falls within a gravitational field. When flow is strictly horizontal, or when the fluid density is very small, or when the pressure gradient is much greater than the gravitational potential gradient, then the second term may be neglected, and Darcy's law becomes:

$$\dot{V} = \frac{-kA}{\mu}\frac{\partial p}{\partial L} \qquad (1.8.2)$$

Notice that the negative sign signifies that flow is positive when the pressure gradient is negative.

permeability   The permeability, k, is usually an average value of the ease with which a fluid will flow through the medium. The higher the permeability, the easier it is for the fluid to flow through the medium. The permeability takes into account all the various flow channels that differ in size, shape, direction, and interconnections. The fluid will often flow preferentially through certain flow channels in the medium. The average permeability ignores this fact.

permeability
coefficient

Hydrologists, dealing only with the flow of water (Figure 1.8.1), will frequently combine the viscosity and permeability to form a permeability coefficient:

$$k_p = k/\mu \tag{1.8.3}$$

where k_p is the permeability coefficient $[(\text{N sec})^{-1}]$.

If $\dot{V}$ in Eq. 1.8.1 is divided by the area A, then the resulting equation is similar in form to the general flow equation 1.7.19. Although Eq. 1.8.2 has been presented for flow in one dimension only, similar equations can be given for flow in the other two spatial dimensions, similar to Eqs. 1.7.20 and 1.7.21.

three
geometries

Since these equations are all of the same form as Eqs. 1.7.19–1.7.21, the results derived in Section 1.7.3.2 are applicable here. In particular, fluid flowing one dimensionally through porous media bounded by surfaces of a flat plane is given by

$$\dot{V} = \frac{-kA}{\mu L}(p_2 - p_1) \tag{1.8.4}$$

similar to Eq. 1.7.36.

Figure 1.8.1. Flow of water through soil is just one example of flow through a porous medium. There is evidence that water moves more quickly through soil than would otherwise be expected, and the reason is preferential channels probably caused by freezing and thawing, earthworms, decay of roots, soilborne animals, drying and wetting, or dissolution of dissolvable minerals. Such nonhomogeneities can also be present in other examples of flow through porous media.

If the flow is one dimensional in a radial direction in a cylinder, then

$$\dot{V} = \frac{-2\pi kL(p_1 - p_2)}{\mu \ln(r_2/r_1)} \tag{1.8.5}$$

similar to Eq. 1.7.43.

If the flow is one dimensional in a radial direction in a sphere, then

$$\dot{V} = \frac{-2\pi k(\phi_1 - \phi_2)}{\mu(1/r_1 - 1/r_2)} \tag{1.8.6}$$

similar to Eq. 1.7.46.

laminar flow necessary   Darcy's law is an adequate representation of flow through porous media only when flow is laminar, or streamlined. The effect of inertia in streamline flow is negligible because friction, or viscous efffects, dominate fluid particle flow paths. At high particle velocities, however, particle momentum (mass times velocity) becomes large enough that particle flow paths no longer easily bend around corners in the porous medium. Small eddies begin to form, and flow is no longer streamlined, or layered.

Reynolds number   A means to determine when inertial effects can no longer be neglected is the calculation of the Reynolds number in a porous medium

$$Re = \frac{dv\rho}{\mu} \tag{1.8.7}$$

where Re is the Reynolds number (dimensionless); d, the effective particle diameter (m); v, the superficial particle speed (m/sec); ρ, the density (kg/m^3); and μ, the viscosity [kg/(sec m)].

Because fluid velocity within a porous medium is usually determined from volumetric flow rate divided by total cross-sectional area (not just pore area), then the Reynolds number can be calculated as:

$$Re = \frac{d\dot{V}\rho}{\mu A} \tag{1.8.8}$$

where $\dot{V}$ is the volumetric flow rate (m^3/sec); and A, the total cross-sectional area (m^2).

When the value of the Reynolds number is greater than 1.0, Darcy's law has been found to fail adequately to describe flow through a porous medium. A correction factor is usually applied in this case.

Klinkenberg flow   The second restriction on the use of Darcy's law applies to gaseous flow at low pressures. When the average distance traveled by gas molecules between molecular collisions becomes comparable to the radii of the medium pore spaces, then flow can no longer be considered to be streamlined. Low pressures are required for this to happen because only at extremely low pressures will the mean free path of the gas lengthen to become as large as the pore radii. This type of flow is known as Knudsen flow in mass transfer and Klinkenberg flow in momentum transfer (fluid flow).

In Klinkenberg flow, gas molecules slip past one another rather than flow with one another. This results in higher flow rates than predicted on the basis of Darcy's law. An apparent permeability can be used so that Darcy's law can be used anyway. This apparent permeability is given by:

$$k_{ap} = k(1 + b/p) \tag{1.8.9}$$

where k_{ap} is the apparent permeability (m^2); k, the true permeability (m^2); b, a constant (m^2/N); and p, the mean absolute pressure of the gas (N/m^2).

The value for b will depend on the natures of both the gas and the medium.

resistance
Aside from these few restrictions on the use of Darcy's law, analysis of flow through porous media fits very well within the general development given earlier. For this reason, it can be used as a conceptual framework for analogically considering other transport processes. Thus resistance to flow through a cylindrical porous structure, with flow occurring radially, is

$$R = \frac{\mu \ln(r_2/r_1)}{2\pi k L} \tag{1.8.10}$$

where R is the fluid resistance $(N \sec/m^5)$, similar to Eq. 1.7.41.

Capacity is obtained from Eq. 1.7.38 as being the product of the volume (V) and a physical property (ξ). We must now determine what physical property is involved.

compliance, or capacity
From Eq. 1.4.8, compliance is given as

$$C_f = \frac{1}{p} \int_0^t \dot{V} \, dt = \frac{\Delta V}{\Delta p} \tag{1.4.8}$$

bulk modulus
The relation between pressure and volume changes in a fluid is also given by its bulk modulus of elasticity:

$$B = \frac{-\Delta p}{\Delta V/V} \tag{1.8.11}$$

where B is the bulk modulus of elasticity (N/m^2); and V, the volume (m^3).

The negative sign indicates that volume decreases as pressure increases. Thus

$$C_f = \frac{V}{B} \tag{1.8.12}$$

where compliance is taken to be a positive value.

Comparing compliance from Eq. 1.7.38 with that in Eq. 1.8.12 shows that the physical property required is the inverse of the bulk modulus. Thus

$$C_f = \frac{V}{B} = \frac{\pi(r_2^2 - r_1^2)Lv}{B} \tag{1.8.13}$$

where C_f is the compliance (m^5/N); r_2, the outside radius of the hollow cylinder (m); r_1, the inside radius of the hollow cylinder (m); L, the cylinder length (m); v, the void fraction of the medium (dimensionless); and B, the bulk modulus (N/m^2).

time constant
A sudden change of filling pressure will result in an exponential increase in flow, characterized by a time constant of

$$\tau = \frac{v \ln(r_2/r_1)(r_2^2 - r_1^2)\mu}{2kB} \tag{1.8.14}$$

where τ is the time constant (sec).

The bulk modulus of an ideal gas is

$$B = p, \quad \text{isothermal process} \tag{1.8.15a}$$

$$B = pc_p/c_v, \quad \text{adiabatic process} \tag{1.8.15b}$$

where p is the total gas pressure (N/m^2); C_v, the specific heat at constant volume [m^2/(sec^2 °C)]; and C_p, the specific heat at constant pressure [m^2/(sec^2 °C)].

At atmospheric pressure, B for an ideal gas is 10^5 N/m^2 for isothermal compression and 1.4×10^5 N/m^2 for adiabatic compression. Other representative values of B are 2.2×10^9 for water, 1.7×10^9 for oil, 2.6×10^{10} for mercury, and 2×10^{11} for steel.

time constant related to speed of sound
Thus an interesting effect should be able to be seen for a cylindrical porous medium filled initially with air, but with water displacing the air from the center of the cylinder (if water were displacing the air from the bottom, rectangular coordinates would be more convenient to use). The moving front of water raises the local pressure on the air and compresses it slightly. The pressure at the outer surface of the cylinder should then rise exponentially toward a higher value (Figure 1.8.2). Once water displaces all the air in the voids, any pressure change occurring at the inner surface of the hollow cylinder is transmitted almost instantaneously to the outer surface. This is because water is so much less compressible than air that its time constant is nearly infinitesimal compared to that for air. If the medium through which the water was flowing were made of steel, pressure changes would be transmitted faster through the steel than through the water, because steel is much less compressible and its time constant is thus much shorter. Thus sound should travel faster through steel than through water, and through water faster than through air.

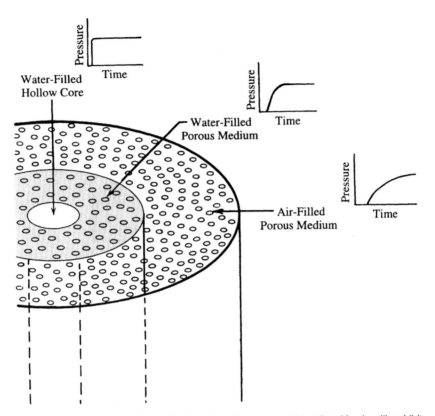

Figure 1.8.2. A hollow cylinder filled partly with water and partly with air will exhibit different capacities depending on the fluid. Since the water is nearly incompressible, its bulk modulus will be much higher than the bulk modulus for air. Pressure changes will thus be transmitted more rapidly through water.

Flow through porous media is an example of the application of the general flow balance equation. Although the application was fully developed only for hollow cylindrical coordinates, the other coordinate systems could be used where geometry makes them most convenient.

1.8.2 Conduction Heat Transfer

Fourier's law   Fourier's law for heat conduction in fluids or solids is given by:

$$\dot{q} = -kA \frac{d\theta}{dx} \qquad (1.8.16)$$

where $\dot{q}$ is the rate of heat transfer (N m/sec); k, the thermal conductivity [N/(°C sec)]; A, the cross-sectional area (m^2); θ, the temperature (°C); and x, the direction of heat flow (m).

The negative sign indicates that heat flows positively from a region of higher temperature to lower temperature.

thermal conductivity The thermal conductivity is a parameter whose value depends on the medium through which the heat is moving. High thermal conductivity values [of the order of $100 \, N/(°C \, \text{sec})$] are characteristic of materials called heat conductors and low conductivity values [of the order of $0.1 \, N(°C \, \text{sec})$] are said to belong to heat insulators.

three geometries Steady-state, one-dimensional heat conduction can be obtained from Eqs. 1.7.36, 1.7.43, and 1.7.46 for rectangular, cylindrical, and spherical coordinates, with the generalized effort variable replaced by temperature

$$\dot{q} = \frac{kA}{L}(\theta_2 - \theta_1) \tag{1.8.17}$$

$$\dot{q} = \frac{2\pi kL(\theta_0 - \theta_i)}{\ln(r_0/r_i)} \tag{1.8.18}$$

$$\dot{q} = \frac{4\pi k(\theta_0 - \theta_i)}{1/r_i - 1/r_0} \tag{1.8.19}$$

Restrictions on the use of Fourier's law of heat conduction are few: There can be no internal heat generation, and the medium through which the heat is flowing must be static. Heat can be transferred by a flowing fluid, but this is convective heat transfer, not conduction, and will be dealt with later.

resistances Resistances to heat flow (Figure 1.8.3) can be obtained from previous analyses to be

$$R = L/ka \tag{1.8.20}$$

$$R = \frac{\ln(r_0/r_i)}{2\pi kL} \tag{1.8.21}$$

$$R = \frac{1/r_i - 1/r_i}{4\pi k} \tag{1.8.22}$$

for rectangular, cylindrical, and spherical coordinate systems.

The rate of heat storage is given by

$$\dot{q} = mc_p \frac{\partial \theta}{\partial t} \tag{1.8.23}$$

where m is the mass of the object (kg); and c_p, the specific heat [at constant pressure, $N \, m/(kg \, °C)$].

The value of specific heat is a physical property of the mass storing heat.

thermal capacity Since mass is the product of density and volume,

$$\dot{q} = \rho c_p V \frac{\partial \theta}{\partial t} \tag{1.8.24}$$

where ρ is the density (kg/m^3); and V, the volume (m^3).

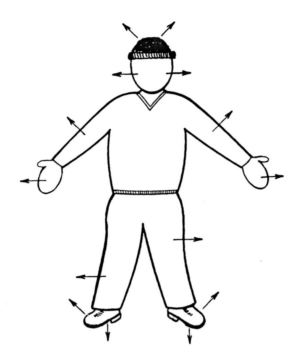

Figure 1.8.3. Heat loss through clothing represents a combination of series and parallel resistances. Conduction resistance to heat flow is given as $R = L/kA$. To use clothes to insulate against heat loss, thicker clothes of lower thermal conductivity material are used. The use of more insulation on one part of the body allows the person to tolerate higher heat loss on another part.

This equation can be seen to be of the form of Eq. 1.7.25, with $\zeta = \rho c_p$. Therefore, thermal capacity is given by

$$C = \rho c_p V = m c_p \tag{1.8.25}$$

time constants and the time constant characterizing responses of heat flow to sudden changes in temperature difference is

$$\tau = \frac{\rho c_p L^2}{k} \tag{1.8.26}$$

as in Eq. 1.7.39,

$$\tau = \frac{\rho c_p (r_o^2 - r_i^2)}{2k} [\ln(r_o/r_i)] \tag{1.8.27}$$

as in Eq. 1.7.48, and

$$\tau = \frac{\rho c_p (r_o^3 - r_i^3)(1/r_i - 1/r_o)}{3k} \tag{1.8.28}$$

as in Eq. 1.7.49.

1.8.3 Binary Diffusion Mass Transfer

Fick's laws   Diffusion mass transfer occurs because of a chemical activity difference of mass between two locations. This usually means a difference in concentration. Fick's law is the equation governing diffusion, and it comes in two forms: the steady-state form

$$\dot{m}_A = -D_{AB} A \, \frac{dc_A}{dz} \tag{1.8.29}$$

and the non-steady-state form

$$\frac{\partial c_A}{\partial t} = D_{AB} \, \frac{\partial^2 c_A}{\partial z^2} \tag{1.8.30}$$

where $\dot{m}_A$ is the mass flux of substance A (kg/sec); D_{AB}, the mass diffusivity of substance A diffusing through substance B (m^2/sec); A, the area perpendicular to the direction of diffusion (m^2); c_A, the concentration of substance A (kg/m^3); z, the Cartesian dimension of length (m); and t, the time (sec).

These forms for Fick's law are appropriate for binary diffusion in one dimension. Binary diffusion means that only two substances are present, and the unidimensionality of these equations can be replaced by multidimensionality with the guidance of Eqs. 1.7.28 and 1.7.29.

mass diffusivity   The mass diffusivity, also called the diffusion coefficient, assumes higher values for gases than for liquids, and higher values for liquids compared to solids. This reflects the fact that gas molecules are very free to move in any given space. Liquid molecules are somewhat more constrained by surrounding molecules, and molecules of solids are bound to each other quite tightly. Values for gas diffusivities are of the order of 10^{-5} m^2/sec, values for liquids are of the order of 10^{-9} to 10^{-11} m^2/sec, and values for solids vary from 10^{-10} m^2/sec for amorphous solids to 10^{-34} m^2/sec for crystalline solids. Diffusivity values are not constant, but depend on temperature, pressure, and molecular weight for gases and liquids. In addition, diffusivities for solutes in liquids depend on solute concentration.

binary mixtures   These forms of Fick's law are to be applied to binary mixtures only. When more than two substances are present, it is frequent that only one of them is of particular interest, and the others can be grouped together and treated as a single substance. An example of this is air, which is itself a mixture of other gases, but which does not usually vary in composition from one place to another. Since the various constituents of air are not subject to diffusion pressure among themselves, air is usually treated as a single gas. Thus the diffusion of water in air and ammonia in air are both treated as binary diffusion.

Diffusion of various solutes in water is of particular biological importance. Many biological systems are more complex, however. There are biological gels, porous media, and membranes to contend with, and each of these must be analyzed as a different diffusion mode.

Steady-state one-dimensional diffusion across a substance of thickness L bounded by two flat planes has a resistance given by

$$R = \frac{L}{D_{AB}A} \tag{1.8.31}$$

resistances

as indicated by Eq. 1.7.37. The resistance for one-dimensional radial diffusion in a cylinder is given by

$$R = \frac{\ln(r_o/r_i)}{2\pi D_{AB}L} \tag{1.8.32}$$

as indicated by Eq. 1.7.44. The resistance for one-dimensional steady-state diffusion within a hollow sphere is given by

$$R = \frac{1/r_i - 1/r_o}{4\pi D_{AB}} \tag{1.8.33}$$

resistance of shells

as indicated by Eq. 1.7.47.

If the diffusion occurs through two spherical shells (Figure 1.8.4), one inside the other, each with its own diffusivity, then the two resistances are in series, because the same mass that flows through one shell must also flow through the other. According to Eq. 1.6.1

$$R_{\text{tot}} = \frac{1/r_1 - 1/r_2}{4\pi D_{AB}} + \frac{1/r_2 - 1/r_3}{4\pi D_{AC}} \tag{1.8.34}$$

where R_{tot} is the total resistance to diffusion (sec/m^3); r_1, r_2, r_3, the radial boundaries of the inner and outer shell (m); and D_{AB}, D_{AC}, the mass diffusivities of substance B in one shell and of substance C in another shell (m^2/sec).

On the other hand, if the hollow sphere were split into hemispheres, each with its own diffusivity, then the two resistances are in parallel, because the path taken by the diffusing mass may split, with part of the mass flowing through each resistance. According to Eq. 1.6.2

$$\frac{1}{R_{\text{tot}}} = \frac{2\pi D_{AB}}{1/r_1 - 1/r_2} + \frac{2\pi D_{AC}}{1/r_1 - 1/r_2} \tag{1.8.35}$$

$$R_{\text{tot}} = \frac{(1/r_1 - 1/r_2)}{2\pi(D_{AC} + D_{AB})} \tag{1.8.36}$$

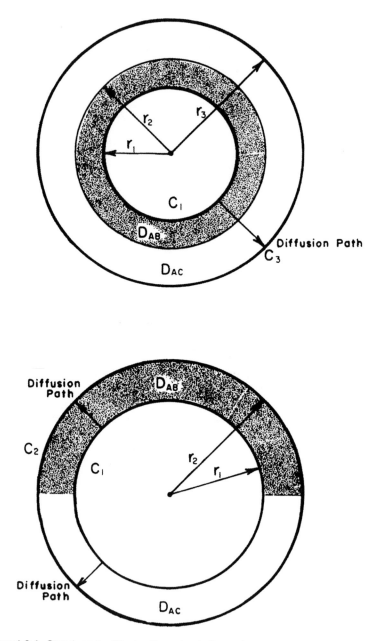

Figure 1.8.4. Steady-state diffusion through a hollow sphere contrasting series (top) and parallel (bottom) diffusion regions. An example of series diffusion is water diffusing through the skin and waxy layer of an apple. An example of parallel diffusion is the apple with part of the skin removed.

The use of 2π, rather than 4π in the numerator accounts for the fact that only $\frac{1}{2}$ of the spherical area is involved in each diffusion region.

mass
capacitance

Notice from Eq. 1.4.9 that mass capacitance has units of m^3 and thus must represent a volume. Since it has been shown in Eq. 1.7.17 that capacity is given by ξV, and V is volume, the parameter ξ must be dimensionless.

There is no normally measured parameter representing mass capacitance. The volume represented by mass capacitance is the volume of the mass that has diffused into the volume of another substance. Thus the quantity ξV represents the original volume of the diffusing material. For a binary system of gases, the total volume after diffusion is expected to be the sum of volumes of the two separate gases; for a binary system where one substance is crystalline, the total volume is not expected to be much different from the volume of the crystalline substance; for liquid solutions, the solute fits among solvent molecules and the total volume does not change. Thus the parameter ξ changes, depending on the type of binary system employed.

time constant

Time constants for the planar, cylindrical, and spherical systems are given by the resistances in Eqs. 1.8.31–1.8.33 multiplied by ξV. These time constants characterize the time it takes for two separate substances to diffuse completely into one another. Any mass inertia present is very small, and entirely negligible except on a very microscopic scale. Thus the time constant is the best way to characterize the transient behavior of a diffusion system.

1.8.4 Conduction of Electricity

Although electricity is mostly thought to be confined to special conduction paths defined by wires, resistors, capacitors, inductors, and other circuit elements, diffuse conduction occurs in many atmospheric, aquatic, and biological media (Figure 1.8.5). The general field Eq. 1.7.26 can be applied to such systems.

diffuse
conduction

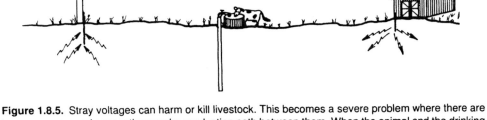

Figure 1.8.5. Stray voltages can harm or kill livestock. This becomes a severe problem where there are two or more ground connections and a conduction path between them. When the animal and the drinking water are at two different voltages, a large enough current can flow through the animal's body to cause severe injury. This is an instance of diffuse electrical conduction, both through the earth and through the animal's body.

We have already seen that the effort variable for electrical conduction is voltage difference and the flow variable is current flow. It makes no difference for our purposes whether current is considered to be the flow of positive or negative charges. As long as the electrical system is linear (current and voltage do not appear with powers greater than 1.0), and components remain constantly valued, superposition is valid, and it does not matter which component is first encountered by the current.

In the general field Eq. 1.7.26 appear two physical properties (k and ξ) that must be defined for electrical systems. The first of these is given by an

electrical conductivity

electrical conductivity

$$k = \sigma = 1/\rho \qquad (1.8.37)$$

where k is the general constant in Eq. 1.7.26; σ, the electrical conductivity (N sec m/coul^2); and ρ, the electrical resistivity, [coul^2/N sec m)] and the second is

$$\xi = C/V \qquad (1.8.38)$$

where ξ is the general constant in Eq. 1.7.26; C, the electrical capacitance [coul^2/(N m)]; and V, the volume (m^3).

Thus the Fourier Eq. 1.7.28 for diffuse electrical conduction is

$$\frac{\partial^2 e}{\partial x^2} + \frac{\partial^2 e}{\partial y^2} + \frac{\partial^2 e}{\partial z^2} = \frac{C\rho}{V} \frac{\partial e}{\partial t} \qquad (1.8.39)$$

where e is the electrical potential (N m/coul, or volts).

resistance

Electrical resistance for the hollow cylinder follows Eq. 1.7.44 as

$$R = \rho \frac{\ln(r_o/r_i)}{2\pi L} \qquad (1.8.40)$$

where R is the electrical resistance (N sec/coul^2, or ohms); r_o, the outer boundary of the hollow cylinder (m); r_i, the inner boundary of the hollow cylinder (m); and L, the length of the cylinder (m).

Resistivity (ρ) rather than conductivity (σ) is used in Eqs. 1.8.38 and 1.8.40 because resistivity is the more commonly tabulated parameter.

time constant

The time constant for this system is

$$\tau = RC = \frac{\rho \ln(r_o/r_i)C}{2\pi L} \qquad (1.8.41)$$

Resistances for planar and spherical geometries follow from Eqs. 1.1.29 and 1.7.47.

1.8.5 Other Transport Systems

mechanical
systems

There are other systems that resemble those already introduced. These include mechanical systems, optical systems, and magnetic systems. In the case of mechanical systems, the effort variable is force, and the flow variable is speed. Power is thus given by the product of effort and flow variables, or force times distance divided by time:

$$P = F\dot{x} \qquad (1.8.42)$$

where P is power (N m/sec); F, force (N); and $\dot{x}$, speed (m/sec).

Mechanical resistance is the quotient of force divided by speed. Mechanical resistors are usually called dashpots (or shock absorbers in automobile suspension systems).

Mechanical capacitors are springs, and since the force developed in a spring is the spring constant times the deflection

$$F = kx = k \int \dot{x}\, dt \qquad (1.8.43)$$

where k is the spring constant (N/m); x, the spring deflection (m); and t, the time (sec), then the mechanical capacitance is the spring constant (see Eq. 1.4.6).

Mechanical inertia is given by mass. We can see this from Newton's second law

$$F = ma = m\,\frac{d\dot{x}}{dt} \qquad (1.8.44)$$

where m is the mass. Comparing Eq. 1.8.44 with 1.4.10 shows that mass is equivalent to electrical inductance.

Systems where mass, elasticity, and resistance are distributed throughout space are not commonly treated by the means developed here, and, therefore, little more will be said about these systems. Application areas do appear, however, in astrophysics and noncontinuum mechanics.

optics

The field of optics could be understood by the general approach used here, but rarely, if ever, is this done. Light (or photon) sources represent effort sources, and flows of photons are the flow variables. Resistance is represented by medium absorption and capacity by internal reflection. There is little likelihood of the existence of optical inertia except, perhaps, on an astrophysical scale, where emitted light travels on despite the quenching of its source.

magnetics

Magnetics also could be considered using this approach. The magnetic field strength (or magnetomotive force) is the effort variable, and magnetic flux is the flow variable. Resistance in a magnetic circuit is called reluctance.

Magnetic capacity is related to magnetic permeability. Inertia for most magnetic systems would be negligible.

The previous processes were all physical. These concepts can be applied, however, to inherently biological, psychological, sociological, political, economic, and other systems. In psychology, for instance, the effort variable could be motivation in higher-order organisms, or attractive biological stimuli for lower-order organisms; and the flow variable could usually be physical movement, or sometimes biochemical reaction. Resistance, being the ratio of the effort variable to the flow variable, determines the vigor of the response. Capacity determines eagerness and anticipation. Inertia represents reticence or habit. Problems ranging from the destruction of the tropical rain forest to the movement of microbes should be able to be treated profitably as relations between appropriate effort and flow variables.

psychology

I once attended a speech on technology transfer, where means to impart knowledge and skills to others were being discussed. The speaker talked about a source, a receiver, and the transfer medium. It soon became apparent that the whole issue of technology transfer could be understood well in the context of effort and flow variables, resistance, capacity, and inertia.

technology transfer

There may be other analogous transport systems that can be conceptualized in an equivalent way. The point here is not, however, to be exhaustive in detail, but to point out that general concepts developed for transport systems are powerful and widely applicable. The engineer need only learn the general approach well, and from then on she/he has the key to understanding of all these systems.

1.9 SYSTEMS APPROACH

We have seen that there are some commonalities among transport processes. Each has at least one effort variable and one flow variable. There are the physical properties of resistance, capacity, and inertia that are defined in terms of appropriate effort and flow variables. Various balances can be written in terms of these effort and flow variables in order to apply the general concepts to specific cases.

do not sweat the small stuff

To use the systems approach, one must learn to conceptualize a problem in the standard means we have shown in this chapter. It helps to be able to conceptualize in terms of electrical symbols, because these are both succinct and thoroughly defined, but it is not necessary as long as the symbolic concepts are well defined. Each symbol used must be "pure," in the sense that resistance, capacity, and inertia are not mixed together in a way that makes it difficult to translate the symbolic form into mathematical form. Done correctly, this step is trivial, and the problem can be solved easily.

summary

To summarize, then, we have seen that, at least abstractly, there are two kinds of sources: effort and flow. Effort sources maintain the effort variable constant, but can supply any needed flow. Flow sources maintain constant flow, but can supply any required effort in order to maintain that constant flow.

Resistance limits flow for an effort source, but has no such effect for a flow source. Resistance can transform a flow into an effort. Resistance also dissipates energy and power.

Capacity stores flow and maintains effort. It does not dissipate energy, but stores it instead.

Inertia maintains flow, and thus it stores kinetic energy.

The time scale for system changes depends on the magnitudes for resistance, capacity, and inertia. The system may either manifest an exponential or oscillatory response. Resistance–capacity systems are most common, and, for these, larger resistance or capacity values lead to slower exponential responses.

EXAMPLE 1.9-1. Systems Diagram in Ecology

Draw the systems diagram and give the systems equation for a large herd of caribou migrating through a constrictive narrow valley.

Solution

The valley can be represented by a resistance that regulates the flow of animals. On the downstream side of the valley, the caribou are presumed to migrate freely without significant impediment. We can thus represent the downstream side as a connection to the zero potential for the effort variable, which is migratory pressure.

Representation of the upstream side of the valley could be one of three possibilities:

1. A capacity element. Using this element would signify that the number of migrating animals is finite, whatever could be stored in the capacity element. Flow of animals from the element would decrease as the migratory pressure, and the number of animals upstream, decreases.

2. A pressure source. Using this element would signify that the migratory pressure of animals attempting to move through the valley does not decrease with the passage of animals. However, the constriction of the valley still has an effect on the flow of animals through it.

3. A flow source. Use of this element would mean that flow of animals would be constant, and not affected by the resistance represented by the valley.

Of these three, the flow source is clearly wrong. We know that a more restrictive valley would slow the flow of animals. There are some properties of the capacity element and the pressure source that make each of these at least partially correct. The capacity element represents the fact that the number of animals is finite; the pressure source, however, indicates that the flow of migrating animals is not likely to decrease over time (except, perhaps, for the very last animals). The best choice for a representation of the upstream side of the valley is thus the pressure source. The systems diagram for caribou passing through a restrictive valley thus appears in Figure 1.9.1; the systems equation, after Eq. 1.7.76, is:

$$\text{flow} = \dot{N} = p_s / R_v$$

EXAMPLE 1.9-2. Systems Diagram in the Environment

Give the systems diagram and give the systems equation for the movement through the soil of a spill of gasoline.

Solution

Gasoline, or any other substance, does not move instantaneously through the soil. One reason is because the soil presents resistance to the flow of liquids. A compacted soil has more resistance, and a light soil has less.

The ultimate destination for the gasoline, probably the water table, can be represented by a point of zero pressure. At this point, the gasoline no longer has a tendency to flow.

The spill itself can be represented as a capacity element; the amount of gasoline is limited, and the flow is higher at the beginning than at the end. This is much more appropriate than either a flow source or pressure source. Thus the systems diagram appeares in Figure 1.9.2; as to the systems equation, Kirchhoff's node equation at the junction of resistor and capacity indicates that:

$$\frac{p}{R_s} + C \frac{dp}{dt} = 0$$

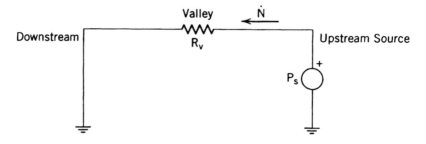

Figure 1.9.1. Systems diagram representing the movement of a herd of caribou through a narrow valley.

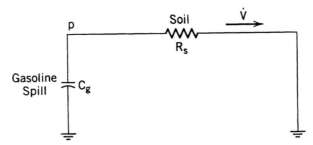

Figure 1.9.2. Systems diagram representing movement of gasoline through the soil from a spill.

Kirchhoff's loop equation gives:

$$\dot{V} R_s + \frac{1}{C} \int \dot{V} \, dt = 0$$

where $\dot{V}$ is the flow rate and p, the pressure.
  Either of these equations suffices.

EXAMPLE 1.9-3. Systems Diagram in a Building

Give the systems diagram and give the systems equaiton for heat in a room containing several human occupants and a window through which sunlight is streaming.

Solution

The walls of the room resist heat transfer to the environment at a constant temperature, designated as a temperature source. Heat accumulated within the room can be designated by a capacity element. There are a number of heat (flow) sources: The humans inside the room represent one, other heat sources such as electrical devices or the furnace output represent another, and the sunlight streaming into the room through the window represents a third. The systems diagram thus appears in Figure 1.9.3; the systems equation can be easily determined using Kirchhoff's node equation at the capacity element

$$\dot{q}_s + \dot{q}_h + \dot{q}_o + \frac{(\theta_r - \theta_o)}{R_w} + C \frac{d\theta_r}{dt} = 0$$

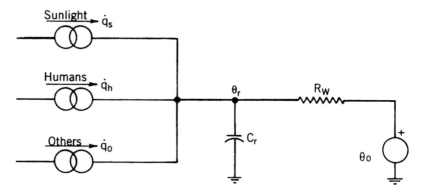

Figure 1.9.3. Systems diagram representing a heat balance in a room with human occupants, sunlight, and other sources.

Problems

1.1-1.   Units. Given the integral equation

$$P = \int_0^\infty e^{-(\dot{V}/K)t} \, dt$$

The units of $\dot{V}$ are m^3/sec, and the units of t, sec. What are the units of K and of P?

1.1-2.   Coulombs in a gas. What is the maximum number of coulombs that could exist in a kg mole of gaseous hydrogen ions at $0°C$? How many coulombs would there be in a kg mole of helium at the same temperature?

1.1-3.   Weight of a tomato soup can. If a can of tomato soup concentrate has a net weight of 305 g, how many newtons does this represent? Why is it preferable to express weight in newtons rather than in grams?

1.2-1.   Uphill and downhill flow. Why does water flow faster when moving downhill and fire move faster when moving uphill? Explain in terms of effort variables.

1.2-2.   Perfume. The perfume industry is based on attractive scents. Would this be an example of an effort variable?

1.3-1.   Eating food. A hungry person eats food. Identify the effort and flow variables.

1.3-2.   Effort and flow variables in cardiac defibrillation. A cardiac defibrillator uses a high-voltage pulse across the chest to momentarily stop all myocardial activity. The term *high-voltage current* has been used to describe the electrical pulse. Why is this term incorrect?

1.4-1. Systems approach to bird flight. Birds flap their wings in flight. Explain where resistance, inertia, and capacity apply to the vertical movement of the wings.

1.4-2. Mass transfer. What are the units of effort times flow in mass transfer, and why are these not equivalent to power?

1.4-3. Collecting and making maple syrup. If sugar maple sap is collected in the spring at the rate of 6 L/day per tree, and if the tap in the tree is located at a height of 6 m above the mean height of the roots, how much power did the tree expend to deliver the sap to the tap? Where does this power come from? If the volume of maple sap is reduced by a ratio of 30–40 in 6 hours to make maple syrup, and heat is used to evaporate the excess water, compare the power requirements of the heater and of the tree in order to make 1 L of syrup.

1.4-4. Thermal resistance. If 55 N m/sec of heat flows through an object with an imposed temperature difference of 5°C, what is the thermal resistance of the object?

1.4-5. Mechanical resistance units. Show that the units of mechanical resistance are N sec/m.

1.4-6. Element types. If the effort and flow variables are in phase, are they related through resistance, capacity, or inertia? What is the simplest combination of element types to give an effort variable that leads a flow variable by 60° (or $\pi/3$ radians)?

1.4-7. Mechanical capacitance. Describe what a mechanical capacity element might look like.

1.4-8. Energy storage. What is the essential difference between energy storage in capacity and inertia elements?

1.4-9. Thermal capacity. Show by means of the phase difference between heat flow and temperature that thermal inertia is really thermal capacity.

1.4-10. Go fly a kite. A youngster pulls on his kite string with a force of $t/10$ newtons, where t is the time in seconds. The kite responds with an $\dot{x} = 0.25t^2$. What is the inertia of the kite? What is the mass of the kite?

1.4-11. Lung compliance. If compliance at any given lung volume is given by the slope of the line defined by Eq. 1.4.15, what is the compliance value at a lung volume of 2.4 L? Vital capacity for adult human males is about 4.8 L.

1.4-12. Nonlinear force—deformation. List biological materials that show the same general shape as given in Figure 1.4.2.

1.4-13. Hysteresis. If the direction of arrows in Figure 1.4.3 were reversed, what would that mean? Why might you like that to be true?

1.4-14. Airways pressure drop. Give the expression for pressure loss across the airways.

1.4-15. Skunks in the basement. Who ya gonna call?

1.4-16. Racing horses. The following is a list of thoroughbred horses in one race. Calculate the average speed and the average kinetic energy.

Horse	Mass	Speed
1	460 kg	14.7 m/sec
2	470	15.6
3	472	16.0
4	480	15.2
5	455	13.8
6	475	16.2
7	472	15.6
8	452	15.9
9	463	14.8
10	470	15.5

1.4-17. Myoelectric signals. Muscle fatigue can be measured by means of the frequency spectrum, related to standard deviation, of electrical signals recorded from the muscle. The following is a list of values (in arbitrary units) for the myoelectric signal during isokinetic knee extension using the rectus femoris muscle. Compute the standard deviation.

0.35	0.15	0.70	−0.65	0.28	−0.10
−0.25	−0.24	−0.80	0.16	−0.35	−0.03
−0.10	0.15	0.04	0.00	0.04	0.00
−0.10	−0.13	0.50	0.25	0.08	0.06
0.08	−0.23	0.22	0.01	0.08	0.01

1.4-18. Licking a chocolate-center. How many licks does it take to get to the center of a chocolate-center pop? Here are data on the number of licks to help you answer the question:

92	133	114	117
66	143	125	135
127	119	137	123
150	183	139	130
115	122	109	128

1.4-19. Antibacterial tests. In developing a countertop spray cleaner effective against *Salmonella* bacteria, candidate antibiotics are tested on agarose plates inoculated with *Salmonella* bacteria. Single drops (0.1 ml) of 1 M antibiotic solutions are introduced to each plate, and the zones of *Salmonella* inhibition are measured after a standard incubation time. The standard for the experiment is a 70% solution of ethanol in water. Diameters (in cm) for each of 16 replications for each antibiotic are given below. Which antibiotic do you recommend?

Sodium chlorite	2-Benzyle 4-chlorophenol	Dehydrobiethylanine	70% Ethanol
1.85	0.95	1.47	1.02
1.58	0.90	1.71	0.80
1.50	1.06	1.25	0.94
1.19	0.86	2.15	0.88
1.87	1.32	2.10	0.85
1.46	1.20	1.57	0.90
1.73	1.08	2.21	0.89
1.58	0.99	2.03	0.76
1.50	1.24	1.63	0.99
1.78	0.93	2.74	0.93
1.76	1.16	1.92	0.85
1.36	1.18	1.87	0.91
0.76	1.07	2.40	0.75
1.02	1.04	0.81	0.81
1.45	1.06	1.58	0.77
1.30	1.27	1.25	0.85

1.4-20. Standard deviation. Calculate the standard deviation for each of the antibiotics in the previous question. What is the number of centimeters that contain 95.4% of the data?

1.4-21. Acidification rate in a bioreactor. If pH can be measured to within 1.5%, flow rate can be measured to 3.5%, and the two measurements are multiplied together to give the rate of acidification, what is the uncertainty in the final acidification rate estimate?

1.4-22. Lake volume. Measuring the diameter of a circular lake can be accomplished by surveying equipment to within 2%. Lake depth can be ascertained sonically to ±5%. What is the uncertainty of the calculated volume value?

1.5-1. Effort and flow sources. Give examples of effort and flow sources not mentioned in the text.

1.5-2. Other characteristics. What are some possible means to distinguish between the two nonideal sources shown in Figure 1.5.3?

1.5-3. **Dual sources.** If $\dot{\omega}_0 = 15$ and $R = 5$ in Figure 1.5.3, what value should be given to ϕ_0 in order that the two sources be duals of each other?

1.5-4. **Real flow sources.** Discuss the ideality of the following flow sources:
   (a) Flow of waste from a city.
   (b) Flow of migrating birds.
   (c) Growth of a mushroom through the top layer of earth.
   (d) Flow of blood in the body.
   (e) Spread of the Mediterranean Fruit Fly in California (an introduced nonindigenous pest).

1.6-1. **Thermal resistance of composite wall.** In the composite wall of Figure 1.6.2, the brick thermal resistance is $0.0085°C\,\text{sec}/(N\,m)$, the concrete block thermal resistance is $0.0030°C\,\text{sec}/(N\,m)$, and the window thermal resistance is $0.0195°C\,\text{sec}/(N\,m)$. What is the thermal resistance of the wall? If the inside temperature is maintained at $22°C$, what is the heat flow through the wall when the outside temperature is $0°C$? How much heat flows through the window? How large is the electric heater that is needed to maintain a temperature inside at $22°C$ when the outside temperature is $0°C$?

1.6-2. **Alternative representation of a wall.** Draw a mechanical analog (springs, masses, and dashpots) of the wall in Figure 1.6.2. Label each component with regard to the part of the wall for which it stands.

1.6-3. **Combination of respiratory impedances.** In the diagram is shown an equivalent schematic of respiratory mechanics in humans. Reduce this complex diagram to a simpler representation with a single resistance, compliance, and inertance. Determine values for these components. What information is lost when the respiratory system is

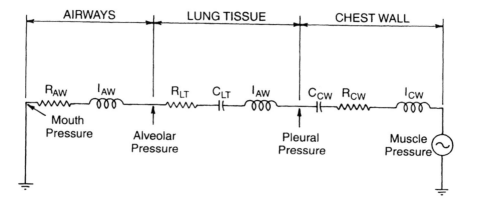

represented by a single resistance, compliance, and inertance?

$$R_{AW} = 157 \text{ kN sec/m}^5$$

$$R_{LT} = 39 \text{ kN sec/m}^5$$

$$R_{CW} = 196 \text{ kN sec/m}^5$$

$$I_{LT} = 774 \text{ N sedc}^2/\text{m}^5$$

$$I_{CW} = 1690 \text{ N sec}^2/\text{m}^5$$

$$I_{AW} = 137 \text{ N sec}^2/\text{m}^5$$

$$C_{LT} = 2.45 \times 10^{-6} \text{ m}^5/\text{N}$$

$$C_{CW} = 2.45 \times 10^{-6} \text{ m}^5/\text{N}$$

1.6-4. Time constant for sterilization. Food sterilization time is based on the decimal reduction time D defined as the time required to reduce the microbial population to 10% of the original number. Thus the number of microbial survivors is $N = N_0 10^{-t/D}$. Express the decimal reduction time in terms of the time constant.

1.6-5. Stiffness of apples. Nondestructive testing of apples and other fruits for ripeness is a goal for biological engineers. As fruits ripen they become less stiff (or firm). Given that inertia in a mechanical system is equal to the mass of the fruit, and stiffness is the inverse of compliance, suggest how acoustic vibrations in an apple can be used to test for ripeness.

1.6-6. Intracellular biopotential microelectrode. A microelectrode filled with 3 M KCl is an ultrafine device used to measure intracellular biopotentials. An equivalent systems diagram of the microelectrode is:

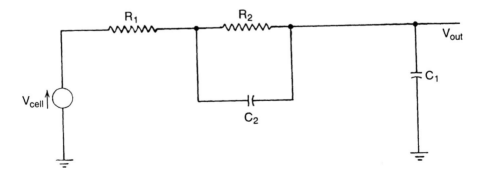

The value of R_1 is typically 13.5×10^6 N m sec/coul^2, and that of C_1 is about 23×10^{-12} coul^2/(N m). Values for R_2 and C_2 are negligible. If a nerve action potential suddenly raises the intracellular voltage by 120 mV, how long will it take the output voltage to rise to 99% of its correct value? Recalculate the time if the electrode is attached to an amplifier with an input capacitance of 15×10^{-12} coul^2/(N m) through a 1 m length of small coaxial cable [88.6 coul^2/(N m^2)]. If the nerve action potential rises for $\frac{1}{2}$ msec and falls thereafter for $\frac{1}{2}$ msec, what would you think the output voltage would appear like?

1.6-7.   **Composting of hydrocarbon contaminated soil.** The owner of a property contaminated with petroleum must completely remediate the site within one year. Approximately 23,000 cubic meters of soil are contaminated with a mean total petroleum hydrocarbons (TPH) concentration of 12,000 mg/kg. The site must be remediated to a mean TPH of not greater than 100 mg/kg. As an alternative to the costs of incineration and off-site disposal, the owner has conducted laboratory tests of composting that show the reductions in TPH as:

Day	
0	18,600 mg/kg
3	16,400
6	13,200
10	11,200

Treat this as an exponential decay and determine the expected time for the mean TPH to be reduced to the allowable level.

1.6-8.   **Seiches.** These are free-standing waves that develop primarily in closed or semiclosed bodies of water. They are not normal shoreline waves, but are standing waves of unusually large magnitude and duration. Merian's formula (Korgen, 1995) gives the maximum time period between seiche peaks as:

$$T = 2L/\sqrt{gd}$$

where T is the time between peaks (sec); L, the length of the enclosed rectangular body of water (m), g, the acceleration of gravity (m/sec^2); and d, the constant depth of water (m). Strong southwest winds often pile up water at the northeastern end of Lake Erie (length of 400 km and effective depth of 26 m) at Buffalo. When the wind abruptly stops, a seiche of up to 6 m height is formed along the length of the lake. If you lived at the southwestern end of the lake at Toledo, how much time after the wind stopped would you have to move your belongings to higher ground?

1.7-1.   **Rate of oxygen consumption.** A sitting human being normally produces about 105 N m/sec of heat. What is the rate of oxygen consumption to satisfy this requirement if carbohydrate is the substrate oxidized?

1.7-2. **Jumping on the earth and moon.** It has been found that the leg muscles can produce a maximum force roughly equal to twice the body weight. An average adult man has a body mass of about 70 kg. An average adult woman has a body mass of about 85% of the average man. What is the maximum rate of vertical acceleration that can be produced during a jump on Earth? If the man and woman were to travel to the moon, where the force of gravity is one-sixth that of the Earth, what would be the maximum vertical accelerations during jumps?

1.7-3. **Bathysphere.** Undersea water exploration can be accomplished in generally spherical vessels with rigid walls made from strong materials. If the pressure inside the vessel is to be maintained at about an atmosphere ($101,300 \, N/m^2$) because of the human occupants, and if the vessel were to be lowered to a depth of 3 km below the surface, what is the pressure that must be resisted by the walls? If the vessel has a diameter of 2 m and the wall material can withstand a tensile stress of up to $138 \, MN/m^2$, what is the minimum wall thickness in order to keep the vessel from collapsing at 3 km depth?

1.7-4. **Capillary bursting pressures.** The bursting pressure of a medium-sized artery 4 mm in diameter is about $26.7 \, kN/m^2$ (200 mm Hg). Its wall thickness is about 1 mm. Calculate the approximate bursting pressure of a capillary that is 0.008 mm in diameter and has a wall thickness of 0.001 mm. Why would real capillaries have about one-half the bursting pressure as you calculated?

1.7-5. **Balances in chemostat.** A chemostat is a growth chamber for the continuous culture of microbes. A medium composed of an excess of all required nutrients except one (called the limiting growth factor) is added at a constant rate while the overflow is removed at the same rate. Write verbal and mathematical forms of balances for: (a) the concentration of the limiting growth factor, and (b) the number of cells in the chemostat. Draw systems diagrams to represent these balances. Explain why you represented it this way.

1.7-6. **Tiger balance.** The rate of tigers migrating to a certain Siberian area is 20/yr. The rate of tigers killed or dying is 100/yr. The rate of tiger reproduction is 50/yr. If the tiger population in 1995 is 250, how long will it be before there are no tigers left in that region of Siberia?

1.7-7. **Captan balance.** An orchardist mixes 0.91 kg of Captan in 760 liters of spray and begins spraying his trees. Noting that rain threatens, he stops spraying with 570 liters of spray left in the tank. Being environmentally conscious, he will reuse the spray after the rain. The next day, he refills the tank with water. If the effectiveness of the Captan decreases exponentially with time after it is mixed with water, and loses 50% of its effectiveness after 8.3 h at pH 7.0, what is your recommendation to the orchardist for the amount of Captan to mix in the refilled tank?

1.7-8.   Carbon balance. This figure of the global carbon cycle appeared in Moore (1995). Arrows give exchanges of carbon between compartments in units of Pg (petagrams) C/year. Values inside the boxes are stored carbon in units of Pg C. From figures in the drawing, write carbon balances for the land, oceans, and atmosphere.

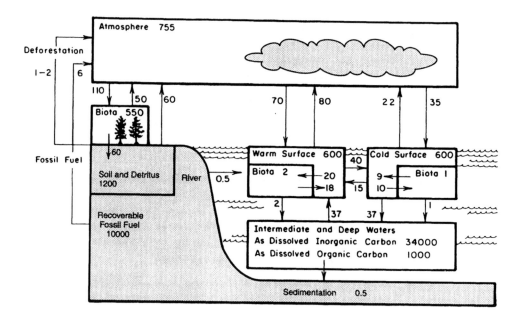

1.7-9.   Oxygen balance. A person runs on level ground at 4.47 m/sec and requires 6.70×10^{-5} m^3 O$_2$/sec. Maximum oxygen consumption for the young male runner is 4.2×10^{-5} m^3 O$_2$/sec. If the maximum oxygen debt is 9×10^{-3} m^3 O$_2$, how long can the running be sustained? Assume that young women have maximum oxygen consumptions and oxygen debts 70% as large as those of their male counterparts. How long can they sustain the same rate of exercise? If the efficiency of repayment of the maximum oxygen debt is 50% (twice as much O$_2$ required as actually repaid), and the normal oxygen consumption at rest is 5.0×10^{-6} m^3/sec, what is the minimum amount of time required to repay the maximum oxygen debt?

1.7-10. Iodine balance. If the rate of uptake of ^{131}I (radioactive iodine) by the thyroid gland is proportional to the concentration of ^{131}I in the blood, and if the only ^{131}I in the blood comes from a bolus of ^{131}I given by injection, determine the concentration of ^{131}I in the blood as time goes on.

1.7-11. Calories in powdered yogurtessee. Yogurtessee is a dairy-based, no-fat, no-cholesterol fat substitute for makers of various food products. It is available in two forms: a wet form with 28 percent solids content and a spray-dried powdered form. The wet form contains 90.8 calories per 100 grams. How many calories would you expect in 100 grams of powdered yogurtessee?

1.7-12. Kirchhoff's node equation. What would be the consequence if all flows that reached a junction (or node) did not sum to zero?

1.7-13. Kirchhoff's loop equation. Write Kirchhoff's general loop equation for a fluid flow system. Repeat for a thermal system.

1.7-14. Ventilation duct. Repeat Example 1.7.5-2, but for the case where the two flows to be delivered are 0.83 and 6.2 L/sec.

1.7-15. Flow source boundary conditions. Describe what happens in the third drawing from the top in Figure 1.7.12.

1.8-1. Resistance to water flowing through soil. The permeability of soil ranges from 0.29 to 14 darcys (Richardson, 1961). Determine the average resistance to the flow of water through soil 15 m deep and 1 hectare in area. Assume the pressure gradient over the 15 m thickness is due to the column of free water, and the pressure difference over this column is $1.47 \times 10^5 \, N/m^2$. The viscosity of water is $9.84 \times 10^{-4} \, N\,sec/m^2$.

1.8-2. Flow through a hemodialysis membrane. A hermodialysis membrane with effective area of $0.6 \, m^2$, thickness of $50 \, \mu m$, and permeability of $2.96 \times 10^{-14} \, m^2$ is used to filter urea and other impurities from the blood. The viscosity of blood plasma is $1.2 \times 10^{-3} \, N\,sec/m^2$. What is the expected filtration rate if the transmembrane pressure is $2.25 \, N/m^2$?

1.8-3. Conduction heat transfer through a tooth. If the outside of a tooth is exposed to an average temperature of $32°C$, and the inside of the tooth is maintained at $37°C$, how much heat flows by conduction through the enamel layer 0.25 mm thick? Thermal conductivity of the enamel is $0.82 \, N/(°C\,sec)$ and tooth surface area is $90 \, mm^2$.

1.8-4. Conduction through a thin-walled cylinder. The resistance to conduction heat flow through a hollow cylinder is properly given by Eq. 1.8.21. Sometimes, however, Eq. 1.8.20 is used, substituting

$(r_o - r_i)$ for thickness L and $2\pi r_o L$ for A (L here is cylinder length). Plot the error of the resistance calculation using the planar approximation against (r_o/r_i).

1.8-5. **Composite membranes.** Membrane systems are used in a wide variety of situations to separate various chemical species. They are used in the pharmaceutical, biotechnology, and food industries, for wastewater treatment, for producing potable water, and for encapsulating artificial organs. Composite membranes are produced where the membrane of choice must be strengthened by the addition of a second material. Assume a hollow fiber membrane of inside diameter 1.2 mm and outside diameter 1.3 mm. Mass diffusivity of the target species through this membrane is $3.0 \times 10^{-10} \, \text{m}^2/\text{sec}$. A second material is layered on the outside of the first material to add strength. Mass diffusivity of the same target species through this material is $18 \times 10^{-10} \, \text{m}^2/\text{sec}$. What thickness of the outside material can be added if the total diffusion resistance does not exceed the resistance of the inner material by more than 25%?

1.8-6. **Impedance plethysmography.** Impedance between electrodes can be measured by injecting an electrical signal into the organ or body and measuring the ratio of interelectrode voltage to current. The frequency of the signal can make a big difference in the results obtained. When dc is used, ions in biological fluids can migrate to the electrodes, causing them to polarize and eventually obstruct current flow. An alternating electrical signal can be used to solve the polarization problem, but a frequency that is too low can cause repeated stimulation of nerves, muscles, and glands.

Impedance plethysmography is often used to measure blood volume changes occurring in various parts of the body, such as the limbs, digits, head, thorax, kidney, and eye. If the body segment of interest is represented by a cylinder and the electrodes are placed at the ends of the cylinder, then the resistance measured between the electrodes is:

$$R = \rho L/A = \rho L^2/V$$

where R is resistance (N m sec/coul^2); ρ, the resistivity ($\text{N m}^2 \, \text{sec/coul}^2$); L, the length of the cylinder (m); A, the cylinder cross-sectional area (m^2); and V, the volume (m^3).

As long as the volume change (ΔV) is small,

$$\Delta R \approx \frac{\rho L^2}{V^2} \Delta V = \left(\frac{R^2}{\rho L^2}\right) \Delta V$$

Determine the impedance plethysmography sensitivity, $\Delta R/\Delta V$, due to volume changes in the human forefinger, as measured by a ring electrode placed at the base of the finger and an adhesive electrode placed at the end of the finger. The length of the forefinger is 90 mm, and its diameter is 25 mm. Assume that electrical resistance measurements reflect mainly blood flow through the included phalangeal bone and surrounding soft tissue. The electrical resistivity of blood is $1.5\,\mathrm{N\,m^2\,sec/coul^2}$.

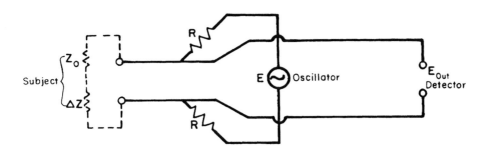

1.9-1.   Systems analysis of worker ant odor. The complete immobility and crumpled posture of a dead ant cause no response by other ants. However, certain long-chain fatty acids and their esters, products of chemical decomposition, accumulate in a day or two, and stimulate the workers to bear the corpse to the garbage pile outside the nest. When other living worker ants are experimentally daubed with these substances, they are dutifully carried to the garbage pile. Upon being dumped in the garbage, these poor creatures scramble to their feet and return promptly to the nest, only to be carried out again. The hapless creatures are thrown back on the pile time and again until most of the death scent has been worn off their bodies by the ritual. Write a balance to describe their body odor.

1.9-2.   Systems analysis of ant food trail. The fire ant worker lays an odor trail to a food source by exuding pheromone as it touches its stinger to the ground. The more dense the food source, the stronger is the pheromone signal laid by returning workers. Old trails no longer used disappear below threshold level in 2 min. Draw a systems diagram representing the food trail. What systems equations apply?

1.9-3.   Systems analysis of neural synapse. Transmission of electrical signals from one neuron to another, or from a neuron to a muscle cell, occurs chemically over a short gap called a synapse. Cholinergic neurons release acetylcholine into the gap. When the acetylcholine reaches the cell membrane of the next neuron, it enhances the permeability of the membrane to small ions, including sodium. Thus begins the action potential in the second neuron. Acetylcholinesterase rapidly removes acetylcholine from the synapse or else the membrane permeability enhancement of acetylcholine would be permanent and not allow more signals to pass. Draw a systems diagram for the process of neuron-to-neuron signal transmission. Write a simple balance equation that describes the ion permeability of the neuronal membrane in terms of quantities of acetylcholine and acetylcholinesterase.

1.9-4. Systems analysis of glucose tolerance test. A glucose tolerance test can be given to test for insulin-dependent diabetes. A constant intravenous infusion of glucose is given to an individual, and blood glucose levels are ascertained at about 10-minute intervals. When plotted as glucose concentrations above the normal resting value, data appear to rise exponentially to an asymptotic value. The higher the asymptotic value, the more glucose intolerant is the individual. Draw a systems diagram that gives the observed response. Identify the effort and flow variables. What anatomical or physiological features are represented by each systems element?

1.9-5. Systems analysis of compost. Composting of organic material can be done either aerobically or anaerobically. Aerobic composting produces very little odor, but requires an adequate supply of oxygen to accomplish. Define effort and flow variables, draw the systems diagram, and give the systems equation (if you can) for the delivery of air to a large compost pile.

1.9-6. Weight training. Lifting weights for strength training often uses a machine consisting of weights attached to cables that are fed over pulleys. The exercising person lies on her/his back and pulls the weights up with downward movement of the arms. Define effort variables, draw the systems diagram, and give the systems equation for the operation of this weight training machine.

1.9-7. Spread of diseases. New strains of Asian flu often begin in China, where genes are exchanged between avian influenza viruses indigenous to ducks and mammalian influenza viruses hosted by pigs. Define effort and flow variables, draw the systems diagram, and give the systems equation for the spread of Asian flu around the world. What factors do you think could be important in the determination of values for various systems parameters?

1.9-8. Metal oxide pH electrode. A metal oxide pH electrode can be used to ascertain the pH of an aqueous solution. They work like this: Hydrogen ions from the solution pass through a membrane into a small chamber, where they combine with electrons and oxygen from the metal oxide electrode to form water. The current drawn from the electrode depends on the number of hydrogen ions oxidized. Define the effort and flow variables, draw the systems diagram, and give the systems equation for a pH electrode measuring pH in a very small container of solution. Give a diagram for the pH electrode in a flowing stream of solution.

1.9-9. Bioluminescence ATP assay. Bacterial activity in environmental samples can be determined by extracting ATP from the bacteria and using this ATP as an energy source to drive the luciferin–luciferase reaction of firefly extract (Krones et al., 1990). The resulting bioluminescence can be detected with a photomultiplier tube. The output from the tube immediately rises to a peak and then falls exponentially toward zero. A curve such as this can be obtained as the effort variable measurement across a capacity element in series with a resistance. Speculate on the meanings of capacity and resistance. If the amount of ATP in the sample is related analogously to the original amount of stuff (flow variable) stored on the capacity element, how would you obtain this information from the time course of data from the photomultiplier tube?

1.9-10. Transpirational moisture loss. Transpiration is the pi)cess of moisture loss from plant tissues. Water can be lost to the atmosphere from the leaves or from the stems of the plant. Water that is lost from the stems goes through either the lenticels of the sten, or through the nearly impermeable nonlenticel surface. Water lost from each of the leaves goes either through the stomata or leaf cuticle. Surrounding each plant surface is a boundary layer that limits moisture movement to the atmosphere and then an aerodynamic resistance that controls gross moisture movement through the atmosphere. Taking this description of moisture loss and assuming that each step in the process exhibits some resistance to moisture movement, draw a systems diagram for moisture movement from the plant to the atmosphere.

REFERENCES

Bates, J. H. T., M. S. Ludwig, P. D. Sly, K. Brown, J. G. Martin, and J. J. Fredberg, 1988, Interrupter Resistance Elucidated by Alveolar Pressure Measurement in Open-Chest Normal Dogs, *J. Appl. Physiol.* **65**: 408–14.

Carslaw, H. S., and J. C. Jaeger, 1959, *Conduction of Heat in Solids*, London, Oxford University Press.

Glass, L., and M. C. Mackey, 1988, *From Clocks to Chaos, the Rhythms of Life*, Princeton, NJ: Princeton University Press.

Ingersoll, L. H., O. J. Zobel, and A. C. Ingersoll, 1954, *Heat Conduction with Engineering, Geological, and other Applications*, Madison, WI: University of Wisconsin Press.

Johnson, A. T., 1986, Conversion between Plethysmograph and Perturbational Airways Resistance Measurements, *IEEE Trans. Biomed. Engr.* **33**: 803–38;6.

Johnson, A. T., 1984, Multidimensional Curve-Fitting Program for Biological Data, *Computer Programs in Biomedicine* **18**: 25–64.

Korgen, B. J., 1995, Seiches, *Amer. Sci.* **83**: 330–41.

Krones, M. J., A. T. Johnson, and O. J. Hao, 1990, A Numerical Technique for Interpreting the Bioluminescence ATP Assay, *Environm. Technol.* **11**: 1107–11.

Moore, B., 1995. Global Carbon Cycle, in *Encyclopedia of Environmental Biology*, W. A. Nierenberg, Ed., New York: Academic Press, pp. 215–23.

Richardson, J. G., 1961, Flow through Porous Media, in *Handbook of Fluid Dynamics*, V. L. Streeter, Ed., New York: McGraw–Hill, pp. 16-1–16-112.

2

FLUID FLOW SYSTEMS

"The human mind is as driven to understand as the body is driven to survive."

Hugh Gilmore

2.1 INTRODUCTION

pressure and height

Fluids flow through conduits due to a pressure gradient along the conduits. The effort variable is thus pressure difference and the flow variable is the flowing fluid. Height differences can be considered as a second effort variable in fluid flow. Although pressure in a fluid sometimes depends on the height of the fluid above it, there are instances, such as when a pump or a container with elastic walls is involved, when pressure and fluid height do not correspond. Thus pressure and fluid height are linked, but to consider these two as separate effort variables is sometimes instructive (Figure 2.1.1).

capillarity

There is a third effort variable that must be considered when analyzing fluid movement in plants or sometimes in fluid-filled tubes having diameters less than 10 mm. This effort variable is capillarity, related to surface tension, which is a combination of adhesive and cohesive forces. When adhesive forces dominate cohesive forces, the interface between the lower fluid (usually a liquid such as water) and the upper fluid (often a gas such as air) rises in the tube above the fluid level outside the tube. The amount of rise is inversely proportional to tube diameter, which is about 0.01 μm for the xylem in plants (Rand and Cooke, 1996). The combination of capillarity in the xylem and liquid cohesion in plant leaves is sufficient to cause movement of water from the roots to the top of the tallest plant. Capillarity is not a major contributor to the movement of water except at a very small geometric scale.

109

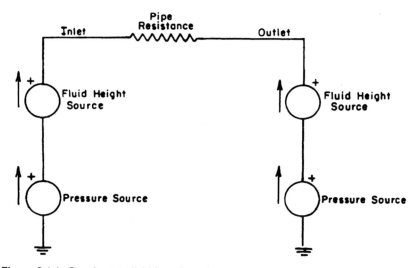

Figure 2.1.1. Steady-state fluid flow through a conduit. The rate of flow depends upon the difference between the sum of pressure and fluid height at the inlet and outlet, divided by the pipe resistance.

Conduit resisitance limits the amount of fluid flowing through it. In this chapter will be presented the means to calculate the value of resistance from a conduct and fluid physical parameters.

Of the three major transport processes in this book (fluid flow, heat transfer, and mass transfer), fluid flow is unique in several respects. We have already seen how these three processes, and others, can be unified in analogous fashion by considering effort variables, flow variables, and the relationships between them. Yet fluid flow is the only transport process where there is a sufficiently large moving mass to make kinetic energy and momentum significant.

We have already seen the applications of the general field equation to flow through porous media (Section 1.8.1). A different approach is taken for fluid flow through defined conduits: pipes, channels, and ducts. Table 2.1.1 presents a summary of the systems approach to fluid flow in conduits. This table can be used to place the material in this chapter in the context that was established in Chapter 1.

Most fluid flow design problems involve the steady state, and rates of change of flow with respect to time are not important in most of these problems. It is true that many biological organisms have some form of pulsatile flow, but problems involving these organisms are often ones for analysis and not design. For these reasons, pipe friction and resistance dominate the material to follow, and compliance and inertance are not given much emphasis.

We begin first with the three basic balances of mass, energy, and momentum that will be used for typical design problems involving fluid flow

moving mass

(Table 2.1.1). Pipe friction determines pipe resistance, which is important for determining steady-state fluid flow. Thus there is a substantial amount of space devoted to pipe frictioin. Other fluid flow cases important to the biological engineer are treated at the end of this chapter.

2.2 CONSERVATION OF MASS

mass balance

The control volume to be considered here is a section of a pipe of variable cross section flowing full (Figure 2.2.1). A mass balance on the pipe section gives:

$$\text{mass flow rate in} - \text{mass flow rate out} = \text{rate of mass storage} \quad (2.2.1)$$

As indicated in Section 1.7.3.1, there is normally no allowable mass generation inside the control volume.

2.2.1 Continuity Equation

For the special case where the conduit flows steadily full (Figure 2.2.1), the rate of mass storage is zero, and Eq. 2.2.1 becomes:

$$\left.\frac{\Delta m}{\Delta t}\right|_1 - \left.\frac{\Delta m}{\Delta t}\right|_2 = 0 \quad (2.2.2)$$

Table 2.1.1 Systems Approach to Fluid Flow in Conduits

Modes of action	only one means for fluid to flow
Effort variable	there are two: pressure and height (a third, capillarity, is important in plants)
Flow variable	volumetric rate of flow
Resistance	due to energy dissipation within the fluid and to changes in velocity profile
Capacity	compliance relates substance volume to differences in either pressure or height
Inertia	inertance can be significant due to movement of the dense fluid
Power	mechanical in nature, usually pressure times volume flow rate
Time	most analyses are steady state
Substance generation	none
General field equation	not used except through a porous medium
Balances written	mass balance (continuity equation) energy balance (Bernoulli equation) momentum balance (including Navier–Stokes equation)
Typical design problem	determine pump specifications: use Bernoulli equation and continuity equation

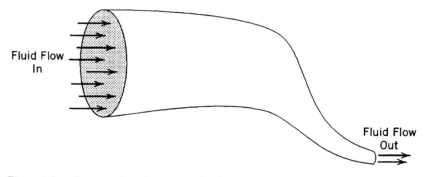

Figure 2.2.1. Conservation of mass in a pipe flowing full. If the fluid is incompressible, the continuity equation results.

where $(\Delta m/\Delta t)|_i$ is the amount of mass (kg) moving during time interval Δt (sec) at section i.

Notice that the condition that the rate of change of mass storage equals zero also requires that the conduit must be rigid and the fluid incompressible. For many biological applications, the conduit is not rigid, and the fluid, if a gas, is not incompressible. Most food and biochemical processes, however, use rigid pipes and Eq. 2.2.2 is correct.

Mass can be considered to be density times volume, and volume is length times area; thus

$$\left.\frac{\Delta m}{\Delta t}\right|_1 = \rho_1 A_1 \left.\frac{\Delta L}{\Delta t}\right|_1 = \left.\frac{\Delta m}{\Delta t}\right|_2 = \rho_2 A_2 \left.\frac{\Delta L}{\Delta t}\right|_2 \qquad (2.2.3)$$

For an incompressible fluid density remains constant ($\rho_1 = \rho_2$) and cancels from both sides of the equation.

The term $\Delta L/\Delta t$ can be recognized as the speed v of the fluid. Although direction is really of little consequence, v is often called "velocity." Thus

$$A_1 v_1 = A_2 v_2 = \dot{V} \qquad (2.2.4)$$

continuity equation

where $\dot{V}$ is the volume rate of flow (m³/sec). This is the so-called continuity equation, which relates velocity with area. Because the cross-sectional area of a circular pipe is related to the square of the diameter, changing the diameter by a factor of two changes the velocity by a factor of four.

When comparing velocities through conduits of different sizes, account must be taken of the source of the fluid. For instance, if there are two pipes,

flow source required

with the diameter of the second being twice that of the first, then the flow velocity when the second pipe replaces the first will be one-quarter of the flow velocity in the first pipe, *only if the fluid is supplied by an ideal flow source.* If an ideal pressure source is used, then the additional pipe resistance for the smaller-diameter pipe will cause the flow velocity to be less than four times the velocity in the larger-diameter pipe, and perhaps even less than the velocity in the larger-diameter pipe. Thus one must be careful to note the circumstances when applying the continuity equation. If both pipes are in series, and no changes are made to the piping and pumping systems, then Eq. 2.2.4 properly relates the relationship between flow velocities.

EXAMPLE 2.2.1-1. Velocity in a Pipe

A 30 m long horizontal pipe 5 cm in diameter flows full of water at a rate of $0.65 \text{ m}^3/\text{sec}$. What is the average velocity of the water?

Solution

Using Eq. 2.2.4

$$\dot{V} = Av$$

or

$$v = \dot{V}/A = \frac{0.65 \text{ m}^3/\text{sec}}{[\pi(0.05 \text{ m})^2/4]} = 331 \text{ m/sec}$$

2.2.2 Elemental Form of Continuity Equation

Although Eq. 2.2.4 is an acceptable form for the continuity equation in one-dimensional flow in a closed conduit on a gross scale, for several dimensions or on an elemental scale, another form for the continuity equation is required. This elemental form is not usually necessary for piping system design, but may be useful for analyzing fluid flow in biological systems, and is most useful for computer modelling.

Refer to Figure 2.2.2, which illustrates an elemental Cartesian control volume of fluid with dimensions Δx, Δy, and Δz. The mass balance of Eq. 2.2.1 is again the starting point.

flow balance in differential element
The rate of mass flowing into the control volume is the velocity in the x-direction, u, times the density times the area ($\Delta y \, \Delta z$):

$$\dot{m}_{x_1} = \rho u \, \Delta y \, \Delta z \tag{2.2.5}$$

The corresponding rate of mass flow out of the control volume is the same $\dot{m}_x$ as above, with the addition of any incremental change of velocity in the x direction

$$\dot{m}_{x_2} = \rho u \, \Delta y \, \Delta z + \left(\frac{\partial(\rho u)}{\partial x} \, \Delta x\right) \Delta y \, \Delta z \tag{2.2.6}$$

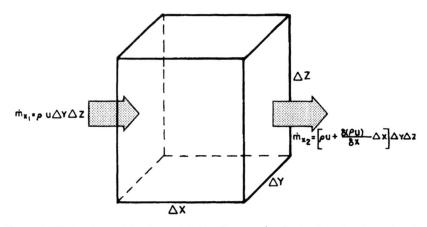

Figure 2.2.2. An elemental volume of fluid with mass $\dot{m}_{x_1}$ flowing into the element and $\dot{m}_{x_2}$ flowing out. The difference between $\dot{m}_{x_1}$ and $\dot{m}_{x_2}$ contributes to the accumulation of mass inside the element. If $\dot{m}_{x_1}$ equals $\dot{m}_{x_2}$, then there is no further accumulation of mass inside.

The net rate of mass flow into the control volume in the x-direction is just $(\dot{m}_{x_1} - \dot{m}_{x_2}) = -[\partial(\rho u)/\partial x]\,\Delta x\,\Delta y\,\Delta z$. Similar expressions exist for net mass influx in the y and z directions, where the velocities are defined as v and w, respectively.

The last term in the mass balance of Eq. 2.2.1 is the rate of change of mass accumulation, which is the time rate of change of density times volume

$$\frac{\partial(\rho\,\Delta x\,\Delta y\,\Delta z)}{\partial t} = \frac{\partial \rho}{\partial t}\,\Delta x\,\Delta y\,\Delta z \tag{2.2.7}$$

Since the elemental volume $\Delta x\,\Delta y\,\Delta z$ is common to all terms, it cancels out, leaving

$$-\frac{\partial(\rho u)}{\partial x} - \frac{\partial(\rho v)}{\partial y} - \frac{\partial(\rho w)}{\partial z} = \frac{\partial \rho}{\partial t} \tag{2.2.8}$$

If the flow is steady, so that ρ does not vary with time

steady-state equation

$$\frac{\partial(\rho u)}{\partial x} + \frac{\partial(\rho v)}{\partial y} + \frac{\partial(\rho w)}{\partial z} = 0 \tag{2.2.9}$$

incompressible fluid For an incompressible fluid, density is constant and Eq. 2.2.9 becomes:

$$\frac{\partial u}{\partial x} + \frac{\partial v}{\partial y} + \frac{\partial w}{\partial z} = 0 \tag{2.2.10}$$

and since the derivative of the product (ρu) is:

$$\frac{\partial(\rho u)}{\partial x} = \rho \frac{\partial u}{\partial x} + u \frac{\partial \rho}{\partial x} \qquad (2.2.11)$$

Eq. 2.2.9 becomes:

$$u \frac{\partial \rho}{\partial x} + v \frac{\partial \rho}{\partial y} + w \frac{\partial \rho}{\partial z} + \rho \left(\frac{\partial u}{\partial x} + \frac{\partial v}{\partial y} + \frac{\partial w}{\partial z} \right) = 0 \qquad (2.2.12)$$

if ρ varies with location.

These are all forms of the continuity equation that may be used on an elemental scale. Velocities, density, and derivatives in Eqs. 2.2.8, 2.2.10, and 2.2.12 are all evaluated at a point.

2.3 CONSERVATION OF ENERGY

energy
balance

An energy balance of the form of Eq. 2.2.1 could be written for a fluid system. Instead, however, steady state with no energy generation will be assumed. Thus the energy balance becomes

total energy into control volume − total energy out from control volume = 0

$$(2.3.1)$$

We must consider the various energy forms that contribute to the overall energy total.

2.3.1 Potential Energy

Potential energy is static, nonmoving. As indicated in Table 2.1.1, there are two effort variables, pressure and height, to contribute to potential energy.

height

The first form of potential energy is represented by a mass of fluid existing at some height above a reference plane. If the mass were to fall to the reference, energy would be liberated and work could be done by the mass. In this case, the gravitational field (as given by the height) is the effort variable, and the flow variable is weight rate of flow. As given in Table 1.2.1, power is the effort variable times the flow variable, or $P = \dot{W}h$. Energy is just power multiplied by time, so

$$PE_g = Wh \qquad (2.3.2)$$

where PE_g is the gravitational potential energy (N m); W, the weight of fluid (N); and h, the height above a reference plane (m).

pressure

The second form of potential energy is that due to fluid pressure. If given the ability, fluid pressure would produce a column of fluid above the height of the container holding the fluid (Figure 2.3.1). The height of this column would be

$$h = p/\gamma = p/\rho g \qquad (2.3.3)$$

where h is the equivalent height for the fluid (m); p, the pressure (N/m^2); γ, the specific weight of fluid (N/m^3); ρ, the fluid density (kg/m^3 or N sec^2/m^4); and g, the acceleration due to gravity (m/sec^2).

equivalence between height and pressure

Considered in another way, the pressure p exists as the pressure beneath a column of fluid h high. This is true if the column exists, as in Eq. 2.3.2 or if only the pressure exists, as in Eq. 2.3.3, and the actual column is prevented from forming by the top of the container holding the fluid.

height, not volume, determines potential energy

The total amount or weight of fluid in the column is of no consequence when calculating the potential energy: Only the fluid height matters. As can be seen in Figure 2.3.2, the fluid columns are the same height regardless of the size of the standpipe. Nor does the standpipe angle matter, because the gravitational field is exerted in only the vertical direction.

Pressure potential energy is calculated as the fluid weight (W) times the equivalent height ($h = p/\gamma$), or

$$\text{PE}_\text{p} = Wp/\gamma \qquad (2.3.4)$$

pump pressure

pressure potential energy

Pressure energy can be added to the fluid system by a device such as a pump. It is not difficult to realize that the pump does not create energy, if considered to be an energy transducer. From the fluid system standpoint, however, the pump does create fluid energy, and this will be given as WE, where E has units of fluid height (m).

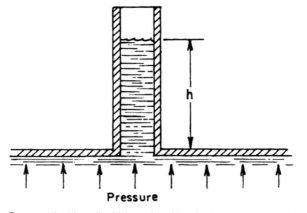

Pressure

Figure 2.3.1. Pressure inside a liquid is resisted by the walls of the container. Where there is a hole in the wall, the pressure is resisted by a column of liquid that rises to a height $h = p/\gamma$. If the tube in which the liquid rises is not high enough, the liquid will form a fountain.

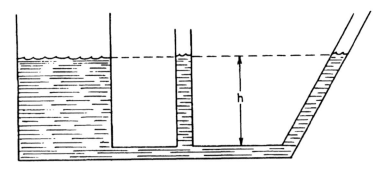

Figure 2.3.2. The height of a fluid in a standpipe depends only on the pressure and not on total weight. Because of the much larger volume of the pipe on the left, there is much more weight of fluid there than in the thinner pipe in the center. Pressure at the base of both pipes is the same because pressure equals the total weight of the fluid in the pipe divided by the cross-sectional area. The pipe on the right forms an angle with the vertical, but the same height is reached by the fluid as in the other two pipes.

Pressure added by a pump is really no different from fluid pressure as given in Eq. 2.3.3. However, pressure added by the pump is treated separately because it is often the variable that must be calculated. Therefore, pump pressure is of special interest.

total potential energy   Thus all potential energy terms are summed to

$$PE = WE + Wp/\gamma + Wh \tag{2.3.5}$$

EXAMPLE 2.3.1-1.   Human Blood Pressure

Human blood pressure measured at the level of the heart is about $16.0 \, \text{kN/m}^2$ (systolic) and $10.7 \, \text{kN/m}^2$ (diastolic). What is the blood pressure at the level of the brain stem for an upright posture?

Solution

Assuming that the brain stem is 0.3 m above the heart, both systolic and diastolic blood pressures are lower than at the heart level by a value of

$$p = \gamma h$$

The density of blood is about $1050 \, \text{kg/m}^3$. Thus the specific weight is $(1050 \, \text{kg/m}^3) \, (9.8 \, \text{m/sec}^2) = 10290 \, \text{N/m}^3$ and $p = (10290 \, \text{N/m}^3) \, (0.3 \, \text{m}) = 3087 \, \text{N/m}^2 = 3.09 \, \text{kN/m}^3$. Systolic pressure at the brain stem is $16.0 - 3.1 = 12.9 \, \text{kN/m}^2$, and diastolic pressure at the brain stem is $10.7 - 3.1 = 7.6 \, \text{kN/m}^2$.

EXAMPLE 2.3.1-2. Potential Energy Storage

The squid *Loligo vulgaris* can propel itself by expelling pressurized water from its mantle cavity. A 350 g squid stores about 50% of its body mass as water pressurized by muscular contraction to about 20 kN/m² (Trueman, 1980). Calculate the potential energy stored as pressurized water.

Solution

Pressure potential energy is given by Eq. 2.3.4 as

$$PE_p = Wp/\gamma$$

The weight of the stored water is

$$\left(\frac{0.350}{2}\,kg\right)\left(9.8\,\frac{N}{kg}\right) = 1.72 \text{ N}$$

$$PE_p = \frac{(1.72 \text{ N})(20{,}000 \text{ N/m}^2)}{(9810 \text{ N/m}^3)} = 3.5 \text{ N m}$$

2.3.2 Kinetic Energy

Kinetic energy is associated with movement. From physics, we know that kinetic energy is given by

$$KE = mv_p^2/2 \qquad (2.3.6)$$

where m is the mass (kg); and v_p, the local velocity of the mass (m/sec).

Fluid kinetic energy is given by Eq. 2.3.6, too. The problem comes in defining which velocity is to be used.

All fluid particles flowing in a conduit normally do not flow at the same rate. Those particles nearer the walls are held back by friction and move more slowly than those particles in the center. Total kinetic energy of a fluid is thus given by the kinetic energy contributions of all the individual particles. But we normally do not measure any velocity except average velocity: graphically, this is the velocity represented by the time it takes to fill a container of known volume. Both slow and fast particles contribute to the full container, and they are mixed very thoroughly in the filling process.

None of this would matter except that kinetic energy is given by the square of the velocity. Average kinetic energy involves the average v_p^2, while the velocity obtained from filling the pail is the average v. In general,

kinetic energy of fluid particles

particles contribute unequally to total kinetic energy

$$(\overline{v_p})^2 \neq (\overline{v_p^2}) \qquad (2.3.7)$$

or the average velocity squared is not equal to the average squared velocity. In addition to this, fluid mass passing an imaginary plane in the pipe will be greater for higher-velocity locations. To account for these differences, Eq. 2.3.6 is modified to give

$$\text{KE} = mv^2/\alpha = Wv^2/\alpha g \qquad (2.3.8)$$

where α is a dimensionless parameter with a value between 1.0 and 2.0; and g, the acceleration due to gravity, 9.8 m/sec^2.

kinetic energy at pipe section In order to determine values of α, a differential approach is required. Thus we return to the definition of kinetic energy for each individual particle and sum over all particles. More practically, an imaginary plane is formed perpendicular to the axis of the pipe, and kinetic energy is integrated over the cross-sectional area (Figure 2.3.3). Local velocities are assumed to be the same over the entire differential area, dA. As long as the flow is steady, velocity does not change with time, and

$$\text{KE} = \int_m \tfrac{1}{2} v_p^2 \, dm \qquad (2.3.9)$$

where KE is the kinetic energy (N m); m, the mass (kg or N sec^2/m); and v_p, the particle velocity (m/sec). But from Eqs. 2.2.3 and 2.2.4,

$$\frac{dm}{dt} = \rho \, d\dot{V} = \rho v_p \, dA \qquad (2.3.10)$$

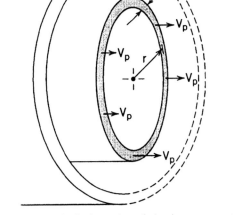

Figure 2.3.3. To find average velocity in a pipe of circular cross section, multiply local velocity by incremental area, integrate over the entire cross section, and divide by total area. Kinetic energy contributions for every annulus are also summed through integration. The incremental area is just the area of an annulus with circular length $2\pi r$ and width dr.

where m is the accumulated mass (kg); $\dot{V}$, the volume rate of flow (m³/sec); ρ, the density (kg/m³); and t, time (sec). Thus

$$KE = \int_t \int_A \tfrac{1}{2}\rho v_p^3 \, dA \, dt \tag{2.3.11}$$

determining
correction
factor
In order to determine the value for α, kinetic energy using average velocity is equated to kinetic energy using particle velocities

$$KE = \int_t \int_A \tfrac{1}{2}\rho v_p^3 \, dA \, dt = \int_t \int_A \frac{\rho v^3}{\alpha} \, dA \, dt \tag{2.3.12}$$

where v is the average conduit velocity (m/sec); and α, a dimensionless parameter introduced earlier.

For steady flow, there is no time dependence of velocities. Thus time cancels from both sides of the equation above. The parameter α can be solved as

$$\alpha = \frac{\displaystyle\int_A \rho v^3 \, dA}{\dfrac{1}{2}\displaystyle\int_A \rho v_p^3 \, dA} = \frac{A v^3}{\dfrac{1}{2}\displaystyle\int_A v_p^3 \, dA} \tag{2.3.13}$$

But average velocity is calculated from

$$v = \frac{1}{A}\int_A v_p \, dA \tag{2.3.14}$$

$$v^3 = \frac{1}{A^3}\left(\int_A v_p \, dA\right)^3 \tag{2.3.15}$$

Thus

$$\alpha = \frac{2\left(\displaystyle\int_A v_p \, dA\right)^3}{A^2 \displaystyle\int_A v_p^3 \, dA} \tag{2.3.16}$$

two velocity
profiles
In Figure 2.3.4 are shown two velocity profiles. For the case where all local velocities in a pipe flowing full are equal, $v = v_p$, and

$$\alpha = \frac{2 v^3 A^3}{v^3 A^3} = 2.0 \tag{2.3.17}$$

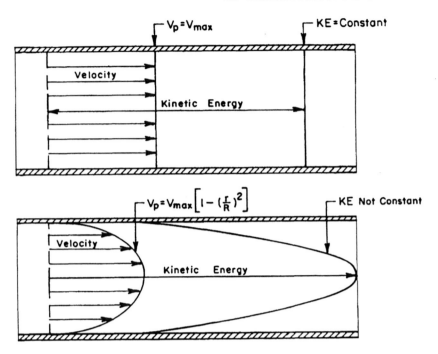

Figure 2.3.4. Two velocity profiles existing in a pipe. The upper profile is uniform: All particles flow at the same rate. Kinetic energies of all particles are identical. The lower profile is parabolic: Particles in the center of the pipe flow faster. Kinetic energies of the center particles are much greater than those near the wall.

For the case of a parabolic velocity profile

$$v_p = v_{max}\left[1 - \left(\frac{r}{R}\right)^2\right] \tag{2.3.18}$$

where v_{max} is the maximum local velocity (m/sec); R, the inner radius of the conduit (m); and

$$
\alpha = \frac{2\left\{v_{max}\displaystyle\int_0^R 2\pi r\left[1 - \left(\frac{r}{R}\right)^2\right] dr\right\}^3}{\pi^2 R^4 \displaystyle\int_0^R 2\pi r v_{max}^3\left[1 - \left(\frac{r}{R}\right)^2\right]^3 dr}
$$

$$
= \frac{16\pi^3 v_{max}^3\left(\dfrac{R^2}{2} - \dfrac{R^2}{4}\right)^3}{2\pi^3 R^4 v_{max}^3\left(\dfrac{R^2}{2} - \dfrac{3R^2}{4} + \dfrac{3R^2}{6} - \dfrac{R^2}{8}\right)}
$$

$$= 1.0 \tag{2.3.19}$$

kinetic energy usually small   When using the energy equation, the kinetic energy term for liquids is usually very small compared to potential energy terms and friction losses (not covered yet). Thus standard engineering practice has usually ignored differences in velocity profiles and used $\alpha = 2$ for all. Later, non-Newtonian fluids (see Section 2.6.2) will be found to require different values of α; so this standard practice does not serve our purposes.

This formulation of kinetic energy is correct only for round pipes flowing full and in steady flow. A large amount of turbulence or eddying in the fluid gives rise to a significant amount of rotational kinetic energy that has not been considered here.

EXAMPLE 2.3.2-1. Steady Flow of Beer

Beer flows through a 10 cm diameter stainless steel pipe 100 m long. The rate of flow is 38 L/min. What is the kinetic energy of the beer in the pipe if the flow velocity profile is parabolic?

Solution

Average velocity can be obtained from the continuity equation

$$v = \dot{V}/A = \left(\frac{38 \text{ L}}{\text{min}}\right)(10^{-3} \text{ m}^3/\text{L})\left(\frac{1 \text{ min}}{60 \text{ sec}}\right)/[\pi(0.1 \text{ m})^2/4]$$

$$= 8.06 \times 10^{-2} \text{ m/sec}$$

The mass of beer in the pipe is $m = \rho V$, where ρ is the beer density and V the volume of the pipe. Since the density of beer is nearly that of water.

$$m = (1000 \text{ kg/m}^3)(100 \text{ m})[\pi(0.1 \text{ m})^2/4]$$

$$= 7.85 \text{ kg}$$

From Eq. 2.3.8, using a value of $\alpha = 1.0$,

$$\text{KE} = mv^2/\alpha = (785 \text{ kg})(8.06 \times 10^{-2} \text{ m/sec})^2/1.0$$

$$= 5.11 \text{ kg m}^2/\text{sec}^2 = 5.11 \text{ N m}$$

2.3.3 Modified Bernoulli Equation

The modified Bernoulli equation is the conservation of energy equation that can include energy added by pumping devices and can also include friction losses. Like the continuity equation 2.2.4, the energy balance is usually considered at two different pipe sections. Summing all potential and kinetic energy terms

energy
balance

$$Wh_1 + Wp_1/\gamma + Wv_1^2/\alpha g = Wh_2 + Wp_2/\gamma + Wv_2^2/\alpha g + WE + Wh_f$$

$$(2.3.20)$$

where W is the weight of fluid (N); h, the fluid of height (m); p, the fluid pressure (N/m²); v, the average velocity (m/sec); γ, the fluid specific weight (N/m³); α, a dimensionless parameter (unitless); g, the acceleration due to gravity (9.81 m/sec²); E, the equivalent height added to fluid by a pump (m); h_f, friction losses (m); and 1, 2 refer to two different pipe locations.

There is no numbered subscript associated with either E or h_f because these actually occur *between* sections 1 and 2 and not *at* those sections.

The common weight factor, W, can be canceled from each term because, from the continuity equation, the weight rate of flow is the same for all sections in the steady state:

Bernoulli
equation

$$h_1 + p_1/\gamma + v_1^2/\alpha g = h_2 + p_2/\gamma + v_2^2/\alpha g + E + h_f \qquad (2.3.21)$$

where h is the height above a reference plane (m); p, the pressure (N/m²); γ, the specific weight (N/m³); v, the average velocity (m/sec); E, the equivalent height added by pump (m); and h_f, friction loss (m).

This is the modified Bernoulli equation to describe energy relations to fluid flow. Correctly, however, each term in this equation is not an energy term, but rather a distance term. The distance referred to is a height of the fluid in question. If the equation is applied to water, each term assumes units of meters of water, if milk is the fluid, each term assumes units of meters of milk. We will return to this point when specifying pumps.

differences
are important

What the modified Bernoulli equation tells us is that it is differences in height, differences in pressure, and differences in average velocity squared that matter. Thus these differences, along with the friction existing between the two pipe locations of interest, can determine the required amount of energy to be added to the fluid by a pump.

EXAMPLE 2.3.3-1. Pressure in the Heart

Typical systolic pressure at the right atrium of the heart is 800 N/m² and at the pulmonary artery is 3330 N/m². How much pump pressure is added by the right heart?

Solution

The Bernoulli equation 2.3.21, when rearranged, becomes

$$(h_1 - h_2) + \frac{(p_1 - p_2)}{\gamma} + \frac{(v_1^2 - v_2^2)}{\alpha g} + E + h_f = 0$$

Since the heart is small, with little height difference between the right atrium and the entrance to the pulmonary artery, $(h_1 - h_2) = 0$. Also, the diameter of the vena cava feeding the right atrium is nearly the same as that of the pulmonary artery. Thus $v_1 \approx v_2$. Now, for the sake of simplicity, two assumptions are made:

1. The flow velocity profile is assumed to be a constant value (plug flow);
2. The heart is assumed to be frictionless.

We can also for now neglect the pulsatile flow. As a result

$$E = \frac{(p_2 - p_1)}{\gamma} = \frac{(p_2 - p_1)}{\rho g}$$

$$= \frac{(3330 - 800) \text{ N/m}^2}{(1050 \text{ kg/m}^3)(9.8 \text{ m/sec}^2)} = 0.24 \text{ m}$$

The pump pressure is about 24 cm of blood.

EXAMPLE 2.3.3-2. Energy in a Branching System

Typical cardiac output in the aorta at rest is $92 \times 10^{-6} \text{ m}^3/\text{sec}$ (5.5 L/min). Flow rate in the carotid artery, 0.3 m above the heart, is $12.5 \text{ m}^3/\text{sec}$ (0.75 L/min). The diameter of the aorta is about 2.5 cm and the diameter of the carotid artery is about 0.4 cm. Systolic/diastolic pressures in the aorta are $16,000/10,700 \text{ N/m}^2$ and in the carotid artery are about $12,000/7,000 \text{ N/m}^2$. What is the friction loss between aorta and carotid artery?

Solution

We could attempt to use the Bernoulli equation to solve this problem. We have all information to calculate velocities in both vessels, pressure differences, height differences, and there is no pump between the aorta and the carotid artery ($E = 0$). However, the Bernoulli equation cannot be used because there are other arteries fed by the aorta, including the brachial artery, superior mesenteric artery, splenic artery, hepatic artery, iliac artery, and vessels that feed the kidneys. Kinetic energy of blood flowing in the aorta becomes partitioned into vessels about which we know little or nothing. This is not an energy conserving system, and the Bernoulli's equation does not apply.

2.3.4 Energy Allocation Within the Fluid

The modified Bernoulli Eq. 2.3.21 accounts for fluid energy gains or losses.

higher velocity means lower pressure

Without external intervention, the gain of one kind of energy (for example, kinetic energy) must be accompanied by a loss of another type (for example, potential energy). This reciprocity is illustrated in Figure 2.3.5, where differences in pipe diameter cause different flow velocities to result. Sections with higher velocities and higher kinetic energies have lower static pressures and lower potential energies.

One might suppose that fluid kinetic energy might also be reduced, but kinetic energy reductions require velocity reductions, and, for an incompressible fluid in a constant diameter pipe the continuity equation 2.2.4 shows that a velocity reduction cannot exist. For incompressible fluids in pipes of

transformation between potential and kinetic energy

noncontant diameters, velocity reductions do, indeed, occur, but Bernoulli's equation 2.3.21 shows that lower velocities accompany higher pressures, and, therefore, kinetic energy is transformed into potential energy. When velocity is again increased, potential energy is transformed into additional kinetic energy. This process of energy transformation is not 100% reversible because some kinetic energy is transformed into heat and is lost.

If the Bernoulli equation predicts that higher velocities are accompanied by lower pressures, and if velocity is higher in the center of the pipe than it is at

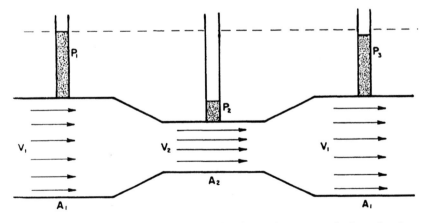

Figure 2.3.5. Reducing the pipe cross-sectional area increases velocity and reduces static pressure. Subsequently increasing the pipe area reduces velocity to its original value, but the original pressure is not attained because of a small amount of energy loss due to friction.

the wall, why is there not a higher pressure at the wall than in the center of the pipe? And, if there is a pressure difference, why does all the fluid not flow radially from the wall to the pipe center? It would appear that such a conclusion could be reached by careful consideration of the Bernoulli equation.

static
pressure is
uniform

We do know that this radial flow within a pipe does not occur. Therefore, there cannot be a radial pressure difference in the pipe. Static pressure inside the pipe at any cross section must be uniform. Thus, except for usually negligible differences in height, potential energy at any point in a pipe cross section must be constant.

Kinetic energy does vary radially within the pipe because of velocity differences. Therefore, the utility of the Bernoulli equation is in comparing pipe cross sections at different locations, but not in comparing radial locations at one pipe cross section.

2.3.5   General Form of Energy Balance Equation

There are many forms of the general energy balance equation. The one presented here relates to the Bernoulli equation already given.

Consider again the differential control volume of Figure 2.2.2. As mass enters the control volume, it brings with it potential energy due to height (mgh) and internal potential energy (mu). This last term can include reversible chemical or electrical potential energies.

additional
energy terms
possible

In addition, there is energy associated with mass movement ($\int_{m_p v \, dt}$). Internal energy and pressure–volume energy are often added together to form enthalpy (h_{en}). Kinetic energy can be translational ($\frac{1}{2}mv^2$) and rotational ($\frac{1}{2}I\omega^2$). To these needs to be added the net amount of heat addition (q) and any shaft (mechanical) work done to the control volume. If

$$e = h_{en} + v^2/2 + I\omega^2/2m + gh \tag{2.3.22}$$

where e is the energy per unit mass, or specific energy (N m/kg); h_{en}, the specific enthalpy (N m/kg); m, the mass (kg or N sec²/m); v, the velocity (m/sec); I, the moment of inertia (kg m² or N sec² m); ω, the radial velocity (rad/sec); g, the acceleration due to gravity (9.81 m/sec²); and h, the distance above a reference plane (m). Then the energy balance

(rate of energy in) − (rate of energy out) + (rate of energy generated)

$$= \text{(rate of change of energy storage)} \tag{2.3.23}$$

becomes

general
energy
balance

$$\dot{m}_{x_1} e_{x_1} - \dot{m}_{x_2} e_{x_2} + \dot{m}_{y_1} e_{y_1} - \dot{m}_{y_2} e_{y_2} + \dot{m}_{z_1} e_{z_1}$$

$$- \dot{m}_{z_2} e_{z_2} + \dot{q} + \dot{W}_s + \dot{e} = \frac{\partial}{\partial t} \int_V \rho e \, dV \qquad (2.3.24)$$

where $\dot{q}$ is the rate of heat transferred (N m/sec); W_s, the rate of mechanical work (N m/sec); and $\dot{e}$, the rate of specific energy generated (N m/sec).

Without some functional dependence of e on x, y, and z, there is not much further this form of the general energy equation can go. However, for one-dimensional flow without energy generation, this equation becomes

$$\int_A e\rho v_p \cos \phi \, dA + \dot{W}_s + \dot{q} = \frac{\partial}{\partial t} e\rho \, dV \qquad (2.3.25)$$

where v_p is the local velocity (m/sec); ϕ, the angle between the velocity vector and a vector normal to the control volume surface (rad); A, the surface area of the face of the control volume (m^2); ρ, the density (kg/m^3); V, the volume (m^3); $\dot{W}_s$, rate of mechanical work (N m/sec); e, the specific energy (N m/kg); $\dot{q}$, the rate of heat transferred (N m/sec); and t, the time (sec).

2.4 MOMENTUM BALANCE

There are useful pieces of information that can be obtained from performing a momentum balance on a control volume in a flowing fluid. These pieces of information include the velocity profile at a cross section in the flow path and an expression for shear stress in the fluid. Unlike the mass and energy balances presented in the previous two sections, momentum is a vector quantity. In the limited conditions presented here, with one-dimensional flow, many requirements of vector mathematics may be relaxed.

momentum
balance
equals force
balance

For our purposes, a momentum rate balance is equivalent to a force balance on an elemental volume in the fluid. Many engineering students do not feel comfortable considering momentum, but do have reasonable facility with forces. For those students, keep in mind that rate of change of momentum, like a force, is an effort variable. As indicated in Section 1.7.2, effort variables cannot accumulate. Hence a momentum balance is

momentum
balance

(rate of momentum in) − (rate of momentum out)

$$+ \text{(rate of momentum generated)} = 0 \qquad (2.4.1)$$

The rate of momentum generated (or added) depends on external forces acting on the control volume. Newton's second law gives this equivalance

$$F = ma = m\frac{dv}{dt} = \frac{d(mv)}{dt} \qquad (2.4.2)$$

where F is the force (N); m, the mass (kg); a, acceleration (m/sec^2); v, velocity (m/sec); and mv, momentum (kg m/sec).

Because momentum can be generated by external forces acting on the system, momentum is said to be "not conserved." This contrasts with mass and energy, both of which do not change their values due to external forces; these are said to be "conservative." The title of this section is thus "Momentum Balance" rather than "Conservation of Momentum."

2.4.1 Viscosity

viscosity measures thickness

Viscosity of a fluid measures its resistance to motion. If two plates containing a thickness r of fluid between them are drawn apart by a force F at rate v (Figure 2.4.1), then the shear stress on the fluid is

$$\tau = F/A \qquad (2.4.3)$$

shear stress

where τ is the shear stress (N/m^2); A, the surface area of flat plate (m^2); and F, the force required to sustain movement (N).

The rate of shear is defined as

rate of shear

$$\gamma = \frac{dv}{dr} \qquad (2.4.4)$$

where γ is the rate of shear [m/(m sec)]; v, the speed of relative plate movement (m/sec); and r, the distance between plates (m). Shear stress is

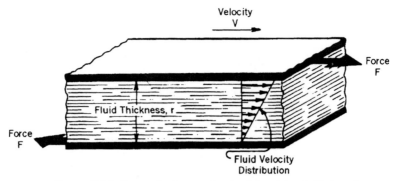

Figure 2.4.1. Conceptual apparatus for determining fluid viscosity.

related to the force required to pump a fluid through a tube, and the rate of shear is related to the rate at which the fluid flows.

Viscosity is the ratio of shear stress to rate of shear

$$\mu = \tau/\gamma \qquad (2.4.5)$$

where μ is the viscosity [$N\,sec/m^2$ or $kg/(m\,sec)$].

The more viscous the moving fluid, the higher is the shear stress and the more difficult the fluid is to pump. Various viscosity values are found in Table 2.4.1.

viscosities for gases   Viscosity values for gases are generally low, on the order of $10^{-5}\,kg/(m\,sec)$ or $10^{-5}\,N\,sec/m^2$ (Tables 2.4.2 and 2.4.3). There is very little difference in viscosity from gas to gas. Liquids, on the other hand, are much more viscous, and the range of viscosities for liquids appearing in Table 2.4.1 is from $2 \times 10^{-4}\,kg/(m\,sec)$ for Freon-12 at $80°C$ to $6.6\,kg/(m\,sec)$ for molasses at $21°C$.

Viscosities for gases increase with the square root of absolute temperature and are nearly independent of pressure below $1000\,kN/m^2$ (10 atm). When interpolation of viscosity from tabled values is necessary, the interpolation formula to use for gases is thus

interpolation of gas viscosities

$$\frac{(\mu_x - \mu_1)}{(\mu_2 - \mu_1)} = \frac{\sqrt{T_x} - \sqrt{T_1}}{\sqrt{T_2} - \sqrt{T_1}} \qquad (2.4.6a)$$

where μ is the viscosity of gas [$kg/(m\,sec)$]; T, the absolute temperature (K); and x, 1, 2 denote the unknown and two known values of viscosity.

Table 2.4.3 gives coefficients for empirical equations developed to estimate viscosity values for single gases at atmospheric pressure and varying temperature. Viscosities of mixtures of gases may be estimated from viscosity values for individual gases multiplied by the mass fraction of each respective gas in the mixture. The result is only approximate, however, because there is interaction that occurs among gases.

viscosities for liquids   Viscosity values for liquids decrease with absolute temperature and, since liquids are practically incompressible, are unaffected by pressure. The decrease of viscosity with temperature can be dramatic: The proverbial "molasses in January" is notable because it flows so slowly. Molasses heated to room temperature or above flows much more easily. Of all the physical properties of materials that must be considered in transport processes, viscosities of liquids are generally the most temperature dependent, and corrections must often be made where simultaneous fluid flow and heat transfer occur. Because water is often the liquid base for many fluids important in biological and food systems, a detailed listing of viscosity values for water at different temperatures is given in Table 2.4.4.

Interpolation of viscosity from tabled values may be necessary. Because the viscosity of water is inversely proportional to absolute temperature, the inverse

Table 2.4.1 Viscosity Values for Various Fluids

Material	Temperature (°C)	Density (kg/m^3)	Viscosity [kg/(m sec)]
Air	0	1.293	1.71×10^{-5}
	21	1.201	1.81×10^{-5}
	100	0.946	2.18×10^{-5}
Beer	0		1.3×10^{-3}
Water	0	1000	1.793×10^{-3}
	21	998	9.84×10^{-4}
	49	987	5.59×10^{-4}
Sucrose,	0	1086	3.818×10^{-3}
20% solution	21	1082	1.916×10^{-3}
	80	1055	5.92×10^{-4}
60% solution	21	1289	6.02×10^{-2}
	80	1252	5.42×10^{-3}
Lubricating oil,			
S.A.E. 10	16	900	1.00×10^{-1}
S.A.E. 30	66	870	1.0×10^{-2}
	16	900	4.00×10^{-1}
	66	870	2.7×10^{-2}
Liquid ammonia	−15	660	2.5×10^{-4}
	27	600	2.1×10^{-4}
Freon-12	−15	144	3.3×10^{-4}
	27	130	2.6×10^{-4}
CaCl$_2$ brine,	−23	1238	1.25×10^{-2}
24% solution	−18	1234	8.8×10^{-3}
	2	1227	3.7×10^{-3}
NaCl brine,	−18	1190	6.1×10^{-3}
22% solution	2	1170	2.7×10^{-3}
Molasses,	21	1430	6.600
heavy dark	38	1380	1.872
	49	1310	9.20×10^{-1}
	66	1160	3.74×10^{-1}
Soybean oil	30	920	4.06×10^{-2}
Olive oil	30	920	8.40×10^{-2}
Cottonseed oil	16	920	9.10×10^{-2}
Milk, whole	0	1035	4.28×10^{-3}
	20	1030	2.12×10^{-3}
Milk, skim	25	1040	1.37×10^{-3}
Cream, pasteurized			
20% fat	3	1010	6.20×10^{-3}
30% fat	3	1000	1.378×10^{-2}
Acetic acid	15		1.31×10^{-3}
	41		1.00×10^{-3}
	100		4.30×10^{-4}

(continued)

Table 2.4.1 (continued)

Material	Temperature (°C)	Density (kg/m^3)	Viscosity [kg/(m sec)]
Ethyl alcohol	0	800	1.773×10^{-3}
	20	789	1.200×10^{-3}
Methyl alcohol	0	810	8.2×10^{-4}
	20		5.97×10^{-4}
Linseed oil	30	942	3.31×10^{-2}
Rapeseed oil	30		9.6×10^{-2}
Salicylic acid	20		2.71×10^{-3}
Benzene	20		6.7×10^{-4}
Glycerol	60	1260	9.8×10^{-1}
Oxygen	37.8	1.255	2.13×10^{-5}
Carbon dioxide	37.8	1.725	1.54×10^{-5}
Blood			
plasma	37		$1.1–1.6 \times 10^{-3}$
whole	37	1050	$4.0–5.0 \times 10^{-3}$
Saliva	37		1.0×10^{-3}

[a] Viscosity values have units of kg/(m sec) or N sec/m^2. One centipoise is equal to 0.001 kg/(m sec).

Table 2.4.2 Variation of Viscosities of Selected Gases with Temperature [10^{-5} kg/(m sec)][a]

Temperature (°C)	Gas				
	Air	N$_2$	O$_2$	CO$_2$	CO
−20	1.61	1.57	1.80	1.27	1.55
−10	1.66	1.62	1.86	1.32	1.60
0	1.72	1.66	1.92	1.37	1.65
10	1.78	1.71	1.97	1.41	1.69
20	1.82	1.75	2.03	1.46	1.74
30	1.87	1.80	2.09	1.50	1.79
37	1.90	1.83	2.13	1.54	1.83
40	1.91	1.84	2.14	1.55	1.84
50	1.96	1.89	2.20	1.60	1.88
60	2.00	1.93	2.25	1.64	1.93
70	2.05	1.98	2.30	1.69	1.97
80	2.09	2.02	2.35	1.73	2.02
90	2.14	2.07	2.39	1.78	2.06
100	2.18	2.11	2.44	1.82	2.11

[a] For example, the density of air at 37°C is 1.90×10^{-5} kg/(m sec).

Table 2.4.3 Empirical Equations to Determine Viscosities [(kg/m sec)] of Gases at Various Absolute Temperatures

Equation: $\mu = A + BT + CT^2$

Gas	A	B	C
Benzene	-1.509×10^{-8}	2.5706×10^{-8}	-8.9797×10^{-13}
Carbon dioxide	1.18109×10^{-6}	4.9838×10^{-8}	-1.0851×10^{-11}
Carbon monoxide	2.38114×10^{-6}	5.3944×10^{-8}	-1.5411×10^{-11}
Chloroform	-4.3915×10^{-7}	3.7309×10^{-8}	-5.1696×10^{-12}
Ethanol	1.4991×10^{-7}	3.0741×10^{-8}	-4.4479×10^{-12}
Ethylene	-3.9851×10^{-7}	3.8726×10^{-8}	-1.1227×10^{-11}
Methane	3.8435×10^{-7}	4.0112×10^{-8}	-1.4303×10^{-11}

Adapted from Yaws (1995).

[a] For example, the viscosity of CO at 300 K is $2.38114 \times 10^{-6} + (5.3944 \times 10^{-8})(300) - 1.5411 \times 10^{-11}(300)^2 = 1.7177 \times 10^{-5}$ kg/(m sec)

Equation: $10^7 \ln \mu = A \ln T + BT^{-1} + CT^{-2} + D, \quad 300 \leq T \leq 1000$ K

Gas	A	B	C	D
Ammonia	8.1181026×10^{-1}	-1.6192541×10^2	1.3635348×10^4	3.8586405×10^{-1}
Carbon Dioxide	5.4330318×10^{-1}	-1.8823898×10^2	8.8726567×10^3	2.4499362
Carbon monoxide	6.0443938×10^{-1}	-4.3632704×10^1	-8.8441949×10^2	1.8972150
Chlorine	5.9042211×10^{-1}	-6.8463350×10^1	1.1486756×10^2	2.0947363
Helium	6.4802751×10^{-1}	4.3051414×10^{-1}	-3.7873123×10^1	1.6131962
Hydrogen sulfide	5.2624074×10^{-1}	-2.4493210×10^2	1.4706252×10^4	2.5104802
Methane	5.5243600×10^{-1}	-1.6260917×10^2	6.4734038×10^3	1.9463233
Nitric oxide (NO)	5.9536071×10^{-1}	-5.7867416×10^1	-3.8658607×10^2	2.0594392

(continued)

Table 2.4.3 (continued)

Equation: $10^7 \ln \mu = A \ln T + BT^{-1} + CT^{-2} + D$, $\quad 300 \le T \le 1000\ \mathrm{K}^a$

Gas	A	B	C	D
Nitrogen	6.0443938×10^{-1}	-4.3632704×10^{1}	-8.8441949×10^{2}	1.8972150
Nitrogen dioxide (NO$_2$)	5.5638659×10^{-1}	-1.5082685×10^{2}	5.4896589×10^{3}	2.3748776
Nitrous oxide (N$_2$O)	5.4648279×10^{-1}	-1.7538256×10^{2}	7.6356925×10^{3}	2.3887157
Oxygen	6.1936357×10^{-1}	-4.4608607×10^{1}	-1.3460714×10^{3}	1.9597562
Sulfur dioxide	1.0275639×10^{-0}	-2.6860106×10^{2}	1.7696352×10^{4}	2.5434068

Adapted from McBride et al. (1993).

^{a}For example, the viscosity of CO at 300 K is $10^7 \ln \mu = 6.0443938 \times 10^{-1} (\ln 300) - 4.3632704 \times 10^1 (300)^{-1} - 8.8441949 \times 10^2 (300)^{-2} + 1.8972150 = 5.18953651275$
$\mu = 1.7939 \times 10^{-5}\ \mathrm{kg/(m\,sec)}$.

Table 2.4.4   Dependence of Viscosity of Liquid Water on Temperature

Temperature (°C)	Viscosity [kg/(m sec)]
0	1.7921×10^{-3}
10	1.3077×10^{-3}
20	1.0050×10^{-3}
30	8.007×10^{-4}
37	6.984×10^{-4}
40	6.560×10^{-4}
50	5.494×10^{-4}
60	4.688×10^{-4}
70	4.061×10^{-4}
80	3.565×10^{-4}
90	3.165×10^{-4}
100	2.838×10^{-4}

interpolation of liquid viscosities

of absolute temperature should be used in the interpolation ratio. *This equation is not valid for gases*

$$\frac{(\mu_x - \mu_1)}{(\mu_2 - \mu_1)} = \frac{\left(\dfrac{1}{T_x} - \dfrac{1}{T_1}\right)}{\left(\dfrac{1}{T_2} - \dfrac{1}{T_1}\right)} \tag{2.4.6b}$$

where μ is the viscosity of water [kg/(m sec)]; T, the absolute temperature (K); and x, 1, 2 denote the unknown and two known values of viscosity.

viscosity of blood

hematocrit

Viscosity values for suspensions of inclusions within liquids depend on the concentrations of particles. An important example of this is the viscosity of human blood, which depends strongly (Figure 2.4.2) on hematocrit level (hematocrit is the fraction of blood volume represented by blood cells). When hematocrit becomes extremely high (greater than 60%), it is hard to describe blood as a liquid at all. If the heart had to be powerful enough to propel throughout the body blood with viscosities corresponding to normal hematocrit values (47% for men and 42% for women), there would be little room in the chest for the lungs and other structures. Fortunately, at the extremely low flows that are present in the very small blood vessels, there is a

axial streaming of red blood cells

tendency for red blood cells to be carried in midstream and away from vessel walls. The blood near the walls is composed almost completely of plasma, which has a much lower viscosity than whole blood (Figure 2.4.3). Because most of the friction in a flowing fluid occurs near the tube wall where the rate of shear (and the rate of change of velocity) is highest, the lower viscosity of blood in this region considerably reduces the power requirement for pumping blood. A second consequence of the axial streaming of red blood cells is that

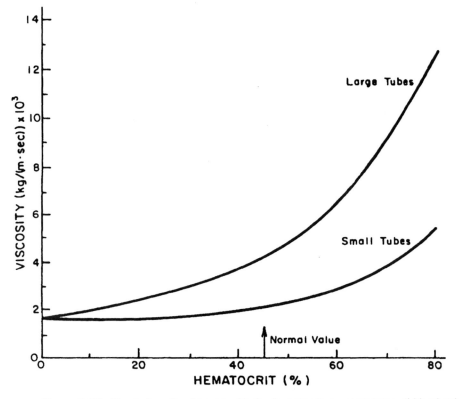

Figure 2.4.2. Blood viscosity changes with the hematocrit, or percentage of blood volume represented by blood cells. Because these cells tend toward the center of a small tube, the fluid near the wall is almost entirely plasma with low viscosity.

the cells have a higher average velocity than the blood as a whole, making more oxygen available to the tissues.

There are other fluids that are composed of particles suspended within liquids. Many foods are among these, and, at least for some of them, separation of the suspending liquid (usually with a water base), near walls of small tubes can substantially change local viscosities. This effect is most pronounced, however, in tubes too small to be of practical significance in food processing.

kinematic
viscosity

Viscosity divided by density is called "kinematic viscosity." Kinematic viscosity is identical to momenum diffusivity (Table 1.7.1), which is analogous to the mass diffusivity and thermal diffusivity to be studied in later chapters. Momentum diffusivity can be used to give a relative thickness of the hydrodynamic boundary layer next to a wall.

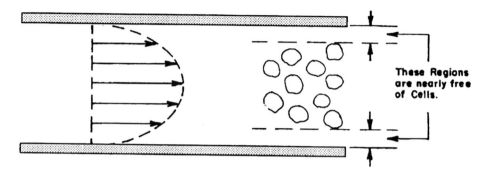

Figure 2.4.3. The higher velocities found on the sides of blood cells closer to the center of the tube are accompanied by lower static pressures. This causes the blood cells to flow in the center of the tube and frees the layer nearest the tube wall of blood cells. Fluid viscosity is thus considerably lower next to the wall.

EXAMPLE 2.4.1-1. Temperature Dependence of Viscosity

From Table 2.4.4, we see that the viscosity of water assumes a value of 1.7921×10^{-3} kg/(m sec) at $0°C$ (273.15 K) and a value of 1.0050×10^{-3} kg/(m sec) at $20°C$ (293.15 K). Calculate the viscosity value at $10°C$ (283.15 K) and compare with the tabled value.

Solution

Using Eq. 2.4.6b,

$$\frac{\mu_{283} - 1.7921 \times 10^{-3}}{1.0050 \times 10^{-3} - 1.7921 \times 10^{-3}} = \left(\frac{\dfrac{1}{283.15} - \dfrac{1}{273.15}}{\dfrac{1}{293.15} - \dfrac{1}{273.15}}\right)$$

$$\mu_{283} = 1.7921 \times 10^{-3} + (0.51766)(-0.7871 \times 10^{-3})$$

$$= 1.3846 \times 10^{-3} \text{ kg/(m sec)}$$

Remark

This value differs from the tabled value of 1.3077×10^{-3} by 6%.

2.4.2 Momentum Balance in a Circular Pipe

shear stress
and rate of
shear related
through
viscosity

We have seen in the previous section that shear stress (force per unit area acting on the fluid in the direction of movement) is related to rate of shear (rate of change of velocity with distance perpendicular to the direction of movement) through the viscosity. The rate of shear inside a pipe, dv/dr, is negative. Thus

$$\tau = -\mu \frac{dv}{dr} = -\mu \gamma \qquad (2.4.7)$$

where τ is the shear stress (N/m^2); μ, the viscosity ($N\,sec/m^2$); γ, the rate of shear (sec^{-1}); v, the velocity (m/sec), and r, the radial distance (m).

shear stress
equals
momentum
flux rate

Shear stress also has equivalent units of $kg\,m/(m^2\,sec^2)$, which, in turn, is just the time rate of change of momentum (kg m/sec) per unit area (m^2). Thus the shear stress τ can be considered to be rate of momentum flux.

This interpretation leads to the concept of effort and flow variables for momentum that some people find useful. The effort variable is momentum concentration ($mv/V = \rho \dot{V}/A$) and the flow variable is the rate of momentum flux (τ). See Table 1.7.1. Manipulation of Eq. 2.4.7 gives

$$\tau = \frac{-\mu}{\rho} \frac{d(\rho \dot{V}/A)}{dr} = \frac{-\mu}{\rho} \frac{d(\rho v)}{dr} = -v \frac{d(mv/V)}{dr} \qquad (2.4.7a)$$

kinematic
viscosity is
momentum
diffusivity

The viscosity (μ) divided by the density (ρ) gives the momentum diffusivity (v), otherwise known as kinematic viscosity.

There is a physical explanation for the generation of shear stress that involves momentum transfer. Fluids in laminar flow can be considered to be flowing in cylindrical shells within the pipe. Shells closer to the pipe center flow faster than those closer to the pipe wall. As fluid molecules from the faster shells bump, collide, and grind against molecules from adjacent slower shells, they add momentum to the slower molecules at the expense of their own momenta. This process tends to retard the faster molecules. We call this shear stress.

flow in shells

We form a control volume around one of these shells from position x to $x + \Delta x$ (Figure 2.4.4). The net rate of momentum flux into the control volume through the cylindrical surface at radius r is just $\tau_{rx} A_r$, where τ_{rx} is shear stress acting in the x direction, and varying in magnitude with radius. A_r is the surface area of the inner cylindrical surface of the control volume. The net rate of momentum flux into the control volume through the cylindrical surface at outer radius $r + \Delta r$ is negative with a value $-\tau_{r+\Delta r,x} A_{r+\Delta r}$. Thus the net efflux of momentum from the cylindrical surfaces of the control volume is

$$(\text{net momentum efflux}) = (\tau_{r+\Delta r,x} A_{r+\Delta r} - (\tau_{rx} A_r)$$
$$= [\tau_{r+\Delta r,x} 2\pi (r + \Delta r)\,\Delta x] - (\tau_{rx} 2\pi r\,\Delta x) \qquad (2.4.8)$$

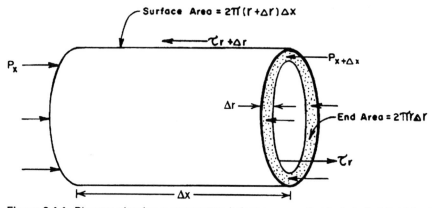

Figure 2.4.4. Diagram showing a momentum balance on a cylindrical shell of fluid flowing within a circular tube. Pressures at the ends are p_x and $p_{x+\Delta x}$. Shear stresses on the walls are τ_r and $\tau_{r+\Delta r}$. Pressure and shear stresses are multiplied by their respective areas to obtain forces acting on the shell.

where τ_{rx} is the shear stress of fluid flowing in the x direction, and that varies with radius $[(\text{kg m/sec})/(\text{m}^2\,\text{sec})$; A_r, the area of cylindrical surface (m^2); r, the radius of inner surface of control volume (m); Δr, the thickness of the control volume (m); and Δx, the length of the control volume (m). Notice that the sign of momentum efflux is opposite to the sign of momentum flux into the control volume.

Momentum fluxes added to the control volume through the two end faces are due to forces acting on those surfaces, as in Eq. 2.4.2. Thus the sum of these forces is

$$\sum F = p_x A_x - p_{x+\Delta x} A_{x+\Delta x}$$
$$= p_x 2\pi r\, \Delta r - p_{x+\Delta x} 2\pi r\, \Delta r \tag{2.4.9}$$

where p_x is the fluid pressure acting on an annular surface at x (N/m^2); and A_x, the area of the end of the cylinder (m^2).

There is no momentum generation by external forces. The momentum balance in Eq. 2.4.1 thus becomes

$$\tau_{r+\Delta r,x} 2\pi(r + \Delta r)\, \Delta x - \tau_{rx} 2\pi r\, \Delta x = p_x 2\pi r\, \Delta r - p_{x+\Delta x} 2\pi r\, \Delta r \tag{2.4.10}$$

Note that all terms in Eq. 2.4.10 are forces. The terms involving p represent the forces on the end of the cylindrical shell, and the terms involving τ represent the forces on the sides of the shell. Thus the above equation is a simple force balance. Rearranging this equation

$$\frac{\tau_{r+\Delta r,x}(r + \Delta r) - \tau_{rx} r}{\Delta r} = \frac{(p_x - p_{x+\Delta x})r}{\Delta x} \tag{2.4.11}$$

In fully developed flow, the pressure drop (Δp) per unit length (L) of pipe is constant, $\Delta p/L$. After this substitution, we next let Δr shrink to zero

$$\frac{d(\tau_{rx}r)}{dr} = \frac{\Delta p}{L}r \qquad (2.4.12)$$

Separating variables gives

$$d(r\tau_{rx}) = \frac{\Delta p}{L}r\,dr \qquad (2.4.13)$$

and integrating gives:

$$\tau_{rx} = \frac{\Delta p}{L}\frac{r}{2} + \frac{C}{r} \qquad (2.4.14)$$

where C is a constant of integration (N/m).

In order that momentum flux not become infinite at $r = 0$, C must be zero. Thus

shear stress in
a pipe

$$\tau_{rx} = \frac{\Delta p}{2L}r \qquad (2.4.15)$$

shear stress
highest at wall

This important result shows that, in a circular pipe with fully developed laminar flow, the shear stress is a linear function of the radius r and is a maximum at the pipe wall. Fluids with greater pressure losses per length of pipe will also have higher values of shear stress.

Please note the meaning of Δp in Eq. 2.4.15. This pressure difference is due solely to friction loss in the fluid flowing in the pipe. In terms of the previous Bernoulli Eq. 2.3.21, we have denoted friction loss in the pipe with the symbol h_f. However, h_f has units of height of fluid; thus Δp in Eq. 2.4.15 is equivalent to $(h_f\gamma)$ or $(h_f\rho g)$ as used previously.

friction loss

EXAMPLE 2.4.2-1. Shear Stress in Fluid

Water flows in a vertical pipe 50 m long and 5 cm in diameter. The water flows at a rate of 0.126 L/sec. The pressure in the fluid at the top end of the pipe is $207\,kN/m^2$, and the pressure at the bottom is $800\,kN/m^2$. What is the shear stress at the wall of the pipe?

Solution

From the continuity equation 2.2.4, the velocity of the water in the pipe is:

$$v = \frac{\dot{V}}{A} = \frac{(0.126 \text{ L/sec})(0.001 \text{ m}^3/\text{L})}{\pi(2.5 \text{ cm})^2(0.01 \text{ m/cm})^2} = 6.43 \text{ m/sec}$$

From the Bernoulli equation 2.3.21,

$$(h_2 - h_1) + \frac{(p_2 - p_1)}{\gamma} + \frac{V_2^2 - V_1^2}{\alpha g} + E + h_f = 0$$

$$50\text{m} + \frac{(p_2 - p_1)}{\gamma} + 0 + 0 + h_f = 0$$

$$(p_2 - p_1) = (-h_f - 50\text{m})\gamma$$

or

$$p_1 - p_2 = (50\text{m} + h_f)v$$

$$(h_f\gamma) = (p_1 - p_2) - (50\text{m})\gamma$$

$$= (800,000 - 207,000) \text{ N/m}^2 - (50\text{m})(9810 \text{ N/m}^3)$$

$$= 102,500 \text{ N/m}^2$$

From Eq. 2.4.15,

$$\tau_r = \frac{(h_f\gamma)r}{2L} = \frac{(102,500 \text{ N/m}^2)(0.05\text{m})}{(2)(50\text{m})} = 51.2 \text{ N/m}^2$$

Remark

If the pipe had been horizontal, rather than vertical, the pressure difference $(p_1 - p_2)$ would have been due entirely to friction.

2.4.3 Flow Velocity Profile

For the relatively simple case where fluid is flowing full in a horizontal circular pipe, and the fluid is incompressible, with constant viscosity, and the flow is one dimensional, steady state, and fully developed laminar (there is no change of velocity profile along the length of the pipe), the results of the momentum balance can lead to insight related to the fluid velocity across the pipe cross section. Intuition probably tells you, correctly, that fluid velocity is greatest in the center of the pipe and slowest next to the pipe wall. We will see in more detail how fluid velocity varies with radial position in the pipe.

From Eqs. 2.4.15 and 2.4.7

$$\tau_{rx} = \frac{p}{2L} r = -\mu \frac{dv}{dr} \qquad (2.4.16)$$

or

$$dv = \frac{-\Delta p}{2L\mu} r \, dr \qquad (2.4.17)$$

Integrating,

$$v = \frac{-\Delta p}{4L\mu} r^2 + C \qquad (2.4.18)$$

Since $v = 0$ at the wall, $r = r_0$

$$C = \frac{\Delta p \, r_0^2}{4L\mu} \qquad (2.4.19)$$

laminar flow
velocity profile

$$v = \frac{\Delta p \, r_0^2}{4L\mu} \left[1 - \left(\frac{r}{r_0} \right)^2 \right] \qquad (2.4.20)$$

The flow velocity profile for fully developed laminar flow is parabolic with decreases in radius (illustrated in the lower part of Figure 2.3.4).

For fluids undergoing a phase change while flowing in a pipe, the phase change modifies not only the flow velocity profile, but also the average velocity. Most liquid densities are higher than gas densities. Thus a gas undergoing condensation within a pipe will have a higher inlet volume flow rate than outlet flow rate. Exactly the opposite is true for liquids evaporating within a pipe. These effects are common in refrigeration systems, but are beyond the scope of this text.

2.4.4 General Form for Momentum Balance

Development of the general form for momentum balance is not extremly difficult, but it is tedious and not necessary to the material to be subsequently presented. In general, three momentum balances are required for the three orthogonal Cartesian directions. Rotational momentum is neglected. The general momentum balance in Eq. 2.4.1 is used to provide a framework, and details are filled in. Momentum carried into and from a differential element due to bulk flows (convection) is added to momentum transfer by molecular means (shear stress). Forces acting on the differential element are fluid pressure and gravity.

The quantity ρv_x is the volume concentration of momentum in the x direction (ρ in kg/m^3 and v_x in m/sec). Momentum flux (related to shear stress) is obtained by multiplying ρv_x by v_x [units of (kg m/sec^2)]. Multiplying by v_y and v_z gives momentum flux in the other two dimensions. The result is a set of three equations

general momentum balances in three directions

$$\frac{\partial(\rho v_x)}{\partial t} = -\left(\frac{\partial(\rho v_x v_x)}{\partial x} + \frac{\partial(\rho v_y v_x)}{\partial y} + \frac{\partial(\rho v_z v_x)}{\partial z}\right)$$

$$-\left(\frac{\partial \tau_{xx}}{\partial x} + \frac{\partial \tau_{yx}}{\partial y} + \frac{\partial \tau_{zx}}{\partial z}\right) - \frac{\partial p}{\partial x} + \rho g_x \tag{2.4.21}$$

$$\frac{\partial(\rho v_y)}{\partial t} = -\left(\frac{\partial(\rho v_x v_y)}{\partial x} + \frac{\partial(\rho v_y v_y)}{\partial y} + \frac{\partial(\rho v_z v_y)}{\partial z}\right)$$

$$-\left(\frac{\partial \tau_{xy}}{\partial x} + \frac{\partial \tau_{yy}}{\partial y} + \frac{\partial \tau_{zy}}{\partial z}\right) - \frac{\partial p}{\partial y} + \rho g_y \tag{2.4.22}$$

$$\frac{\partial(\rho v_z)}{\partial t} = -\left(\frac{\partial(\rho v_x v_z)}{\partial x} + \frac{\partial(\rho v_y v_z)}{\partial y} + \frac{\partial(\rho v_z v_z)}{\partial z}\right)$$

$$-\left(\frac{\partial \tau_{xz}}{\partial x} + \frac{\partial \tau_{yz}}{\partial y} + \frac{\partial \tau_{zz}}{\partial z}\right) - \frac{\partial p}{\partial z} + \rho g_z \tag{2.4.23}$$

where v_x is the velocity component in the x direction (m/sec); τ_{xy}, the shear stress in the x direction, and varying in the y direction (N/m^2); p, the pressure in the fluid (N/m^2); ρ, the density (kg/m^3); and g_x, the component of gravity in the x direction (m/sec^2). These equations would be helpful especially for constructing a numerical model of elemental flow in a fluid flow system. The model would be programmed to be solved by computer.

2.4.5 Navier–Stokes Equations

The Navier–Stokes equations follow directly from the general momentum equations just given. The Navier–Stokes equations are used to solve for various pressure and flow distributions in flowing fluids (Kiris et al., 1993), and are also used for simultaneous heat transfer and fluid flow.

These equations arise from cases where the density (ρ) and viscosity (μ) are constant. The resulting three equations are

Navier–Stokes equations

$$\rho\left(\frac{\partial v_x}{\partial t} + v_x \frac{\partial v_x}{\partial x} + v_y \frac{\partial v_x}{\partial y} + v_z \frac{\partial v_x}{\partial z}\right)$$

$$= \mu\left(\frac{\partial^2 v_x}{\partial x^2} + \frac{\partial^2 v_x}{\partial y^2} + \frac{\partial^2 v_x}{\partial z^2}\right) - \frac{\partial p}{\partial x} + \rho g_x \tag{2.4.24}$$

$$\rho\left(\frac{\partial v_y}{\partial t} + v_x\frac{\partial v_y}{\partial x} + v_y\frac{\partial v_y}{\partial y} + v_z\frac{\partial v_y}{\partial z}\right)$$

$$= \mu\left(\frac{\partial^2 v_y}{\partial x^2} + \frac{\partial^2 v_y}{\partial y^2} + \frac{\partial^2 v_y}{\partial z^2}\right) - \frac{\partial p}{\partial y} + \rho g_y \qquad (2.4.25)$$

$$\rho\left(\frac{\partial v_z}{\partial t} + v_x\frac{\partial v_x}{\partial x} + v_y\frac{\partial v_z}{\partial z} + v_z\frac{\partial v_z}{\partial z}\right)$$

$$= \mu\left(\frac{\partial^2 v_z}{\partial x^2} + \frac{\partial^2 v_z}{\partial y^2} + \frac{\partial^2 v_z}{\partial z^2}\right) - \frac{\partial p}{\partial z} + \rho g_z \qquad (2.4.26)$$

At very low flow rates, at Reynolds numbers below approximately 1, the flow is called creeping flow. This type of flow describes the settling of small particles through a fluid. It has been found that inertial effects are unimportant in creeping flow, which leads to Stokes' laws for incompressible fluids

Stokes laws

$$\frac{\partial p}{\partial x} = \mu\left(\frac{\partial^2 v_x}{\partial x^2} + \frac{\partial^2 v_x}{\partial y^2} + \frac{\partial^2 v_x}{\partial z^2}\right) \qquad (2.4.27)$$

$$\frac{\partial p}{\partial y} = \mu\left(\frac{\partial^2 v_y}{\partial x^2} + \frac{\partial^2 v_y}{\partial y^2} + \frac{\partial^2 v_y}{\partial z^2}\right) \qquad (2.4.28)$$

$$\frac{\partial p}{\partial z} = \mu\left(\frac{\partial^2 v_z}{\partial x^2} + \frac{\partial^2 v^z}{\partial y^2} + \frac{\partial^2 v_z}{\partial z^2}\right) \qquad (2.4.29)$$

In many applications, these equations can be reduced to one dimension by proper assignment of direction.

EXAMPLE 2.4.5-1. Ventricular Filling

Doppler-shift ultrasonic flow sensors can be used to measure blood velocities in the heart. From these the atrium to ventricle pressure drop can be estimated. If mitral blood velocity and acceleration have been measured as 1 m/sec and 20 m/sec², respectively, and a polynomial has been fit to the ratio of blood velocity to mitral velocity along a central streamline between the atrium and ventricle ($x = 0$ to 6 cm), to give: $v(x)/v_{mv} = x^2/9 + 2x/3$, find the pressure drop between the atrium and ventricle.

Solution

The Navier–Stokes equation 2.4.24 will be used for one-dimensional flow ($v_y = v_z = 0$). Gravitational and viscous effects are considered negligible compared to inertial effects

$$\rho \frac{\partial v}{\partial t} = \frac{-\partial p}{\partial x} - \rho v \frac{\partial v}{\partial x}$$

Multiplying by dx and integrating in the x direction from atrium (A) to ventricle (V)

$$\rho \int_A^V \frac{\partial v}{\partial t}\, dx = -\int_A^V dp - \rho \int_A^V v\, dv = p_A - p_V - \frac{1}{2}\rho(v_V^2 - v_A^2)$$

Next we assume that the left-hand integral is composed of the product of time-varying and position-varying components

$$\rho \int_A^V \frac{\partial v(x, t)}{\partial t}\, dx = \left(\rho \int_A^V \frac{A_{mv}}{A(x)}\, dx\right)\left(\frac{dv_{mv}(t)}{dt}\right)$$

where the subscript mv denotes "mitral valve."

From the continuity equation 2.2.4, the ratio of areas $A_{mv}/A(x)$ can be replaced by the velocity ratio $v(x)/v_{mv}$ as long as the fluid is incompressible and not accumulating anywhere.

Finally, atrial velocity is usually small, and v_V can be replaced by mitral valve velocity (Thomas and Weyman, 1992). Thus

$$\frac{dv_{mv}}{dt}\left(\rho \int_A^V \frac{v(x)}{v_{mv}}\, dx\right) = p_A - p_V - \frac{1}{2}\rho v_{mv}^2$$

or

$$p_A - p_v = \frac{1}{2}\rho v_{mv}^2 + \frac{dv_{mv}}{dt}\left(\rho \int_A^V \frac{v(x)}{v_{mv}}\, dx\right)$$

$$= \frac{1}{2}\left(1050\frac{kg}{m^3}\right)\left(\frac{1\ m}{sec}\right)^2 + \left(20\frac{m}{sec^2}\right)\left(1050\frac{kg}{m^3}\right)\int_{0\ cm}^{6\ cm}\left(-\frac{x^2}{9} + \frac{2x}{3}\right) dx$$

$$= 525\frac{kg}{m\ sec^2} + 21{,}000\frac{kg}{m^2\ sec^2}\left(-\frac{x^3}{27} + \frac{x^2}{3}\right)_{0\ m}^{0.06\ m}$$

$$= 525\frac{kg}{m\ sec^2} + \left(21{,}000\frac{kg}{m^2\ sec^2}\right)(0.00119\ m)$$

$$p_A - p_V = 550\frac{kg}{m\ sec^2} = 550\frac{N}{m^2}$$

Remarks

One difficulty with polynomial curve fitting is that the coefficients often have units that are not specified. Such was the case here where the necessary units on $x^3/27$ and $x^2/3$ had to be meters.

Although there was a considerable amount of manipulation necessary, this example does show some utility to the Navier–Stokes equations.

2.4.6 Drag Coefficient and Settling Velocity

For the special case of slow flow past a sphere, the Navier–Stokes equations have been used to determine velocity and pressure distributions. The pressure distribution, integrated over the sphere, has then been used to determine the drag force on the sphere:

drag force on
sphere

$$F_D = C_D \frac{\rho \pi d^2}{8} v^2 \tag{2.4.30}$$

where F_D is the drag force (N); C_D, the drag coefficient (dimensionless); d, the sphere diameter (m); v, the relative velocity of the sphere in the fluid (m/sec); and ρ, the fluid density (kg/m^3 or N sec^2/m^4).

The drag coefficient for spheres with very low Reynolds numbers is given as

drag
coefficient

$$C_D = 24/\text{Re} \tag{2.4.31}$$

where Re is the Reynolds number of the sphere (dimensionless; see Eq. 2.5.4).

For higher values of the Reynolds number, the drag coefficient assumes other values, and is approximately constant at 0.44 unless the Reynolds number exceeds 10^5 (Figure 2.4.5).

A simple force balance on a falling sphere can lead to an estimate of settling velocity for the particle. Causing the sphere to fall is its weight; retarding its fall is the drag force. The sphere will accelerate in the downward direction until the terminal (or settling) velocity is reached:

settling
velocity of
particles

$$\sum F = ma = 0 = \text{weight} - \text{drag force} \tag{2.4.32}$$

$$0 = V(\rho_s - \rho_f)g - \left(\frac{24}{\text{Re}}\right)\frac{\rho \pi d^2}{8} v^2$$

where ρ is the density of a sphere (kg/m^3); and ρ_f, the density of a fluid (kg/m^3).

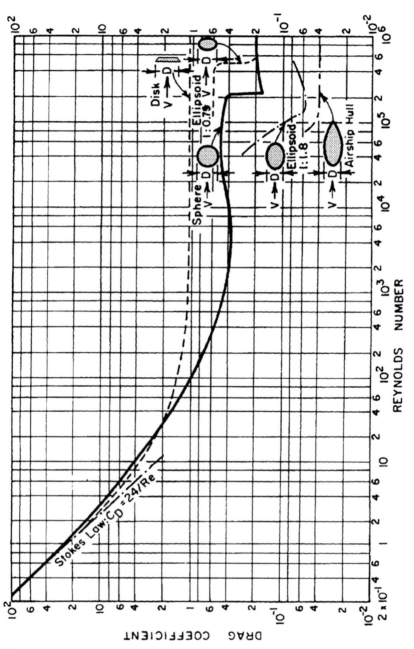

Figure 2.4.5. Drag coefficient for bodies of revolution (redrawn with permission of the McGraw-Hill Companies from Daugherty and Ingersoll, *Fluid Mechanics*, 1954).

From Eqs. 2.4.30 and 2.4.31

$$0 = \frac{\pi d^3}{6}(\rho_s - \rho_f)g - \left(\frac{24\mu_f}{dv\rho_f}\right)\frac{\rho_f \pi d^2}{8}v^2 \tag{2.4.33}$$

Thus

Stokes law for
spheres

$$v = \frac{d^2(\rho_s - \rho_f)g}{18\mu_f} \tag{2.4.34}$$

This equation is called Stokes law for spherical particles settling through a fluid medium in streamline (or laminar) flow.

Spherical particles moving in a fluid stream develop turbulence at Reynolds settling of dust numbers much lower than 2000 (see Section 2.5.1). Thus Stokes law has limited usefulness to determine settling velocities over a wide range of particle sizes. In Table 2.4.5 are shown equations for settling velocities for dust particles in air. These equations have been developed for spherical particles only; for other shapes, coefficient values must be changed. Although they apply to the settling of dusts in air, they can be used to estimate the velocity of settling of particles in other fluids, such as water. Notice that one means to hasten the settling of particles is to increase the apparent acceleration of gravity, as in a centrifuge.

EXAMPLE 2.4.6-1. Settling of Bacteria

Droplets of various sizes are emitted during a sneeze. How much time would it take for a droplet the size of an individual bacterium (about 2 μm diameter) to fall to the ground, about 1.7 m below the mouth?

Solution

For the solution to this problem, we assume the bacterium to be a coccus, spherical in shape. Using Stokes law, Eq. 2.4.34,

$$v = \frac{g(\rho_s - \rho_f)d_s^2}{18\mu_f}$$

The density of the sphere will be assumed to be that of water, or 1000 kg/m^3. Density and viscosity values for air are found in Table 2.4.1.

$$v = \frac{(9.81 \text{ m/sec}^2)(1000 \text{ kg/m}^3 - 1.2 \text{ kg/m}^3)(2 \times 10^{-6} \text{ m})^2}{18[1.81 \times 10^{-5} \text{ kg/(m sec)}]}$$

$$= 1.20 \times 10^{-4} \text{ m/sec}$$

Table 2.4.5 Settling Velocities for Dust Particles in Air

Particle Size Range (10^{-6} m)	Flow Regime	Equation for v	Examples of Particles in Size Range
>2000	Turbulent	$\sqrt{\dfrac{2gd_s\rho_s}{0.275\rho_f}}$	Raindrops
100–1000	Intermediate	$0.536\,d_s^3\sqrt{\dfrac{g(\rho_s - \rho_f)^2}{\rho_f\mu_f}}$	Mist, pulverized coal
2–50	Laminar	$\dfrac{g(\rho_s - \rho_f)d_s^2}{18\mu_f}$	Fog, fly ash, pigments, foundary dust, cement, silicosis-causing dusts, pollen, spores, bacteria
0.1–1	Ultralaminar	$\dfrac{g(\rho_s - \rho_f)d_s^2}{18\mu_f}\left(1 + \dfrac{0.172}{d_s}\right)$	Oil smoke, metallurgical fumes, dust in outdoor quiet air
<0.1	Brownian motion	(none—doesn't fall)	Tobacco smoke, carbon black

Source: Tigges and Karlsson (1967).

The time it takes for the droplet to fall 1.7 m is just

$$t = \frac{1.7 \text{ m}}{1.20 \times 10^{-4} \text{ m/sec}} = 14131 \text{ sec} = 3.9 \text{ h}$$

Remark

A droplet this size may not fall to the ground by its own weight. Small air currents may buoy the droplet, or it may dry before it reaches the ground.

2.5   FRICTION LOSSES IN PIPES

Friction losses usually account for the numerically largest term in the energy balance equation in normal piping systems. Friction generates heat, and this may either be dissipated to the environment or added to the fluid. There are two types of friction losses in a piping system: First there is friction of fluid particles moving against other particles. In piping systems these are called pipe losses. Second, any changes in direction or magnitude of fluid velocity due to entrances, bends, valves, and so on, requires the dissipation of energy. These are called minor losses, although they may not be minor in value.

friction equals energy loss

2.5.1   Pipe Losses

laminar flow

There are two distinct regions of flow that determine friction losses. Laminar flow is characterized by fluid particles moving in straight lines (Figure 2.5.1). There is little mixing between these stream lines, and the steady velocity

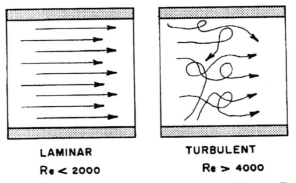

LAMINAR   **TURBULENT**

Re < 2000   **Re > 4000**

Figure 2.5.1. The differences between laminar and turbulent flow are illustrated here. Laminar flow occurs slowly, with little mixing across the diameter of the tube. Turbulent flow is relatively fast, with much mixing.

profile in a pipe is parabolic in shape (Figure 2.3.4, bottom, and Eq. 2.4.20). In

pressure
proportional to
flow

$$\Delta p \propto v \tag{2.5.1}$$

turbulent flow   Turbulent flow is characterized by much mixing across the pipe. Fluid particles flow in pseudorandom or eddying paths, and there is much energy dissipation and more fluid heating than with laminar flow (Figure 2.5.1). Because any given fluid particle could begin in any position and end up located at any other position within the pipe cross section, the net velocity profile is nearly flat (Figure 2.3.4, top). In turbulent flow, pressure drops along the pipe have been measured to be proportional with $v^{1.7}$ to v^3. Normally,

pressure
proportional to
flow squared

$$\Delta p \propto v^2 \tag{2.5.2}$$

transitional
flow

Reynolds
number

Between these two regions is a transitional zone, where flow cannot be counted upon to be either fully laminar or fully turbulent. Disturbances in the line may cause some turbulence that attenuates some distance downstream. Boundaries of the transition zone are not precisely defined, and accurate predictions of fluid behavior within the zone are difficult to make.

The type of flow can be determined once the value of the Reynolds number is known. The Reynolds number is a dimensionless quantity relating inertial (or gravitational) force to viscous (friction) force. From Newton's law, the inertial force is

$$F = ma = \rho L^3 v/t = \rho L^2 (L/t)v = \rho L^2 v^2 \tag{2.5.3}$$

where F is the gravitational force (N); ρ, the density (kg/m^3 or N sec^2/m^4); L, the significant length (m); v, the average velocity (m/sec); and t, the time (sec).

Friction force is μLv, where μ is the viscosity [N sec/m^2 or kg/(m sec)]. The Reynolds number is thus

$$\text{Re} = \frac{\rho L^2 v^2}{\mu Lv} = \frac{\rho Lv}{\mu} \tag{2.5.4}$$

For pipe flow, the Reynolds number uses pipe diameter as the significant length. Thus

$$\text{Re} = \frac{\rho Dv}{\mu} \tag{2.5.5}$$

critical
Reynolds
numbers

When the Reynolds number value is 2000 or below, laminar flow is normally assumed to exist. When the Reynolds number value is 4000 or above, turbulence is present. Between the values of 2000 and 4000, flow is not quite laminar and not fully turbulent. Of course, local conditions, such as internal protrusions, can cause some turbulence to exist at low Reynolds numbers, but turbulent eddies or vortices dampen very quickly at low Reynolds numbers. Likewise, flow straightening vanes can cause flow to be nearly laminar at high Reynolds numbers, but the flow quickly returns to turbulence away from the vanes.

There are cases where the pipe is either not round or not flowing full. In both of those cases, the diameter value to use in the Reynolds number calculation must be modified. The hydraulic diameter has been introduced for this reason.

hydraulic diameter

The hydraulic diameter is defined as four times the wetted area divided by the wetted perimeter

$$D = 4 \text{ (wetted area)/(wetted perimeter)} \tag{2.5.6}$$

And this value is used in the Reynolds number calculation.

For example, a circular pipe flowing full has a wetted area of $\pi d^2/4$ and a wetted perimeter of πd. Thus

$$D = 4(\pi d^2/4)/(\pi d) = d \tag{2.5.7}$$

circular pipe

A circular pipe flowing one-half full has a wetted area one-half as large as the full pipe, or $\pi d^2/8$ (Figure 2.5.2). It has a wetted perimeter of one-half of that for a full pipe, or $\pi d/2$. Thus

$$D = 4(\pi d^2/8)/(\pi d/2) = d \tag{2.5.8}$$

It should not be inferred from this that the hydraulic diameter of a circular pipe is always d. A circular pipe flowing one-quarter full has a wetted perimeter of $1.16d$ and a hydraulic diameter of $0.68d$.

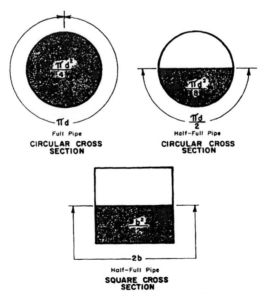

Figure 2.5.2. Hydraulic diameters for two shapes with two degrees of filling.

square pipe

A square pipe has a wetted area of b^2 and a wetted perimeter of $4b$. Thus

$$D = 4(b^2)/(4b) = b \tag{2.5.9}$$

Calculation of pipe friction losses is not usually undertaken for various regions inside the pipe. Friction is usually calculated for an entire pipe segment. Pipe friction losses can be calculated from

$$h_f = f(L/D)(v^2/2g) \tag{2.5.10}$$

where h_f is the pipe friction loss (m); f, the friction factor (dimensionless); L, the pipe length (m); D, the pipe hydraulic diameter (m); v, the average velocity (m/sec); and g, the acceleration due to gravity (9.8 m/sec^2).

friction factor

In this form, the friction factor is known as the Moody friction factor. Another form for pipe friction loss is

$$h_f = f_f(L/D)(2v^2/g) \tag{2.5.11}$$

where the Fanning friction factor f_f is used. The Fanning friction factor is four times smaller than the Moody friction factor. Unless otherwise stated, the Moody friction factor and Eq. 2.5.10 will be used here.

Friction factor values have been charted in Moody diagrams (Figure 2.5.3).

laminar flow
friction factor

For laminar flow the determination of friction factor is straightforward, since friction factor is inversely related to Reynolds number:

$$f = 64/\mathrm{Re} \tag{2.5.12}$$

Pipe friction losses for turbulent flow also use Eq. 2.5.10. The friction factor value, however, is no longer a simple function of Reynolds number.

The flow velocity profile for turbulent flow is not parabolic but is instead

mixing in
turbulent flow

flat, as in the top of Figure 2.3.4. This is because fluid particles mix very thoroughly across the pipe cross section. A particle in the center with a forward velocity of v_1 can easily find itself tossed by eddying to the edge of the pipe lumen and then tossed to the central zone again. Viscous friction tends to accelerate slower particles and decelerate faster particles until they all have the same forward velocity v_1. A very thin boundary layer at the pipe wall acts as the transition between zero velocity at the wall to maximum velocity over most of the pipe cross section.

In laminar flow this wholesale mixing does not occur. Particles next to the

wall
roughness
important in
turbulent flow

pipe wall are not likely to move away from the wall, and pipe wall roughness (ε) has little consequent effect on overall pipe friction loss. However, many more fluid particles contact the pipe wall in turbulent flow. Pipe wall roughness has considerable influence on pipe friction factor in turbulent flow.

Moody
diagram

The Moody diagram includes wall roughness as relative roughness (ε/D). Relative roughness is used because (1) wall roughness should matter less as

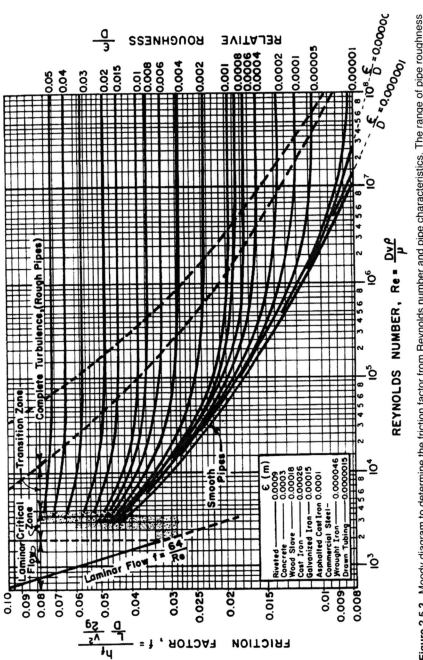

Figure 2.5.3. Moody diagram to determine the friction factor from Reynolds number and pipe characteristics. The range of pipe roughness values is given in the box; relative roughness is pipe roughness divided by hydraulic diameter.

the pipe becomes larger and (2) nondimensionalizing the wall roughness makes its appication more general.

Notice that the turbulent friction factor lines on the Moody diagram are curved. The procedure for locating a turbulent friction factor is:

1. Calculate the relative roughness, ε/D;
2. Locate the ε/D value on the right-hand side of the Moody diagram;
3. Follow the constant ε/D line to the left, curving as the line curves, until the proper Reynolds number is located on the abscissa;
4. Look horizontally to the left to obtain the value for f.

Alternatively, the Colebrook and White equation can be used to calculate f (Daugherty and Ingersoll, 1954)

$$1/\sqrt{f} = -2.01 \log_{10}[\varepsilon/3.7D + 2.51/(\text{Re}\sqrt{f})] \qquad (2.5.13)$$

Friction factor is not explicitly given in this equation, but a friction factor value may be solved by iteration on computer, which begins by assuming a value for f to use on the right-hand side of the equation, then solving for the value of f on the left-hand side of the equation, and using this new f value on the right-hand side. The procedure for this is given in Figure 2.5.4.

EXAMPLE 2.5.1-1. Reynolds Number in an Open Channel

Waste water flows at a rate of $8 \times 10^{-4}\,\text{m}^3/\text{sec}$ in an open channel of rectangular cross section. The base of the channel is 2 m wide and the height of the water is 1 m. Is the flow laminar or turbulent?

Solution

Water velocity can be determined from the continuity equation 2.2.4

$$v = \dot{V}/A = (8 \times 19^{-4}\ \text{m}^3/\text{sec})/(2\ \text{m})(1\ \text{m}) = 4 \times 10^{-4}\ \text{m/sec}$$

Hydraulic diameter is calculated using Eq. 2.5.6 as

$$D = 4(2\ \text{m})(1\ \text{m})/(1\ \text{m} + 2\ \text{m} + 1\ \text{m}) = 2\text{m}$$

Reynolds number comes from Eq. 2.5.5

$$\text{Re} = \frac{Dv\rho}{\mu} = \frac{(2\ \text{m})(4 \times 10^{-4}\ \text{m/sec})(1000\ \text{kg/m}^3)}{9.84 \times 10^{-4}\ \text{kg/(m sec)}}$$

$$= 813$$

Since the Reynolds number is less than 2000, the flow is laminar.

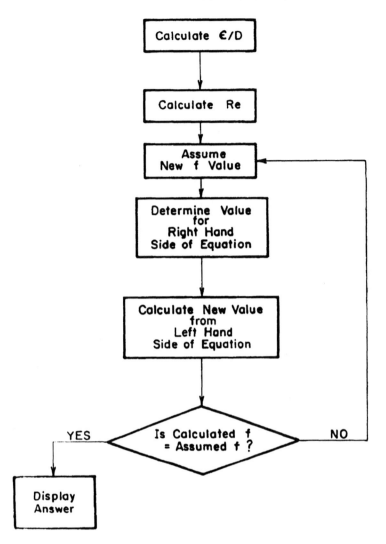

Figure 2.5.4. Schematic of the iterative procedure to calculate friction factor in turbulent flow.

EXAMPLE 2.5.1-2. Pipe Friction Loss

Soybean oil flows through a 15-cm-diameter smooth stainless steel pipe 125 m long with an average velocity of 0.15 m/sec. Determine the friction factor and calculate pipe friction loss.

Solution

Reynolds number can be calculated as

$$\mathrm{Re} = \frac{Dv\rho}{\mu} = \frac{(0.15 \text{ m})(0.15 \text{ m/sec})(920 \text{ kg/m}^3)}{4.06 \times 10^{-2} \text{ kg/(m sec)}}$$

$$= 510$$

Since $\mathrm{Re} < 2000$, the flow is laminar. Thus, from Eq. 2.5.12,

$$f = 64/\mathrm{Re} = 0.126$$

Friction loss can be calculated from Eq. 2.5.10 as

$$h_f = f(L/D)(v^2/2g)$$

$$= 0.126(125 \text{ m}/0.15 \text{ m})(0.15 \text{ m/sec})^2/(2)(9.8 \text{ m/sec}^2)$$

$$= 0.121 \text{ m}$$

Remark

The friction loss just calculated has units of meters of soybean oil.

EXAMPLE 2.5.1-3. Pipe Turbulent Friction

The soybean oil given in Example 2.5.1-2 flows through the same pipe with a velocity of 1.0 m/sec. Determine the friction factor and calculate pipe friction loss.

Solution

The value for the Reynolds number is

$$\mathrm{Re} = \frac{Dv\rho}{\mu} = \frac{(0.15 \text{ m})(1.0 \text{ m/sec})(920 \text{ kg/m}^3)}{4.06 \times 10^{-2} \text{ kg/(m sec)}}$$

$$= 3399$$

Since $\mathrm{Re} > 2000$, the flow is turbulent. Consulting the Moody diagram, Figure 2.5.3, the friction factor value for smooth pipes is found to be

$$f = 0.0415$$

Friction loss can be calculated from Eq. 2.5.10 as:

$$h_f = f(L/D)(v^2/2g)$$

$$= 0.0415(125 \text{ m}/0.15 \text{ m})(1 \text{ m/sec})^2/(2)(9.8 \text{ m/sec}^2)$$

$$= 1.76 \text{ m}$$

Remarks

The friction factor value obtained from the Moody diagram can be checked by means of the Colebrook and White equation 2.5.13

$$\frac{1}{\sqrt{f}} = -2.01 \log_{10}\left(\frac{\varepsilon}{3.7D} + \frac{2.51}{\text{Re}\sqrt{f}}\right)$$

thus

$$\frac{1}{\sqrt{f}} = 2.01 \log_{10}\left(\frac{(0 \text{ m})}{(3.7)(0.15 \text{ m})} + \frac{2.51}{(3399)\sqrt{0.0415}}\right)$$

and

$$f = 0.0416$$

This is essentially the same value obtained from the Moody diagram. If there had been a discrepancy, the Colebrook and White equation could have been used to give an accurate value for friction factor.

2.5.2 Minor Losses

minor losses

In addition to pipe losses, there are the so-called "minor" friction losses. These may not be so minor and may be the numerically dominant friction terms. Minor losses come about as a result of changing the flow velocity profile in the pipe. Any change in either direction or magnitude of the velocity of the fluid in the pipe results from an acceleration, which, in turn, means a force is present. A force through a distance gives work that must be done by either the pipe or the fluid. Since the fluid is the medium that can change energy level, a change in fluid velocity profile results in a loss of energy to the fluid. This is the fluid minor friction loss.

2.5.2.1 Loss Coefficients There are several techniques used to calculate pipe minor losses. All of them require experimental values and an empirical approach. The technique used here employs loss coefficients.

Minor losses are calculated similarly to pipe losses by multiplying measured loss coefficients by average velocity squared and dividing by 2 times the acceleration due to gravity. Total minor loss is the sum of minor losses from each pipe fitting used

minor loss
calculation

$$h_f = \sum K_m v^2 / 2g \tag{2.5.14}$$

where h_f is the friction loss of pipe fittings (m); K_m, the loss coefficient (dimensionless); v, the average fluid velocity (m/sec); and g, the acceleration due to gravity (m/sec^2).

Table 2.5.1 Loss Coefficient K_m for Various Fittings

Globe valve (fully open)	10.0
Angle valve	5.0
Gate valve	0.19
Standard tee	1.8
Standard elbow	0.9
Medium sweep elbow	0.75
Long sweep elbow	0.60
Entrance	
Square-cornered, flush with wall	0.5
Rounded	0.1
Square-cornered, projecting inward	0.9
Discharge nozzles	
Plain	0.013–0.04
Ring	0.013–0.053
Spray	varies; see manufacturer specifications
Sudden contraction, $d_1/d_2 =$	
0.1	0.48
0.3	0.40
0.5	0.29
0.7	0.14
0.9	0.02
Sudden enlargement ($d_2 > d_1$)	$\dfrac{(v_1 - v_2)^2}{1.5g}$

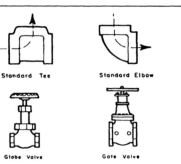

Standard Tee   Standard Elbow   Medium Sweep Elbow   Long Sweep Elbow

Globe Valve   Gate Valve   Angle Valve

Loss coefficient values for some standard pipe fittings are given in Table 2.5.1.

The use of the 2 in the denominator of the equation above is preferred over the value for $\alpha = 1$ or 2, as in the kinetic energy equation, for several reasons: (1) Fittings usually cause at least some eddying that flattens the flow velocity profile, (2) values of K_m are not so precise that the value of $\alpha = 1$ or 2 makes the minor loss value more accurate, and (3) the procedure is simplified if a value of $\alpha = 2$ is used all the time.

Vennard (1961) indicates that for pipe bends the friction loss given in Eq. 2.5.14 is not complete. Using the tabulated K_m value for bends gives friction loss over and above friction loss that would exist for the same length of pipe if the bend did not exist. This length is θr, where θ is the bend angle in radians and r is the bend radius in meters. Thus an additional term $f(L/D)(v^2/2g) = f(\theta r/D)(v^2/2g)$ should be added to the value for h_f calculated from Eq. 2.5.14. However, calculated minor friction losses are sufficiently imprecise that little penalty is paid by ignoring the additional term.

minor losses imprecise

2.5.2.2 *Entrance Length* Development of velocity profiles in pipe cross sections is important for a number of reasons:

1. Average fluid kinetic energy,
2. Pressure differences along the pipe,
3. Pipe friction losses.

All depend on the velocity profile. As the velocity profile changes, energy is consumed.

energy loss and velocity profile

Fluid velocity at the walls of the pipe is very small and usually considered to be zero (called a no-slip condition). Thus, with no velocity, the pressure existing at the wall is the static pressure of the fluid. Pressures measured anywhere else in the pipe cross section will be different from the static pressure as long as the fluid moves. Correct measurements of static pressure are nearly always made at the wall.

static pressure

The region next to the wall is called the boundary layer. It is in this region that most viscous friction is present and, therefore, most energy dissipation occurs. The boundary layer is usually considered to be the thickness of the layer of fluid from the solid wall surface to the point in the fluid where velocity reaches a value of 99% of the free-stream velocity. Thus fluid velocity changes very rapidly with distance from the wall in the boundary layer.

boundary layer

Most boundary layer theory is developed for flow external to, and across, immersed bodies. For flow inside pipes and ducts the development of a boundary layer at the pipe or duct entrance leads to the concept of entrance length.

entrance
length

The entrance length is defined as the distance from the pipe entrance to the point where the velocity profile is fully developed. Within the entrance length friction losses are very high. For fluids with a viscosity that does not depend on the rate of shear (Newtonian fluids), Mishra and Singh (1976) have found entrance length to be

$$L_e = 0.02875 \text{ Re } D \qquad (2.5.15)$$

where L_e is the entrance length (m) Re, the Reynolds number; and D, the pipe hydraulic diameter (m).

friction loss

Friction loss for the entrance length is

$$h_f = 9.8v^2/g \qquad (2.5.16)$$

where h_f is the friction loss (m); v, the velocity (m/sec); and g, the acceleration due to gravity (9.81 m/sec^2).

To use entrance length in design calculations, the value for L_e is first calculated and then the friction loss for the entrance length. The length of pipe used to calculate pipe losses becomes the total pipe length with entrance length subtracted. Otherwise, friction losses in the entrance length would be calculated twice.

It has been reported (Lightfoot, 1974) that entrance lengths for arteries are relatively long. Of course, these vessels are not straight, are elastic, are of

entrance
lengths in
airways and
arteries

nonuniform dimensions, and are subject to pulsatile flow. The entrance lengths for airways segments in the respiratory system are often longer than the segments themselves (Johnson, 1991), thus indicating that fully developed flow velocity profiles never are achieved.

2.5.2.3   Pipe Discharge

kinetic energy
is lost

There is a loss of energy at the point where an open-ended pipe empties into the atmosphere (Figure 2.5.5). Kinetic energy present in the fluid at the end of the pipe is completely dissipated into the atmosphere a short distance from the pipe. From Bernoulli's equation

$$h_1 + p_1/\gamma + v_1^2/\alpha g = h_2 + p_2/\gamma + v_2^2/\alpha g + h_f \qquad (2.5.17)$$

where condition 1 is located at the end of the pipe and condition 2 is located in the atmosphere. But, for a horizontal pipe, $h_1 = h_2$, and $p_1 = p_2 = 0$. The velocity of the fluid (v_2) is zero after it comes to rest. Thus

$$h_f = \frac{v_1^2}{\alpha g} \qquad (2.5.18)$$

and the equivalent resistance of the end of the pipe is

$$R = \frac{\gamma h_f}{Av} = \frac{\gamma v_1}{\alpha A g} \qquad (2.5.19)$$

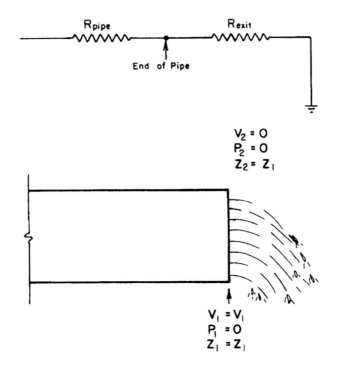

Figure 2.5.5. An equivalent resistance at the mouth of an open-ended pipe accounts for the loss of kinetic energy of the fluid.

EXAMPLE 2.5.2-1. Pipe Friction Loss

Ethyl alcohol at 20°C is being pumped through a 15-cm-diameter smooth stainless steel pipe 125 m long. Velocity of the liquid is 0.2 m/sec. The source end of the pipe forms a square-cornered entrance flush with the wall. The downstream end empties into a large tank where the liquid level is below the pipe. There are a globe valve and two medium sweep elbows in the pipe. What are the minor losses for this pipe? What is the total friction loss?

Solution

From Table 2.5.1, the following loss coefficient values are found:

square entrance	0.5
globe valve	10.0
medium sweep elbow	0.75

$$\sum K_m = 0.5 + 10.0 + 2(0.75) = 12.0$$

The minor loss (Eq. 2.5.14) due to these components is

$$h_f = \sum K_m v^2/2g = (12.0)(0.2 \text{ m/sec})^2/(2)(9.8 \text{ m/sec}^2)$$
$$= 2.4 \times 10^{-2} \text{ m}$$

Friction loss for the entrance length can either be obtained from the K_m value in Table 2.5.1 or from Eq. 2.5.16. The value from Table 2.5.1 is used in this example.

To obtain the total friction loss, the value for minor loss must be added to the exit loss and pipe losses. First the exit loss: From Eq. 2.5.18

$$h_f = \frac{v_1^2}{\alpha g}$$

The value for α will depend on the velocity profile. Since the nearness of the valve or elbows to the end of the pipe will influence the velocity profile, and thus the value for α, we use a value of $\alpha = 1$ as the most conservative choice (giving the largest friction loss).

$$h_f = (0.2 \text{ m/sec})^2/(9.8 \text{ m/sec}^2) = 0.4 \times 10^{-2} \text{ m}$$

To calculate pipe losses, Reynolds number is first calculated

$$\text{Re} = \frac{dv\rho}{\mu} = \frac{(0.15 \text{ m})(0.2 \text{ m/sec})(789 \text{ kg/m}^3)}{[1.200 \times 10^{-3} \text{ kg/(m sec)}]} = 19{,}725$$

Flow is thus turbulent, and this reinforces the choice of $\alpha = 1$ above. From the Moody diagram, $f = 0.026$.

Entrance length (Eq. 2.5.15) is

$$L_e = 0.02875 \text{ Re } D = 85 \text{ m}$$

Pipe losses are thus

$$h_f = f \frac{L}{D} \frac{v^2}{2g} = \frac{(0.026)(125 \text{ m} - 85 \text{ m})(0.2 \text{ m/sec})^2}{(0.15 \text{ m})2(9.8 \text{ m/sec}^2)}$$
$$= 1.4 \times 10^{-2} \text{m}$$

Total friction loss is, therefore,

$$h_f = 2.4 \times 10^{-2} \text{ m} + 0.4 \times 10^{-2} \text{ m} + 1.4 \times 10^{-2} \text{ m} = 4.2 \times 10^{-2} \text{ m}$$

2.5.3 Fluid System Impedance

Pipes exhibit resistive, capacitive, and inertial properties collectively called impedance.

2.5.3.1 Compliance As long as the conduit is rigid and flowing full, and the fluid is nearly incompressible, fluid compliance is negligible. These conditions do not often exist for biological systems subject to analysis, but are often true for fluid systems to be designed by the engineer for various applications.

compliance definitions
Fluid capacity is called compliance, and compliance has been defined in Chapter 1 as

$$C_f = \frac{1}{p} \int_0^t \dot{V} \, dt \qquad (1.4.8)$$

where C_f is the fluid compliance (m^5/N); p, the pressure (N/m^2); $\dot{V}$, the volumetric rate of flow (m^3/sec); and t, time (sec).

If the rate of flow is steady, or if an average rate of flow is used, then

$$C_f = \frac{1}{p} \int_0^t \dot{V} \, dt = \frac{V}{p} \qquad (2.5.20a)$$

where V is the accumulated volume of fluid (m^3).

Compliance can also be defined over part of the p–V curve as

$$C_f = \frac{\Delta V}{\Delta p} \qquad (2.5.20b)$$

where ΔV is the change of volume (m^3); and Δp, the corresponding change of pressure (N/m^2).

Three possibilities for compliance are shown in Figure 2.5.6. The rigid pipe flowing full has a curve similar to curve number (3), whereas a typical biological structure (such as the lung) has a nonlinear curve similar to number (2). An ideal Windkessel vessel (Section 2.5.5.1) has a characteristic similar to (1).

We noted earlier (Table 2.1.1) that there are two effort variables in fluid flow. Fluid height is the second, after pressure. If the fluid is pumped into the

open-top container
bottom of a storage container open at the top to ambient pressure, then pressure above the fluid is constant, and pressure at the bottom of the fluid is

$$p = \rho_f g h \qquad (2.3.3)$$

where p is the pressure referenced to ambient (N/m^2); ρ_f, the accumulating fluid density (kg/m^3 or N sec^2/m^4); g, the acceleration due to gravity (9.81 m/sec^2); and h, the height of fluid above the bottom (m).

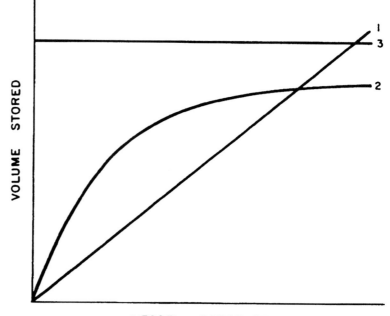

VESSEL   PRESSURE

Figure 2.5.6. Diagrams of the relationships between vessel pressure and stored volume. The straight line (1) has a constant compliance. The downward-curved line (2) shows a typical compliance curve for biological and many nonbiological materials, where the compliance decreases to zero at some volume. Increasing the pressure beyond this point will not increase the volume stored until the vessels burst. The horizontal line (3) shows the very low compliance for a rigid vessel flowing full.

Fluid height is fluid volume divided by cross-sectional area of the container (for a container of constant cross-sectional area)

$$h = V/A \qquad (2.5.21)$$

where V is the fluid volume (m³); and A, the container cross-sectional area (m²).

Thus the fluid compliance of the container is

$$C_f = \frac{V}{p} = \frac{V}{\rho_f g h} = \frac{V}{\rho_f g V/A} = \frac{A}{\rho_f g} \qquad (2.5.22)$$

which is constant.

closed container with gas
Another case of interest occurs with a closed storage container occupied by a compressible gas originally at some pressure p_0. An incompressible fluid is then pumped into this container and compresses the gas. Household water systems often use containers such as this to maintain pressure on the water

even when the pump is not running. Pressure in the gas is obtained from the ideal gas law as

$$p_g = \frac{n_0 R T}{V_g} \tag{2.5.23}$$

where p_g is the pressure in the gas (N/m^2); n_0, the number of moles of gas present (kg mol); R, the universal gas constant $[8314.34\,N\,m/(kg\,mol\,K)]$; T, the absolute temperature (K); and V_g, the gas volume (m^3).

The numer of moles of the gas (n_0) can be obtained from Eq. 2.5.23 when the gas completely occupies the tank (V_{tot}) at a pressure of p_0 and temperature T.

Pressure at the bottom of the fluid is the pressure of the liquid due to gravitation effects plus the pressure exerted on the top of the fluid by the gas

$$p_f = \rho_f g V_f / A + p_g = \frac{\rho_f g V_f}{A} + \frac{n_0 R T}{V_g} \tag{2.5.24}$$

where P_f is the pressure at the bottom of the fluid (N/m^2); and V_f, the fluid volume (m^3).

Since $V_{tot} = V_g + V_f$,

$$p = \frac{\rho_f g V_f}{A} + \frac{n_0 R T}{(V_{tot} V_f)} = \frac{\rho_f g V_f}{A} + \frac{p_0 V_f}{(V_{tot} - V_f)} \tag{2.5.25}$$

$$C_f = \frac{V_f}{p_f} = \frac{1}{\dfrac{\rho_f g}{A} + \dfrac{p_0}{(V_{tot} - V_f)}} \tag{2.5.26}$$

series combination

The first term in the denominator is the inverse of the fluid compliance, and the second term is the inverse of the gas compliance. The resultant of two compliances in series is the inverse of the inverses of the two compliances (see Section 1.6.3). Thus this compliance can be recognized as the series combination of fluid compliance (Eq. 2.5.22) and gas compliance (Figure 2.5.7).

If the gas compression were not isothermal, there would be an additional gas pressure effect of the change in temperature. What usually happens in the case of water pressure tanks is that water pumped from a well is often at a lower temperature than ambient. This may be reversed in the winter. Since the heat capacity of the water is much greater than that of the air in the tank, the air temperature practically becomes that of the water.

adiabatic gas combustion

If gas compression occurred adiabatically (no transfer of heat), then (Millard, 1953)

$$c_v \ln\left(\frac{T_2}{T_1}\right) = -R \ln\left(\frac{V_2 n_1}{V_1 n_2}\right) = R \ln\left(\frac{p_2}{p_1}\right) \tag{2.5.27}$$

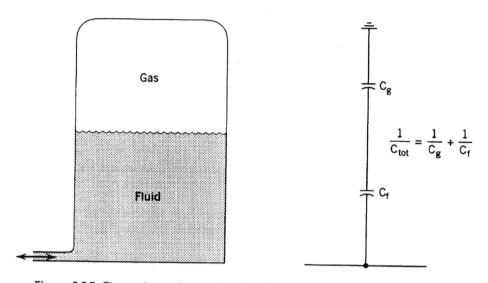

Figure 2.5.7. The total compliance of a closed tank filled with gas and fluid is the series combination of the gas compliance and the fluid compliance.

where c_v is the specific heat at constant volume (see Section 3.6.1); T, the absolute temperature (K); R, the universal gas constant [8314.34 N m/(kg mol K)]; V, the gas volume (m^3); n, the number of kg moles (kg mol); p, the pressure (N/m^2); and 1, 2 denote two conditions of the gas.

Since the gas is trapped above the fluid in the tank, $n_1 = n_2$, and

$$T_2 = T_1 \, e^{(R/c_v) \, \ln(V_1/V_2)} \tag{2.5.28}$$

Thus

$$p_2 = \frac{nRT_1}{V_2} e^{(R/c_v) \, \ln(V_1/V_2)} \tag{2.5.29}$$

This value can be substituted for p_0 in Eq. 2.5.26.

Another case of interest concerns a rigid container being filled with compressible gas. It does not matter whether the elastic force required to push gas from the container comes from the walls of the container or from the compressed gas itself. If gas in the container begins at atmospheric pressure, then the number of moles of gas contained therein is

$$n_0 = \frac{p_0 V}{RT} \tag{2.5.30}$$

where n_0 is the number of moles in the container at reference pressure p_0 (kg mol); p_0, the reference pressure (N/m^2); V, the container volume (m^3); R, the universal gas constant [8314.34 N m/kg mol k)]; and T, the absolute temperature (K).

compressed
gas

If the gas in the rigid container is compressed, then the number of gas moles inside is

$$n_1 = \frac{p_1 V}{RT} \qquad (2.5.31)$$

where n_1 is the number of moles in the container at pressure p_1 (kg mol); p_1, the compression pressure (N/m^2). Isothermal conditions have been assumed.

When the gas is allowed to escape from the tank, the pressure declines toward p_0. The number of kg moles released is $(n_1 - n_0)$. This represents a volume released to the atmosphere of

$$V_g = \frac{(n_1 - n_0)RT}{p_0} \qquad (2.5.32)$$

where V_g is the volume of gas, at reference pressure p_0, escaping from the container (m^3).

The fluid compliance is

$$C_f = \frac{\Delta V}{\Delta p} = \frac{V_g}{p_1 - p_0} = \frac{(n_1 - n_0)RT/p_0}{p_1 - p_0}$$
$$= \frac{p_1 V/p_0 - V}{(p_1 - p_0)} = \frac{V}{p_0} \qquad (2.5.33)$$

Or fluid compliance for a compressible gas in a rigid-walled container, assuming isothermal conditions, is just the volume of the container divided by atmospheric pressure.

The pressure required to store fluid in a compliance element is just the volume of fluid to be stored in that element divided by the value of compliance.

EXAMPLE 2.5.3.1-1. Compliance in a Tank

A tank contains 22% sodium chloride brine solution 5 m deep. The tank size is 6 m high by 4 m diameter. What is the difference between the compliance of the brine in the tank if the top is open to the atmosphere or if the tank is closed and the 1 m difference between the brine level and the top of the closed tank is filled with air at 1 atmosphere pressure?

Solution

Compliance is the relationship between volume and pressure. From Eq. 2.5.22, compliance of the open tank is

$$C_f = \frac{A}{\rho_f g}$$

Brine density is found from Table 2.4.1 as $1170\,\text{kg/m}^3$. Thus

$$C_f = \frac{\pi(4\ \text{m})^2}{(4)(1170\ \text{kg/m}^3)(9.81\ \text{N/kg})} = 1.09 \times 10^{-3}\ \text{m}^5/\text{N}$$

We are interested in the compliance of the brine in the tank, with pressure referenced to the atmosphere. The pressure at the bottom of the brine is the same in both open and closed tanks as long as the air pressure at the top of both is atmospheric. With volumes and pressures equal, compliances for both open and closed tanks are equal.

If the closed tank began with no brine, but filled with air at atmospheric pressure, then the pressure of the compressed air above the brine would be (assuming isothermal conditions)

$$p_2 = p_1 V_1/V_2 = p_1 h_1/h_2$$

where h is the height of the volume of air.

$$p_2 = (101.5\ \text{kN/m}^2)(6\ \text{m}/1\ \text{m}) = 609\ \text{kN/m}^2$$

The total pressure at the bottom of the brine is

$$p = \rho_f g h_f + p_2 = (1170\ \text{kg/m}^3)(9.81\ \text{N/kg})(5\ \text{m}) + 609\ \text{kN/m}^2$$
$$= 666\ \text{kN/m}^2$$

Subtracting atmospheric pressure, this pressure is measured as $565\,\text{kN/m}^2$. Compliance for the closed tank in this case is

$$C = \frac{V}{p} = \frac{\pi(4\ \text{m})^2(5\ \text{m})}{4(565\ \text{kN/m}^2)} = 0.111 \times 10^{-3}\ \text{m}^5/\text{N}$$

Remarks

Notice that the closed tank has become much less compliant. That means that a higher amount of pressure is required to fill with more liquid volume than if the tank were open at the top.

Compliance of the air above the brine in the closed tank with $609\,\text{kN/m}^2$ pressure is, from Eq. 2.5.26

$$C_{\text{air}} = \frac{V_{\text{air}}}{p_{\text{air}}} = \frac{\pi(4\ \text{m})^2(1\ \text{m})}{4(609\ \text{kN/m}^2)} = 0.0206 \times 10^{-3}\ \text{m}^5/\text{N}$$

The series combination of the brine compliance and the air compliance is

$$C_{\text{tot}} = \frac{1}{\dfrac{1}{c_{\text{brine}}} + \dfrac{1}{C_{\text{air}}}} = \frac{1}{\dfrac{1}{0.111 \times 10^{-3}} + \dfrac{1}{0.0206 \times 10^{-3}}}$$

$$= 0.0174 \times 10^{-3}\ \text{m}^5/\text{N}$$

The value of the total compliance (given as $0.111 \times 10^{-3}\,\text{m}^5/\text{N}$) is not equal to the series combination of brine and air compliances because of the reference of the total compliance to atmospheric pressure rather than to absolute zero pressure.

---

pipe inertia

2.5.3.2 Inertance Inertia of fluid flowing in pipes (called inertance) can sometimes be important to consider, and cases where inertance is important almost always involves pipes flowing full. Momentum of fluid flowing in a pipe is given as

$$mv = \rho A L v \qquad (2.5.34)$$

where m is the fluid mass (kg); v, the fluid velocity (m/sec); ρ, the fluid density (kg/m^3); A, the pipe cross-sectional area (m^2); and L, the pipe length (m).

The force required to accelerate or decelerate the fluid is

$$F = m\,\frac{dv}{dt} = \rho A L\,\frac{dv}{dt} \qquad (2.5.35)$$

where F is the force (N or kg m/sec^2).

From the definition of fluid inertance in Eq. 1.4.12,

$$I_f = p\left/\left(\frac{d\dot{V}}{dt}\right)\right. = \left(\frac{F}{A}\right)\left/A\left(\frac{dv}{dt}\right)\right. = \rho L/A \qquad (2.5.36)$$

where I_f is the fluid inertance ($\text{N sec}^2/\text{m}^5$).

The same conditions that cause compliance to be small usually result in significant fluid inertia. Therefore, incompressible flow in rigid tubes flowing full will have considerable amounts of inertia. Under steady flow conditions, inertia will have no effect on flow rates, but if the flow is nonsteady, as in pulsatile or reversing flow, inertial effects should be accounted for.

acceleration pressure

Multiplying inertance by the rate of change of volume flow rate yields the pressure required to accelerate or decelerate the fluid.

Most bioprocesses (food processing, biochemical extractions, etc.) operate in the steady state, and inertial effects are largely ignored. Many biological systems, however, do not produce steady flow. Examples are vertebrate respiratory flow, blood flow, and water flow in the invertebrate species *Crassostrea virginica* (Chesapeake Oyster). In these instances, extra pressure must be developed by the muscles to propel fluid initially at rest or even flowing in the opposite direction.

biological systems (margin note)

EXAMPLE 2.5.3.2-1. Inertia of Water Flowing in a Pipe

Water flows through a 5-cm-diameter pipe 100 m long at a rate of 25 L/min. The faucet at the end of the pipe takes 2 sec to turn off. Calculate the pressure produced at the faucet due to inertia of the flowing water.

Solution

Inertance of the water in the pipe can be calculated from Eq. 2.5.36 as:

$$I_f = \rho L/A$$

The density of water is about 1000 kg/m^3; so

$$I_f = (1000 \text{ kg/m}^3)(100 \text{ m})/[\pi(0.05 \text{ m})^2/4]$$

$$= 5.1 \times 10^7 \text{ kg/m}^4 \text{ or N sec}^2/\text{m}^5$$

The rate of change of velocity with time is

$$\ddot{V} = (25 \text{ L/min})(0.001 \text{ m}^3/\text{L})(1 \text{ min}/60 \text{ sec})/2 \text{ sec}$$

$$= 2.08 \times 10^{-4} \text{ m}^3/\text{sec}^2$$

The pressure produced is thus

$$p = I_f \ddot{V} = (5.1 \times 10^2 \text{ N sec}^2/\text{m}^5)(2.08 \times 10^{-4} \text{ m}^3/\text{sec}^2)$$

$$= 10.6 \text{ kN/m}^2, \text{ or about } 1/10 \text{ atm}$$

---

resistance represents friction (margin note)

2.5.3.3 *Resistance* More than any other factor, pipe resistance often determines the size of the pump required to move fluid in the pipe. The resistance, another way to represent pipe friction, is given in Eq. 1.4.5 as the ratio of effort variable (pressure) to flow variable (volume flow rate). We have seen that friction in a pipe is given by pipe losses and minor losses

$$h_f = f\left(\frac{L}{D}\right)\left(\frac{v^2}{2g}\right) + \sum \frac{K_m v^2}{2g} \tag{2.5.37}$$

where h_f is the pipe friction loss (m). Conversion of h_f to pressure loss is accomplished by multiplying by specific weight of the fluid

$$p = \gamma h_f \tag{2.5.38}$$

where p is the pressure loss in the pipe (N/m^2); and γ, the specific weight (N/m^3).

Thus pipe resistance is given by

$$R = p/\dot{V} = \gamma h_f/Av$$

$$= \frac{\gamma[f(L/D)(v/2g) + \sum K_m v/2g]}{Av}$$

$$= \frac{\rho v[f L/D + \sum K_m]}{2A} \tag{2.5.39}$$

Because of the relationship between Reynolds number and friction factor in laminar flow, a round pipe without fittings (no minor losses) has a resistance of

Hagen–
Poiseuille
formula

$$R = \frac{fLv^2\gamma}{AD2gv} = \frac{64Lv^2\gamma}{Re\ AD2gv} = \frac{128\mu L}{\pi D^4} \tag{2.5.40}$$

where μ is the fluid viscosity [kg/(m sec) or $N\,sec/m^2$]; and Re, Reynolds number (dimensionless).

This resistance is only properly applied to the section of the pipe where fluid flow is fully developed and laminar. That is, the steady-state velocity profiles no longer change with distance along the pipe. As a rule of thumb, this occurs at a distance of ten pipe diameters from the entrance to the pipe.

The Hagen–Poiseuille formula (Eq. 2.5.40) shows that tube resistance is related directly to tube length, but is inversely related to tube diameter to the fourth power. Thus a narrow tube has much more resistance to fluid flow than does a wider tube. This fact can be used to understand neural control in both respiratory and cardiovascular systems.

biological
applications

Arterioles, for instance, are innervated by sympathetic autonomic fibers. Resistance of these vessels is regulated to control tissue blood flow and arterial blood pressure. A reduction in arteriolar diameter by 16% is sufficient to double resistance and thus halve blood flow. Contrarily, the presence of inhaled or locally produced nitric oxide (NO) reduces vascular resistance.

Respiratory airways also constrict when irritants are inhaled, an apparent protection mechanism.

The Hagen–Poiseuille formula can be used to estimate resistance to fluid flow in plant phloem, but underestimate the measured resistance by a factor of two (Rand and Cooke, 1996).

EXAMPLE 2.5.3.3-1. Pipe Resistance

What is the fluid reistance of a 15-cm-diameter pipe 100 m long flowing full of SAE 10 lubricating oil at 50°C? Flow is fully developed and laminar.

Solution

The viscosity for SAE 10 oil is found from Table 2.4.1 to be: 0.100 at 16°C and 0.010 at 66°C. Using Eq. 2.4.6b to find the viscosity at 50°C,

$$
\mu_{50} = \mu_{16} + (\mu_{66} - \mu_{16}) \left(\frac{\dfrac{1}{T_{50}} - \dfrac{1}{T_{16}}}{\dfrac{1}{T_{66}} - \dfrac{1}{T_{16}}} \right)
$$

$$
= 0.100 + (0.010 - 0.100) \left(\frac{\dfrac{1}{323.15} - \dfrac{1}{289.15}}{\dfrac{1}{339.15} - \dfrac{1}{289.15}} \right)
$$

$$
= 3.58 \times 10^2 \text{ kg/(m sec)}
$$

From Eq. 2.5.40,

$$
R = \frac{(128)[3.58 \times 10^{-2} \text{ kg/(m sec)}](100 \text{ m})}{\pi(0.15 \text{ m})^4} \left(\frac{1 \text{ N sec}^2}{\text{kg m}} \right)
$$

$$
= 2.88 \times 10^5 \text{ N sec/m}^5
$$

2.5.3.4 Time Relationships As discussed in Section 1.6.5, combinations of resistance, compliance, and inertance can lead to pressures and flows that vary with time. The example of water hammering has already been given in Figure 1.6.4, and the solution to this problem by adding a compliance to the system has been discussed in Section 1.6.5. There are other important cases where time-varying flows and pressures are important in biological systems. Pendelluft is a condition in the lungs where air flows between the lungs in an oscillatory manner because the time constants (resistance times compliance) are not the same

examples

in both lungs. A stiff lung is usually a sign of some respiratory disease. Ocean wave motion can be analyzed as the result of compliance and inertance of sea water. The propagation of sound in the atmosphere is similar.

The larger the compliance and inertance of a fluid system, the more slowly the system responds to external and internal forces. The more that resistance dominates both compliance and inertance, the less the system varies with time.

2.5.4 Nonisothermal Flow

If heat transfer occurs to or from a fluid flowing within a pipe, its temperature changes throughout the pipe. This is called nonisothermal flow. Viscosity values often change dramatically with temperature (see Section 2.4.1), and, as viscosity changes, so do the Reynolds number and friction factor. An approximate means to deal with the effect of heat transfer on friction factor is to evaluate viscosity at the two temperatures associated with (1) the tube wall and (2) the mean bulk fluid. The latter is calculated as the average between inlet and outlet fluid temperatures.

viscosity effects

After determining Reynolds number based on the mean bulk temperature, calculate an initial friction factor estimate in the manner described in previous sections. The corrected friction factor can be calculated from

$$f_\theta = f_0 \left(\frac{\mu_0}{\mu_\infty} \right)^m \tag{2.5.41}$$

where f_θ is the corrected friction factor (dimensionless); f_0, the initial friction factor (dimensionless); μ_0, the viscosity evaluated at the pipe wall temperature (N sec/m^2); μ_∞, the viscosity evaluated at bulk fluid temperature (N sec/m^2); and m, an exponent (dimensionless).

Values of m for different heating/cooling and Reynolds number conditions appear in Table 2.5.2. When the fluid is being heated, the corrected friction factor is lower than its initial value. When the fluid is being cooled, the friction factor becomes greater.

Table 2.5.2 Values of Exponent _m_ for Various Conditions

	Reynolds number	
	<2100 (laminar)	>2100 (turbulent)
Fluid being heated	0.38	0.17
Fluid being cooled	0.23	0.11

EXAMPLE 2.5.4-1. Nonisothermal Resistance

Corn syrup initially at 35°C is being heated inside a shell and tube heat exchanger to 90°C. Calculate the resistance to laminar flow in a straight tube 3 m long and 5 cm in diameter. The syrup is flowing at a rate of 25 L/min, and the pipe wall is maintained at 100°C.

Solution

Temperature dependence of resistance will be caused by viscosity changes in the water that acts as the base fluid for the corn syrup. The bulk fluid temperature for this short pipe will be estimated as the average of inlet and outlet temperature, or

$$\theta_{bulk} = \frac{35°C + 90°C}{2} = 62.5°C$$

The value for water viscosity at 62.5°C is, according to Eq. 2.4.6b and values given in Table 2.4.4,

$$\mu_{62.5} = 4.588 \times 10^{-4} + (4.061 \times 10^{-4} - 4.688 \times 10^{-4})$$

$$\times \left(\frac{\dfrac{1}{335.65} - \dfrac{1}{333.15}}{\dfrac{1}{343.15} - \dfrac{1}{333.15}} \right)$$

$$= 4.528 \times 10^{-4} \text{ N sec/m}^2$$

Pipe laminar resistance at 62.5°C is, from Eq. 2.5.40,

$$R_{62.5} = \frac{(128)(4.528 \times 10^{-4} \text{ N sec/m}^2)(3 \text{ m})}{\pi(0.05 \text{ m})^4}$$

$$= 8855 \text{ N sec/m}^5$$

From Eq. 2.5.41, and a value of viscosity of $\mu_{90} = 3.165 \times 10^{-4}$ N sec/m², the corrected resistance is

$$R = R_{62.5} \left(\frac{\mu_{90}}{\mu_{62.5}} \right)^{0.38}$$

$$= \left(8855 \ \frac{\text{N sec}}{\text{m}^5} \right) \left(\frac{3.165 \times 10^{-4} \text{ N sec/m}^2}{4.528 \times 10^{-4} \text{ N sec/m}^2} \right)^{0.38}$$

$$= 7728 \text{ N sec/m}^5$$

Remarks

If pipe flow had been turbulent, we could not have used the Hagen–Poiseuille formula to calculate resistance.

Because of the local viscosity reduction at the wall for this pipe, resistance to flow is about 13% less than otherwise expected.

2.5.5 Elastic Tubes

Unlike the rigid tubes usually used in designs by biological engineers, many tubes encountered in biological systems are distensible. This is certainly true for tubes in the cardiovascular, respiratory, and alimentary systems. Elastic tubes behave differently from rigid tubes and give different relations between pressure and flow.

2.5.5.1 Pulsating Flow For example, the aorta is the blood vessel most proximal to the heart. It exhibits a great deal of compliance that tends to reduce pressure fluctuations and to steady blood flow. Such a vessel is called a "Windkessel" (Fung, 1990).

Windkessel vessel   The Windkessel vessel is modelled as an elastic element and a rigid resistance element (Figure 2.5.8). The electrical analog to the Windkessel is also shown. Notice that the layout for the vessel shows the elastic chamber apparently in series (the same fluid slows first through the elastic element and

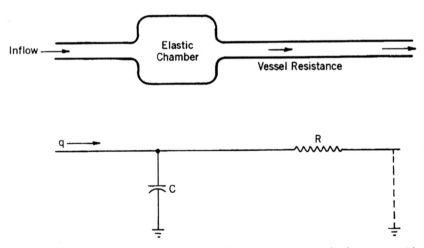

Figure 2.5.8. The Windkessel vessel is considered to possess an elastic component in parallel with a resistance component. The equivalent electrical schematic is given below. Kirchhoff's junction equation can be used to give the relation between pressure and flow into the vessel.

then through the resistance element), but the electrical analog shows the compliance in parallel with the resistance (the flow splits at the junction between the two). The apparent difference between the two representations is caused by the nature of the flows through them. In the vessel model, the flow is assumed to be steady, but the electrical analog is clearly meant to apply to pulsatile flow. With a little thought, it should be apparent that the elastic chamber, as well as the electrical capacitor, both allow pulsations to escape to the outside (or to ground). The other path for pulsations to reach the outside (or ground) is through the rigid resistance element. Since the pulsations can reach the outside through two parallel paths, the resistance and compliance elements are in parallel for pulsating flow. For steady flow, the elastic chamber initially inflates somewhat and subsequently changes no more. Thus the entire chamber is of no consequence for steady flow. Again, the electrical analog reflects this fact. If the capacitor were in series with the resistor, no steady flow would be possible, because capacitors block direct current.

We have already seen the general Kirchhoff relation between pressure and flow at a junction between parallel elements (Eq. 1.7.76). For this vessel the junction equation is

Kirchhoff relation

$$\dot{\omega} + \frac{\phi}{R} + C\,\frac{d\phi}{dt} = 0 \qquad (2.5.42a)$$

where $\dot{\omega}$ is a generalized flow variable; ϕ, a generalized effort variable; R, the resistance; C, the capacity; and t, the time.

Or, in terms of the pressure and flow in fluid systems

$$\dot{V} + \frac{p}{R} + C\,\frac{dp}{dt} = 0 \qquad (2.5.42b)$$

where $\dot{V}$ is the volume flow rate delivered to the vessel by the source (m^3/sec); p, the pulsatile pressure at the compliance–resistance junction (N/m^2); R, the resistance ($N\,sec/m^2$); C, the compliance (m^5/N); and t, the time (sec).

Before proceeding farther, we must decide what type of source drives the flow through the Windkessel vessel. If it is an ideal flow source, then the flow rate is determined entirely by the source and Eq. 2.5.42b is the correct one to use. On the other hand, if the source is an ideal pressure source, then the pressure is determined by the source, and the pressure would be the same across the parallel combination of compliance and resistance. No pressure attenuation would be possible in this case. The most likely scenario is a nonideal source composed of an ideal pressure source in series with a resistor (Figure 2.5.9). The junction equation for this case is

source type determines response

nonideal flow or pressure source

$$\sum \dot{V} = 0 = \frac{p_s - p}{R_s} + \frac{p}{R} + C\,\frac{dp}{dt} \qquad (2.5.42c)$$

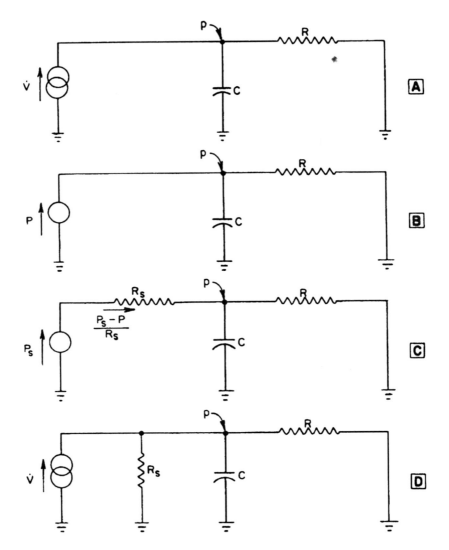

Figure 2.5.9. Four different kinds of sources can supply the Windkessel vessel. An ideal flow source appears in A, an ideal pressure source appears in B, a nonideal pressure source appears in C, and a nonideal flow source appears in D. The response of the vessel is determined in large part by the type of source.

where p_s is the pulsatile pressure at the source (N/m^2); R_s, the source resistance (N sec/m^5); p, the pulsatile pressure at the compliance–resistance junction (N/m^2); R, the vessel resistance (N sec/m^5); C, the vessel compliance (m^5/N); and t, the time.

Equivalently, the source can be considered to be a nonideal flow source composed of an ideal flow source in parallel with a source resistor. From

Thevenin's theorem, the parallel resistance R_s for the nonideal flow source is the same value as the series resistor R_s for the nonideal pressure source. The junction equation for this case is

$$\sum \dot{V} = 0 = \dot{V} + \frac{p}{R_s} + \frac{p}{R} + C\frac{dp}{dt} \qquad (2.5.42d)$$

Notice that the resistors R and R_s are in parallel and can easily be combined into one equivalent resistor using Eq. 1.6.3.

As implicitly stated earlier, the effect of the Windkessel vessel depends on the type of source supplying it. If an ideal flow source is assumed, and Eq. 2.5.4b results, then a solution for pressure p as a function of time can be obtained. Dividing the equation by C and multiplying by $e^{t/RC}$ gives

$$\frac{\dot{V}e^{t/RC}}{C} + \frac{pe^{t/RC}}{RC} + e^{t/RC}\frac{dp}{dt} = 0 \qquad (2.5.43)$$

But

$$\frac{pe^{t/RC}}{RC} + e^{t/RC}\frac{dp}{dt} = \frac{d}{dt}(pe^{t/RC}) \qquad (2.5.44)$$

Thus

$$\frac{1}{C}\int_0^t \dot{V}e^{t/RC}\,dt = -\int_0^{pe^{t/RC}} d(pe^{t/RC}) \qquad (2.5.45)$$

$$\frac{1}{C}\int_0^t \dot{V}e^{t/RC}\,dt = -pe^{t/RC}\Big|_0^{pe^{t/RC}}$$

$$= -pe^{t/RC} + p_0 \qquad (2.5.46)$$

where p_0 is the initial pressure (N/m^2). Therefore,

$$p = p_0 e^{-t/RC} - \frac{1}{C}e^{-t/RC}\int_0^t \dot{V}e^{t/RC}\,dt \qquad (2.5.47)$$

which shows that the pressure at the junction between compliance and resistor has a value that begins with initial value p_0 and decays exponentially with time constant RC, and a second component that is attenuated depending on the value of C, but that could possibly grow large over time because of the $e^{t/RC}$ term in the integral.

EXAMPLE 2.5.5.1-1. Pressure in a Windkessel Vessel

A distensible tube with resistance of 4.00×10^6 N sec/m^5 and compliance of 2.10×10^{-9} m^5/N is being driven by an ideal flow source $16 \times 10^3 \sin$

$9.42t \, \text{m}^3/\text{sec}$. Determine the magnitude of the steady-state pressure in the tube.

Solution

Equation 2.5.47 pertains. The $e^{t/RC}$ in the integral and the sinusoidal flow source mean that there really will not be a steady-state pressure solution. However, the first term in Eq. 2.5.47 can be ignored after a long period of time. Hence, using the second term only,

$$p = \frac{1}{C} e^{-t/RC} \int_0^t \dot{V} e^{t/RC} \, dt$$

$$= -\frac{\dot{V}_0}{C} e^{-t/RC} \int_0^t \sin \omega t \, e^{t/RC} \, dt$$

The value for this integral can be looked up to be

$$p = -\frac{\dot{V}_0}{C} e^{-t/RC} \frac{e^{t/RC}[(RC)^{-1} \sin \omega t - \omega \cos \omega t]}{(RC)^{-2} + \omega^2}$$

$$RC = \left(4.00 \times 10^6 \frac{\text{N sec}}{\text{m}^5}\right)(2.10 \times 10^{-9} \, \text{m}^5/\text{N}) = 8.4 \times 10^{-3} \, \text{sec}$$

Thus

$$p = \frac{16 \times 10^3 \, \text{m}^3/\text{sec}}{2.10 \times 10^{-9} \, \text{m}^5/\text{N}} \frac{(9.42 \cos 9.42t - 119 \sin 9.42t)/\text{sec}}{(119^2 + 9.42^2)/\text{sec}^2}$$

$$= 534 \times 10^6 (9.42 \cos 9.42t - 119 \sin 9.42t) \, \text{N/m}^2$$

Remarks

If the $e^{-t/RC}$ term were canceled with the $e^{t/RC}$ term in the integral, before the integration process, the wrong result would have been obtained.

Example 1.4.5-1 shows that compliance in biological systems often decreases as the tissue is stretched. This, then, limits the amount of fluid that can be stored inside.

---

steady flow analyzed differently

2.5.5.2 Steady Flow Elastic tubes can distend with increasing pressure. Steady flow can affect the sizes of these tubes because upstream pressures are higher than downstream pressures. Diameters of these tubes are nonuniform (Figure 2.5.10).

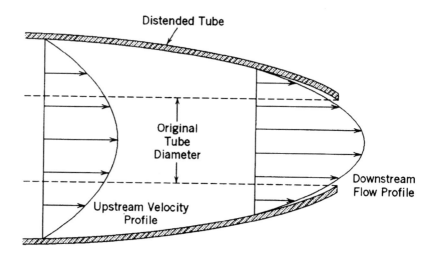

Figure 2.5.10. An elastic tube distends nonuniformly due to pressure differences within a flowing fluid.

The simplest relationship between pressure and diameter is a linear one

$$D = D_0 + D_1 p \qquad (2.5.48)$$

where D is the tube hydraulic diameter (m); D_0, the tube diameter at zero pressure (m); D_1, the tube compliance constant (m^3/N); and p the pressure inside the tube relative to pressure outside the tube (N/m^2).

Pulmonary arteries and veins are known to exhibit this linear relationship (Fung, 1990).

Equation 2.5.40 presents the resistance of a rigid tube with laminar flow as

$$R = \frac{128\mu L}{\pi D^4} = \frac{\Delta p}{\dot{V}} \qquad (2.5.49a)$$

This equation can be rewritten as the relation between pressure loss in the tube and distance along the tube

$$\frac{dp}{dx} = \frac{128\mu \dot{V}}{\pi D^4} \qquad (2.5.49b)$$

Following the analysis given in Fung (1990)

$$D^4 \, dp = \frac{128\mu \dot{V}}{\pi} \, dx \qquad (2.5.50)$$

Using the chain rule

$$D^4 \frac{dp}{dD} \, dD = D^4 \, dp = \frac{128\mu\dot{V}}{\pi} \, dx \qquad (2.5.51)$$

But, from Eq. 2.5.48,

$$\frac{dD}{dD} = 1 = 0 + D_1 \frac{dp}{dD} \qquad (2.5.52)$$

So

$$\int_{D(0)}^{D(L)} \frac{D^4}{D_1} \, dD = \int_0^L \frac{128\mu\dot{V} \, dx}{\pi} \qquad (2.5.53)$$

$$\frac{D^5(L) - D^5(0)}{D_1} = \frac{640}{\pi}\mu\dot{V}L \qquad (2.5.54)$$

$$\dot{V} = \frac{\pi}{640\mu L D_1}[D^5(L) - D^5(0)]$$

$$= \frac{\pi}{640\mu L D_1}\{[D_0 + D_1 p(0)]^5 - [D_0 + D_1 p(L)]^5\} \qquad (2.5.55)$$

where the terms 0 and L in parentheses correspond to values evaluated at the beginning and end of the tube.

For blood flow in large vessels with Reynolds numbers much greater than 1.0, inertial force terms must be considered. For this analysis, the reader is referred to Fung (1990). The solution leads to multiple values of possible flow rate, with implied instability as a result.

inertial forces can destablize

EXAMPLE 2.5.5.2-1. Flow in the Pulmonary Vein

The pulmonary vein is a very distensible blood vessel. The vein has a zero-pressure diameter of about 0.5 cm and a length about 15 cm. If the mean blood pressure is about 1200 N/m² where blood enters the pulmonary vein, and is about 0 N/m² where the vein empties into the left cardiac atrium, determine the tube compliance constant value (D_1) when the vein carries a volume flow rate of 83 mL/sec.

Solution

Because the pressure at the distal end of the pulmonary vein is about 0 N/m², and the term D_0^5 is small, we can obtain an initial estimate for D_1 by approximating Eq. 2.5.55 as

$$\dot{V} \approx \frac{\pi}{640\mu L D_1}[D_1 p(0)]^5 = \frac{\pi D_1^4 p^5(0)}{640\mu L}$$

Thus

$$D_1^4 = \frac{(83 \times 10^{-6} \text{ m}^3/\text{sec})(640)(4.5 \times 10^{-3} \text{ N sec/m}^2)(0.15 \text{ m})}{\pi(1200 \text{ N/m}^2)^5}$$

$$= 8.2 \times 10^{-6} \text{ m}^3/\text{N}$$

We check this result by calculating $\dot{V}$ from Eq. 2.5.55 (as written) and comparing against the known value of $8.3 \times 10^{-5} \text{ m}^3/\text{sec}$.

$$\dot{V} = \frac{\pi}{640(4.5 \times 10^{-3} \text{ N sec/m}^2)(0.15 \text{ m})(8.2 \times 10^{-6} \text{ m}^3/\text{N})}$$

$$\times \left\{\left[0.005 \text{ m} + \left(8.2 \times 10^{-6}\frac{\text{m}^3}{\text{N}}\right)\left(1200\frac{\text{N}}{\text{m}^2}\right)\right]^5 - (0.005 \text{ m})^5\right\}$$

$$= 6.4 \times 10^{-4} \text{ m}^3/\text{sec}$$

By trial and error, we generate the following values

$D_1(10^{-6} \text{ m}^3/\text{N})$	$\dot{V}(10^{-4} \text{ m}^3/\text{sec})$
8.2	6.4
5	2.3
3	1.1
2	0.69
2.5	0.86
2.4	0.83

The value for D_1 is thus about $2.4 \times 10^{-6} \text{ m}^3/\text{N}$.

Remark

The diameter of the pulmonary vein varies from 0.5 cm at the left atrium to 0.788 cm ($2.4 \times 10^{-6} \text{ m}^3/\text{N}$ times 1200 N/m^2), where the pressure is 1200 N/m^2.

2.5.6 Bifurcations

Real tubes found in biological systems also have more complications. They may not be smooth, and they often have branching, or bifurcations. Pulsatile flow in such tubes leads to propagated pressure waves, which, upon encountering a branch, are reflected back upstream. The characteristic impedance of a tube with reflected waves is given as

$$Z = \frac{\rho v_{\text{son}}}{A}$$
(2.5.56)

where Z, is the characteristic impedance (N sec/m^5); ρ, the density of fluid (kg/m^3 or N sec^2/m^4); v_{son}, the speed of sound in the fluid (m/sec); and A, the cross-sectional area of tube (m^2).

The characteristic impedance is the ratio of pulsating, or oscillatory, pressure to oscillatory flow when the wave goes in the direction of the flow.

The parabolic flow velocity profile for fully developed laminar flow is split symmetrically when it reaches a bifurcation, but forms two asymmetric flow velocity profiles in the branches past the bifurcation (Figure 2.5.11). The flow velocities are higher at the branch walls joined at the bifurcation because the flow velocity was highest in the center of the tube prior to the bifurcation. The same forces that shaped the parabolic velocity profile in the original tube also reshape the velocity profiles into parabolas in the two branch tubes, but the reshaping process requires energy that is manifested by a pressure loss. The entrance length after the bifurcation can be calculated from Eq. 2.5.15.

velocity
profiles must
be reformed

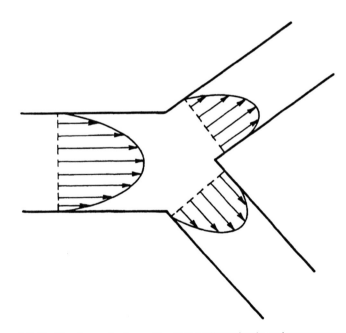

Figure 2.5.11. The flow velocity profile after a bifurcation is no longer symmetrical.

2.5.7 Compressible Flow

It is not uncommon that flow systems analyzed or designed by biological engineers use air or other gases as the working fluid. Ventilation systems for glass houses, animal housing facilities, or spaces meant for human occupancy

air systems
common

are examples of systems designed by biological engineers that use air as the moving fluid. Air entering and leaving the respiratory system is another example. Normally, as long as pressure changes from one part of the system to another are no greater than about 10%, there is no need to treat the gas any different than an incompressible fluid. When the gas must be pumped against a high pressure, however, compressibility effects can be significant and pipe resistance equations may be in error. Such an example might occur when pumping air under water to supply the needs of fish.

2.5.7.1 Sonic Velocity

One of the most important things to remember about compressible flow is that the flow velocity within a tube cannot exceed

sonic velocity
limits flow rate

the sonic velocity. To demonstrate this, consider a piston accelerating within a cylinder (Figure 2.5.12). When the piston moves slowly, there is hardly any pressure increase in the gas ahead of the piston, and the entire column of gas moves within the tube at the same speed as the piston. As the piston speed increases, there is a modest increase in gas compression ahead of the cylinder, but the pressure wave moves ahead of the compressed gas until the entire column of gas moves. If the piston approaches the speed of sound, the gas compresses more and more, but the pressure wave at the front of the compressed gas travels no faster than the speed of sound. If the piston exceeds the speed of sound, it will eventually overtake the pressure wave front, gas immediately in front of the piston will be highly compressed, but no pressure waveform is ahead of the piston, and the rest of the gas remains motionless until it interacts locally with the piston.

isothermal
sonic velocity

The sonic velocity of a gas in isothermal flow is

$$v_{son} = \sqrt{\frac{RT}{M}} \qquad (2.5.57)$$

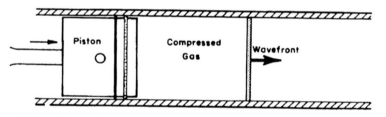

Figure 2.5.12. When a piston moves within a cylinder, it compresses the gas ahead of it. The pressure wavefront travels at the speed of sound.

where v_{son} is the sonic velocity (m/sec); R, the universal gas constant [8314.34 N m/(kg mol K)]; T, the absolute temperature (K); and M, the molecular mass [kg/kg mol or N sec^2/(m kg mol)].

And for adiabatic flow it is

$$v_{son} = \sqrt{\frac{c_p R T}{c_v M}} \qquad (2.5.58)$$

adiabatic
sonic velocity

where c_p is the specific heat at constant pressure [N m/(sec kg °C)]; and c_v, the specific heat at constant volume [N m/(sec kg °C)].

The value of c_p/c_v for air is 1.4. The sonic velocity for adiabatic flow of air is about 20% greater than for isothermal flow.

2.5.7.2 Pressure Drop and Maximum Flow Rate

To find the pressure drop in a tube with a flowing compressible fluid, we begin with the Bernoulli equation 2.3.21

$$\Delta h + \frac{\Delta p}{\gamma} + \frac{\Delta v^2}{\alpha g} + E + h_f = 0 \qquad (2.5.59)$$

Assuming a horizontal tube with $\Delta h = 0$, and no addition of mechanical work ($E = 0$),

$$\frac{\Delta p}{\gamma} + \frac{\Delta v^2}{\alpha g} + h_f = 0 \qquad (2.5.60)$$

where p is the relative pressure (N/m^2); γ, the specific weight (N/m^3); v, the velocity (m/sec); α, a velocity correction factor (unitless); g, the acceleration due to gravity (9.81 m/sec^2); and h_f, the friction loss (m).

Putting this equation into differential form,

$$\frac{dp}{\gamma} + \frac{2v\,dv}{\alpha g} + dh_f = 0 \qquad (2.5.61)$$

Using the result for pipe friction losses in Eq. 2.5.10

$$dh_f = f \frac{v^2}{2gD}\,dL \qquad (2.5.62)$$

and rearranging, gives:

$$\frac{g\,dp}{\gamma v^2} + \frac{2\,dv}{\alpha v} + \frac{f\,dL}{2D} = 0 \tag{2.5.63}$$

As long as the flow is isothermal, the ideal gas equation can be used to specify the relation between pressure and flow rate

$$pV = \frac{mRT}{M} \tag{2.5.64}$$

where p is the absolute pressure (N/m²); V, the volume (m³); m, the mass (kg); R, the universal gas constant [8314.34 N m/(kg mol K)]; T, the absolute temperature (K); and M, the molecular mass (kg/kg mol).

Density is equal to

$$\rho = \frac{m}{v} = \frac{pM}{RT} \tag{2.5.65}$$

and, since $1\,\text{N} = (1\,\text{kg})\,(1\,\text{m/sec}^2)$,

$$\gamma = \frac{pM}{RT} \tag{2.5.66}$$

Velocity can be expressed as:

$$v = \frac{\dot{W}}{\gamma A} \tag{2.5.67}$$

where $\dot{W}$ is the weight rate of flow (N/sec); γ, the specific weight (N/m³); and A, the cross-sectional area of the tube (m²).

And so

$$\left(\frac{gA^2 M}{\dot{W}^2 RT}\right) p\,dp + \frac{2\,dp}{\alpha p} + f\,\frac{dL}{2D} = 0 \tag{2.5.68}$$

This equation can be integrated to give

$$p_i^2 - p_o^2 = \frac{\dot{W}^2 RT}{gA^2 M}\left[f\frac{L}{D} + \frac{4}{\alpha}\ln\left(\frac{p_i}{p_o}\right)\right] \tag{2.5.69}$$

where p_i and p_o denote absolute pressures at the inlet and outlet of the pipe (N/m²).

weight rate of flow

Solving for the weight rate of flow

$$
\dot{W} = \left[\frac{\dfrac{gA^2M}{RT}(p_i^2 - p_o^2)}{f\dfrac{L}{D} + \dfrac{4}{\alpha}\ln\left(\dfrac{p_i}{p_o}\right)} \right]^{1/2}
\tag{2.5.70}
$$

outlet pressure does not always control flow

The weight rate of flow becomes zero at two places: when $p_i = p_o$, and when $p_o = 0$ [so that $\ln(p_i/p_o) = \infty$]. A maximum flow rate occurs when $d\dot{W}/dp = 0$, which occurs at the sonic velocity. Decreasing p_o beyond the point of maximum flow does not result in a flow increase (Figure 2.5.13).

solving for flow rate

When using Eq. 2.5.69 to determine the pressure drop along a tube, use the same value for friction factor, f, that would have been used for a noncompressible fluid. For laminar flow this would be $f = 64/Re$, and for turbulent flow use the Moody diagram. Because the pressures p_i and p_o appear in two places in the equation, a solution must be obtained by the process of iteration. That is, solve for the value of the unknown pressure on the left side of the equation using an estimate on the right side. Then use the new pressure value to update the value on the right side and repeat the procedure.

Data for the graph in Figure 2.5.13 were generated for a smooth tube 18 mm in diameter and 1 m long. Air was assumed to flow through the tube at 38°C and entered at atmospheric pressure (101.3 kN/m²). Two iterative loops had to be incorporated: (1) Because the flow regime turned out to be turbulent, the Colebrook and White equation 2.5.13 was used to determine friction factor, and (2) because friction factor depends on flow velocity, which, in turn, depends on weight rate of flow (Eq. 2.5.67), the weight rate of flow appears implicitly on both sides of Eq. 2.5.70. Iteration loops were used to obtain the required answers.

EXAMPLE 2.5.7.2-1. Flow of Air in a Rigid Tube

Air at 311 K flows through a smooth tube 1 m long and 18 mm diameter with a pressure of 101.3 kN/m² at the upstream end and 40.52 kN/m² at the downstream end. Determine the weight rate of flow of air.

Solution

This is the same set of conditions used for Figure 2.5.13, with $p_o/p_i = 40.52/101.3 = 0.4$. Average pressure in the tube is $(40.52 + 101.3)/2 = 70.91$ kN/m². The density of the air is, from the ideal gas law

$$
\rho = \frac{m}{v} = \frac{p_{avg}M}{RT} = \frac{(70{,}910 \text{ N/m}^2)(29 \text{ kg/kg mol})}{(8314.34 \text{ N m/kg mol K})(311 \text{ K})}
$$

$$
= 0.795 \text{ kg/m}^3
$$

The viscosity of air at 311 K is $\mu = 1.9 \times 10^{-5}$ kg/(m sec).

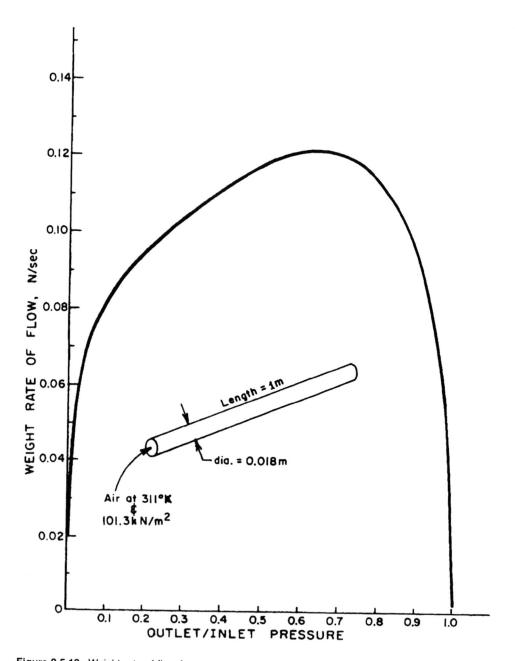

Figure 2.5.13. Weight rate of flow for a compressible fluid, air, for the pipe shown in the diagram. The curve shows the limited control exerted by outlet pressure on flow through the tube.

In order to calculate the Reynolds number, an estimate of velocity must be made. This requires a starting estimate of weight rate of flow, using Eq. 2.5.67. We start with an estimate of $\dot{W} = 0.01\,\text{N/sec}$

$$v = \frac{\dot{W}}{\rho g A} = \frac{0.01\ \text{N/sec}}{\left(0.795\ \dfrac{\text{kg}}{\text{m}^3}\right)\left(9.81\ \dfrac{\text{N}}{\text{kg}}\right)\left(\pi\dfrac{0.018^2}{4}\,\text{m}^2\right)}$$

$$= 5.04\ \text{m/sec}$$

The Reynolds number is

$$\text{Re} = \frac{Dv\rho}{\mu} = \frac{(0.018\ \text{m})(5.04\ \text{m/sec})(0.795\ \text{kg/m}^3)}{[1.9 \times 10^{-5}\ \text{kg/(m sec)}]}$$

$$= 3800$$

This flow is turbulent, with a friction factor obtained from the Colebrook–White equation 2.5.13 (with $\varepsilon/D = 0$ because the pipe was specified as smooth)

$$\frac{1}{\sqrt{f}} = -2.01\ \log_{10}(0 + 2.51/\text{Re}\sqrt{0.1})$$

This equation must also be solved by iteration, with an initial estimate of $f = 0.1$ to inject on the right-hand side.

$$\frac{1}{\sqrt{f}} = -2.01\ \log_{10}[2.51/(3800\sqrt{0.1})]$$

$$f = 0.0345$$

This iteration continues until the same value of f is obtained on the left-hand side, as is assumed on the right-hand side. This value is $f = 0.0402$.

The weight rate of flow can now be calculated using Eq. 2.5.70

$$\dot{W} = \left[\frac{\dfrac{(9.81\ \text{m/sec}^2)(2.54\times10^{-4}\ \text{m}^2)^2(29\ \text{kg/kg mol})(101{,}300^2 - 40{,}520^2)\ \text{N}^2/\text{m}^4}{(8314.34\ \text{N m/kg mol K})(311\ \text{K})}}{(0.0402)\left(\dfrac{1\ \text{m}}{0.018\ \text{m}}\right) + \dfrac{4}{1}\ \ln\left(\dfrac{101{,}300\ \text{N/m}^2}{40{,}520\ \text{N/m}^2}\right)}\right]^{1/2}$$

$$= 0.102\ \text{N/sec}$$

Since this value for weight rate of flow does not agree with the initial estimate of 0.01 N/sec, the entire process must be repeated with the new estimate of $\dot{W} = 0.102$ used to obtain velocity.

Results of the last four iterations in the process are:

$\dot{W}$	v	Re	f
0.1020	5.04	3,795	0.0402
0.1121	51.40	38,724	0.0219
0.1124	56.48	42,555	0.0215
0.1124	56.63	42,668	0.0215

A final value of $\dot{W} = 0.1124$ N/sec is obtained.

2.5.7.3 Viscosity and Density Dependence

Flow regimes in biological systems can be very complicated. As an example we consider the respiratory system. Flow in the trachea is very high, usually high enough to be turbulent. Flow in the very smallest airways is much lower and is normally considered to be in the laminar range (the flow would be laminar except that the lengths of the lower airways are so short that the laminar velocity profile is not fully developed). Total cross-sectional area of the airways in any particular generation generally declines from the peripheral airways to the trachea (a generation refers to the number of bifurcations in the conducting air passages closer to the mouth). Therefore, flow velocity decreases from the central (largest) to the peripheral (smallest) airways during inhalation and increases from the peripheral airways to the central airways during exhalation. Therefore, the flow in the airways is accelerating, splitting, and mixing all at the same time.

respiratory airways

Knudsen and Schroter (1983) have described the types of flow for these different conditions. The resistance of tubes with fully developed laminar flow is, from Eq. 2.5.40, related directly to fluid viscosity. Resistance of tubes containing accelerating fluids is inertial in nature and depends on fluid density, but not viscosity.

generalized tube resistance

In general, then, tube resistance can be given as

$$R = R_i \mu^{2-a} \rho^{a-1} \dot{V}^{a-1} \qquad (2.5.71)$$

where R is the tube resistance (N sec/m^5); R_i, a coefficient (sec/m^{5+a}); μ, the viscosity [kg/(m sec)]; ρ, density (kg/m^3); $\dot{V}$, the volume flow rate (m^3/sec); and a, an exponent (dimensionless).

The parameter a assumes different values depending on the type of flow. These are summarized in Table 2.5.3. Since each of these can exist simultaneously in different parts of a complex arrangement of fluid passages, overall resistance of the tubes can be very complex. Note that in the respiratory system, resistance can be viscosity dependent in the smallest airways, density dependent in the middle-sized airways, and dependent on both viscosity and

density and viscosity dependence

Table 2.5.3   Flow Resistance Dependence on Viscosity and Density

Flow type	Parameter a Value	Equation for R
Fully developed laminar	1.00	$R_1 \mu$
Boundary layer growth (developing velocity profile)	1.50	$R_2 \mu^{0.5} \rho^{0.5} \dot{V}^{0.5}$
Turbulent[a]	1.75	$R_3 \mu^{0.25} \rho^{0.75} \dot{V}^{0.75}$
Fluid acceleration	2.00	$R_4 \rho \dot{V}$

[a] Pipe resistance in turbulent flow is often assumed to be dependent on the first power of volume flow rate. The actual exponent depends on particular pipe conditions.

density elsewhere. The same could be true for a complex arrangement of ventilation ducts or irrigation pipes.

Density-dependent resistance is of most importance during compressible flow, when density can change along the length of the tube. Because upstream pressure is always higher than downstream pressure, and compressible fluid density is directly related to pressure, tube upstream resistance, for all but fully developed laminar flow, is always higher than tube downstream resistance, even for the same size tube and uniform velocity. Higher resistances cause higher pressure drops; so the pressure–flow relationship of the tube becomes somewhat to highly nonlinear, with resistance being location dependent.

resistance can depend on location

In addition, because the Reynolds number includes density (Eq. 2.5.5), upstream Reynolds number values are higher than downstream values, even for a tube of uniform cross section and velocity. It is conceivable that upstream Reynolds numbers could be high enough to indicate turbulent flow while downstream Reynolds numbers indicate laminar flow. The effects of these differences would be extremely difficult to separate from effects attributable to entrance length, however, so they would not be noticed of themselves.

laminar downstream, turbulent upstream

2.5.7.4 Compression Heating Another consequence of compressible flow is that abruptly slowing the flow causes a rise in temperature. Compression heating occurs, especially where the flow is brought to a halt. This is called the stagnation temperature. For a thermometer to read the correct temperature in a stream of flowing gas, it should travel at the same velocity as the gas. Otherwise, it reads an elevated temperature. Equating kinetic energy of the flowing gas with thermal energy of the stagnant gas gives the stagnation temperature as

stagnation temperature

$$\theta_s = \theta_a + \frac{v_p^2}{2c_p} \qquad (2.5.72)$$

where θ_s is the stagnation temperature (°C); θ_a, the ambient temperature (°C); v_p, the local velocity in the flowing stream (m/sec); and c_p, the specific heat at constant pressure [N m/(kg °C) or m^2/(sec^2 °C)].

EXAMPLE 2.5.4-1. Stagnation Temperature

Air flowing in turbulence in a conduit at 56.6 m/sec encounters a tee. If the temperature of the free-flowing air is 311 K, what is the temperature at the tee? The specific heat of air at 311 K is 1.0048 kN m/(kg K).

Solution

Air at the tee comes to a standstill before splitting to flow to either side. The temperature of the air at the tee can thus be calculated using Eq. 2.5.72 as

$$\theta_s = 311 \text{ K} + \frac{(56.5 \text{ m/sec})^2}{2(1004.8 \text{ m}^2/(\text{sec}^2 \text{ K})}$$

$$= 312.6 \text{ K}$$

Remark

If the flow had been laminar at the same velocity, then the peak velocity would have been twice the average velocity, and θ_s would have been 317.4 K.

2.5.8 Fluid Flow in Plants

Water flow in plants is diagrammed schematically in Figure 2.5.14. There are several significant differences between plant water flow and flow in other biological systems. The first is that a phase change occurs at one end of the system (in the leaves), and the resistance to flow at that point involves diffusion of gaseous water vapor (see Section 4.3). The second is that vertical movement of water in the xylem of plants, as indicated earlier, involves capillarity. Capillarity (the pressure that draws a column of liquid though an extremely small tube) can be calculated from:

evaporation in leaves

capillarity

$$p = \frac{2\sigma}{r} \tag{2.5.73}$$

where p is the capillary pressure (N/m^2); σ, the surface tension coefficient (N/m); and r, the radius of curvature of the meniscus (m).

The value for σ is about 0.073 N/m for an air–water interface at 20°C. Using a xylem radius of 0.005 μm, the calculated capillary pressure is 29.4 × 10^6 N/m^2, enough to draw water to a height of 3000 m. If capillarity were not present, water could not be drawn to a height of any more than about 10 m (the height supported by one atmosphere of pressure).

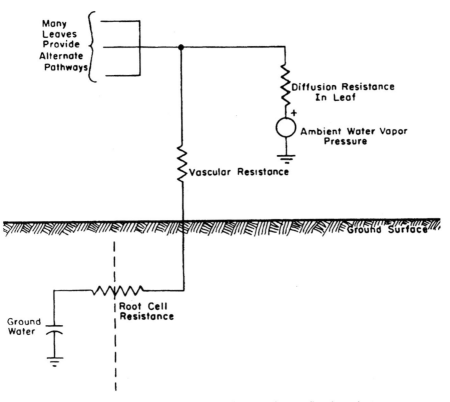

Figure 2.5.14. Steady-state diagram of water flow in a plant.

A third difference that occurs with water movement in a plant is that osmotic pressure significantly affects movement. Osmosis involves the concentration of solutes in solution, and is covered in Section 4.3.3.3. Thus plant physiologists often talk in terms of water potential, which is the sum of hydrostatic pressure, gravitational potential, and osmotic potential.

Water in the xylem is usually under tension, meaning that pressures measured inside the water column are more negative than an absolute vacuum (-5 to -50 atmospheres). Perhaps because of this, the water column in the xylem breaks very often, forming voids called emboli. Turgid living cells surrounding the xylem communicate with the interiors of the xylem passages through tiny pores in the xylem walls. When an embolus appears in the water column, water seeps into the xylem from these surrounding cells and refills the void. This maintains the ability of the plant to draw water to its highest level (Canny, 1998).

plant water
potential

breaks in
water column

2.5.9 Deposition of Suspended Particles

deposition
narrows tubes

Sedimentation of suspended particles flowing in a closed conduit is a very real possibility if the flow rate is slow enough. If deposition occurs, then the cross-sectional area of the pipe available for fluid flow decreases. If the volume flow rate can be maintained, then flow velocity can increase because area decreases (from the continuity equation 2.2.4). Contrarily, a decrease in open area increases pipe resistance, and velocity may slow further. In the former case where velocity increases, an equilibrium will be established and eventually no more deposition will occur. In the latter case, the pipe will quickly fill with sediment until no more flow occurs.

To determine whether particles transported by a fluid will or will not settle is a complex problem that is best answered by empirical means. Particle weights tend to cause settling; higher velocities away from pipe walls tend to push particles to the center of the pipe (see Section 2.4.1); drag forces move the particles along with the flow. The result is a dynamic that depends on fluid density, fluid velocity, particle size, and particle density, among other things.

Fluid streams that have additional particle transport capacity can cause erosion. Fluid streams for which the transport capacity is exceeded cause deposition. This is true whether the situation is open-channel flow (see Section 2.7), pipe flow of waste water, or settling of dusts or smokes (see Section 2.4.5).

No permanent deposit of sediment suspended in water occurs if

deposition
velocity

$$ v > \frac{10}{\pi} \sqrt{d_s g \left(\frac{\rho_s - \rho_w}{\rho_w} \right)} \qquad (2.5.74) $$

where v is the average fluid velocity (m/sec); d_s, the diameter of suspended particles (m); g, the acceleration due to gravity (m/sec^2); ρ_w, the density of water (kg/m^3); and ρ_s, the density of suspended solids (kg/m^3).

When dealing with natural sediments, the specific gravity of quartz is often used (Table 2.5.4).

Because the fluid velocity varies across the conduit diameter, and fluids with slower velocities can carry fewer or smaller particles, it is recommended that water carrying waste particles be kept in turbulent flow. Particles that would otherwise tend to settle will be thoroughly mixed in the fluid stream.

Table 2.5.4 Densities of Selected Materials

Material	Density (kg/m^3)
Minerals	
Basalt	2400 – 3100
Bauxite	2550
Chalk	1900 – 2800
Clay	1800 – 2600
Coal, bituminous	1200 – 1600
Dolomite	2840
Earth, dry	1200 – 1500
Earth, moist	1300 – 1600
Mud	1700 – 1800
Granite	2650 – 2700
Gravel	1400 – 1900
Gypsum	2310 – 2330
Lava, basaltic	2800 – 3000
Limestone	2000 – 2900
Magnetite	4900 – 5200
Marble	2600 – 2840
Mica	2600 – 3200
Porcelain	2300 – 2500
Pumice	370 – 900
Quartz	2650
Rock salt	2180
Salt	780 – 1250
Sand, dry	1440 – 1760
Sand, wet	1890 – 2070
Sandy clay, compacted	2200
Shale	2600 – 2900
Slate	2600 – 3300
Food Materials	
Apples (Lodi)	780
Applesauce	1114
Beer	1000
Butter	998
Carrot	1040
Gelatin	1270
Grapefruit sections	1070
Grapes, seedless	1090
Honey	1400
Marashino cherries	1150
Meat	1060
Milk, whole	1030
Milk, skim	1040

(*continued*)

Table 2.5.4 (*continued*)

Material	Density (kg/m^3)
Molasses, 21°C	1430
Molasses, 37.8°C	1380
Molasses, 49°C	1310
Molasses, 66°C	1160
Oil, castor	969
Oil, corn	920
Oil, olive	920
Oil, soybean	920
Peach	1010
Peanut butter	1096
Pear	1010
Pineapple chunks	1060
Plum	1040
Prune	1290
Potatoes, frozen	980
Salt	2165
Starch	1530
Sugar	1610
Yogurt	1051

Waste Materials

Fish feces	1190
Commercial fish food	1130 – 1200
Manure, swine and beef	1050

Natural Materials

Antler	1866
Cow leg bone	2060
Ivory	1830 – 1920
Leather	860 – 1020
Paper	700 – 1150
Rubber	1000 – 2000
Whale tympanic bulla	2470

Gases

Acetylene	1.1708
Air	1.2929
Ammonia	0.7710
Argon	1.7837
Butane	2.637
Carbon dioxide	1.9769
Carbon monoxide	1.2504

(*continued*)

Table 2.5.4 (*continued*)

Material	Density (kg/m³)
Chlorine	3.214
Ethylene	1.2604
Helium	0.1785
Hydrogen	0.08988
Hydrogen chloride	1.6392
Hydrogen sulfide	1.539
Methane	0.717
Nitric oxide	1.3402
Nitrogen	1.2506
Nitrous oxide	1.978
Oxygen	1.42904
Phosgene	4.531
Propane	2.020
Sulfur dioxide	2.9269

Liquids

Acetaldehyde	806
Acetic acid	1050
Acetone	792
Benzene	879
Butyric acid	954
Castor oil	960
Ethyl alcohol	789
Ethylene chloride	1246
Gasoline	660 – 690
Glycerine	1261
Kerosene	780 – 820
Linseed oil	934
Methyl alcohol	792
Naphthalene	1152
Nitric acid (100%)	1513
Petroleum	878
Turpentine	873
Water	1000

Insulators

Fire brick	706 – 882
Mineral wool blanket	128 – 193
Glass wool blanket	48

(*continued*)

Table 2.5.4 (*continued*)

Material	Density (kg/m^3)
Woods	
Ash	450 – 540
Basswood	320 – 370
Cedar, red	440 – 470
Fir	360 – 380
Hemlock	380 – 400
Locust, black	660 – 690
Maple, red	490 – 540
Oak	570 – 670
Pine	470 – 510
Poplar	380 – 400
Walnut, black	510 – 550
Cypress	460
Metals	
Aluminum	2700
Arsenic	6618
Barium	3500
Brass	8520
Cadmium	8648
Calcium	1550
Chromium	7140
Cobalt	8900
Copper	8940
Gold	19300
Iridium	22420
Iron	7870
Lead	11342
Magnesium	1740
Mercury	13546
Nickel	8850
Platinum	2145
Silver	10500
Sodium	9712
Steel	7800
Strontium	2600
Tin	7300
Titanium	4500
Tungsten	19300
Zinc	7140

(*continued*)

Table 2.5.4 (*continued*)

Material	Density (kg/m^3)
Building Materials	
Asbestos, loose	470–570
Asbestos	2000–2800
Brick, common	1600
Brick, fireclay	2000
Cement, Portland (dry)	1500
Cement, Portland (wet)	3050–3150
Concrete	2311
Cork, regranulated	45–120
Corkboard	160
Diatomaceous earth	320
Glass, window	2700
Glass wool	24
Glass	2500–2750
Limestone	2500
Masonry	2240–2560
Plaster, gypsum	1440
Sandstone	2160–2300
Human Materials	
Human skin	1100
Human muscle	1080
Human fat	850
Blood	1050
Enamel	2200
Dentin	1900
Bone, cortical	1700–2000
Bone, trabecular	1600–1900
Food, As Stored in Bulk	
Apples	310–450
Beans, green, frozen	570
Beef	360
Beef, boneless	1280
Blueberries, frozen	450
Broccoli, frozen	340
Celery	480
Cheese, solid	520–650
Cheese, Swiss	640
Chicken, whole fryers	410
Chicken, fryer parts	620
Chicken, fryer parts, frozen	670

(*continued*)

Table 2.5.4 (*continued*)

Material	Density (kg/m^3)
Chili peppers	260
Citrus juice, concentrate, frozen	850
Cranberries	360
Dried fruit	720
Eggs	310
Frozen fish	460 – 960
Frozen asparagus	380
Grapefruit	560
Grapes	470
Lamb, boneless	980
Lard	840
Lemons	600
Lettuce, head	230 – 400
Milk, condensed	730
Nuts, shelled almonds	440
Nuts, almonds in shell	210
Nuts, English walnuts in shell	330
Nuts, English walnuts shelled	350
Nuts, peanuts, shelled	620
Oranges	370 – 610
Peaches	420 – 650
Peaches, frozen	580
Pears	760
Peas, frozen	440 – 450
Pork	530
Pork, boneless	960
Potatoes	440
Potatoes, frozen french fries	380 – 460
Spinach, frozen	500
Strawberries	570 – 670
Strawberries, frozen ($-20°$C)	960
Strawberries, unfrozen	1040
Tomatoes	500 – 620
Turkeys	320 – 400
Veal, boneless	980

Food Bulk Densities

Cocoa beans, bulk	1073
Coconut, shredded, bulk	320 – 352
Coffee beans, green, bulk	673
Coffee beans, ground, bulk	400
Coffee beans, roasted, bulk	368
Corn, ear, bulk	448
Corn, shelled, bulk	720

(*continued*)

Table 2.5.4 (*continued*)

Material	Density (kg/m^3)
Milk, whole dried	320
Mustard seed	720
Peanuts, hulled	480 – 720
Peas, dried	800
Rapeseed	770
Rice, clean	770
Rice, hulls	320
Soybeans, whole, bulk	800
Sugar, granulated	800
Wheat	770

Solid Basic Food Materials

Ash (minerals in food)	1650 – 1740
Carbohydrate	1330 – 1430
Cellulose, solid	1270 – 1610
Citric acid, solid	1540
Fat, solid	900 – 950
Glucose, solid	1560
Protein, solid globular	1400
Salt, solid	2160
Starch, solid	1500
Sucrose, solid	1590

Temperature-Dependent Food

Buffalo milk	$\rho = 923.84 - 0.4\theta$
Cow's milk	$\rho = 923.51 - 0.430\theta$
Cream	$\rho = 1038.2 - 0.17\theta - 0.003\theta^2 - (133.7 - 475.5/\theta)X_f$
Skim milk	$\rho = 1036.0 - 0.146\theta + 0.0023\theta^2 - 0.0016\theta^3$
Whole milk	$\rho = 1035.0 - 0.358\theta + 0.0049\theta^2 - 0.000100\theta^3$

$\theta = {}^\circ C$
X_f = fat content (mass fraction)

EXAMPLE 2.5.9-1. Flushing Manure

Hog manure with a mean particle diameter of 450 μm and standard deviation of 95 μg (Patni, 1980) is to be washed from a piggery. What must the water velocity be in order to remove virtually all the manure? The density of the manure is 1050 kg/μm^3 (Tunney, 1980).

Solution

We will use Eq. 2.5.74 to solve for velocity, since the particle size distribution and density are known. To remove virtually all the manure (99.9%) requires consideration of particles at least 3.1 standard deviations larger than the mean, or

$$d_s = 450 \times 10^{-6} \text{ m} + (3.1)(95 \times 10^{-6} \text{ m}) = 744 \times 10^{-6} \text{ m}$$

$$v > \frac{10}{\pi} \sqrt{(744 \times 10^{-6} \text{ m})\left(9.81 \frac{\text{m}}{\text{sec}^2}\right)\left(\frac{1050 - 1000 \text{ kg/m}^3}{1000 \text{ kg/m}^3}\right)}$$

$$= 6.08 \times 10^{-2} \text{ m/sec}$$

Remarks

Assuring that this flow will be turbulent requires that

$$\text{Re} = \frac{Dv\rho}{\mu} > 2000$$

$$D > \frac{2000\,\mu}{v\rho} = \frac{2000[9.84 \times 10^{-4} \text{ kg/(m sec)}]}{(6.08 \times 10^{-2} \text{ m/sec})(998 \text{ kg/m}^3)}$$

$$= 3.24 \times 10^{-2} \text{ m}$$

$$\dot{V} = Av = \left(\frac{\pi D^2}{4}\right)v = \frac{\pi}{4}(3.24 \times 10^{-2} \text{ m})^2(6.08 \times 10^{-2} \text{ m/sec})$$

$$= 5.02 \times 10^{-5} \text{ m}^3/\text{sec}$$

The water to flush the manure from the hog house will probably come from the lagoon, where, because the water is still, all particles but the very smallest will settle from the wash water.

2.6 NON-NEWTONIAN FLUID FLOW

non-Newtonian fluids have variable viscosities

The nondimensional Reynolds number was introduced in Eq. 2.5.4, and one physical parameter appearing in the Reynolds number is the viscosity. While there are many fluids that have constant viscosities, many biological fluids do not. Fluids possessing a constant viscosity are called Newtonian; those with nonconstant viscosity are called non-Newtonian. Actually, even Newtonian fluids have nonconstant viscosities; nearly all fluids have viscosities that vary with temperature (viscosities of gases increase and viscosities of liquids decrease with increasing temperature). Additionally, non-Newtonian fluids

viscosity
varies with
flow rate

have viscosities that vary with the rate of flow, and this property makes them more complicated, but more interesting to work with.

2.6.1   Rheological Properties

Rheology is the science of deformation and flowing materials. Rheological properties of non-Newtonian fluids are those properties that characterize the flow of these types of fluids and that distinguish between Newtonian and non-Newtonian types.

If the apparatus diagrammed in Figure 2.4.1 were used to obtain viscosity data for various kinds of fluids, then the shear stress and rate of shear curves in Figure 2.6.1 could result. Newtonian fluids are seen to have a constant slope and a straight line. Bingham plastics are nearly Newtonian but require a yield

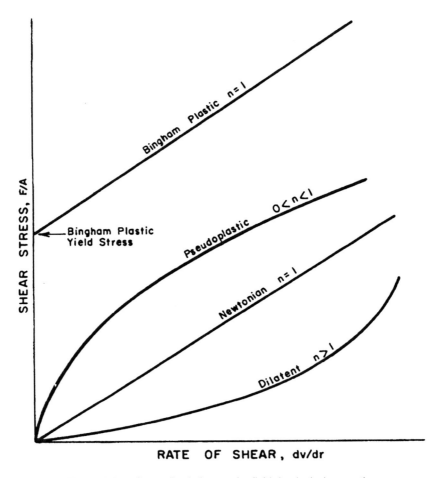

Figure 2.6.1. Generalized diagram for fluid rheological properties.

stress to be overcome before they will move. Pseudoplastic materials generally have decreasing viscosity (slope of the line) with increasing shear rate. Dilatent fluids generally have increasing viscosity with increasing rate of shear.

The simplest mathematical description used to describe the characteristics seen in Figure 2.6.1 is the power law model

power law
model

$$\tau = K\gamma^n + \tau_0 \tag{2.6.1}$$

where K is the consistency coefficient ($\text{N sec}^n/\text{m}^2$); n, the flow behavior index (dimensionless); τ_0, the yield stress (N/m^2); and γ, the rate of shear [m/m sec)].

In the case of a Newtonian fluid, $n = 1$, $\tau_0 = 0$, and $K = \mu = $ viscosity. If the yield stress is ignored, then mathematical manipulation of Eq. 2.6.1 becomes much easier, and thus Bingham fluids are often approximated as pseudoplastics. Pseudoplastic fluids are characterized by curves concave down, and thus have flow behavior indexes less than 1.0; dilatent fluids with concave-up curves have flow behavior indexes greater than 1.0. The power law model has been found useful over two to four decades of shear rate.

pseudoplastic
model

Pseudoplastics may be conceptualized as long-chain molecules suspended in a fluid bed (Figure 2.6.2). Upon standing they tangle and intertwine. Resistance to movement, and thus viscosity, is very high. Once moved, however, the long chains begin to untangle, and line up parallel to each other. Eventually they slide past each other with relative ease. Fluid viscosity has decreased and tends toward Newtonian. Expect pseudoplastic behavior in

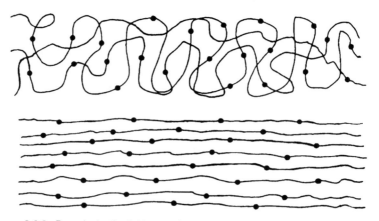

Figure 2.6.2. Pseudoplastic fluids can be considered to be composed of long-chain molecules suspended in a liquid. At rest (above) the molecular chains entwine, causing high viscosity. When moved (below), the chains align, with minimal interference and low viscosity.

suspensions containing polysaccharides, proteins, and other long macro-molecules.

dilatent model   Dilatent fluids can be thought of as densely packed hard spheres with just enough fluid between them to fill the voids (Figure 2.6.3). Resistance to movement of spheres past each other is relatively low because the fluid lubricates movement. As they move faster, their dense packing becomes disrupted, and there is insufficient fluid to lubricate their motions completely. Thus they become harder to move faster. Dilatent behavior should be expected in materials containing gritty material without excess liquid.

A list of non-Newtonian fluid examples appears in Table 2.6.1. It can be seen that the types of fluids appearing in each category generally fits the conceptual description just given. A list of fluid parameter values appears in Table 2.6.2. This list was compiled from values appearing in the literature and

flow behavior index   shows the extreme variability present in these measurements. In general, values of flow behavior index are more reliable than those of the consistency coefficient, which is fortunate, since the flow behavior index appears in more places than consistency coefficient in the equations to follow.

There is very little change in the flow behavior index (n) with temperature and concentration. If there is any trend, then as temperature increases, n tends toward 1.0 (the fluid moves closer to Newtonian); as concentration increases, n tends toward 0 (the fluid moves farther from Newtonian).

Values of the consistency coefficient (K) are much more sensitive to temperature and concentration. As concentration increases, K increases

consistency coefficient variations   (becomes more viscous); as temperature increases, K tends to decrease by an amount that very much depends on the suspending fluid

$$K_\theta = K_0 \mu_\theta / \mu_0 \qquad (2.6.2)$$

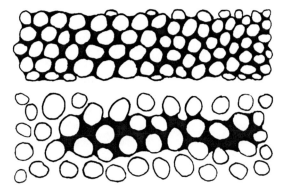

Figure 2.6.3. Dilatent fluids can be considered to be composed of hard spheres with a minimum of fluid. When at rest (above), surfaces of the densely packed spheres are lubricated when they contact each other. During flow (below), however, there is insufficient liquid to lubricate the spheres, and the resulting friction appears as an increase in viscosity.

Table 2.6.1 Examples of Non-Newtonian Fluids

Bingham fluids	$\tau_0 \neq 0$
Damp clay	$K \neq 0, K > 0$
Concentrated slurries	$n = 1$
Window putty	
Cheese	
Pseudoplastics	$\tau_0 = 0$
Paints	$K \neq 0, K > 0$
Mayonnaise	$0 < n < 1$
Heavy slurries	
Human blood	
Some honeys	
Tomato paste	
Purees	
Solutions of gums	
Dilatent	$\tau_0 = 0$
Moist sand	$K \neq 0, K > 0$
Quick sand	$1 \leqslant n < \infty$
Heavy starch suspensions	
Many honeys (buckwheat, sumac, orange, tar weed,	
Hawaiian honeydew, etc.)	

where K_θ is the consistency coefficient at temperature θ ($N \sec^n/m^2$); K_0, the consistency coefficient at reference temperature ($N \sec^n/m^2$); μ_0, the viscosity of the suspending fluid at reference temperature ($N \sec/m^2$); and μ_θ, the viscosity of the suspending fluid at temperature θ ($N \sec/m^2$).

Many food and biological materials use water as the suspending fluid. Viscosity values for water at various temperatures are well known (Table 2.4.2). The use of Eq. 2.6.2 is highly recommended to extend the temperature range of tabulated experimental values.

There are additional kinds of non-Newtonian fluid behavior as well as the ones just described. There are some fluids that show elastic or viscoelastic properties. They particles suspended in an elastic fluid act as tiny springs. These fluids tend to respond with stresses related to the time rate of change of the shear rate (Figure 2.6.4). Viscoelastic fluids combine responses of viscous fluids (as previously described) and elastic fluids. Treating viscoelastic fluids as viscous fluids has been considered to be a conservative engineering approach.

viscoelastic
fluids

Some non-Newtonian fluids exhibit time-dependent properties. Thixotropic viscosities decrease with time. When motion ceases, the fluid often will eventually return to its initial viscosity. Examples of thixotropic fluids are bread dough, some soil suspensions, and some paints (especially those formulated to be used on ceilings without running down the paint brush).

thixotropic
fluids

Table 2.6.2 Rheological Properties of Some Fluids

Product	Temperature (°C)	Composition	Consistency Coefficient $\left(\dfrac{N\,sec^n}{m^2}\right)$	Flow Behavior Index (n)
Aerated poultry waste slurry	10–25	v% = volume percent solids	$1.12 \times 10^{-11}(v\%)^{2.59}$	$1.81 - 0.161 \ln v\%$
Ammonium alginate	24	3.37%	12.5	0.500
Ammonium alginate	24	3.37%	13.5	0.477
Apple juice	27	20° Brix	0.00021	1.0
Apple juice	27	60° Brix	0.0003	1.0
Applesauce	24	unknown	0.66	0.408
Applesauce	25	31.7% T.S.	22.0	0.4
Applesauce	27	11.6% T.S.	12.7	0.28
Applesauce	24	unknown	0.50	0.645
Applesauce	24	unknown	0.66	0.408
Applesauce	unknown	unknown	5.63	0.47
Apricot purée	27	17.7% T.S.	5.40	0.29
Apricot purée	25	19% T.S.	20.0	0.3
Apricot purée	27	13.8% T.S.	7.20	0.41
Apricot concentrate	25	26% T.S.	67.0	0.3
Banana purée	24	unknown	6.50	0.458
Banana purée	24	unknown	10.7	0.333
Banana purée	20	unknown	6.89	0.46
Banana purée	42	unknown	5.26	0.486

(continued)

Table 2.6.2 *(continued)*

Product	Temperature (°C)	Composition	Consistency Coefficient $\left(\dfrac{N\ sec^n}{m^2}\right)$	Flow Behavior Index (n)
Banana purée	49	unknown	4.15	0.478
Chicken minced	23	unknown	911	0.088
Chocolate	30	unknown	0.685	0.500
Chocolate milk	unknown	unknown	0.39	0.26–0.42
Cocoa butter	40	unknown	0.041	1.0
Corn syrup	27	48.4% T.S.	0.0053	1.0
Cream	3	20% fat	0.0062	1.0
Cream	3	30% fat	0.0138	1.0
Egg white	62	unknown	0.45	1.0
Egg white, fresh	2	unknown	0.19	0.56
Egg yolk	5–60	42.6% T.S.	0.17–0.77	0.87–0.89
Grape juice	27	20° Brix	0.0025	1.0
Grape juice	27	60° Brix	0.11	1.0
Guava purée	23.4	10.3% T.S.	4.36	0.49
Honey	22	unknown	1914	0.89
Honey	24	normal	5.60	1.0
Honey	24	normal	6.18	1.0
Human blood	27		0.00384	0.890
Ketchup	22	unknown	33.2	0.242

(continued)

Table 2.6.2 (*continued*)

Product	Temperature (°C)	Composition	Consistency Coefficient $\left(\dfrac{N\ sec^n}{m^2}\right)$	Flow Behavior Index (n)
Ketchup	25	unknown	2.0–9.4	0.38–0.61
Mango pulp	30–70	16–30° Brix	2.8–10.3	0.28–0.30
Mayonnaise	25	unknown	4.2–4.7	0.54–0.59
Mayonnaise	20	80% oil	127.4	0.69
Meat batter, raw	15	unknown	14–639	0.16–0.72
Mustard	25	unknown	3.4–37.0	0.21–0.56
Olive oil	20	normal	0.084	1.0
Orange Juice	0	unknown	1.89	0.680
Peach purée	27	10.0% T.S.	4.50	0.34
Peach purée	27	10.0% T.S.	0.94	0.44
Pear purée	27	14.6% T.S.	5.30	0.38
Pear purée	27	15.2% T.S.	4.25	0.35
Pear purée	32	18.31% T.S.	2.25	0.486
Pear purée	32	45.75% T.S.	35.5	0.479
Sewage sludge	20	w% = weight percent solids w% = 1.7 to 6.2	0.00113 (w%)$^{3.40}$	0.817 – 0.843 w%
Skim milk	25	normal	0.0014	1.0
Soybean oil	30	normal	0.04	1.0
Tomato concentrate	32	5.8% T.S.	0.223	0.59

(*continued*)

Table 2.6.2 *(continued)*

Product	Temperature (°C)	Composition	Consistency Coefficient $\left(\dfrac{N\ sec^n}{m^2}\right)$	Flow Behavior Index (n)
Tomato concentrate	32	30% T.S.	18.7	0.4
Tomato paste	unknown	unknown	15.0	0.475
Tomato purée	unknown	unknown	0.92	0.554
Whole egg	61	unknown	0.67	1.0
Whole milk	20	normal	0.00212	1.0
Yogurt	20	unknown	25.4	0.545

Source: Heldman (1975), Rao (1994).

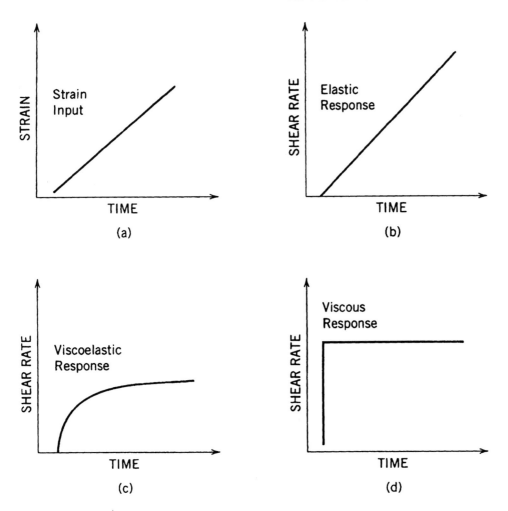

Figure 2.6.4. Different types of material response to a step change in rate of shear. For a Newtonian (viscous) fluid, shear rate is directly proportional to stress; for an elastic solid, strain is directly proportional to stress; for a viscoelastic material, the response is between the viscous and elastic responses.

rheopectic
fluids

Rheopectic viscosities increase with time. Examples of rheopectic substances are jelly, glue, gelatin, and clotting blood. When designing with either thixotropic or rheopectic substances, it is best to design for the maximum viscosities expected to be encountered.

EXAMPLE 2.6.1-1. Consistency Coefficient of Blood

Human blood is listed in Table 2.6.2 with a consistency coefficient of $0.00384 \, \text{N sec}^n/\text{m}^2$ and a flow behavior index of 0.890 at a temperature of 27°C. What are the values of consistency coefficient and flow behavior index to be used at a human body temperature of 37°C?

Solution

Flow behavior index values are nearly unaffected by temperature, and the value of 0.890 will suffice. Consistency coefficient values, however, depend strongly on temperature, as given by Eq. 2.6.2.

Water is the suspending fluid. Using values listed in Table 2.4.4 and Eq. 2.4.6b, the viscosity of water was calculated to be $8.507 \times 10^{-4} \, \text{kg}/(\text{m sec})$ at 27°C and $6.984 \times 10^{-4} \, \text{kg}/(\text{m sec})$ at 37°C. The corrected value for the consistency coefficient is thus

$$K_{37} = (0.00384 \, \text{N sec}^n/\text{m}^2)(6.984 \times 10^{-4}/8.507 \times 10^{-4})$$

$$= 0.00315 \, \text{N sec}^n/\text{m}^2$$

2.6.2 Pipe Flow

The design of pumping and piping systems for non-Newtonian fluids is based upon the same procedures as previously given for Newtonian fluids, with certain modifications. The continuity equations in Sections 2.2.1 and 2.2.2 can be used unmodified. The energy balance equations in Section 2.3, however, must be modified to account for different velocity profiles that depend on the flow behavior index n. These velocity profiles are diagrammed in Figure 2.6.5. As the flow behavior index varies from 0 to ∞ (pseudoplastic to dilatent), the flow velocity profiles vary from plug flow (all local velocities equal) to triangular.

2.6.2.1 Velocity Profiles   To show that these are, indeed, the flow velocity profiles obtained, we begin by assuming laminar flow in a pipe of round cross section by a fluid with shear stress rate of shear properties described by a power law equation:

$$\tau = K\gamma^n = K\left(-\frac{dv_p}{dr}\right)^n \tag{2.6.3}$$

where v_p is the local velocity (m/sec); and r, the radial coordinate (m).

The reason for the negative sign on the rate of shear, dv_p/dr, is that the radius r increases with distance from the center of the pipe. As r increases, the

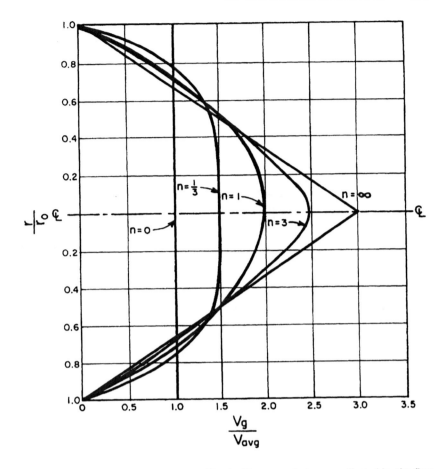

Figure 2.6.5. Laminar flow velocity profiles inside a round pipe are affected by the flow velocity index of the flowing fluid. Here, relative particle velocities are shown as they vary with location in the pipe.

rate of shear decreases for the same change of velocity across the section. Boundary values for this problem are

$$v_p = \begin{cases} 0, & r = r_0 \text{ (outer radius)} \\ v_{\max}, & r = 0 \text{ (center)} \end{cases}$$

Manipulating and integrating Eq. 2.6.3

$$\int_{r_0}^{r} \left(\frac{\tau}{K}\right)^{1/n} dr = -\int_0^{v_p} dv_p \qquad (2.6.4)$$

From momentum transfer, Eq. 2.4.15, we saw that

$$\tau = \frac{r\,\Delta p}{2L} \tag{2.6.5}$$

Thus

$$v_{\mathrm p} = -\left(\frac{\Delta p}{2LK}\right)^{1/n}\left(\frac{n}{n+1}\right)\left(r^{n+1/n} - r_0^{n+1/n}\right) \tag{2.6.6}$$

To obtain the average velocity, annuli of area $2\pi r\,dr$ are multiplied by local velocity $v_{\mathrm p}$ and integrated. The result is divided by the total cross-sectional area, πr_0^2

$$v = \frac{1}{\pi r_0^2}\int_0^{r_0} 2\pi r v_{\mathrm p}\,dr \tag{2.6.7}$$

where v is the average velocity (m/sec). Using Eq. 2.6.6,

$$v = -\frac{2}{r_0^2}\left(\frac{\Delta p}{2LK}\right)^{1/n}\left(\frac{n}{n+1}\right)\int_0^{r_0}\left(r^{2+1/n} - r_0^{n+1/n}r\right)dr \tag{2.6.8}$$

$$v = \left(\frac{\Delta p}{2LK}\right)^{1/n}\left(\frac{n}{n+1}\right)\left(\frac{n+1}{3n+1}\right)r_0^{n+1/n} \tag{2.6.9}$$

Thus

$$\frac{v_{\mathrm p}}{v} = \left(\frac{3n+1}{n+1}\right)\left[1 - \left(\frac{r}{r_0}\right)^{n+1/n}\right] \tag{2.6.10}$$

As the fluid becomes an ultradilatent, $n \to \infty$, and

$$\frac{v_{\mathrm p}}{v} = \left(\frac{3+1/n}{1+1/n}\right)\left[1 - \left(\frac{r}{r_0}\right)^{1+1/n}\right]$$

$$= 3\left(1 - \frac{r}{r_0}\right) \tag{2.6.11}$$

which is linear in r and forms a triangular velocity distribution: When the fluid is an extreme pseudoplastic, $n = 0$, and

$$\frac{v_{\mathrm p}}{v} = \left(\frac{3n+1}{n+1}\right)\left[1 - \left(\frac{r}{r_0}\right)^{1+1/n}\right] = 1 \tag{2.6.12}$$

which is constant and identical to a turbulent velocity profile. When the fluid is Newtonian, $n = 1$, and

$$\frac{v_p}{v} = \left(\frac{4}{2}\right)\left[1 - \left(\frac{r}{r_0}\right)^2\right] = 2\left[1 - \left(\frac{r}{r_0}\right)^2\right] \tag{2.6.13}$$

which forms a parabolic velocity profile.

2.6.2.2 *Kinetic Energy* We have seen in Section 2.3.2 that a correction factor α had to be incorporated within the kinetic energy term for Bernoulli's equation to account for different velocity profiles within the pipe. This must also happen for the velocity profiles associated with non-Newtonian fluids.

Using the definition of α in Eq. 2.3.13,

$$\alpha = \frac{Av^3}{\frac{1}{2}\displaystyle\int_A v_p^3\, dA} = \frac{2\pi r_0^2 v^3}{\displaystyle\int_A v_p^3\, dA} \tag{2.6.14}$$

Inserting the results from Eq. 2.6.10,

$$\alpha = \frac{2\pi r_0^2 v^3}{\displaystyle\int_A \left(\frac{3n+1}{n+1}\right)^3\left[1 - \left(\frac{r}{r_0}\right)^{1+1/n}\right]^3 v^3\, dA} \tag{2.6.15}$$

$$dA = 2\pi r\, dr \tag{2.6.16}$$

Thus

$$\alpha = \frac{r_0^2\left(\dfrac{1+n}{1+3n}\right)^3}{\dfrac{r_0^2}{2} - 3r_0^2\left(\dfrac{n}{3n+1}\right) + 3r_0^2\left(\dfrac{n}{4n+2}\right) - r_0^2\left(\dfrac{n}{5n+3}\right)}$$

$$= \frac{(4n+2)(5n+3)}{3(1+3n)^2} \tag{2.6.17}$$

For extreme pseudoplastic fluids, $n = 0$, and $\alpha = 2.0$; for Newtonian fluids, $n = 1$, and $\alpha = 1.0$; for ultradilatant fluids, $n = \infty$, and $\alpha = 20/27 = 0.74$.

Hence the Bernoulli equation already developed for Newtonian fluids flowing in a pipe (eq. 2.3.21) can be used for non-Newtonian fluids provided that the value for α in the kinetic energy term is chosen correctly. In laminar flow, the value for α is that in Eq. 2.6.17. In turbulent flow, there is a great deal of mixing, as there is for Newtonian fluids. There are no differences in velocity profiles for Newtonian or non-Newtonian fluids in turbulence, and $\alpha = 2.0$.

2.6.2.3 Friction Losses The friction factor for Newtonian fluids depends on whether there is laminar or turbulent flow in the pipe. So it is also with non-Newtonian fluids.

The Reynolds number value was taken to indicate the flow regime. However, there is no constant viscosity value from non-Newtonian fluids to insert into the expression for the Reynolds number.

generalized
Reynolds
number

A generalized Reynolds number has been defined for non-Newtonian fluids flowing in pipes

$$\mathrm{GRe} = \frac{8D^n v^{2-n} \rho}{(3 + 1/n)^n 2^n K} \tag{2.6.18}$$

where GRe is the generalized Reynolds number (dimensionless); D, the hydraulic diameter (m); v, the fluid velocity (m/sec); ρ, the fluid density (kg/m^3 or N sec^2/m^4); K, the consistency coefficient (N sec^n/m^2); and n, the flow behavior index (dimensionless).

Inserting the value of $n = 1$ for Newtonian fluids causes the generalized Reynolds number to reduce to the Reynolds number defined previously in Eq. 2.5.5.

Moody
diagram

friction factor

A Moody diagram for non-Newtonian fluids flowing in smooth pipes is given in Figure 2.6.6. It generally has the same shape as the Moody diagram for Newtonian fluids appearing in Figure 2.5.3. The friction factor for laminar flow is a straight line equal to 64/GRe; lines for the turbulent region friction factor also appear to have the same general shape as similar lines on the Moody diagram in Figure 2.5.3. Notice, however, that the parameter distinguishing the family of curves for Newtonian fluids in turbulent flow is the relative roughness, ε/d. The parameter distinguishinig the family of curves for non-Newtonian fluids in turbulent flow is the flow behavior index, n. The non-Newtonian Moody diagram cannot be used for rough conduits.

The transition from laminar to turbulent flow in pipes occurs at higher generalized Reynolds numbers for lower values of the flow behavior index. For very highly pseudoplastic fluids, this transition does not begin before

laminar to
turbulent
transition

Reynolds number values of 5,000–10,000. Fully developed turbulent flow does not occur until Reynolds numbers of about 70,000. Highly viscous non-Newtonian Reynolds number values of 70,000 or more are not likely to be achieved.

For ultradilatent fluids, the laminar-to-turbulent transition can occur at very low Reynolds numbers. Values of several hundred are possible.

Friction factor values in Figure 2.6.6 were obtained assuming that: (1) the velocity profile depends only on the flow behavior index, and (2) the laminar sublayer next to the pipe wall is relatively thin. The latter is true for Newtonian

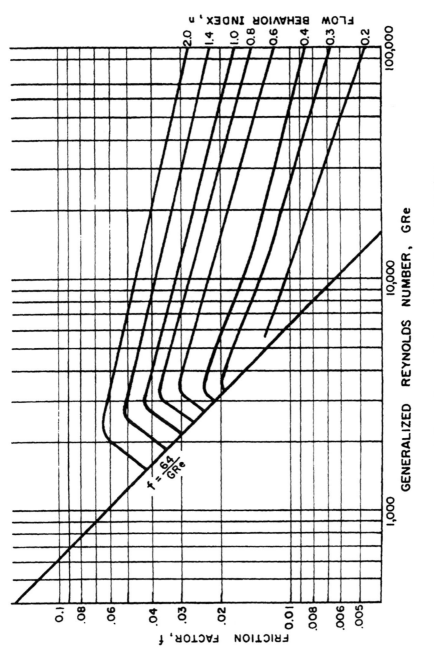

Figure 2.6.6. The Moody diagram for non-Newtonian fluids flowing in smooth pipes.

217

and pseudoplastic fluids and not true when $n \gg 1$. The equation used to calculate the lines in the non-Newtonian Moody diagram is

$$\frac{1}{\sqrt{f}} = \frac{2}{n^{0.75}} \log_{10}[GRe(f/4)^{1-n/2}] - \frac{0.2}{n^{1.2}} \qquad (2.6.19)$$

smooth pipes   When $n = 1$ and $\varepsilon = 0$, for Newtonian fluids flowing in smooth pipes, this equation reduces to the Colebrook and White equation 2.5.13. This equation has been found to underestimate the friction fractor for $n < 0.4$.

Calculation of pipe losses for rough pipes containing non-Newtonian fluids is often not necessary because many fluids exhibiting non-Newtonian properties must be kept sanitary. Smooth pipes do not contain the small pockets where stagnant fluids can accumulate and spoil as rough pipes do. Thus fluid foods, blood, and many biological suspensions are required to be pumped through smooth pipes.

rough pipes   Many biological waste products, however, exhibit non-Newtonian properties and do not merit the added expense of smooth pipes. These fluids may contain suspended particles that should be pumped with turbulence to inhibit settling. Hence there is a need at times to pump non-Newtonian fluids in turbulence through rough pipes.

Equation 2.6.19 can be modified for relative roughnesses different from zero. Using the form for the Colebrook and White equation 2.5.13 as a guide, turbulent friction factor   an estimate of the turbulent friction factor can be obtained from:

$$\frac{1}{\sqrt{f}} = 2 \log_{10} \left(\frac{GRe \, f^{(1-0.5n)}}{10(0.1/n^{1/2} + \log_{10} 4^{(1-n/2)})} + \frac{\varepsilon}{3.7D} \right) \qquad (2.6.20)$$

This equation becomes equal to the Colebrook and White equation when $n = 1$ (Newtonian fluid). An iterative procedure must be used to calculate f.

minor losses   Pipe minor losses for fittings may be calculated in a manner similar to those for Newtonian fluids, with the exception that the value for α should be that given in Eq. 2.6.17. The same loss coefficients referred to in Eq. 2.5.14 are used

$$h_f = \sum K_m v^2 / \alpha g \qquad (2.6.21)$$

entrance length   There are several ways to determine entrance length mathematically. Mishra and Singh (1976) have defined it as the point where the laminar boundary layer approaches a value equal to the tube radius. A graph of their calculated entrance length appears in Figure 2.6.7, and indicates that as the fluid becomes more pseudoplastic, entrance length decreases. This is to be expected, considering that the ending velocity profile for highly pseudoplastic fluids is very flat, and not much different in shape from the velocity profile entering a pipe.

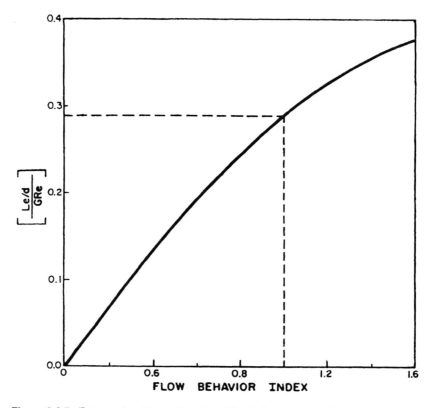

Figure 2.6.7. Entrance length as a function of flow behavior index. Entrance length for a Newtonian fluid ($n = 1.0$) is given by Eq. 2.5.25.

Total pressure drop for the entrance region for laminar flow is

$$h_f = \left[\frac{32}{GRe} \frac{L_e}{D} + 0.4 \left(\frac{(3n+1)^2}{(n+1)^2} - 1 \right) \right] \frac{v^2}{2g} \qquad (2.6.22)$$

To avoid accounting twice for the friction loss in the entrance section, the entrance length is subtracted from the total pipe length before pipe losses are calculated.

heat
exchangers

Non-Newtonian fluids are sometimes pumped through heat exchangers (Section 3.7.1). One example of this is with cardiopulmonary bypass, where blood is cooled while it is oxygenated (Galletti and Colton, 1995). Sometimes the physical embodiment of this type of heat exchanger is a smaller pipe concentrically located within a larger pipe. Friction loss calculations for the fluid in the inner pipe are no different than just discussed; friction loss for the fluid in the annulus between the pipes may be calculated from Hanks and

Larsen (1979), or may be estimated the same as for a circular pipe with hydraulic diameter calculated using a wetted perimeter including both inside and outside pipes.

EXAMPLE 2.6.2.3-1. Friction Factor in Smooth Pipes

Applesauce with 11.6% (by weight) total solids and at 27°C is to be pumped at a rate of 0.0846 m³/sec through a 15-cm-diameter smooth glass pipe. Determine the friction factor value.

Solution

We obtain values for consistency coefficient (12.7 N sec^n/m²) and flow behavior index (0.28) from Table 2.6.2. There is need also for a density value for applesauce, but this does not appear in Table 2.5.4. Thus density must be estimated from

$$\rho_{tot} = \sum X_i \rho_i$$

where ρ_{tot} is the total density (kg/m³); X_i, the mass fraction of constituent i (kg/kg); and ρ_i, the density of constituent i (kg/m³).

For this approximation, we assume 5% of total solids are sucrose and 7% are cellulose. The rest (88%) is water. We also assume that mass fractions are equal to the percentages of total solids. With these assumptions, there is no use correcting densities for temperature. From Table 2.5.4, we find the densities of sucrose and cellulose to be 1590 and about 1500 kg/m³, respectively. Thus

$$\rho_{tot} = (0.05)(1590 \text{ kg m}^3) + (0.07)(1500 \text{ kg/m}^3)$$

$$+ (0.88)(1000 \text{ kg/m}^3)$$

$$= 1065 \text{ kg/m}^3$$

The cross-sectional area of the pipe is

$$A = \frac{\pi D^2}{4} = \frac{\pi (0.15 \text{ m})^2}{4} = 0.0177 \text{ m}^2$$

and velocity is

$$v = \frac{0.0846 \text{ m}^3/\text{sec}}{0.0177 \text{ m}^2} = 4.78 \text{ m/sec}$$

giving a generalized Reynold's number of

$$\text{GRe} = \frac{8(0.15 \text{ m})^{0.28}(4.78 \text{ m/sec})^{2-0.28}(1065 \text{ kg/m}^3)}{(3+1/0.28)^{0.28}(12.7 \text{ N sec}^n/\text{m}^2)2^{0.28}} = 2830$$

From the Moody diagram, Figure 2.6.6, we see that

$$f = \frac{64}{GRe} = 0.023$$

Remark

If this had been a Newtonian fluid, a Reynolds number above 2000 would have indicated that the flow was turbulent and the friction factor would not have been equal to 64/GRe. A highly pseudoplastic fluid such as applesauce remains in laminar flow for high velocities.

EXAMPLE 2.6.2.3-2. Friction Factor in Rough Pipes

Tomato pomace is the waste left from filtering ground tomatoes to make juice and paste. This material, which is "slightly pseudoplastic," can be pumped as a slurry to a waste-handling facility. Determine the friction factor if the value of the generalized Reynolds number is 50,000 in a 15-cm-diameter cast iron pipe.

Solution

Biological materials are all too often described with nonprecise language such as "slightly pseudoplastic." A flow behavior index value will have to be assumed, but there should be no illusions about the absolute accuracy of the final result. After studying Table 2.6.2, a value of $n = 0.7$ is assumed. This problem does not require a consistency coefficient value because GRe is already specified, but a value of $K = 10 \, N \, sec^n/m^2$ could probably be justified.

From Figure 2.5.3, we see that the roughness value to use with cast iron pipe is 0.00026 m. The relative roughness is thus

$$\frac{\varepsilon}{D} = \frac{0.00026 \, m}{0.15 \, m} = 1.73 \times 10^{-3}$$

From Figure 2.6.6 we obtain a value for friction factor for smooth pipes as $f = 0.018$. This value will be used as an initial estimate in Eq. 2.6.20.

$$\frac{1}{\sqrt{f}} = 2 \log_{10} \left(\frac{(50,000)(0.018)^{1-0.35}}{10 \, (0.1/0.7^{1/2} + \log_{10} (4^{1-0.35})} + \frac{1.73 \times 10^{-3}}{3.7} \right)$$

$$= 5.66$$

$$f = 0.0312$$

Using $f = 0.0312$ as our new estimate on the right-hand side of the equation gives $f = 0.0280$. Repeating the procedure gives $f = 0.0286$, $f = 0.0285$, $f = 0.0285$. Thus a final friction factor value of $f = 0.028$ can be used.

Remark

Some biological waste products, such as tomato pomace and municipal solid waste (MSW), are highly variable in nature and do not deserve a great deal of precision when designing systems with them.

EXAMPLE 2.6.2.3-3. Entrance Length for Tomato Pomace

Determine the entrance length and the pressure drop for tomato pomace entering the 15-cm-diameter cast iron pipe from the preceding example. For this example, let the velocity be 9.5 m/sec and density, 1100 kg/m^3. Other values are as given in the previous example.

Solution

First we calculate the generalized Reynolds number from Eq. 2.6.18

$$\text{GRe} = \frac{8(0.15 \text{ m})^{0.7} \ (9.5 \text{ m/sec})^{2-0.7} \ (1100 \text{ N sec}^2/\text{m}^4)}{\left(3 + \dfrac{1}{0.7}\right)^{0.7} \ 2^{0.7}(10 \text{ N sec}^{0.7}/\text{m}^2)}$$

$$= 945$$

From Figure 2.6.7, we see that $[\text{Le}/D)/\text{GRe}] = 0.19$ corresponding to a flow behavior index of 0.7. Thus

$$\text{Le} = 0.19 \ (\text{GRe})D = 0.19(945)(0.15 \text{ m})$$

$$= 27 \text{ m}$$

Friction loss is calculated from Eq. 2.6.22 as

$$h_f = \left[32(0.19) + 0.4\left(\frac{(2.1+1)^2}{(0.7+1)^2} - 1\right) \frac{(9.5 \text{ m/sec})^2}{2(9.8 \text{ m/sec}^2)} \right]$$

$$= 32.3 \text{ m of tomato pomace}$$

Remarks

If the tomato pomace had been pumped such that turbulence existed, the entrance length would have been zero, because the flow velocity profiles at the entrance and for fully developed turbulent flow would have been nearly identical. A second remark is that a velocity of 9.5 m/sec is extremely fast and not likely to be practical.

2.7 OPEN-CHANNEL FLOW

Flow of water or fluid wastes in open channels can be the object of biological engineering design. The types of fluids can be Newtonian or non-Newtonian. Concepts and equations in the preceding sections were, for the most part, general enough to cover both pipe flow and open-channel flow conditions. The continuity equation, Bernoulli equation, momentum balance, Reynolds number, and equations for friction loss are all valid for open-channel flow. About the only modification necessary is the flow velocity distribution determination, which is different from pipe flow.

effort variable is height

The value for α, as defined in Eq. 2.3.13, and used in the Bernoulli equation and friction loss expressions, is usually taken to be 2.0 because the flow velocity distribution is assumed to be uniform in open channels. In channels of complex geometry, the value for α can decrease to as low as 1.25, and can vary abruptly from one section to another.

uniform flow velocity distribution

If the flow velocity distribution is not known, it is hard to determine the average velocity in the channel, and, consequently, hard to determine the volume flow rate in the channel. The Bernoulli equation can be used for this. Friction loss is calculated by similar means to pipe flow. The hydraulic diameter, as given in Eq. 2.5.6, is used to determine the Reynolds number, and from this the flow regime, laminar or turbulent, is ascertained in the same way as with pipe flow. If flow is turbulent, as it often is because the large hydraulic diameter value of an open channel results in a large Reynolds number value, then the friction factor value is nearly independent of Reynolds number (found on the right of the Moody diagram, Figure 2.5.3). Open channels are frequently rougher than pipes, causing relatively high friction factor values.

An alternative procedure is to use either the Chézy or Manning formula to determine average velocity. Both of these equations are empirical in nature, and suffer from strange units for the coefficients. Nevertheless, they continue to be popular means to determine velocity.

The Chézy formula is given as

Chézy formula

$$v = \frac{43.5 D \sqrt{S}}{\sqrt{D} + 2m} \qquad (2.7.1)$$

where v is the average velocity (m/sec); D, the hydraulic diameter (m); m, the absolute roughness parameter ($m^{1/2}$); and S, the hydraulic slope (dimensionless).

The hydraulic slope is calculated as the change in height of the fluid surface divided by the length along the channel. For uniform flow in the channel, the hydraulic slope is nearly the same as the slope of the channel.

The roughness parameter m assumes a measured value of $0.0674 m^{1/2}$ for smooth concrete or planed wood, $0.506 m^{1/2}$ for rubble masonry, and 1.44 $m^{1/2}$ for earth channels.

Manning's formula is similar to the Chézy formula

Manning's equation

$$v = \frac{0.127}{n} D^{2/3} S^{1/2} \sqrt{g} \qquad (2.7.2)$$

Table 2.7.1   Values of Manning's Coefficients

	n	
Surface	Min	Max
Neat cement surface	0.010	0.013
Wood-stave pipe	0.010	0.013
Plank flumes, planed	0.010	0.014
Vitrified sewer pipe	0.010	0.017
Metal flumes, smooth	0.011	0.015
Concrete, precast	0.011	0.013
Cement mortar surfaces	0.011	0.015
Plank flumes, unplaned	0.011	0.015
Common-clay drainage tile	0.011	0.017
Concrete, monolithic	0.012	0.016
Brick with cement mortar	0.012	0.017
Cast iron	0.013	0.017
Cement rubble surfaces	0.017	0.030
Riveted steel	0.017	0.020
Canals and ditches, smooth earth	0.017	0.025
Metal flumes, corrugated	0.022	0.030
Canals		
Dredged in earth, smooth	0.025	0.033
In rock cuts, smooth	0.025	0.035
Rough beds and weeds on sides	0.025	0.040
Rock cuts, jagged and irregular	0.035	0.045
Natural streams		
Smoothest	0.025	0.033
Roughest	0.045	0.060
Very weedy	0.075	0.150

where v is the average velocity (m/sec); n, the Manning coefficient ($m^{1/6}$); D, the hydraulic diameter (m); S, the hydraulic slope (dimensionless); and g, the acceleration due to gravity ($9.81 \, m/sec^2$).

Values of Manning's coefficients are found in Table 2.7.1.

Formulas for the Manning's coefficients, channel roughness (as in Eq. 2.5.13), and pipe friction factor have been combined to correlate these parameters (Vennard, 1961). The result is

$$f = \frac{5.74gn^2}{D^{1/3}} \qquad (2.7.3)$$

where f is the friction factor (dimensionless); D, the hydraulic diameter (m); g, the acceleration due to gravity ($9.81 \, m/sec^2$); and n, the Manning's coefficient ($m^{1/6}$).

Values of friction factor calculated using this equation are not very accurate, and there is no dependence on hydraulic slope, but this equation does illustrate the mathematical connection between the empirical Manning coefficients, channel roughness, and friction factor. The value of n is very insensitive to large changes in ε, however; so the correlation has limited effectiveness.

EXAMPLE 2.7-1. Open-Channel Flow

Compare the three methods (Bernoulli, Chézy, Manning) to calculate flow velocity of water in a very steep open channel diagrammed in Figure 2.7.1.

Solution

The hydraulic slope of the channel is

$$S = \frac{5}{\sqrt{100^2 + 5^2}} = 0.05$$

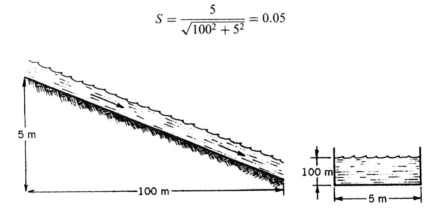

Figure 2.7.1. An open channel with hydraulic slope of 0.05 and hydraulic diameter of 5.45 m lined with smooth concrete gives a flow velocity of 19 m/sec by the Chézy formula, 23 m/sec by the Manning equation, and 26 m/sec from the Bernoulli equation.

The hydraulic diameter of this channel is

$$D = \frac{4(3 \text{ m})(5 \text{ m})}{(3 \text{ m} + 5 \text{ m} + 3 \text{ m})} = 5.45 \text{ m}$$

From the Bernoulli equation 2.3.21.

$$h_f = (h_1 - h_2) + \left(\frac{p_1 - p_2}{\gamma}\right) + \frac{(v_1^2 - v_2^2)}{\alpha g} - E$$

As long as there are no pressure and flow differences from one end of the channel to the other,

$$h_f = f \frac{L}{D} \frac{v^2}{2g} = (h_1 - h_2) = 5 \text{ m}$$

The value for friction factor is unknown, but may be obtained by iteration. The relative roughness can be obtained for concrete from Figure 2.5.3 as

$$\frac{\varepsilon}{D} = \frac{0.0003 \text{ m}}{5.45 \text{ m}} = 0.000055$$

Rather than use the Colebrook–White equation 2.5.13, we will use a graphical iteration. Let us assume that $f = 0.015$. From the Bernoulli equation

$$h_f = 5 \text{ m} = f \frac{L}{D} \frac{v^2}{2g} = (0.015)\left(\frac{100 \text{ m}}{5.45 \text{ m}}\right) \frac{v^2}{2(9.81 \text{ m/sec}^2)}$$

$$v = 18.88 \text{ m/sec}$$

From this, the Reynolds number is

$$\text{Re} = \frac{Dv\rho}{\mu} = \frac{(5.45)(18.88 \text{ m/sec})(1000 \text{ kg/m}^3)}{9.84 \times 10^{-4} \text{ kg/(m sec)}}$$

$$= 1.04 \times 10^8$$

Consulting the Moody diagram (Figure 2.5.3) again, a Reynolds number of 1.04×10^8 corresponds to a friction factor of about $f = 0.011$. We need to guess a lower value of friction factor to increase the velocity and increase the Reynolds number. Let us guess $f = 0.011$. Repeating the procedure, we get $v = 22.05 \text{ m/sec}$ and $\text{Re} = 1.22 \times 10^8$. The friction factor corresponding to this Reynolds number is 0.011, the same as our estimate. If we had used the Colebrook and White equation, our final estimates would have been $f = 0.0113$, $v = 21.8 \text{ m/sec}$, and $\text{Re} = 1.20 \times 10^8$. Since the Reynolds number is so high, the friction factor value is very insensitive to Reynolds number differences, and also to velocity differences.

The Chézy formula, Eq. 2.7.1 gives

$$v = \frac{43.5(5.45 \text{ m})\sqrt{0.05}}{\sqrt{5.45 \text{ m}} + 2(0.0674 \text{ m}^{1/2})}$$

$$= 21.5 \text{ m/sec}$$

Manning's formula, Eq. 2.7.2, with a Manning coefficient value of 0.012, from Table 2.7.1 gives

$$v = \frac{0.127}{0.012 \text{ m}^{1/6}}(5.45 \text{ m})^{2/3}(0.05)^{1/2}\sqrt{9.81 \text{ m/sec}^2}$$

$$= 23.0 \text{ m/sec}$$

Remarks

The three methods give very close agreement for the values of velocity. The channel is very steep, and the flow velocity is very high. Notice, also, that the channel has a large Reynolds number due mainly to its large size.

2.8 DESIGN PROCEDURE FOR PUMP SPECIFICATION

Engineering designs use a mix of quantitative and qualitative information and techniques. Such is the case with the design of piping systems. The size of the pipe will be determined from considerations dealing with requirements for laminar or turbulent flow, or health regulations (for food materials), or sizes commercially available, or specific information about certain sizes, etc. The length of the pipe will be determined by the physical layout of the location where they will be installed, including items such as routing pipes along walls, ceilings, or floors, and distances from fluid supplies to points of use. Types and numbers of fittings depend on health regulations, type of pipe, standard pipe section lengths, physical pipe layout, and points of control. In piping system design, the most common item requiring calculations is a determination of pump size to assure adequate flow, and adequate flow is usually determined from machinery specifications, economic considerations, or time constraints. Thus piping system design is largely determined by nonquantitative inputs.

piping
systems

Pumps are normally specified in terms of flow capacity, discharge pressure developed, suction pressure required, power requirement, and special considerations such as: sanitary pumps (for food or biotech materials), self-priming pumps (can pump air until liquid arrives at the pump), pumps to handle abrasive liquids or liquids with suspensions of abrasive solid particles (for waste systems), or pumps especially gentle on fluids (for blood and

kinds of
pumps

suspensions of microorganisms). Pump flow capacity is determined usually by the problem specification; discharge pressure is determined largely by pipe friction and required pressure at the outlet of the pipe; suction pressure is determined mostly by the pump location (it is even possible that no pump is

pump power

necessary if gravity can provide sufficient flow); pump power requirement is determined by the product of combined suction and discharge pressure with flow rate. As long as volumetric flow rate is used, this product has dimensions of pressure times volumetric flow, which has dimensions of $\triangleq (F/L^2)(L^3/T)$ $\triangleq FL/T$; these are the dimensions of power. If Bernoulli's equation 2.3.21 is used to calculate pump energy requirements, a different approach must be taken because the terms in Eq. 2.3.21 have units of height of a fluid. Thus weight rate of flow must be used in the power calculation to give dimensions of height times weight rate of flow $\triangleq (L)(F/T) \triangleq FL/T \triangleq$ power. This process is diagrammed in Figure 2.8.1.

water power

Pumps are often specified in terms of water power. For fluids different from water the power calculated in the procedure above must be modified by multiplying by the specific gravity of the fluid being pumped. Thus less dense liquids would use less power to pump than water, all else being equal.

efficiency

Pumps, like other electrical and mechanical components, are less than 100% efficient. There is inefficiency associated with a conversion of primary power (for instance, electricity) to mechanical power, and there is inefficiency involved in a conversion of mechanical power to pumping power. The efficiency of electrical motors is about 75–95%, and the mechanical efficiency of pumps is about 60–80%. Since efficiencies less than 100% translate into greater input power requirements for the same output power production, the ideal pump power must be divided by efficiency to determine true power requirement.

Specification of pump power requires the mechanical efficiency only (unless the pump and electrical motor are supplied as a unit), whereas finding electrical power input requires both efficiencies to be used. Engineers are frequently required to specify both pump power and electrical power.

pump over-specification

Some pump overspecification is usually justified on the basis that conditions inside pumps and pipes may not always remain as pristine as assumed for the original design. The penalty paid for specifying higher delivered pressure, flow, or power than the minimum required comes in the cost of the pump and some increment in operating costs. Contrarily, a pump that does not perform to the minimum requirements must be replaced and represents wasted resources.

check transient conditions

Be sure to check transient conditions when specifying pump power. These are especially likely to occur during startup, and, sometimes, shutdown. For instance, a section of pipe consisting of two vertical risers and a horizontal section does not require any special consideration for height as long as the pipe is filled (Figure 2.8.2). This section of pipe acts as a siphon, and the only height difference that matters is the difference between the two lower parts of the pipe. Before the pipe is full, however, it does not act as a siphon, and the

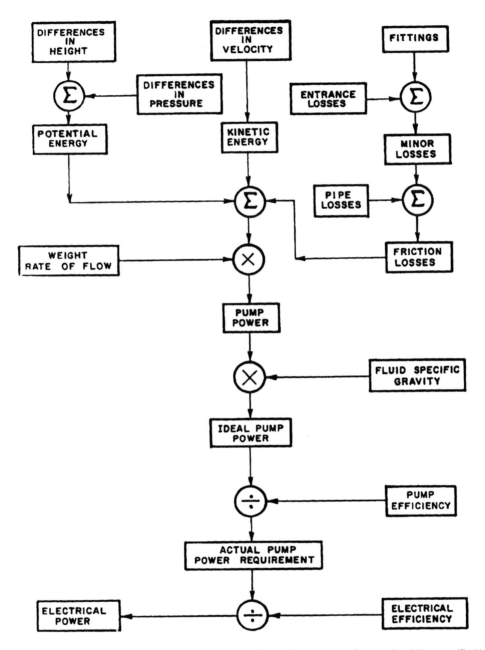

Figure 2.8.1. Schematic of the pump specification design process. Pumps should be specified by volume flow rate, delivery pressure, and power, along with other special considerations such as suction ability, cleanliness, and ability to deal with suspended particles.

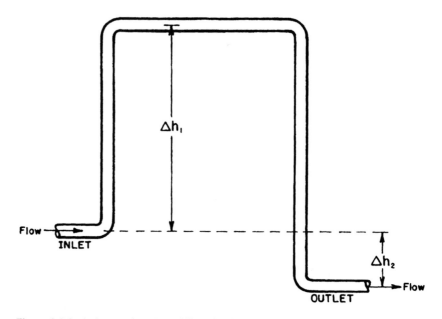

Figure 2.8.2. A pipe section shaped like this will act as a siphon as long as the pipe is full. Liquid will flow out the right end even without a pump because Δh_2 is negative. A pump is required on the inlet side in order to fill the pipe because Δh_1 is positive.

pump discharge pressure requirement may be based entirely on the initial requirement to fill the pipe.

pumping non-Newtonian fluids

The design procedure for pumps carrying non-Newtonian fluids (Figure 2.8.3) is basically the same as for pumps carrying Newtonian fluids (Figure 2.8.1). Bernoulli's equation is used to determine the required amount of pressure and power that must be delivered by and to the pump in order to deliver the correct flow rate. Fluid physical parameters K and n are used to determine the generalized Reynolds number, from which comes a determination of laminar or turbulent flow. Smooth or rough pipes must also be decided if flow is turbulent. Friction factors are either graphically or mathematically determined, and friction losses are calculated. Eventually, pump specifications emerge.

Different pump types have different characteristics that determine their specific applications (Table 2.8.1). According to one classification, pumps may be either reciprocating or rotary. According to another classification, they may be either positive displacement or variable displacement. We will follow the latter.

piston pump

Conceptually the simplest positive displacement pump is a piston pump, which is formed by a piston within a cylinder. When the piston evacuates the cylinder, fluid is drawn into the cylinder; when the piston invades the cylinder, fluid is forced out (Figure 2.8.4). Flow direction is controlled by valves timed

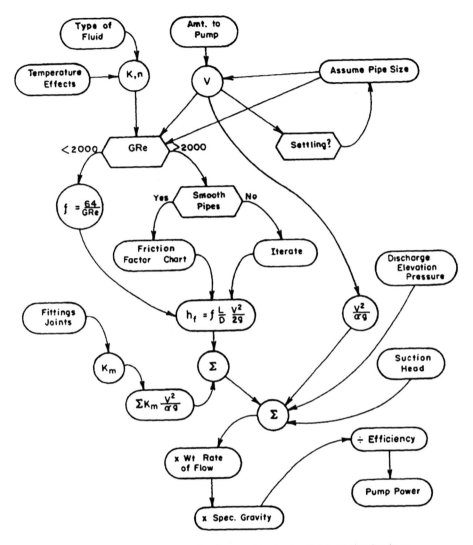

Figure 2.8.3. Design procedure for non-Newtonian fluids flowing in pipes.

so that the intake valve opens as the piston is drawn away from the cylinder head and the discharge valve is open as the piston moves toward the cylinder head. These valves often take the form of flaps that open when pressure on one side is lower than on the other side. Both valves cannot be open simultaneously. An action somewhat similar to a piston pump occurs in a gasoline automobile engine.

As long as the viscosity of the pumped fluid is not so great as to prevent the cylinder from filling completely during each stroke of the piston, the piston pump will deliver a fixed amount of fluid during each stroke (the rate of filling

Table 2.8.1 Pump Characteristics

Type	Characteristics	Fluids Pumped	Percent Pumping Efficiency
Piston	Positive displacement High pressure Pulsatile Fixed volume pumped High efficiency	Nonabrasive Low viscosity Clean, no large suspended particles	60–80%
Rotary gear	Positive displacement Relatively high pressure Nearly steady flow	Nonabrasive Relative high viscosity Small included particles	50–70%
Centrifugal	Variable displacement Low to intermediate pressure Steady flow	Abrasive Large suspended particles Viscous fluids	30–40%
Jet	Variable displacement Better suction than centrifugal Low efficiency Steady flow	Nonabrasive Low viscosity Clean fluids	20–30%
Regenerative Turbine	Variable displacement High suction High pressure Steady flow	Nonabrasive Low viscosity Clean	25–35%

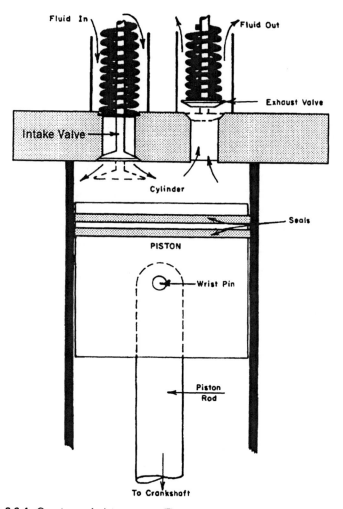

Figure 2.8.4. One type of piston pump. The amount of fluid delivered for each stroke depends on the piston area and the stroke length.

of the cylinder depends upon the pressure drop across the inlet valve: inlet valves must be larger to pump more viscous fluids). Thus the piston pump acts as an ideal flow source, delivering a certain amount of fluid almost independent of the pressure at the pump outlet. In fact, the pressure at which flow can be delivered by the piston pump is limited only by mechanical factors such as the breaking strengths of connecting rod, cylinder walls, or piston, or by the power source driving the pump. The outlet of a piston pump can never be shut off without a spring-loaded safety valve somewhere in the pipe between the pump and the shutoff valve. To violate this is to invite certain disaster.

intermittent
flow

Strong piston pumps can therefore be used to pump fluids at very high pressures. However, the intermittent flow from the pump is sometimes extremely undesirable. Some piston pumps are designed with two connected opposed pistons and cylinders so that pumping occurs from one cylinder while fluid is being drawn into the other. Such a pump is called *double acting*. Other pumps are constructed with pistons having pumping strokes offset from one another. Simplex, duplex, triplex, quadruplex, etc., pumps are those that pump once, twice, three times, four times, etc., for each revolution of the crankshaft. The resultant flows from each of the pistons is summed in the pipe to result in steadier flow rates.

piston fillers

Because piston pumps deliver a known volume of fluid on each stroke, they are often used to fill containers with a measured amount of liquid. Many food containers are filled this way. Containers of oil, detergent, liquid medicines, beer, and virtually all other containers carrying a known volume of liquid are probably filled with piston fillers.

no abrasive
liquids

A piston pump depends upon good sealing between the valves and valve seats (where the valves rest when they are closed), and between the piston and cylinder wall. Thus the piston pump is not used with abrasive liquids or with liquids containing large suspended particles.

rotary gear
pump

An example of a rotary positive-displacement pump is the rotary gear pump diagrammed in Figure 2.8.5. As the two gears rotate together, they form a set of fixed volumes in the spaces between the teeth that move fluid from the low-pressure inlet side to the high-pressure outlet. Seals are made between the ends of the gear teeth and the outer casing, and between mating gear teeth to contain the fluid and prevent leakage. This type of pump is used for the oil pumps in most automobiles. It has the advantage of being both positive displacement and rotary (simpler mechanics and fairly steady flow), but it becomes less like an ideal flow source as it wears.

centrifugal
pump

The centrifugal pump is the workhorse of the pump world. Its workings are a study in energy transformations. Fluid enters the center of a rotating impeller that contains radial vanes (Figure 2.8.6). Kinetic energy is added to the fluid as it accelerates along the vanes in response to the centrifugal force. This kinetic energy is then transformed into potential energy (pressure) when it hits the stationary vanes of the diffuser. and rapidly decelerates.

variable
displacement

Unlike the positive-displacement pumps previously described, there is no imperative to fluid movement in a centrifugal pump. Hence the outlet pipe may be shut off without fear of immediate breakage of the pump. The flow would stop, but the impeller would continue rotating. The energy added to the fluid would not become pressure energy, but would instead become thermal energy. After a while, the pump and fluid would become hot. This is an example of a variable-displacement pump.

Centrifugal pumps can be used with any kind of fluids, including abrasive liquids, liquids containing large suspended particles, and viscous liquids. They

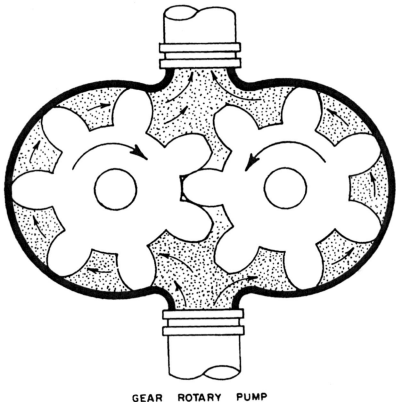

GEAR   ROTARY   PUMP

Figure 2.8.5. Diagram of an external-gear rotary pump. This pump is a positive displacement pump as long as parts fit tightly.

are extremely tolerant of operating conditions, automatically adjusting flow to pressure at the outlet side. They are found in many applications.

jet pump   A centrifugal pump is an essential component of a jet pump, which well illustrates Bernoulli's equation in action (Figure 2.8.7). The part that does the pumping is a narrow section of pipe called a Venturi. The centrifugal pump recycles part of the liquid output of the Venturi section and injects it through a nozzle into the throat of the Venturi. According to Bernoulli's equation, high velocities accompany low pressures, and the low pressure in the Venturi throat pulls liquid from the source and sends it to the outlet.

Because part of the output fluid is recycled, the pumping efficiency (equivalent power output, weight rate of flow times pressure difference between output and input, divided by shaft power input) is low, only 20–30%. This is the penalty paid for the complication of the jet added to the centrifugal pump. The advantage obtained is higher suction than the centrifugal pump is usually capable of. A very inexpensive centrifugal pump may be used as a jet pump component.

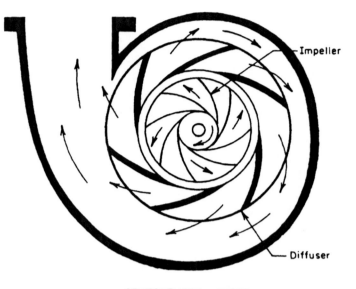

CENTRIFUGAL   PUMP

Figure 2.8.6. A typical diffuser-type centrifugal pump. This type of pump is the most common because it can be used in many situations.

Jet pumps work best for low-viscosity fluids that attain high velocities in the Venturi. They are most often used for pumping water in home wells. Wells deeper than about 6.7 m require the jet to be placed away from the centrifugal pump and near the level of the water.

turbine pump

The regenerative turbine pump has an interesting action (Figure 2.8.8). Both the moving impeller and the stationary stator have small vanes that help form many small pockets; the stator and impeller are sealed together by carefully machined tongues and grooves. The fluid enters at the side, not in the center as with the centrifugal pump. Fluid in contact with the impeller is thrown outward by centrifugal force, and then reaches the stator, where its direction is turned so that it is reintroduced to the part of the impeller closest to the shaft. Again the fluid is forced outward by centrifugal force and turned by the stator. This cycle is repeated many times as the fluid travels from the inlet to the outlet in a circular corkscrew pattern. There are thus many opportunities to impart energy to the fluid.

This type of pump can produce high pressures, and can produce high amounts of suction. The turbine pump can be used to compress air, and is

self-priming

therefore self-priming. Because they depend very much on the integrity of their seals, they cannot be used with abrasive liquids.

other pumps

There are other types of pumps used for specialized applications. Peristalic pumps and roller pumps squeeze flexible tubes containing the fluids for those applications where the fluids cannot contact the pumps (Figure 2.8.9). Screw-

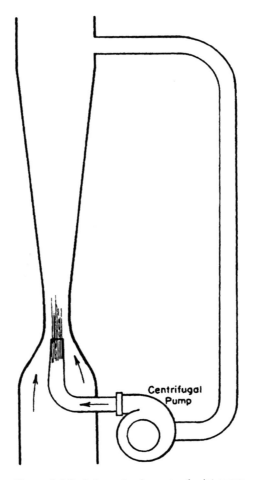

Figure 2.8.7. Schematic elements of a jet pump.

type pumps and specialized piston pumps can even pump granulated solids. Diaphragm pumps are used often in sprayers. Pumps must sometimes be used with suspensions of cells and cannot disrupt the cells. Whole blood is particularly sensitive to disruption of the red blood cells because when the contents of the red blood cells are mixed with plasma, the resulting fluid thickens to the point where it can strain the heart to pump it. Whole blood is thus pumped either with peristaltic pumps or propelled by compressed gas. For specialized applications it is best to consult with pump manufacturers before a pump selection is made.

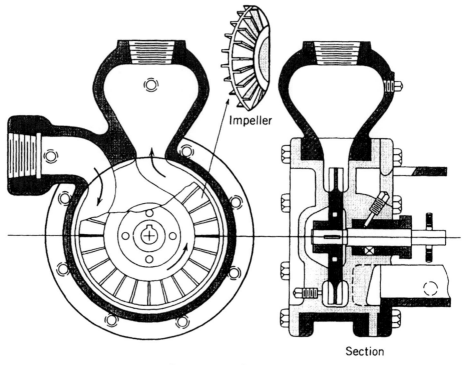

Impeller

Section

Regenerative Turbine Pump

Figure 2.8.8. Diagram of a regenerative turbine pump. This pump can be used to pump air as long as its seals are intact.

EXAMPLE 2.8-1. Pumping Applesauce

Specify the pump required for applesauce flowing full through a 15-cm-diameter insulated stainless-steel pipe 50 m long containing 4 elbows and discharging horizontally into a tank with its bottom 1 m below the level of the pump and its top 2 m above the pump level. The applesauce (32% total solids) flows at a rate of $0.102 \, \text{m}^3/\text{sec}$. Its temperature is 80°C.

Solution

Because this is a food material, a sanitary centrifugal pump will be used. We are not told specifically about the suction requirements of the pump, but normally these should be checked. The pump must provide enough power to the fluid to overcome pipe resistance and pressure in the tank at the end of the pipe.

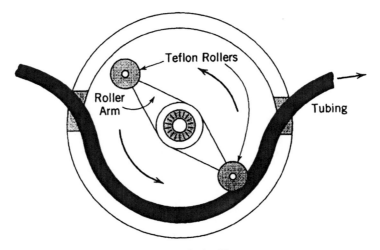

Figure 2.8.9. Peristaltic pump.

From Table 2.6.2, the flow behavior index is found to be 0.4 and the consistency coefficient (at 25°C) is 22.0 N sec^n/m^2. This must be corrected for temperature. From Table 2.4.4, the viscosity of water at 80°C is 3.565×10^{-4} kg/(m sec). Water viscosity is 9.011×10^{-4} kg/(m sec) at 25°C. Thus from Eq. 2.6.2

$$K_{80} = K_{25}(\mu_{80}/\mu_{25}) = (22.0 \text{ N sec}^n/\text{m}^2)(3.565 \times 10^{-4}/9.011 \times 10^{-4})$$

$$= 8.70 \text{ N sec}^n/\text{m}^2$$

Velocity in the pipe is, by Eq. 2.2.4

$$v = \dot{V}/A = (0.102 \text{ m}^3/\text{sec})/[\pi(0.15 \text{ m})^2/4] = 5.75 \text{ m/sec}$$

From Eq. 2.6.18, the generalized Reynolds number is

$$\text{GRe} = \frac{8D^n v^{2-n}\rho}{(1/n + 3)^n 2^n K} = \frac{8(0.15 \text{ m})^{0.4}(5.75 \text{ m/sec})^{1.6}(1100 \text{ kg/m}^3)}{\left(\dfrac{1}{0.4} + 3\right)^{0.4} 2^{0.4}(8.70 \text{ N sec}^n/\text{m}^2)}$$

$$= 2980$$

This is still laminar, despite the generalized Reynolds number above 2000, because of the low flow behavior index (Figure 2.6.6). Also, if the pipe had not been insulated, we would have had to account for nonisothermal flow in the pipe.

The pipe friction factor for laminar flow is

$$f = 64/GRe = 0.0215$$

Pipe resistance is given by Eq. 2.5.39 as

$$R = \frac{\rho v(f L/D + \sum K_m)}{\alpha A}$$

The value for α is obtained from Eq. 2.6.17 as

$$\alpha = \frac{(4n + 2)(5n + 3)}{3(1 + 3n)^2} = \frac{(1.8 + 2)(2.0 + 3)}{3(1 + 1.2)^2} = 0.964$$

The value for each K_m is 0.9 (Table 2.5.1), giving a $\sum K_m = 3.6$. Pipe resistance is

$$R = \frac{(1100 \text{ kg/m}^3)(5.75 \text{ m/sec})[0.0215(50 \text{ m})/(0.15 \text{ m}) + 3.6]}{(0.964)[\pi(0.15 \text{ m})^2/4]}$$

$$= 400 \times 10^6 \text{ N sec/m}^5$$

Pipe friction loss is

$$R\dot{V} = (4.00 \times 10^6 \text{ N sec/m}^5)(0.102 \text{ m}^3/\text{sec}) = 408 \text{ kN/m}^2$$

We assume the pump must accelerate a standing pool of applesauce at its source. Thus pressure developed by the pump to sustain kinetic energy is

$$KE = \frac{\gamma v^2}{\alpha g} = \frac{\rho v^2}{\alpha} = \frac{(1100 \text{ kg/m}^3)(5.75 \text{ m/sec})^2}{0.964} = 37.7 \frac{\text{kN}}{\text{m}^2}$$

Pressure energy to be added at the tank varies as applesauce level in the tank varies. The maximum energy corresponds to the applesauce level 2 m above the pump. Pressure added by the pump to raise the level of applesauce is thus

$$PE = \gamma h = \rho g h = (1100 \text{ kg/m}^3)(9.8 \text{ m/sec}^2)(2 \text{ m})$$

$$= 21.6 \text{ kN/m}^2$$

A pump with 100% efficiency would thus be required to have a power rating of

$$(408 + 37.7 + 21.6 \text{ kN/m}^2)(0.102 \text{ m}^3/\text{sec}) = 47.6 \text{ kN m/sec}$$

The efficiency of a centrifugal pump is about 60–80%. To be conservative, we shall use the 60% figure. Thus pump power is $(47.6\,\text{kN}\,\text{m}/\text{sec})/0.6 = 79.4\,\text{kN}\,\text{m}/\text{sec}$.

The pump specifications are thus:

Pressure: $467\,\text{kN}/\text{m}^2$

Volume flow rate: $0.102\,\text{m}^3/\text{sec}$

Power: $79.4\,\text{kN}\,\text{m}/\text{sec}$

Type: centrifugal, sanitary

EXAMPLE 2.8.2. Power Required for an Artificial Heart

Artificial hearts and ventricular assist devices may be used temporarily to assist the heart during cardiogenic shock or as a bridge to transplant. Calculate the external electrical power requirement for operating an artificial heart.

Solution

Determination of power required to operate an artificial heart can be made in several ways. The simplest way is to multiply maximum systemic blood pressure delivered by the heart (systolic pressure) by the maximum flow rate (cardiac output). This gives a very rough approximation to the power requirement. Systolic pressure is usually about $16\,\text{kN}/\text{m}^2$ and cardiac output at rest is about $107 \times 10^{-6}\,\text{m}^3/\text{sec}$. Thus the hydraulic pumping power requirement is

$$\text{Power} = (16 \text{ kN/m}^2)(107 \times 10^{-6} \text{ m}^3/\text{sec})$$

$$= 1.71 \text{ N m/sec}$$

The overall mechanical and electrical efficiency (η) of the artificial heart is about 20%. Thus the electrical power that must be available for heart operation is

$$\text{Electrical power} = (1.71 \text{ N m/sec})/0.20$$

$$= 8.6 \text{ N m/sec}$$

During exercise, cardiac output can increase to $250 \times 10^{-6}\,\text{m}^3/\text{sec}$, and systolic pressure can increase to $21\,\text{kN}/\text{m}^2$. Total power requirements under this condition would be

$$\text{Electrical power} = (21 \text{ kN/m}^2)(250 \times 10^{-6} \text{ m}^3 \text{ sec})/0.2$$

$$= 26.3 \text{ N m/sec}$$

These figures are about the same as required by natural hearts, which have muscular efficiencies of 3–20%.

The above calculation does not account for any of the details of the power requirements. To show some of the simplest of these, the vascular system is assumed to be able to be represented by a compliance in parallel with a resistance (Figure 2.8.10). The heart delivers blood flow to the vascular system at the aortic pressure, which is assumed to be one-half a sine wave with maximum amplitude of 18 kN/m^2 during ejection (Figure 2.8.11). During diastole (the time between contractions when the heart is refilling) the pressure is assumed to be zero.

Aortic pressure is given by

$$p_a = R_p \dot{V}_r + V_c/C_p = 18 \sin(2\pi t/2T)$$

where p_a is the aortic pressure (kN/m^2); R_p, the peripheral vascular resistance (about 577×10^6 N sec/m^5); C_p, the peripheral vascular compliance (about 1.44×10^{-9} m^5/N); $\dot{V}$, the blood flow through resistance (m^3/sec); V_c, the volume of blood stored in compliance (m^3); T, the period of ejection (assumed to be 0.5 sec); and t, the time (sec).
Since

$$\dot{V} = \dot{V}_r + \dot{V}_c$$

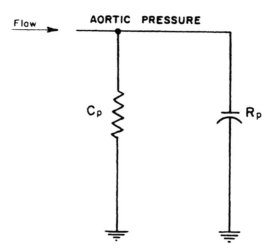

Figure 2.8.10. Diagram of a simplified model of the human vascular system.

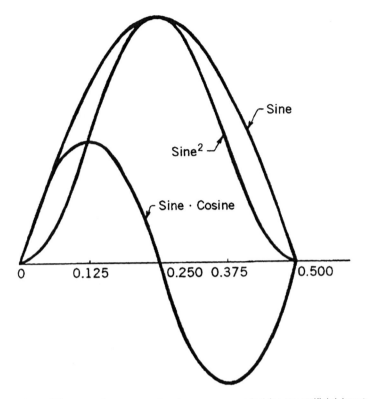

Figure 2.8.11. Diagram of components of power expended by an artificial heart with sinusoidal ejection.

where $\dot{V}$ is the total blood flow rate (m³/sec); and $\dot{V}_c$, the blood flow through compliance (m³/sec).
Then

$$p_a = R_p(\dot{V} - \dot{V}_c) = R_p\dot{V} - R_p\dot{V}_c = R_p\dot{V} - R_pC_p\frac{dp_a}{dt}$$

$$\dot{V} = p_a/R_p + C_p\frac{dp_a}{dt}$$

Power is thus

$$\text{Power} = p_a\dot{V} = p_a^2/R_p + C_pp_a\frac{dp_a}{dt}$$

$$\frac{dp_a}{dt} = (2\pi)(18)\cos(2\pi t) = 113\cos 2\pi t \text{ kN/(m}^2 \text{ sec)}$$

Thus

Electrical power

$$= p_a \dot{V}/\eta = \frac{18^2 \sin^2(2\pi t)(\text{kN/m}^2)^2}{(577 \times 10^6 \text{ N sec/m}^5)(0.2)}$$

$$+ \frac{(1.44 \times 10^{-9} \text{ m}^5/\text{N})[18 \sin(2\pi t) \text{ kN/m}^2][113 \cos(2\pi t) \text{ kN/(m}^2 \text{ sec})]}{0.2}$$

$$= 2.80 \sin^2(2\pi t) + 14.6 \sin(2\pi t) \cos(2\pi t) \text{ N m/sec}$$

Thus, unlike the previous simple analysis, using an elementary vascular model to estimate required power demonstrates that there are two power components. The sine-squared term is unipolar, but the sine–cosine term is bipolar. Over the time of ejection there is no net power delivered to the system compliance: Power delivered by the heart at the beginning of ejection is returned to the system at the end of ejection. That is the nature of energy-storage components.

The peak power requirement, however, occurs when power delivered to the resistance and power delivered to the compliance are both positive during the first half of ejection. The maximum value of power occurs sometime after the point where the sine–cosine reaches its maximum because the sine-squared term rises faster than the sine–cosine falls. To find this maximum power point, we take the derivative and set it to zero

$$\frac{d(\text{power})}{dt} = (2)(2\pi)(2.80)(\sin 2\pi t) \cos(2\pi t)$$

$$+ (2\pi)(14.6)[\cos^2(2\pi t) - \sin^2(2\pi t)] = 0$$

Using the trigonometric identities

$$\sin 2\phi = 2 \sin \phi \cos \phi$$

$$\cos 2\phi = \cos^2 \phi - \sin^2 \phi$$

then the above equation becomes

$$2.80 \sin(4\pi t) + 14.6 \cos(4\pi t) = 0$$

or

$$\tan(4\pi t) = -5.21$$

$$4\pi t = 1.38 \pm \pi k/4 \text{ rad}, \qquad \text{where } k = 1, 2, 3, 4, \ldots$$

Because the tan 2ϕ repeats with a period of $\pi/4$, there are many potential solutions at intervals of $\pi/4$ radians. The correct one is found by numerical means to be at $(\pi - 1.38)$ radians. Thus

$$4\pi t = (\pi - 1.38)$$

$$t = 0.14 \text{ sec}$$

and the power at that point is 8.83 N m/sec. Hence, if the required power were to be calculated only as Power $= p_a^2/R_p$, the maximum required power would be underestimated by nearly 4 times.

PROBLEMS

2.1-1. **Pressure tank walls.** A cylindrical tank contains a pressurized gas. The ends of the tank are rounded outwardly convex. Why must the side walls be made thicker than the ends?

2.1-2. **Sump pump.** Draw a system diagram of a sump pump discharging through a hose to the ground. The pump can be considered to be a nonideal source. Be sure to include the sump in your diagram.

2.1-3. **Plants without capillarity.** If there were no capillarity in plants, what would be the maximum height that water could be drawn in plant xylem?

2.1-4. **Equations for resistance, capacity, inertia.** Write the definitional equations for fluid resistance, capacity, and inertia. Give a biological example of each of these elements.

2.1-5. **Circulatory system.** Give a systems diagram for the flow of blood in the circulatory system. What is the flow variable? What is the effort variable? What kind of element is the heart? Where are resistance, capacity, and inertia located?

2.1-6. **Fish gills.** Give a systems diagram for flow of water through the gills of a finfish. What considerations are there to calculating the water flow rate?

2.1-7. **Hemodialysis.** There are two general schemes for external hemodialysis: continuous arteriovenous hemodiafiltration (CAVHD) and continuous venous–venous hemodiafiltration (CVVHD). The latter uses an external pump to propel the blood. Under what conditions do you think each type would be used?

2.1-8. **Tractor tire inflation pressure.** The critical compaction pressure on a certain soil is $55 \, kN/m^2$. The average contact pressure on the soil produced by a tire is always 7 to $14 \, kN/m^2$ higher than the inflation pressure in the tire (Wiley, 1995). For a tractor weighing 100 kN, what is your recommendation for tire pressure to avoid soil compaction? What is the approximate contact area between each tire and the soil?

2.2-1. **Fluid velocity.** A 30-m-long horizontal pipe with $6.5 \, m^3/sec$ flowing full of liquid has a diameter of 5 cm. If the pipe suddenly expands to

25 cm in a second 30-m section, what is the average velocity in the second section?

2.2-2. **Stents.** A stent is a cylindrical piece of expandable metal mesh for insertion into an atherosclerotic artery with an angioplasty balloon. The stent keeps the artery from recoiling and partially collapsing after intraluminal dilatation of atherosclerotic arterial stenosis. Assume the coronary artery is supplied by an ideal flow source. What are the blood flow velocities before and after dilatation and stenting if the artery lumen has an effective diameter of 1 mm before and 3 mm after the procedure? Cardiac blood flow is about 250 mL/min.

2.2-3. **Aortic blood velocity.** If the cardiac output at rest in a typical human male is 5 L/min, determine the average blood velocity in the aorta that has a lumen diameter of 2.5 cm. Why is the velocity in the aorta not likely to be this value?

2.2-4. **Air flow in the trachea.** The tracheal diameter in a normal human is about 18 mm. Normal resting minute volume is 6 L/min. What is the average air velocity in the trachea? Use your knowledge about breathing to estimate air velocity.

2.2-5. **Use of elemental continuity equation.** Suggest a biological situation where the use of the elemental continuity equation would be appropriate.

2.2-6. **Ventilation systems.** Supply of fresh air to humans, animals, plants, or microbes in closed spaces can be very important to maintaining health. Why must caution be used when applying the continuity equation to ventilation systems?

2.2-7. **Flood plain design.** A small stream has a channel 10 m wide by 0.6 m deep, and normally carries an acceptable flow of 0.86 m³/sec. During storms the volume flow rate in the stream can increase by 8 times. Suggest a width and depth of a constructed flood plain (bounded by levees on both sides) to carry the additional water without an increase in the flow velocity.

2.2-8. **Zebra mussels.** Zebra mussels have become an economically expensive and ecologically damaging nuisance in many waterways. They can completely clog intake pipes up to 600 cm in diameter. Zebra mussels cannot attach to water surfaces if flow velocities attain or exceed 1.5 m/sec. For a water intake that must supply 12 m³/sec, what is the maximum pipe diameter that prevents Zebra mussel attachment?

2.3-1. **Giraffe blood pressure.** A giraffe is about $5\frac{1}{2}$ m high. If the mean arterial blood pressure at the level of the heart at 3.5 m above the ground is 28.7 kN/m² (215 mm Hg), what is the blood pressure in the giraffe's feet? What is the brain blood pressure when it is standing up tall? What adaptations by the giraffe minimize physiological effects of the high pressure?

2.3-2. **Pressure and flow in a pipe.** A 30-m-long horizontal pipe with $6.5\,cm^3/sec$ flowing full of water has a diameter of 5 cm. If the pipe suddenly expands to 25 cm in a second 30-m section, how does the average velocity change? If the pressure in the first section of pipe is $5 \times 10^{-3}\,N/m^2$, what is the pressure in the second section (neglecting friction)?

2.3-3. **Velocity profile correction.** Calculate the value for α if the flow velocity profile in a round pipe is triangular, where

$$v = v_{max}\left(1 - \frac{r}{R}\right)$$

2.3-4. **Squid friction.** The squid in Example 2.3.1-2 is propelled at a speed of 2 m/sec by the water jet. Calculate the energy lost due to friction through the water.

2.3-5. **General energy balance.** Give a biological example where the use of the general energy balance would be preferable to the use of the Bernoulli equation.

2.3-6. **Two cracks.** Liquid flows in a tube at subatmospheric pressure. There are two cracks in the tube, one where a crimp in the tube wall constricts the liquid flow and the other where the tube is its original diameter. Through which crack will more air leak into the tube?

2.4-1. **Temperature dependences of viscosities.** Are viscosities of liquids or of gases more temperature dependent? Why do you say this?

2.4-2. **Bubbles rising through water.** Bubbles 1 mm in diameter are being formed at the bottom of a fish culture tank 2 m deep. The water is maintained at 22°C. How much air at atmospheric pressure is in each bubble? What is the terminal velocity of the bubbles as they rise through the tank?

2.4-3. **Ventricular filling.** Repeat Example 2.4.5-1 with the relationship between central streamline velocity and mitral valve velocity given by: $v(x)/v_{mv} = -x^2/7 + x/2$. Find the pressure drop between atrium and ventricle.

2.4-4. **Settling of particles.** Consult Table 2.5.4 for densities and Table 4.7.2 for sizes of particles and calculate the times it would take for the following particles to fall a distance of 2 m

raindrops through air

pollen through air

insecticide dust through air

viruses through air

red blood cells in the most streamlined orientation through plasma

2.4-5. Viscosity of methane. Calculate the value of viscosity of methane at a temperature of 100°C.

2.4-6. Viscosity of water. Calculate the value of viscosity of water at a temperature of 32°C. What is the kinematic viscosity value?

2.4-7. Viscosity of air. Calculate the value of the viscosity of air at a temperature of 15°C.

2.4-8. Holding a balloon. Calculate the force required to hold a helium-filled balloon in a wind of 5 m/sec. The balloon is spherical with a diameter of 20 cm. Densities can be obtained from Table 2.5.4. Determine the angle of the string.

2.4-9. Homogenized milk. Graph droplet size against the time for a droplet of butterfat to rise to the top of a gallon container of milk. What is the maximum droplet size that you would choose for homogenized milk?

2.4-10. Rotational viscometer. A Brookfield rotational viscometer consists of two concentric cylinders of diameters 2.515 and 2.762 cm and effective length of 9.239 cm. The fluid to be tested is placed in the annulus between the cylinders, and the outer cylinder is made to rotate while the inner cylinder remains stationary. The torque required to rotate the outer cylinder is

$$T = 4\pi\mu L\omega r_o^2 r_i^2 / (r_o^2 - r_i^2)$$

where T is torque (N m); L, the cylinder height (m); ω, the rotational speed (rev/sec); r_o, the outer cylinder radius, (m); and r_i, the inner cylinder radius (m). The average shear rate in the fluid is

$$\gamma = \frac{\omega r_o^3 (r_o + 2r_i)}{2(r_o^2 - r_i^2)^2}$$

The following table of values was obtained for a non-Newtonian fluid. Determine values for flow behavior index and consistency coefficient.

rpm	0.5	1	2.5	5	10	20	50
torque $(10^{-7}$ N m)	86.2	168.9	402.5	754	1365	2379	4636

2.4-11. Shear stress in pipe. Calculate the shear stress at the wall in a 27-m length of pipe 5 cm in diameter, if the pressure drop in the pipe section is $80,000 \, N/m^2$. What is the shear stress in the center of the pipe?

2.5-1. **Hydraulic diameter for an annulus.** Give the hydraulic diameter for fluid flowing in an annulus formed by two concentric cylindrical pipes.

2.5-2. **Reynolds number and friction factor.** Find the Reynolds number for waer at 20°C flowing in a square pipe one-third full. The pipe side is 15 cm, and the water flows at a rate of 2 m/sec. Is the flow laminar or turbulent? If the pipe inside is smooth, what is the friction factor for the pipe?

2.5-3. **Tracheal hydraulic diameter.** Below is a drawing of the human trachea during normal breathing and during a cough (Comroe, 1965).

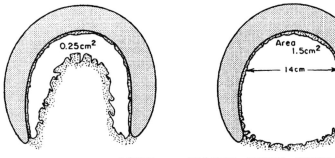

TRACHEA DURING COUGH   TRACHEA DURING NORMAL BREATHING

Positive intrathoracic pressure inverts the noncartilaginous portion of the intrathoracic trachea and decreases its cross-sectional area from 1.5 to 0.25 cm^2. There is also an increase of volumetric flow rate from about 1 to 7 L/sec. Estimate the hydraulic diameter of the trachea in each case. Calculate the Reynolds numbers for each condition and determine if flow is laminar or turbulent.

2.5-4. **Natural stream bed.** A natural stream bed is circularly shaped with a diameter of 1.2 m. The height of the water is normally about one-third of the radius of the circle. After rain the water rises to two-thirds of the radius. Calculate hydraulic diameter in each case. See Example 4.6.1-2 for equations to calculate areas and lengths of circular segments.

2.5-5. **Pipe friction loss.** Repeat Example 2.5.2.3-1 for a velocity of 0.4 m/sec.

2.5-6. **Flowing molasses.** Molasses is usually pumped at a high temperature to avoid moving the material when too thick. Compare straight pipe friction losses for pumping molasses at 21 and 66°C. The pipe is 39 m long and 15 cm in diameter. Volumetric flow rate is 1.8×10^{-3} m^3/sec.

2.5-7. **Pipe pressure loss.** A fluid with a viscosity of 1.0 kg/(m sec) flows in a pipe 8 cm in diameter and 30 m long. The rate of flow is 0.02 m^3/sec. What is the pressure loss in the pipe?

2.5-8. **Glaucoma pressure.** The ciliary body normally secretes aqueous humor into the eye at a rate of about 20×10^{-12} m^3/sec. Excess fluid drains from the eye through a tiny channel called the canal of Schlemm that has a normal resistance of 1.3×10^{14} N sec/m^5. The glaucomatous eye has a reduced aqueous humor flow rate of 1×10^{-12} m^3/sec but a canal of Schlemm resistance of up to 3.2×10^{15} N sec/m^5. Compare intraocular pressures for the normal eye, an eye with glaucoma, and a glaucomatous eye that does not have reduced aqueous humor flow rate.

2.5-9. **Nasal strips.** The nasal canal is not round, but has an equivalent diameter of 5 mm. The popular nasal strips include elastic bands that tend to pull the nostrils open when they are stuck to the outside of the nose just below the bridge. If a nasal strip increases the diameter of the nasal passage by 1 mm, estimate the reduction in nasal resistance as a result of using the strip. Although nasal flow rate is nearly always turbulent, sometimes a laminar flow assumption must be made.

2.5-10. **Arteriolar resistance.** Arterioles are innervated by sympathetic autonomic nerve fibers to control blood flow. If the arteriole is assumed to have fully developed laminar flow, how much reduction in vessel diameter is required to triple vessel resistance?

2.5-11. **Snake pulmonary blood flow.** Pulmonary vascular resistance in the aquatic snake *Acrochordus granulatus* has been measured to range from 6.46×10^{12} N sec/m^5 during diving to 5.92×10^{10} N sec/m^5 during lung ventilation (Lillywhite and Donald, 1989). If pressure in the anterior pulmonary artery is about 2400 N/m^2 during breathing and 2000 N/m^2 during diving, compare pulmonary blood flows for each of these conditions.

2.5.-12. **Coughing.** Coughing begins with a deep inspiration followed by forced expiration against a closed glottis. The glottis is suddenly opened, producing an explosive outflow of air at velocities of up to 270 m/sec. A normal adult trachea has a diameter of about 14 mm, but high intrathoracic pressure inverts the noncartilaginous part of the trachea and reduces its area to 17% of its normal value. Using the respiratory model given in Problem 1.6-3 of Chapter 1, determine the alveolar pressure that must be present to produce flows of this magnitude.

2.5-13. Pipe resistance. If the resistance of the first section of pipe in Problem 2.2-1 is $10 \, \text{N sec/m}^5$, what is the resistance in the second section?

2.5-14. Dyspnea while wearing mask. It has been reported that dyspnea (the feeling of breathlessness) occurs whenever a total pressure swing (inhalation to exhalation) at the mouth exceeds $1670 \, \text{N/m}^2$. Inhalation and exhalation resistances of the U.S. Army M17 air-purifying respirator mask are not constant, but can be given as (Johnson, 1992)

$$R_i = 3.227 \times 10^5 + 5.609 \times 10^7 \dot{V}_i$$

where R_i is inspiratory resistance (N sec/m^5) and $\dot{V}_i$ is inspiratory flow rate (m^3/sec);

$$R_e = 59.93/\dot{V}_e + 6.629 \times 10^4 + 1.376 \times 10^7 \dot{V}_e$$

where R_e is expiratory resistance (N sec/m^5) and $\dot{V}_e$ is expiratory flow rate (m^3/sec). Assuming peak inhalation and exhalation flows are equal, what is the minimum volume rate of airflow through the mask to cause dyspnea? How likely is it that a wearer will encounter these flow rates?

2.5-15. Flow in distensible tube. For the fluid of Problem 2.5-7, flowing in a distensible tube with the pressure–diameter relationship of $D = (8 + 0.2p)$ cm, what is the pressure loss in the tube if the tube is surrounded by atmospheric pressure and empties to the atmosphere?

2.5-16. Respiratory resistance. The resistance to airflow in the respiratory upper airways of humans is $39.2 \, \text{kN sec/m}^2$ (Johnson, 1991). What would be the resistance if the air was not warmed from ambient (assume $20°C$) to $37°$ (body temperature), but remained at $20°C$ as it entered the respiratory system? Neglect density effects.

2.5-17. Resistance of a coronary artery. Repeat Problem 2.2-2 with the assumption that the coronary artery is supplied by an ideal pressure source of $12 \, \text{kN/m}^2$ and flow is laminar. The cardiac blood flow of $250 \, \text{mL/min}$ applies to the case of the 3-mm arterial lumen, and the entire flow resistance is located in the artery.

2.5-18. Bioreactor flow. Water contaminated with hydrocarbons is being pumped from a pond to a bioreactor containing nutrients, oxygen, pH control, cometabolites, and microorganisms. The bioreactor inlet pipe is 50 m long and has an entrance at the pond, discharge into the bioreactor, and three additional elbows. The discharger pipe from the bioreactor is also 50 m long with two additional elbows and a discharge into the atmosphere. All pipe diameters are 15 cm, and the resistance of the bioreactor is $2.5 \times 10^5 \, \text{N/m}^5$. What is the fluid resistance of this system? What is the total pressure drop from the pond to the pipe discharge?

2.5-19. **Syringe injections.** The internal diameter of a 20 gauge hypodermic needle is 0.5 mm. The internal diameter of the syringe is 12 mm. What is the ratio of fluid velocity in the needle to that in the syringe? Estimate the plunger force required to inject 3 mL of liquid in 12 sec.

2.5-20. **Airflow perturbation device (APD).** The APD is a medical device that can be used to assess the level of resistance to airflow in the respiratory system (Johnson et al., 1984). Such a measurement is important for a number of respiratory conditions, including asthma, emphysema, cystic fibrosis, and others. The APD consists of a Fleisch pneumotach to measure airflow and a pressure transducer to measure mouth pressure. A rotating wheel with a screened segment periodically perturbs the airflow and mouth pressure by intermittently adding resistance external to the person breathing through the device. This results in transient mouth pressure and flow perturbations. Assume the respiratory system can be modelled simply by a resistance in series with a pressure source. Show that respiratory resistance can be determined simply by dividing the pressure perturbation magnitude by the flow perturbation magnitude.

2.5-21. **Flow rates in a series of blood vessels.** Repeat Problem 2.2-2 with an ideal pressure source of $12 \, \text{kN/m}^2$ (from the heart) and total blood vessel resistance of $2.88 \times 10^9 \, \text{N sec/m}^5$. Of this, the healthy (nonstenotic) artery contributes 50% of the resistance, the capillary bed contributes 10%, and the vein contributes 40%. Draw a systems diagram of this arrangement. What assumptions are necessary to estimate blood flow values with and without a stent? What blood velocities and volume flow rates are present with and without a stent?

2.5-22. **Windkessel vessel.** Show that, if the nonideal pressure source in Figure 2.5.9c is a dual of the nonideal flow source in Figure 2.5.9d, the Kirchhoff equation for both cases is the same.

2.5-23. **Pressure inside a distensible ventricle.** Robinson formulated a model (Johnson, 1991) of ventricular contraction in a dog weighing 98 N. He approximated the ventricle by a pressure source (p_s) in series with a systolic resistance (R_s) of contracting myocardium. The pressure source was given by

$$p_s = p_o + (42.6 \times 10^3) \left[1 - \left(1 - \frac{V}{20 \times 10^{-6}} \right)^2 \right]$$

where p_o is diastolic pressure (of $1000 \, \text{kN/m}^2$), V the ventricular volume (m^3), and p_s is in N/m^2. The systolic resistance was $333 \times 10^6 \, N \, sec/m^5$.

The distensible ventricle was represented by a compliance (C) of $19.2 \times 10^{-9} \, m^5/N$ and an aortic valve resistance (R) of $4.40 \times 10^6 \, N \, sec/m^5$. Arrangement of the elements of the model is identical to that in Figure 2.5.9c. If ventricular volume decreases with time according to the relation

$$V = 21 \times 10^{-6} + (14 \times 10^{-6})e^{-t/0.16}$$

Use Eq. 2.5.47 to plot ventricular pressure over the systolic duration of 0 to 0.8 sec.

2.5-24. **Pulmonary blood flow during exercise.** Calculate the volume rate of blood flow in the pulmonary vein during exercise when the source pressure is $2202 \, N/m^2$ and the outlet pressure is $-800 \, N/m^2$.

2.5-25. **Slowing speech.** Compare the speed of speech from the mouth to the speed of the same speech when the exhaled air has cooled to room temperature.

2.5-26. **Compressible flow in biological systems.** Suggest some situations involving biological systems where compressible flow occurs in rigid conduits.

2.5-27. **Stagnation temperature.** Plot stagnation temperature across the 10 cm diameter of a pipe for water in laminar flow at a volume flow rate of $0.01 \, m^3/sec$.

2.5-28. **Velocity of water in xylem.** The Reynolds number for flow in the xylem is about 0.02. Calculate the flow velocity.

2.5-29. **Cleaning of raceways.** Brook trout can be raised in raceways, usually rectangular canals through which water flows (Wheaton, 1977; Youngs and Timmons, 1991). Particles from fish grown in aquacultural settings have size distributions, based on volume, given by a mean of about $220 \, \mu m$ and a standard deviation of about $210 \, \mu m$ (McMillan, 1996). How fast must the water flow in the raceway in order to remove all particles?

2.5-30. **Fruit in gelatin.** Grapefruit, apples, and pears will float in liquid gelatin; prunes and maraschino cherries will not. Estimate the density of gelatin.

2.5-31. **Deposition in a constructed wetland.** A channel filled with water carrying natural sediment is 4.1 m wide by 0.9 m deep. The volume flow rate is $6.2 \, m^3/sec$. The channel empties into a constructed wetland 76 m wide by 0.3 m deep by 150 m long. If the channel carries a suspended load of 1000 mg/L, and if you assume an equal distribution of suspended mass at all particle diameters, how long will it be before the wetland is filled completely with sediment?

2.5-32. **Deposition with Gaussian distribution of particle sizes.** Repeat Problem 2.5-31 with the particle size distribution being normal with a mean of 0.8×10^{-2} m and the standard deviation being 4.1×10^{-3} m.

2.5-33. **Intubated neonates.** Neonates are often intubated [given endotracheal (ET) tubes] before oxygen is administered to their lungs. The normal airways resistance of neonatal respiratory airways is about ten times the normal adult resistance values of $160\,\text{kN sec/m}^5$ (mainly because of small airway diameters). Compliance for 2-kg infants is about $4.7 \times 10^{-8}\,\text{m}^5/\text{N}$. The ET tubes are 9 French (3 mm inside diameter) and 15 cm long. Assume sinusoidal inhalation of 15 mL oxygen by the ventilator at 60 breaths/min. At what pressure level should the alarm be set to avoid damaging the baby's lungs?

2.5-34. **Blood-sucking mosquitos.** Mosquitos suck blood through feeding tubes (proboscises) $10\,\mu\text{m}$ in diameter and 3 mm long. Their heads contain cavities that expand when the insects contract surrounding muscles. In this way, they can produce low pressures to draw blood from their victims. If a mosquito can drink $4\,\mu\text{L}$ of blood in 20 sec, what is the pressure difference between the source of the blood and the mosquito air cavity?

2.5-35. **Flow cytometry.** Flow cytometers are used to count, classify, and sort individual cells suspended in a liquid medium. The sample flows past a point where one or more lasers shine on the suspended cells. Signals proportional to cell diameter, cell concentration, and number of antibody-labeled cells are possible. It is even possible to separate healthy from diseased cells when proper labeling can be achieved.

Cytometers are usually configured as in the diagram. The sample is injected into the center of a stable fluid called the sheath. Both sheath and sample nozzles taper from larger to smaller diameter to stabilize flow. The positive velocity gradient in the sample nozzle stretches cells, thus causing them to orient with their long dimensions aligned with the fluid streamlines. Nozzle diameter is $20\,\mu\text{m}$ where the sample leaves the nozzle. The capillary tube containing the sheath fluid is $250\,\mu\text{m}$. Both sample and sheath fluids exit with a velocity of $10\,\text{m/sec}$. For the diagrammed flow chamber, what are the flow velocities of sample and sheath fluid at the top entrance? What is the volumetric flow rate of sample? Estimate the source pressures required to produce sheath and sample exit velocities of $10\,\text{m/sec}$.

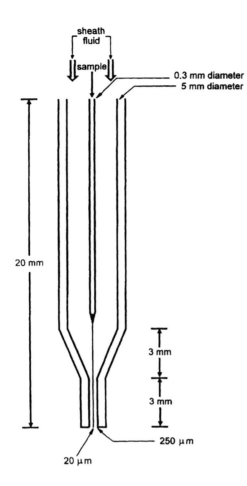

sheath fluid

sample

0.3 mm diameter

5 mm diameter

20 mm

3 mm

3 mm

250 μm

20 μm

2.6-1. Temperature dependence of consistency coefficient. Given

Temperature (°C)	Viscosity of water [10^{-3} kg/(m sec)]	Viscosity of oil [10^{-3} kg/(m sec)]	Consistency coefficient
10	1.3077	5.67	—
20	1.0050	5.47	—
30	0.8007	5.29	62.5

Calculate the consistency coefficient of a water-based non-Newtonian fluid at a temperature of 20°C.

2.6-2. Pumping a non-Newtonian fluid. A non-Newtonian fluid is being pumped through a horizontal steel pipe, 13 cm inside diameter, 14 cm outside diameter, and 100 m long and spilling into the atmosphere.

Draw the equivalent effort and flow variable schematics for fluid flow through the pipe. Label all resistances with equations to determine their magnitudes. The properties for the non-Newtonian fluid are:

$$\text{consistency coefficient} = 0.685 \text{ N sec}^n/\text{m}^2$$
$$\text{flow behavior index} = 0.500$$
$$\text{density} = 919 \text{ kg/m}^3$$

Each of these was measured at 20°C. Determine corrected values for a temperature of 10°C. The properties for the air are:

$$\text{density} = 1.043 \text{ kg/m}^3$$
$$\text{viscosity} = 2.03 \times 10^{-5} \text{ N sec/m}^2$$

Each of these was measured at 65.6°C. Determine corrected values for a temperature of 10°C.

2.6-3. **Pumping a non-Newtonian fluid.** For Problem 2.6-2, determine the pressure at the pump required to maintain a flow inside the pipe of 5 L/sec. Temperature of the fluid is 10°C.

2.6-4. **Non-Newtonian laminar flow.** Between two fluids:

Type	Consistency Coefficient	Flow Behavior Index
Applesauce	56.3	0.47
Apricot purée	54.0	0.29

In which fluid will laminar flow exist at higher Reynolds numbers?

2.6-5. **Density of applesauce.** Estimate the density of applesauce containing 8% sucrose, 7% cellulose, and 85% water.

2.6-6. **Flowing applesauce.** Repeat Example 2.6.2.3-1 with a flow rate of $2.0 \text{ m}^3/\text{sec}$. Calculate the friction loss per meter length of pipe.

2.6-7. **Tomato pomace in a concrete pipe.** Repeat Example 2.6.2.3-2 using a concrete pipe.

2.6-8. **Pumping blood.** Human blood is being pumped through a plastic tube of 4 mm inside diameter. The tube is 0.3 m long. The source of the blood is a pressurized bag, and the tube empties into a hermetically sealed container at atmospheric pressure. How much pressure must be applied to the bag in order to transfer blood at the rate of 0.25 mL/sec? State all assumptions

2.7-1. **Open-channel flow.** For an open channel of trapezoidal cross section, with a bottom width of 1 m and a top width of 2 m, calculate the volume flow rate if the channel is constructed of concrete, is 700 m long, and has a uniform slope along its length of 9 m height difference between one end and the other. The height of the channel is 1 m, and it flows 2/3 full. Use Bernoulli's equation, the Chézy formula, and Manning's equation to obtain three different estimates of the flow rate. Which do you think is correct?

2.7-2. **Moving tomatoes.** Water-filled flumes are often used to convey fresh fruits and vegetables in processing plants. For a flume fabricated from steel, determine how much higher than the downstream end must the upstream end be, if the water is to flow by gravity at a rate of 0.55 m/sec. The flume is 0.61 m (height) × 0.61 m (width) × 30 m (length).

2.8-1. **Greenhouse pest barriers.** Biosecurity in greenhouses can be extended to screening to exclude insects, resulting in reduced pesticide use. Recommended volume flow rate is one air change per minute. Maximum pressures developed by the ventilation fans are about $37 \, N/m^2$, with an additional $12 \, N/m^2$ reserved as a safety margin in case the screen clogs. Static pressure loss through the greenhouse is about $7.5 \, N/m^2$ (without considering screens). Performance data for fans and screening are given below. For a $30 \times 9 \times 3$ m greenhouse, how many fans need to be used, and what areas of screen must be used?

Fan Performance Data

Power (N m/sec)	Volume Flow Rate (m^2/sec) at Static Pressure (N/m^3)			
	0	25	31	37
250	4.72	4.0	3.8	3.5
370	5.5	4.9	4.8	4.6
560	6.3	5.8	5.6	5.5

Screen Pressure Drop

Pressure drop (N/m^2)	25	20	16	12	9	6	5
Air velocity (m/sec)	2.8	2.5	2.3	2.0	1.8	1.5	1.3

2.8-2. **Pumping water into a tank.** Specify the pump required to fill the tank with water from the pond. The pipe is 15-cm-diameter aluminum pipe that comes in 2.4-m lengths. The tank is to be filled in 8 h. (See the diagram on page 258.)

2.8-3. **Pumping through nozzles.** A 0.39-m^3 tank with an outlet in the

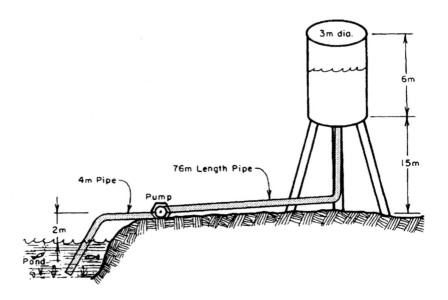

bottom delivers water to two nozzles through 15 m of high-pressure rubber tubing, 1.9 cm inside diameter. The nozzles have the following pressure–flow specifications:

Pressure (kN/m^2)	Flow rate $(10^{-4} \, m^3/sec)$
207	1.87
276	2.04
345	2.57
414	2.91
483	3.28

A pump is to be placed at the tank outlet. Specify the pump flow rate, pressure, power, and type in order to deliver $5.0 \times 10^{-4} \, m^3/sec$ from the nozzles. The nozzles are located 3 m below the bottom of the tank.

2.8-4. **Pumping molasses.** Following is the pressure–flow characteristic of a centrifugal pump used to move molasses through a 30-m-long straight section of pipe 15 cm in diameter. Compare the maximum achievable volumetric flow rates at 21 and 66°C.

2.8-5. **Conveying tomatoes.** Water-filled flumes are often used to convey fruits and vegetables in processing plants. Estimate the pump power required to move water at a rate of 0.55 m/sec through a horizontal flume of dimensions 0.61 (height) × 0.61 (width) × 30 m (length). How much additional power is necessary if tomatoes fall vertically

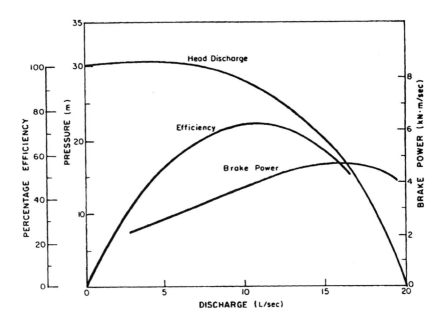

into the flume at a rate of 12 per second and are to be conveyed by the water at the same 0.55 m/sec velocity? The tomatoes are 8 cm in diameter.

REFERENCES

Canny, M. J., 1998, Transporting Water in Plants, *Amer. Sci.* **86**: 152–159.

Comroe, J. H., 1965, *Physiology of Respiration*. Year Book Medical Publishers, Chicago, IL.

Daugherty, R. L., and A. C. Ingersoll, 1954, *Fluid Mechanics with Engineering Applications*, McGraw-Hill, New York, pp. 180–81.

Fung, Y. C., 1990, *Biomechanics: Motion, Flow Stress, and Growth*, Spring-Verlag, New York, pp. 162–78.

Galletti, P. M., and C. K. Colton, 1995, "Artificial Lungs and Blood-Gas Exchange Devices," in *Biomedical Engineering Handbook*, J. D. Bronzino, Ed., CRC Press, Boca Raton, FL, pp. 1879–97.

Gupta, S. N., and P. Mishra, 1974, Turbulent Flow of Inelastic Non-Newtonian Fluids in Pipes, *Indian J. Technol.* **12**: 181–85.

Hanks, R. W., and K. M. Larsen, 1979, The Flow of Power-Law Non-Newtonian Fluids in Concentric Annuli, *Indus. Engr. Chem. Fund.* **18**, 33–35.

Heldman, D. R., 1975, *Food Process Engineering*, AVI, Westport, CT.

Johnson, A. T., 1991, *Biomechanics and Exercise Physiology*, John Wiley & Sons, New York.

Johnson, A. T., 1992, Modelling Metabolic and Cardiorespiratory Effects of Respirator

Mask Wear, Final Report U.S. Army Chemical Research, Development, and Engineering Center Contract DLA900-2045-86-C, Task 188, Battelle, Columbus, OH.

Johnson, A. T., C.-S. Lin, and J. N. Hochheimer, 1984, Airflow Perturbation Device for Measuring Airways Resistance of Humans and Animals, *IEEE Trans. Biomed. Eng.* **31**: 622–26.

Kiris, C., S. Rogers, D. Kwak, and I.-D. Chang, 1993, Computation of Incompressible Viscous Flows through Artificial Heart Devices with Moving Boundaries, in *Contemporary Mathematics (V. 141): Fluid Dynamics in Biology*, A. Y. Cheer and C. P. van Dam, Eds., American Mathematical Society, Providence, R. I., pp. 237–47.

Knudson, R. J., and R. C. Schroter, 1983, A Consideration of Density Dependence of Maximum Expiratory Flow, *Respir. Physiol.* **52**: 125–36.

Lightfoot, E. N., 1974, *Transport Phenomena and Living Systems*, John Wiley & Sons, New York, pp. 100–104.

Lillywhite, H. B., and J. A. Donald, 1989, Pulmonary Blood Flow Regulation in An Aquatic Snake, *Science* **245**: 293–95.

McBride, B. J., S. Gordon, and M. A. Reno, 1993, *Coefficients for Calculating Thermodynamic and Transport Properties of Individual Species*, NASA TM-4513, National Aeronautics and Space Administration, Washington, DC 20546-001.

McMillan, J., 1996, Personal communication, graduate research assistant, University of Maryland, College Park, MD.

Millard, E. B., 1953, *Physical Chemistry for Colleges*, McGraw-Hill, New York, p. 92.

Mishra, P., and L. N. Singh, 1976, Non-Newtonian Flow in the Entrance Region of Tubes, *Indian J. Technol.* **14**: 328–31.

Patni, N. K., 1980, "Pipeline Transport of Liquid Manure" in *Livestock Waste: A Renewable Resource*, Proceedings of the 4th International Symposium on Livestock Wastes, American Society of Agricultural Engineers, St. Joseph, MI, pp. 387–91.

Rand, R. H., and J. R. Cooke, 1996, Fluid Mechanics in Plant Biology, in *Handbook of Fluid Dynamics and Fluid Machinery*, J. A. Schetz and A. E. Fuhs, Eds., John Wiley and Sons, New York, pp. 1921–89.

Rao, M. A., 1994, "Food Process Engineering: Storage and Transport of Fluid Foods in Plants," *Encyclopedia of Agricultural Science*, Vol. 2, Academic Press, San Diego, pp 369–80.

Thomas, J. D., and A. E. Weyman, 1992, Numerical Model of Ventricular Filling, *Ann. Biomed Engr.* **20**: 19–35.

Tigges, A. J., and H. Karlsson, 1967, Classification of Solid Pollutants, in *Marks' Standard Handbook for Mechanical Engineers*, T. Baumeister, Ed., McGraw-Hill, New York, pp. 7–83.

Trueman, E. R., 1980, "Swimming by Jet Propulsion," in *Aspects of Animal Movement*, H. Y. Elder and E. R. Trueman, Eds., Cambridge University Press, Cambridge, U.K., pp. 93–105.

Tunney, H., 1980, "An Overview of the Fertilizer Value of Livestock Waste," in *Livestock Waste: A Renewable Resource*, Proceedings of the 4th International Symposium on Livestock Wastes, American Society of Agricultural Engineers, St. Joseph, MI, pp. 181–84.

Vennard, J. K., 1961, One-Dimensional Flow, in *Handbook of Fluid Dynamics*, V. L. Streeter, Ed., McGraw-Hill, New York, pp. 3-1, to 3-30.

Wheaton, F. W., 1977, *Aquacultural Engineering*, John Wiley and Sons, New York.

Wiley, J. C., 1995, Down to Earth Breakthrough, *Resource*, **2**(9): 9–12.

Yaws, C. L., 1995, *Handbook of Transport Property Data*, Gulf Publishing Co., Houston, TX.

Youngs, W. D., and M. B. Timmons, 1991, A Historical Perspective of Raceway Design, in *Engineering Aspects of Intensive Aquaculture*, Northeast Regional Agricultural Engineering Service, Ithaca, NY, pp. 160–67.

3

HEAT TRANSFER
SYSTEMS

"We are engineers, and we should remember two things: first, there are diminishing returns in trying to get past the 90% point instead of just doing the job, and second, we can always work to change the constraints."

David J. Dewhurst

3.1 INTRODUCTION

heat is kinetic energy   Heat is molecular kinetic energy that is transferred from one body to another when their temperatures are different. Heat itself cannot be measured or observed directly, but the effects of heat transfer can be measured. Those effects usually observed are: (1) changes of temperature, (2) melting or evaporation of a substance, or (3) production of physical work.

thermo-dynamics   There is an equivalence between heat and physical work that is stated as the first law of thermodynamics. There is also the second law of thermodynamics that states that heat (or any other physical substance, for that matter) does not spontaneously concentrate itself. Thus heat flows from higher temperatures (where it is concentrated) to lower temperatures (where it is sparse).

empirical study   The dependence of heat transfer principles upon experimental observations will become clear farther into this chapter. Empirical relationships used to describe mathematically the dependencies of certain heat transfer parameters on other quantities were originated because this was simply the best way for engineers to convey design information. Because heat, unlike fluids or precipitating masses, cannot be directly observed, the problem of quantitative description is not easily reduced to a combination of simple principles.

262

Not that simple experimental principles cannot exist in heat transfer, because they do. The first and second laws of thermodynamics mentioned earlier first existed as sets of experimental observations and were eventually transposed into simple postulates. At first blush, conduction and radiation heat transfer will appear to be simple, as well, until one realizes that the really complex description of experimental results is hidden in the guise of physical properties of the materials being studied. No heat transfer calculations can be made without knowing values for these physical properties, and their values are available almost exclusively from tables of experimental observations. Other modes of heat transfer are even more complex and not reducible to simple mathematical description; the mathematical descriptions themselves are available in tables of experimental observations. Keeping in mind the experimental basis for much of this information can keep the engineer from thinking of calculated results as more precise than they really are.

based on experimental observation

What types of information do engineers seek when performing a heat transfer calculation? There are usually two types of problems. Steady-state calculations are performed to determine the rate of heat transfer, either to be sure that it falls within limits or to be able to supply or remove enough heat. Non-steady-state calculations are often performed to determine the times for heating or cooling.

two types of problems

Heat transfer is similar to fluid flow and mass transfer because it obeys the same general forms of equations as the other two transport processes (Table 3.1.1). These equations and concepts were developed in Chapter 1. Heat is different from both fluid flow and mass transfer in several respects: (1) Heat is energy, and rate of heat transfer is power; thus no physical substance must move for heat to be transferred; (2) there is no fluid flow or mass transfer process analogous to radiation heat transfer; (3) time is more important in heat transfer than in the other two processes, especially because of the times and temperatures necessary to kill microbes in food, medical devices, and tissue culture. Like fluid flow and mass transfer, heat may be transferred by convection. Unlike fluid flow, there is no heat inertia. Unlike mass transfer, there are no common structures that can impede the movement of certain types of heat compared to others. And, importantly, friction within fluid and mass systems can form heat; heat, however, cannot degenerate further.

similarities and differences with other processes

Heat is thus energy flowing due to a temperature difference. The flow variable is heat and the effort variable is temperature. Heat may be transferred from one object or substance to another by four basic mechanisms: (1) conduction, (2) convection, (3) radiation, and (4) change of state.

four heat transfer mechanisms

We have already seen in Section 1.8.2 how the general field equations can be applied to conduction heat transfer using Fourier's law (not to be confused with the Fourier equation 1.7.28). We will return to conduction heat transfer after reviewing the general flow balance for heat transfer.

Table 3.1.1   Systems Approach to Heat Transfer

Modes of action	Multiple, including conduction, convection, radiation, and change of state
Effort variable	Temperature
Flow variable	Heat
Resistance	Depends on material physical properties
Capacity	Amount of heat storage depends on material properties
Inertia	Heat has no mass; thus inertia has no real meaning
Power	Rate of heat flow is power
Time	Conduction analyzed for both transient and stead-state conditions; convection and radiation are considered to be steady state only
Substance generation	Happens by many mechanisms, including mechanical, chemical, and electrical
General field equation	Useful for conduction only
Balances written	Usually just one, a heat balance, which is really a power balance
Typical design problem	1. Determine insulation thickness: use heat balance with conduction, convection, and radiation equations. Determine convection coefficient from equations for Nusselt number.
	2. Determine freezing times: Use Heisler charts for chilling time, Planck equation for freezing time, and Heisler charts for cooling below freezing.
	3. Determine required refrigeration capacity: Use heat balance, including all sources and condensation, if appropriate.
	4. Determine size of heat exchanger: Use heat balance, effectiveness curves, and fouling factors. Alternatively, work from convection and conduction equations to determine heat transfer.

A general flow balance has been given in Eq. 1.7.15. Since heat is the specific substance of interest here, this equation becomes

basic heat balance

$$\text{(rate of heat in)} - \text{(rate of heat out)} + \text{(rate of heat generated)}$$
$$= \text{(rate of heat stored)} \tag{3.1.1}$$

This equation is so basic to heat transfer calculations that it should be the starting point for nearly all engineering designs involving heat transfer.

3.2 CONDUCTION

unmoving object

Conduction occurs when the substance taking part in the heat transfer is unmoving. Thus conduction is a very important mechanism for heat transfer in solids. Liquids and gases must be confined to extremely small spaces such that they cannot move significantly in order to consider conduction to be the dominant mode of heat exchange. Otherwise, heat exchange in fluids occurs by convection.

As discussed in Sections 1.7.3.2 and 1.8.2, the general field equation for conduction heat transfer is

general field
equation

$$\frac{\partial^2 \theta}{\partial x^2} + \frac{\partial^2 \theta}{\partial y^2} + \frac{\partial^2 \theta}{\partial z^2} + \frac{\dot{q}_{gen}'''}{k} = \frac{1}{\alpha} \frac{\partial \theta}{\partial t} \tag{3.2.1}$$

where θ is the temperature (°C); x, y, z, coordinate directions (m); $\dot{q}_{gen}'''$, the rate of heat generation per unit volume [N m/(sec m^3)]; k, the thermal conductivity [N/(°C sec)]; α, the thermal diffusivity (m^2/sec); and t, the time (sec).

Also

thermal
diffusivity

$$\alpha = \frac{k}{\rho c_p} \tag{3.2.2}$$

where ρ is the density (kg/m^3); and c_p, the specific heat at constant pressure [N m/(kg °C)]. Resistances to heat flow and thermal capacities are given in Section 1.8.2. The resistance for heat transfer perpendicular to a planar surface is

thermal
resistance

$$R = L/kA \tag{3.2.3}$$

where R is the thermal resistance [°C sec/(N m)]; L, the plate thickness (m); and A, the area (m^2).

Thermal capacity for the same system is

thermal
capacity

$$C = mc_p \tag{3.2.4}$$

where C is the thermal capacity (N m/°C); and m, the mass (kg).

3.2.1 Thermal Conductivity

Thermal conductivities are measured parameters that determine resistance to heat flow. Thermal conductivities of solids can range from about 8.65 to 415 N/(°C sec) for metals to 0.0346 to 2.60 N/(°C sec) for nonmetallic

conductors

solids. Materials with high thermal conductivities are called conductors and are almost exclusively metals. These materials conduct heat by free electrons moving through the metal lattice and by vibration of adjacent atoms. Nonmetallic solids conduct heat only by molecular vibration. Thermal conductivities for many solids are nearly constant over several tens of degrees of temperature. A large exception to this is solids containing large amounts of frozen food water, such as in foods; since ice has a thermal conductivity [2.25 N/(°C sec)] nearly four times as high as water [0.569 N/(°C sec)], frozen foods usually have much higher thermal conductivities than do unfrozen foods.

Liquids have thermal conductivities that vary moderately with temperature. If large ranges of temperatures are to be encountered, then a linear approximation to thermal conductivity variation is usually used

linear
temperature
dependence

$$k = a + b\theta \qquad (3.2.5)$$

where a, b are constants [N/(°C sec) and N/(°C^2 sec), respectively], before equations similar to 3.2.1 are integrated. Since a linear approximation is used, thermal resistance can be evaluated using the thermal conductivity evaluated at the mean temperature, $(\theta_{hot} + \theta_{cold})/2$.

Thermal conductivity values for nonmetallic liquids range from 0.0865 to 0.692 N/(°C sec). The thermal conductivity of water ($k = 0.569 + 0.00119\theta$, $70 \leq \theta \leq 93$°C) is much higher than those for other common organic-type liquids such as benzene or oil. Thermal conductivities of unfrozen foods and biological materials containing high percentages of water are nearly that of water (Table 3.2.1).

Thermal conductivities of gases at atmospheric pressure are in the range of 0.00692 to 0.173 N/(°C sec). Thermal conduction in gases is relatively

gas
conductivities

simple: heat is transported by collisions between nearly independent molecules. The lighter molecules of hydrogen gas can move relatively faster and hydrogen should, therefore, have a relatively high thermal conductivity [0.175 N/(°C sec)].

Thermal conductivity values for gases increase approximately as the

temperature
dependence

absolute temperature to the 0.85 power; they are nearly independent of

Table 3.2.1   Thermal Conductivities of Water and Selected Gases

Temperature (°C)	Thermal Conductivity [N/(°C sec)]				
	Water	Air	Carbon dioxide	Oxygen	Nitrogen
0	0.558	0.0239	0.0146	0.0248	0.0240
10	0.577	0.0247	0.0153	0.0256	0.0247
20	0.597	0.0255	0.0160	0.0264	0.0254
30	0.615	0.0263	0.0168	0.0272	0.0261
40	0.633	0.0271	0.0175	0.0279	0.0268
50	0.647	0.0279	0.0183	0.0287	0.0275
60	0.658	0.0287	0.0191	0.0295	0.0282
70	0.668	0.0295	0.0198	0.0303	0.0289
80	0.673	0.0302	0.0206	0.0311	0.0296
90	0.678	0.0310	0.0214	0.0318	0.0302
100	0.682	0.0318	0.0222	0.0326	0.0309

pressure up to a few atmospheres. Estimation of thermal conductivity values at temperatures and pressures other than tabulated values should account for both of these facts.

dry air

The low thermal conductivities of gases makes them very good insulators. The thermal conductivity of dry air ($\approx 0.0001477T^{0.907}$, $200 \leq T \leq 400$ K) is very low. The trick to using this low conductivity to advantage is to keep the air from moving and carrying heat with it as it moves. Many of the most common insulating materials contain voids in which air is trapped. For even better insulators, spaces inside the materials are evacuated completely. A perfect vacuum cannot conduct heat at all, but offers no resistance to heat transfer by thermal radiation.

porous
materials

Porous materials are often encountered while working with foods, and these can be considered to be two-phase mixtures of solids and (usually) air. It is usually sufficient to know the thermal conductivity of the bulk of the material, without attempting to predict thermal conductivity values from those of the constitutent phases. A number of thermal conductivity values are found in Table 3.2.2. Others may be found in ASHRAE (1977), Polley et al. (1980), Charm (1971), and Singh and Heldman (1993).

anisotropy

In some mineral materials and more biological materials, the structural organization is different in different directions. Such a material is called anisotropic, and the thermal conductivity value may be much larger in one direction compared to another. As seen in Section 1.7.3.2, Eq. 1.7.26 must be used instead of Eq. 1.7.27 for an anisotropic medium. Since manipulation of this form of the equation requires some relationships to be found between thermal conductivities in the different directions, the solution of the equation to obtain thermal resistance and capacity values may not be easy. For example, if $k_x = 2k_y = 5k_z$, then Eq. 3.2.1 becomes

$$\frac{\partial^2 \theta}{\partial x^2} + 0.5\,\frac{\partial^2 \theta}{\partial y^2} + 0.2\,\frac{\partial^2 \theta}{\partial z^2} + \frac{\dot{q}_{gen}'''}{k_x} = \frac{1}{\alpha}\,\frac{\partial \theta}{\partial t} \qquad (3.2.6)$$

Fortunately, thermal conductivity differences are usually not too pronounced, or they are so pronounced that heat is effectively transferred in only one direction, or the problem geometry can be formulated in such a way that anisotropic behavior can be neglected. As long as a constant thermal conductivity can be rightfully assumed, steady-state heat transfer can be obtained from

$$\dot{q} = \frac{\theta_1 - \theta_2}{R} \qquad (3.2.7)$$

EXAMPLE 3.2.1-1. Thermal Resistance of a Cantaloupe

What is the thermal resistance offered by a cantaloupe 20 cm in diameter with flesh 5 cm thick?

Table 3.2.2 Some Thermal Conductivity Values [N/(°C sec)]

Human and animal materials	
Animal skin	0.50
Beeswax	0.40
Bone	2.0
Cat blood	0.533
Dental amalgam	23.0
Dentine	0.59
Enamel	0.82
Fleece (4.5 cm deep)	0.14–0.21
Fish, poultry, or animal muscle	0.43–0.50
Human blood	0.507
Human blood plasma	0.578
Human skin	0.21–0.63
(varies, depending on blood perfusion and fat content)	
Human fat	0.21–0.33
Human muscle	0.41–0.50
Kidney or liver	0.498
Porcelain	1.00
Pork fat	0.187
Seal blubber	0.190
Oyster shell	1.95–2.27
Wool	0.036
Animal coats	
Arctic mammals	0.036–0.106
Various wild animals	0.038–0.051
Marino sheep	0.037–0.048
Newborn sheep	0.065–0.107
Cattle	0.076–0.147
Rabbit	0.038–0.100
Kangaroo	0.043–0.064
Harp seal pups	0.047–0.065
Grouse feathers	0.029–0.058
Penguin	0.031–0.046
Gosling	0.036–0.046
Artificial fur	0.040–0.067
Chicken	0.050
Waste materials	
Beef manure slurry	
10% total solids	0.58
50% total solids	0.37
100% total solids	0.11
Fruits and vegetables	
Apples	0.39
Cantaloupe	0.57
Cassava	0.50
Corn	0.17

(*continued*)

Table 3.2.2 (Continued)

Eggplant	0.24–0.38
Ginger	0.56
Grapefruit	0.40
Oranges	0.43
Peach	0.65
Pears	0.60
Peas, black-eyed	0.31
Pepper	0.47–0.52
Potatoes	0.55–1.1
Radish	0.56
Squash	0.50
Strawberries	0.68–1.1
Zucchini	0.40–0.58
Food materials, unfrozen	
Applesauce	0.69
Banana purée	0.69
Beet	0.54
Butter, margarine	0.20
Cake	0.12
Carrot purée	1.3
Chicken, dark meat	0.50
Cream	0.18
Fruit juice	0.55
Ground beef	0.35–0.37
Honey	0.50
Mashed potatoes	1.1
Meat emulsion	0.43
Milk	0.54–0.64
Oats	0.064
Olive oil	0.097
Peanut butter	0.56
Peanut oil	0.17
Potato salad	0.48
Sausage	0.36–0.43
Surimi	0.50
Tomato juice (29% solids)	0.55
Food materials, frozen	
Fish	0.019–0.028
Fruits	0.54–1.1
Meat	1.1–1.5
Orange juice	2.4
Peas	0.50
Food materials, dried	
Apple	0.016–0.43
Beef	0.038–0.066
Egg albumen	0.013–0.042

(continued)

Table 3.2.2 (Continued)

Milk	0.18–0.23
Peach	0.016–0.43
Tomato solids	0.20–0.30
Wheat flour	0.45–0.69
Ecological materials	
Air, 20°C	0.0255
Clay soil, 4% H_2O	0.57
Granite	1.7–4.0
Ice, 0°C	2.2
Limestone	1.3
Marble	2.1–2.9
Mica, perpendicular to cleavage plane	0.84
Ocean floor	0.65–0.98
Quartz, parallel to optic axis	8.4
Sandstone	1.8
Sandy soil	
4% H_2O	1.51
10% H_2O	2.16
Silicate rocks	2.1–4.2
Snow, 0°C	0.47
Water, 20°C	0.60
Gases at 273 K (373 K)	
Acetone	0.0099 (0.0171)
Ammonia	0.0218 (0.0332)
Butane	0.0135 (0.0234)
Carbon monoxide	0.0232 (0.0305)
Chlorine	0.00744
Ethane	0.0183 (0.0303)
Ethyl alcohol (293 K)	0.0154 (0.0215)
Ethyl ether	0.0133 (0.0227)
Ethylene	0.0175 (0.0279)
Helium	0.141
Hexane	0.0125
Hydrogen	0.175
Sulfur dioxide	0.0087 (0.0119)
Plant materials	
Cork	0.045
Cotton	0.061
Cypress wood, across grain	0.097
Fir wood, across grain	0.17
Leaves	
Acer tubrum (red maple)	0.32
Cercis canadensis (redbud)	0.32
Cornus florida (dogwood)	0.35
Ilex opaca (American holly)	0.47
Ligustrum volgare (privet)	

(*continued*)

Table 3.2.2 (Continued)

Present year leaf	0.50
Previous year leaf	0.44
Liquidambar styraciflua (sweet gum)	0.41
Magnolia grandiflora (magnolia)	0.48
Quercus alba (white oak)	
Shade leaf	0.25
Sun leaf	0.24
Q. marilandica (blackjack oak)	0.30
Q. phellos (willow oak)	0.27
Q. stellata (post oak)	0.24
Salix nigra (black willow)	0.28
Oak wood, across grain	0.17–0.21
Pine wood, across grain	0.11–0.15
Paper	0.130
Sawdust	0.059
Plastics	
Acrylic	0.17–0.25
Cellulose acetate	0.17–0.33
Epoxy	0.17–1.26
Nylon	0.24–0.29
Phenolic	0.15
Polycarbonate	0.19
Polyethylene	
Low density	0.33
High density	0.46–0.52
Polypropylene	0.12
Polystyrene	0.10–0.13
Polyvinylchloride	0.13
Rubber	0.15
Silicone	0.15–0.31
Metals	
Aluminum	206
Brass	104
Cast iron	52
Copper	377
Lead	33
Steel	
1% C	45
Stainless	16–21
Tin	59
Building and insulating materials	
Asbestos	0.17
Brick	
Building	0.69
Fireclay	1.00
Concrete	0.76

(*continued*)

Table 3.2.2 (Continued)

Corkboard	0.043
Fiber insulation board	0.048
Glass	0.52–1.1
Glass wool	0.041
Rock wool	0.032–0.039
Liquids	
Acetic acid, 100%	0.17
Benzene	0.16
Ethyl alcohol	0.18
Ethylene glycol	0.27
Glycerol	0.28
Kersoene	0.15
Methyl alcohol	0.22
Salt brine	
25%	0.57
12%	0.59
Sulfuric acid	
90%	0.36
60%	0.43
Vaseline	0.18

Solution

Thermal resistance of a hollow sphere is, from Eq. 1.8.22,

$$R = \frac{1/r_i - 1/r_o}{4\pi k}$$

Thermal conductivity of a cantaloupe is, from Table 3.2.2, 0.57 N/(°C sec). Values for r_o and r_i are 0.10 and 0.05 m, respectively. Thus

$$R = \frac{\dfrac{1}{0.05 \text{ m}} - \dfrac{1}{0.10 \text{ m}}}{4\pi[0.57 \text{ N}/(°\text{C sec})]}$$

$$= 1.40°\text{C sec}/(\text{N m})$$

EXAMPLE 3.2.1-2. Thermal Conductivity of Gases

It is proposed to monitor ammonia levels in a closed tank by measuring thermal conductivity of the mixture of air and ammonia in the tank. If the measured thermal conductivity at 25°C is 0.0250 N/(°C sec), what is the fraction of ammonia in the tank?

Solution

As long as the gas mixture contains only air and ammonia, then we can determine the fraction of ammonia by

$$Xk_{NH_4} + (1 - X)k_{air} = 0.0250$$

where X is the fraction of ammonia in the mixture. The problem is now to find thermal conductivities of air and ammonia at $\theta = 25°C$.

The thermal conductivity of air can be obtained from

$$k_{298} = 0.0001477 \ (298.15 \ K)^{0.907}$$
$$= 0.02592 \ N/(°C \ sec)$$

We can also interpolate from values in Table 3.2.1

$$k_{298} = k_{293} + (k_{303} - k_{293})\left(\frac{T_{298}^{0.85} - T_{293}^{0.85}}{T_{303}^{0.85} - T_{293}^{0.85}}\right)$$
$$= 0.0255 + (0.0263 - 0.0255)\left(\frac{298.15^{0.85} - 293.15^{0.85}}{303.15^{0.85} - 293.15^{0.85}}\right)$$
$$= 0.02590 \ N/(°C \ sec)$$

The thermal conductivity for ammonia can be obtained by interpolation of values in Table 3.2.2 as

$$k_{298} = k_{273} + (k_{373} - k_{273})\left(\frac{T_{298}^{0.85} - T_{273}^{0.85}}{T_{373}^{0.85} - T_{273}^{0.85}}\right)$$
$$= 0.218 + (0.0332 - 0.0218)\left(\frac{298.15^{0.85} - 273.15^{0.85}}{273.15^{0.85} - 273.15^{0.85}}\right)$$
$$= 0.0247 \ N/(°C \ sec)$$

Thus

$$0.0247X + (1 - X)(0.0259) = 0.0250 \ N/(°C \ sec) - 0.0012X = -0.0009$$

$X = 0.75$.

Remarks

As long as there are only two gases involved, a single thermal conductivity measurement can determine the fractions of each. Since the thermal conductivities of each gas have nearly the same values, numerical differences are small, and accuracy suffers.

3.2.2 Thermal Conductance

standard
thickness

Tabulated values of thermal conductance are sometimes encountered for materials of standard thicknesses. Examples of these are building materials, such as bricks, window glass, and plywood, or film materials such as plastic sheets. Thermal conductance is just thermal conductivity divided by thickness $(C = k/L)$; so that thermal resistance is just the inverse of the product of conductance and area $(R = 1/CA)$.

3.2.2.1 Clothing Clothing represents a special case of conductive heat transfer. There are social, legal, and moral aspects attributable to clothing, aspects of which the biological engineer should at least be aware, but the most important clothing characteristic to be considered here is its modifying effect on heat loss from the body.

The obvious effect of clothing is its insulative property. Not so obvious is the dead air space formed between the skin and clothes, the reduction in subtle effects permeability to water vapor transmission, and the increase of surface area when wearing clothes (Johnson, 1991a).

The insulation value for clothes is normally found in units of *clo*. A clo was suggested in 1941 as the insulation value of a normal business suit worn comfortably by sedentary workers in an indoor climate of 21°C. Since then, clo the clo has been standardized as the amount of insulation that would allow 6.45 N m/sec of heat from a 1 m^2 area of skin of the wearer to transfer to the environment by radiation and convection when a 1°C difference in temperature exists between the skin and the environment (Goldman, 1967). The higher the clo value of clothing, the less heat will be transferred from the skin.

Thermal conductivity values for clothing would not be very useful because of the variable and unknown thicknesses. Hence the use of thermal conductance is more useful.

Heat conduction through clothing is given by

conduction
through
clothing

$$\dot{q} = \frac{6.45A}{\text{clo}}(\theta_{sk} - \theta_\infty) = C_{cl}A(\theta_{sk} - \theta_\infty) \tag{3.2.8}$$

where $\dot{q}$ is the heat transferred through clothing (N m/sec); clo, the clothing insulation (clo units); C_{cl}, the clothing conductance [N m/(m^2 sec °C)]; A, the surface area (m^2); θ_{sk}, the skin temperature (°C); and θ_∞, the environmental temperature (°C).

This equation shows the normal means to calculate conductive heat transfer using conductance, and shows the equivalence of clo to the normal units in this text.

In Table 3.2.3 are values of conductance for selected items of clothing. Overall thermal resistance for a clothing ensemble should be related to the parallel and serial combination of resistances of all pieces worn (Figure 1.8.3). Because of potential complexities in this process, the following empirical relationships have been developed (ASHRAE, 1985)

clothing
resistance

$$R_{cl} = \left(0.494 \sum C_{cl}^{-1} + 0.0119\right)/A, \qquad \text{men} \qquad (3.2.9a)$$

$$R_{cl} = \left(0.524 \sum C_{cl}^{-1} + 0.0053\right)/A, \qquad \text{women} \qquad (3.2.9b)$$

where R_{cl} is the conduction thermal resistance of a clothing ensemble [°C sec/(N m)]; and A, the surface area (m^2).

The sum of the inverted conductance values $\left(\sum C_{cl}^{-1}\right)$ is calculated by inverting tabulated conductance values for each item of clothing in the ensemble and summing them. Conduction heat loss through the clothing is then calculated as the temperature difference $(\theta_{sk} - \theta_{\infty})$ divided by thermal resistance

$$\dot{q} = (\theta_{sk} - \theta_{\infty})/R_{cl} \qquad (3.2.10)$$

In order to complete this discussion, a few things need to be explained about surface area and about skin temperature. The following relationships were empirically produced.

Clothing surface area is based upon the DuBois formula for calculation of surface area of nude humans

DuBois body
surface area

$$A = 0.07673 W^{0.425} H_t^{0.725} = 0.2025 m^{0.425} H_t^{0.725} \qquad (3.2.11)$$

where A is the nude skin surface area (m^2); W, the body weight (N); H_t, the height (m); and m, the body mass (kg).

The increase of surface area due to clothing has been found to be between 15 and 25% (ASHRAE, 1977). An average increase of 20% is used in the following equations.

Mean skin temperature is also affected by clothing. Temperature calculated by means of the following is essentially independent of metabolic rate up to four or five times the rate expended at rest. For a clothed human

mean skin
temperature

$$\theta_{sk} = 25.8 + 0.267\theta_{\infty} \qquad (3.2.12)$$

where θ_{sk} is an estimate of mean skin temperature (°C); and θ_{∞}, the mean environmental temperature (°C).

Table 3.2.3   Conductance Values for Individual Items of Clothing

Men		Women	
Clothing	[N m/ (m² sec °C)]	Clothing	[N m/ (m² sec °C)]
Underwear			
Sleeveless T shirt	110	Bra and panties	130
T shirt	72	Half slip	50
Briefs	129	Full slip	34
Long underwear upper	18	Long underwear upper	18
Long underwear lower	18	Long underwear lower	18
Torso			
Shirt[a]		Blouse	
Light, short sleeve	46	Light	32
Light, long sleeve	29	Heavy	22
Heavy, short sleeve	26		
Heavy, long sleeve	22	Dress	
		Light	29
		Heavy	9.2
Vest		Skirt	
Light	43	Light	65
Heavy	22	Heavy	29
Trousers		Slacks	
Light	25	Light	25
Heavy	20	Heavy	15
Sweater		Sweater	
Light	32	Light	38
Heavy	17	Heavy	17
Jacket		Jacket	
Light	29	Light	38
Heavy	13	Heavy	17
Footwear			
Socks		Stockings	
Ankle length	161	Any length	640
Knee length	65	Panty hose	640
Shoes		Shoes	
Sandals	320	Sandals	320
Oxfords	160	Pumps	160
Boots	81	Boots	81

[a] 5% lower conductance or 5% higher clo for tie or turtleneck.
Source: Johnson (1991a).

Convection and radiation also contribute to heat loss from clothing. Convection and radiation are parallel heat loss processes, and, likewise, the thermal resistances corresponding to convection and radiation are in parallel with one another. The combination of these parallel resistances is in series with the conduction thermal resistance of the clothing. Thus the overall thermal resistance limiting heat loss through clothing is

total resistance

$$
\begin{aligned}
R_{\text{tot}} &= R_{\text{cl}} + R_{\text{c}} \| R_{\text{r}} \\
&= R_{\text{cl}} + \frac{(1.2Ah_{\text{c}})^{-1}(1.2h_{\text{r}}A)^{-1}}{(1.2h_{\text{c}}A)^{-1} + (1.2h_{\text{r}}A)^{-1}} \\
&= R_{\text{cl}} + [1.2A(h_{\text{r}} + h_{\text{c}})]^{-1}
\end{aligned}
\tag{3.2.13}
$$

where R_{tot} is the total thermal resistance [°C sec/(N m)]; R_{cl}, the conduction thermal resistance of clothing [°C sec/(N m)]; R_{c}, the convection thermal resistance [°C sec/(N m)]; R_{r}, the radiation thermal resistance [°C sec/(N m)]; h_{c}, the convection coefficient [N m/(sec m^2 °C)]; h_{r}, the radiation coefficient [N m/(sec m^2 °C)]; and A, the nude body surface area (m^2).

still air

radiation coefficient

Convection coefficients are discussed in Section 3.3, and values may be obtained from Tables 3.3.3, 3.3.4, and 3.3.5. The velocity of "still" air is usually taken to be 0.15 m/sec. Radiation coefficients are discussed in Section 3.4.6, and a value of 4.7 N m/(sec m^2 °C) is often used for people inside a room.

radiation

Clothed people subjected to radiant heat gain or loss at radiant temperatures different from mean ambient temperature must be treated somewhat differently than those with ambient and radiant temperatures essentially the same (Johnson, 1991a).

moisture transmission

Moisture can also be transmitted through clothing and produce heat loss (McCullough, 1993). Some discussion of this is given in Sections 3.8.1.2 and 4.8.2.3. For more details see ASHRAE (1977) or Johnson (1991a).

EXAMPLE 3.2.2.1-1. Heat Loss through Clothing

In a 1972 study, Van Cott and Kinkade found the average body mass of 19-year-old college students in the eastern United States to be 72.1 kg. If one of these students is 185 cm tall and sits quietly in a classroom at 22°C, what will be his heat loss if he wears briefs, a T shirt, blue jeans (assumed to be light trousers), no socks, and sandals?

Solution

The body weight of the student is (72.1 kg) (9.8 N/kg) = 707 N. The nude skin area of the student is, from Eq. 3.2.11,

$$A_{nude} = 0.07673 W^{0.425} H_t^{0.725} = 0.070673(707)^{0.425}(1.85)^{0.725}$$
$$= 1.79 \text{ m}^2$$

The clothing conductance values are, from Table 3.2.3,

Briefs	129 N/(m sec °C)
T shirt	72
Trousers	25
Sandals	320

Clothing thermal resistance is found from Eq. 3.2.9a as

$$R_{cl} = \{0.494[0.00775 + 0.01389 + 0.04000 + 0.00312] + 0.0119\}/1.79$$
$$= 0.0428°C \text{ sec}/(\text{N m})$$

Using values of $h_c = 3.1$ N m/(sec m^2 °C) and $h_r = 4.7$ N m/(sec m^2 °C), total thermal resistance is, from Eq. 3.2.13

$$R_{tot} = 0.0428 + [(1.2)(1.79 \text{ m}^2)(3.1 + 4.7 \text{ N m}/(\text{sec m}^2 \text{ °C}))]^{-1}$$
$$= 0.103°C \text{ sec}/(\text{N m})$$

Skin temperature can be calculated from Eq. 3.2.15 to be

$$\theta_{sk} = 25.8 + 0.257(22°C) = 31.7°C$$

Heat loss is just

$$\dot{q} = (\theta_{sk} - \theta_\infty)/R_{th} = (31.7°C - 22°C)/0.103 \text{ sec } °C/(\text{N m})$$
$$= 94 \text{ N m/sec}$$

Remark

This amount of heat loss is slightly more than the 86 N m/sec indicated in Table 3.5.3 for basal heat production in a 20-year-old male, and would probably be acceptable to maintain comfort.

3.2.2.2 Fur and Feathers   "Clothing" worn by animals normally takes the form of fur or feathers. The process of heat transfer from the skin of these animals is similar to that of heat through clothing from the skin of humans. Indeed, the thermal resistance of Arctic clothing is nearly the same as fur or feathers of the same thickness (Hammel, 1955).

Values for thermal conductivity of fur and feathers [such as the ones from Cena and Clark (1979) that appear in Table 3.2.2] are not as easily measured as their conductances. Thus values in Table 3.2.4 are conductance values that

Table 3.2.4 Thermal Conductance of Fur and Feathers, Includes Skin and Still Air Layer

Animal	Condition	Body Mass (kg)	Fur Thickness (mm)	Conductance $[N \ m/(m^2 \ sec \ °C)]$
Rabbit	back	2.5	9	2.4
Horse	flank	650	8	4.1
Horse	belly	650	4	4.2
Pig	flank	100	5	2.8
Pig	belly	100	2.5	3.4
Chicken	well-feathered	2	25	1.6
Raven	—	1.25	—	0.75
Cow	—	600	10	3.2
Dog	husky, in winter	50	44	1.2
Polar bear	winter, in air	400	63	1.2
Polar bear	in water	400	57	27
Beaver	winter, in air	26	44	1.2
Beaver	in water	26	40	14
Squirrel	summer	—	3	9.0
Monkey	summer	—	8	3.2

Sources: Tregear (1965), Blaxter et al. (1959), Scholander et al. (1950), Wathes and Clark (1981), Taylor (1981), Webster (1974).

give total rate of heat transfer per unit area. Total surface area of an animal can be estimated from (Cena, 1974)

animal surface area

$$A = 0.09m^{2/3} \tag{3.2.14}$$

where A is the skin surface area (m²); and m, the body mass (kg).

Values in Table 3.2.4 also include convection and radiation from the surface of the fur or feathers. Most of these measurements were made under conditions where radiation was not important, but convection resistance was sometimes a significant portion of the total thermal resistance. For very sparse fur or hair *convection* (number of hairs per unit area is small), convection resistance represents nearly *contribution* 100% of the total thermal resistance from the skin to the air; for very thick fur, the proportion of total thermal resistance represented by convection falls to about 10%. Thus the thermal resistance for a pig flank is about 40% due to convection resistance, that for a horse flank is about 30%, and that for rabbit fur is about 10% (Tregear, 1965).

Natural convection within fur is an important mover of heat (Hammel, 1955). Thus the thermal resistance of fur and feathers depends strongly on the wind speed and direction. Values tabulated here are those made in still air. To *wind* determine values under other wind conditions, either consult original *correction* references, or estimate the values of convection coefficients in still air and wind, and replace the still air convection resistance with that for wind.

The parallel combination of convection and radiation resistance was measured by Wathes and Clark (1981) as 0.12 m² sec °C/(N m); the resistance of the surface layer of skin was measured as 0.12 m² sec °C/(N m); and the resistance of the deep layer of tissues was measured as 0.08 m² sec °C/(N m). This latter value can approximately halve during vasodilation.

EXAMPLE 3.2.2.2-1. Heat Loss Through Feathers

Estimate the heat loss from a domestic chicken of 2 kg body mass and 41°C body temperature to its environment at 10°C.

Solution

From Table 3.2.4, we obtain a value for body conductance of 1.6 N m/(m² sec °C). This includes allowance for the skin and still air layer. Body surface area of the chicken is, from Eq. 3.2.14

$$A = 0.09(2 \text{ kg})^{2/3} = 0.143 \text{ m}^2$$

Total conductance is

$$C = \left(1.6 \frac{\text{N m}}{(\text{m}^2 \text{ sec } °\text{C})}\right)(0.143 \text{ m}^2) = 0.229 \frac{\text{N m}}{\text{sec } °\text{C}}$$

Thermal resistance is

$$R = 1/C = 4.37 \frac{°\text{C sec}}{\text{N m}}$$

The resistance of the deep layer of tissue is

$$R = \left(0.08 \frac{\text{m}^2 \text{ sec } °\text{C}}{\text{N m}}\right)\left(\frac{1}{0.143 \text{ m}^2}\right) = 0.559 \frac{°\text{C sec}}{\text{N m}}$$

Total thermal resistance from the core of the chicken's body to the environment is

$$R_{\text{tot}} = 4.37 + 0.559 = 4.93 \frac{°\text{C sec}}{\text{N m}}$$

This assumes that mean radiant temperature is the same as the ambient temperature of 10°C. Heat loss through the feathers is

$$\dot{q} = \frac{\theta_{\text{body}} - \theta_{\text{amb}}}{R_{\text{tot}}} = \frac{(41°\text{C} - 10°\text{C})}{4.93°\text{C sec}/(\text{N m})}$$

$$= 6.28 \frac{\text{N m}}{\text{sec}}$$

There is approximately 15% heat loss by evaporation and convection through the respiratory system. In addition, heat loss through the legs can be important unless active thermoregulatory mechanisms (vasoconstriction and counter-current heat exchange) are being used to conserve heat. The estimate for total heat loss from the bird is thus about 20% higher than the above figure, or $\dot{q} = 7.5$ N m/sec.

Remark

We will see from Eq. 3.5.7 that calculated heat production from a 2 kg animal is about $\dot{q} = 3.39 \ (2 \ \text{kg})^{0.75} = 5.7$ N m/sec. The value obtained here is higher than that because 10°C is below the thermoneutral zone for the chicken; it generates additional heat to maintain its 41°C body temperature.

3.2.3 Multidimensional Conduction

Until now we have considered only one-dimensional heat conduction, where heat flow was always perpendicular to the faces of the object. Two- and three-dimensional conduction is much more complicated, and our simple thermal resistance techniques are not sufficient to calculate heat transfer. Multidimensional heat conduction is no longer perpendicular to the faces of the object because of interaction between flows of heat in the different directions. Most of these problems are solved numerically for specific solutions matching specific sets of conditions. Analytic solutions are much more generally applied, however, and thus the technique will be illustrated.

analytical solutions sought

Again, the analytical solution must satisfy the heat conduction equation as well as boundary conditions matching the physical problem. Consider a flat rectangular plate insulated on the top and bottom so that no heat flows in the vertical direction. Heat flows in two dimensions (x and y) only. The temperature distribution is uniform in the third (z) direction. There is no heat generation in the plate, and the four exposed faces are subject to the following boundary conditions (Figure 3.2.1)

two-dimensional conduction

boundary conditions

$$\theta = 0, \quad y = 0$$
$$\theta = 0, \quad x = 0$$
$$\theta = 0, \quad x = 2L$$
$$\theta = \theta_o \sin \frac{\pi x}{2L}, \quad y = 2b$$

This last boundary condition is used because it is maximum in the center ($y = b$), but zero at both ends ($y = 0$ and $2b$). Thus there is no conflict with the other boundary conditions at the corners.

From Eq. 1.7.21, for uniform thermal conductivity,

$$\frac{\partial^2 \theta}{\partial x^2} + \frac{\partial^2 \theta}{\partial y^2} = 0 \tag{3.2.15}$$

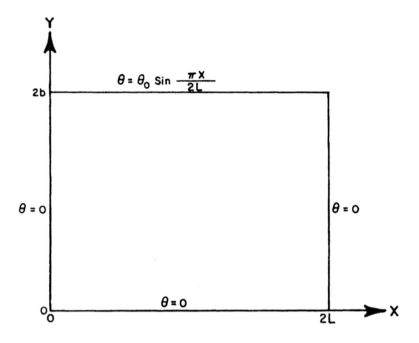

Figure 3.2.1. Rectangular plate with two-dimensional heat conduction.

This is a linear and homogeneous partial differential equation that can usually be integrated by assuming a product solution for $\theta(x, y)$

separation of
variables

$$\theta(x, y) = X(x)Y(y) \qquad (3.2.16)$$

Since $X(x)$ is a function only of x, $dX(x)/dy = 0$, and, likewise, $dY(y)/dx = 0$. Thus from Eq. 3.2.15

$$-\frac{1}{X}\frac{d^2X}{dx^2} = \frac{1}{Y}\frac{d^2Y}{dy^2} \qquad (3.2.17)$$

The variables are now separated, with the left-hand side a function only of x and right-hand side a function only of y. How can a function only of x be equal to a function only of y? This can only happen if both are equal to a constant.

$$-\frac{1}{X}\frac{d^2X}{dx^2} = \lambda^2 = \frac{1}{Y}\frac{d^2Y}{dy^2} \qquad (3.2.18)$$

Thus

$$\frac{d^2X}{dx^2} + \lambda^2 X = 0 \tag{3.2.19}$$

$$\frac{d^2Y}{dy^2} - \lambda^2 Y = 0 \tag{3.2.20}$$

The functions that when differentiated twice return functions of the original forms with opposite signs are sines and cosines.

$$X = A \cos \lambda x + B \sin \lambda x \tag{3.2.21}$$

The functions that when differentiated twice return a function of the original forms with the same signs are exponentials. Thus

$$Y = Ce^{-\lambda y} + De^{\lambda y} \tag{3.2.22}$$

Therefore,

$$\theta = XY = (A \cos \lambda x + B \sin \lambda x)(Ce^{-\lambda y} + De^{\lambda y}) \tag{3.2.23}$$

Since $\theta = 0$ at $x = 0$,

evaluating
constants

$$A(Ce^{-\lambda y} + De^{\lambda y}) = 0 \tag{3.2.24}$$

Thus $A = 0$. Since $\theta = 0$ at $y = 0$,

$$B \sin \lambda x \, (C + D) = 0 \tag{3.2.25}$$

Thus $C = -D$. Since $\theta = 0$ at $x = 2L$,

$$BC \sin 2\lambda L(e^{-\lambda y} - e^{\lambda y}) = 2BC \sin 2\lambda L \sinh \lambda y = 0 \tag{3.2.26}$$

Thus (1) $B = 0$, (2) $C = 0$, or (3) $\sin 2\lambda L = 0$. The first two choices violate the first two boundary conditions. Thus $2\lambda L = n\pi$, or $\lambda = n\pi/2L$, where $n = 0, 1, 2, 3, 4, \ldots$. As long as any of these values satisfies the equation, so must the sum of them. Thus

$$\theta = \sum_{n=0}^{\infty} G_n \sin \frac{n\pi x}{2L} \sinh \frac{n\pi y}{2L} \tag{3.2.27}$$

where G_n is a constant coefficient different for each n (°C). The last boundary condition requires that $\theta = \theta_0 \sin(\pi x/2L)$ at $y = 2b$. Thus, at the last boundary,

$$\theta_0 \sin\frac{\pi x}{2L} = \sum_{n=0}^{\infty} G_n \sin\frac{n\pi x}{2L} \sinh\frac{2n\pi b}{2L} \qquad (3.2.28)$$

The only way that this could be true is if $n = 1$. Thus

$$G_1 = \theta_0 / \sinh(\pi b/2L) \qquad (3.2.29)$$

and, finally

$$\theta(x, y) = \theta_0 \frac{\sinh(\pi y/2L)}{\sinh(\pi b/2L)} \sin\left(\frac{\pi x}{2L}\right) \qquad (3.2.30)$$

The rate of heat transfer in any direction s is just

$$\dot{q} = -kA_s \frac{d\theta}{ds} \qquad (3.2.31)$$

where A_s is the area perpendicular to direction s (m²). Isotherms and lines of constant heat flow for this problem appear in Figure 3.2.2.

In general, more complicated boundary conditions require that the entire sum of solutions in Eq. 3.2.27 be retained. Sometimes these sums recur so often that they are given special names. Hence, certain sums that arise in cylindrical coordinates are called Bessel functions, and others arising in spherical or spheroidal coordinates are called Legendre functions.

Bessel and Legendre functions

Although this example was developed for two-dimensional conduction, the same general approach is also used for three-dimensional conduction.

EXAMPLE 3.2.3-1. Two-Dimensional Heat Flow in Cooking Beef

A piece of beef $0.5 \times 10 \times 40$ cm is being cooked over a grill with a flame temperature of 2300°C. Determine the rate of heat transfer into the beef at the surface in the center of the piece of meat.

Solution

We will assume that the boundary conditions given for the previous development apply here. The boundary condition for $\theta = 0$ on three surfaces is no obstacle, because with conduction we can change the reference temperature at will. The sinusoidal temperature distribution on the fourth side may be questionable. We also assume that the meat is long enough that end effects can be neglected and heat transfer is two dimensional. To visualize this

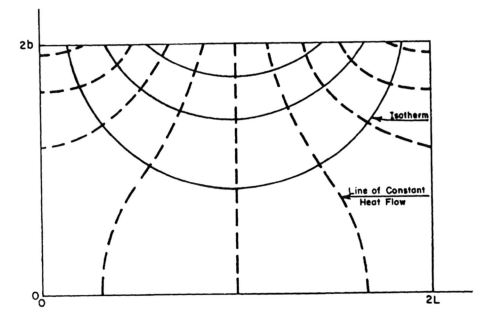

Figure 3.2.2. Isotherms and lines of constant heat flow for the rectangular object in Figure 3.2.1.

problem, imagine that the drawing in Figure 3.2.1 is upside down, with the flame being applied at the surface $y = 2b$.

From Eq. 3.2.30,

$$\dot{q} = -kA_s \frac{d\theta}{ds}$$

and, in this case, $s = -y$ in order to obtain the heat transferred from the top surface in Figure 3.2.1.

Thus, utilizing Eqs. 3.2.30 and 3.2.31

$$\dot{q} = -kA_s\theta_o\left(\frac{-\pi}{2L}\right)\frac{\cosh(\pi y/2L)}{\sinh(\pi b/2L)}\sin\left(\frac{\pi x}{2L}\right)$$

From Table 3.2.2, we estimate the thermal conductivity for meat as $k = 0.45$ N/(°C sec). The value for x in the center of the slab of beef is $x = L$, and $A_s = (0.1 \text{ m})(0.4 \text{ m}) = 0.04 \text{ m}^2$. Making use of the identity that

$$\cosh(\pi y/2L) = \tfrac{1}{2}(e^{-\pi y/2L} + e^{\pi y/2L})$$

and $y = 2b$,

$$\cosh(2\pi b/2L) = \cosh(0.05\pi)$$
$$= \tfrac{1}{2}(e^{-0.05\pi} + e^{0.05\pi})$$
$$= \tfrac{1}{2}(0.855 + 1.170) = 1.01$$

Also

$$\sinh(\pi b/2L) = \sinh(0.025\pi)$$
$$= \tfrac{1}{2}(e^{+0.025\pi} - e^{-0.025\pi})$$
$$= \tfrac{1}{2}(1.08 - 0.924 = 0.00876$$
$$\dot{q} = -\left(0.45 \; \frac{N}{{}^\circ C \; sec}\right)(0.04 \; m^2)(2300^\circ C)\left(\frac{-\pi}{0.10 \; m}\right)\left(\frac{1.01}{0.00786}\right)$$
$$\times (\sin \pi/2)$$
$$\dot{q} = 16{,}700 \; \frac{N \; m}{sec}$$

3.2.4 Non-Steady-State Conduction

We have already seen (Sections 1.7.3 and 1.8.2) that non-steady-state conduction can be considered to be equivalent to the action of thermal capacity acting through a thermal resistance (Figure 3.2.3). The analogy works best when the resistance to heat flow appears mainly outside the object being heated or cooled. When this condition is not true, then a distributed-parameter model rather than a lumped-parameter model must be used.

Given the condition that negligible internal resistance exists, then temperature, and heat flow, decrease exponentially (Eq. 1.6.8 and Figure 1.6.5). Further consideration will be given to non-steady-state heat transfer in Section 3.7.2.

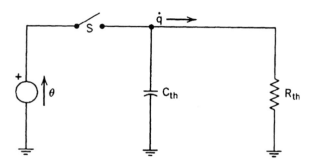

Figure 3.2.3. Transient heat conduction analogue. Thermal capacity is charged to a temperature value θ. When switch S is opened, the capacity discharges exponentially through resistance R_{th}.

3.3 CONVECTION

flowing fluid

Convection heat exchange differs essentially from conduction heat transfer in that heat moves with a flowing fluid. This is a very complicated process, involving the steps of: (1) heat transferred to or from a surface to the fluid mass, (2) the fluid mass either heating or cooling, and (3) the fluid moving away from the surface and being displaced by new fluid that begins the process anew. Thus there is simultaneous heat transfer and fluid flow and at least some interaction between the two.

complex process

To describe accurately the processes involved in convection heat transfer requires a great deal of knowledge about geometry, fluid boundary layer development, transient heat transfer modes, temperature effects on fluid physical properties, and many other considerations. Engineers have long avoided such complications by circumventing the need to formulate specific mathematical descriptions of basic processes contributing to convection heat transfer. Sometimes these complicated processes are included in numerical solutions obtained by computer; more often the complications are hidden by the introduction of a convection coefficient

simplification

convection equation

$$\dot{q} = h_c A(\theta_o - \theta_\infty) \tag{3.3.1}$$

where $\dot{q}$ is the heat transferred by convection (N m/sec); h_c, the convection coefficient [N/(sec m °C)]; A, the surface area (m²); θ_o, the surface temperature (°C); and θ_∞, the temperature of the fluid far away from the surface (°C).

In many cases this approach is sufficient because the exact temperature distribution inside the fluid is of little to no concern (Table 3.3.1).

forced convection

There are two types of convection to discuss. The first type is *forced* convection, wherein the fluid is motivated by a force unrelated to the temperature difference that exists between the surface and the fluid. Examples of forced convection are fluids moved by fans and pumps, and, on a local scale, wind.

natural convection

The second type of convection is *natural* convection, produced when heated surfaces cause the fluid to expand or when cold surfaces cause the fluid to contract. The differences in fluid buoyancy cause fluid movement to occur. Natural convection occurs in rooms with windows, in cooling cups of coffee, and in the oceans on a global scale. Natural convection is sometimes called free convection.

3.3.1 Convection Coefficients

forced convection

Convection coefficient values for forced convection are usually larger than values for natural convection due to the higher velocities achieved during forced convection. Most of the resistance to heat flow in convection occurs at the still boundary layer located at the interface of the surface and the fluid. As

Table 3.3.1   Differences between Forced and Natural Convection

Forced	Natural
Heat is swept along flow stream	Heat rises opposite to gravity or inertial force
Flow patterns are determined primarily by some external flow source	Flow patterns are determined by the buoyant effect of the heated fluid
Velocity profiles are determined first; then they are used to find temperature profiles	Velocity and temperature profiles are intimately connected
Nusselt number depends on Reynolds number and Prandtl number ($\mathrm{Nu} \propto \sqrt{\mathrm{Re}}\sqrt[3]{\mathrm{Pr}}$)	Nusselt number depends on Grashof number and ($\mathrm{Nu} \propto \sqrt[3]{\mathrm{Gr}}\sqrt[3]{\mathrm{Pr}}$)

greater than natural convection

fluid movement increases, the thickness of the boundary layer generally decreases (Figure 3.3.1). Turbulent fluid flow results in higher convection coefficient values than does laminar fluid flow.

3.3.1.1 Dimensionless Numbers

Convection coefficient values cannot generally be derived from basic principles, but instead are usually obtained from experimental results. Since convection coefficient values depend on geometry, temperature differences, fluid physical properties, and other considerations, the number of experiments necessary to determine convection coefficient values under all possible conditions is much too large to be practical. Early experimenters were thus driven to find a more general experimental approach.

They found that experimental results correlating dimensionless numbers gave the generality desired. By properly formulating dimensionless numbers, many sources of variation could be combined to reduce significantly the number of experiments necessary.

Nusselt number

Four such numbers are required: (1) Nusselt number, (2) Reynolds number, (3) Grashof number, and (4) Prandtl number. The Nusselt number expresses the ratio of heat loss from the surface by convection to the heat gain to the fluid by conduction

$$\mathrm{Nu} = \frac{h_c L}{k} \tag{3.3.2}$$

where Nu is the Nusselt number (dimensionless); h_c, the convection coefficient [N m/(sec m² °C)]; L, the significant length (m); and k, the thermal conductivity of fluid [N m/(sec m °C)].

significant length

The length (L) in the Nusselt number is a significant length of meaning for the particular geometry involved. Depending on the path of the fluid across the

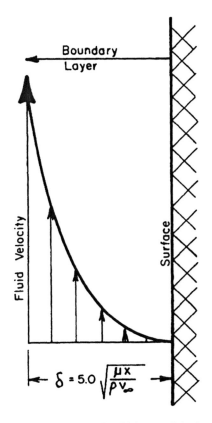

Figure 3.3.1. As fluid velocity increases, the thickness of the boundary layer decreases and convection heat transfer is enhanced. The Blasius equation for laminar flow boundary layer thickness (δ) indicates that fluid viscosity (μ), distance from the leading edge of the surface (x), fluid density (ρ), and free-stream velocity (v_∞) are all important parameters. Thermal and fluid boundary layers are similar, although not usually identical in thickness.

surface, L may be given by the length (L) or replaced by a diameter (d). As an example, consider convection along a flat plate. The value of L is the dimension of the plate in the direction of the flow. For fluid flowing inside tubes or ducts, the significant length is usually taken to be the diameter. If the fluid is flowing across a cylinder, then the significant length would be the diameter of the cylinder; if the fluid is flowing along the cylinder, then the significant length would be the length of the cylinder (Figure 3.3.2). If the fluid is flowing along a very thin cylinder or if the length of the surface is small relative to the boundary layer thickness, then the above generalizations do not apply, and other texts should be consulted. The same significant length used in the Nusselt number should also be used in the Reynolds and Grashof numbers.

Reynolds number

The Reynolds number has been introduced previously in Section 2.5.1 as the ratio of inertial force to viscous force

$$\text{Re} = \frac{\rho \, dv}{\mu} \tag{3.3.3}$$

where Re is the Reynolds number (dimensionless); ρ, the fluid density (kg/m^3); v, the fluid velocity (m/sec); μ, the fluid viscosity [kg/(m sec)]; and d, the pipe diameter (m).

generalized Reynolds number

The generalized Reynolds number for non-Newtonian fluids flowing in a pipe was introduced in Section 2.6.2.3 as

$$\text{GRe} = \frac{8d^n v^{2-n} \rho}{(3 + 1/n)^n 2^n K} \tag{3.3.4}$$

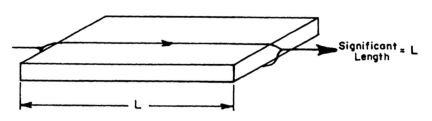

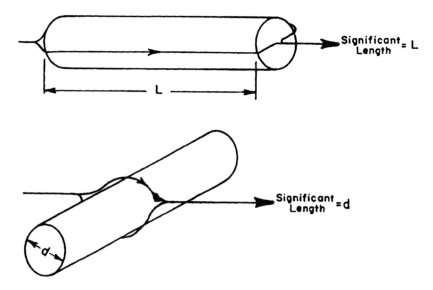

Figure 3.3.2. Determination of the significant length to use in dimensionless numbers depends on the path taken by the fluid.

where n is the flow behavior index (dimensionless); and K, the consistency coefficient (N $\sec^n/m^2$).

forced convection uses Reynolds number

The Reynolds number is used to correlate convection data for forced convection. In general, the Nusselt number is a function of the square root of the Reynolds number (Nu $\propto \sqrt{Re}$).

Grashof number

The Grashof number represents the ratio of buoyancy forces to viscous forces. It is defined as

$$Gr = \frac{\rho^2 g \beta d^3 (\theta_o - \theta_\infty)}{\mu^2} \qquad (3.3.5)$$

where Gr is the Grashof number (dimensionless); ρ, the fluid density (kg/m^3); g, the acceleration of gravity (9.8 m/sec^2); β, the coefficient of thermal expansion (1/°C); d, the pipe diameter (m); θ_o, the surface temperature (°C); θ_∞, the bulk fluid temperature (°C); and μ, the fluid viscosity [kg/(m sec)].

For fluid flowing across a planar surface, the diameter is replaced by a significant length, given as L.

generalized Grashof number

For non-Newtonian fluids flowing inside tubes, the generalized Grashof number (GGr) is the same as in Eq. 3.3.5, except that the viscosity (μ) is replaced by

equivalent viscosity

$$\mu = \left(\frac{v}{d}\right)^{n-1} \frac{(3 + 1/n)^n 2^n K}{8} \qquad (3.3.6)$$

It can be seen that this viscosity has been obtained from the generalized Reynolds number. Again, the diameter term can be replaced by a significant length, if appropriate.

natural convection uses Grashof number

The Grashof number is used to correlate natural convection data. Close inspection of both Reynolds and Grashof numbers might hint that in some respects the Reynolds number squared (for forced convection) is analogous to the Grashof number (for natural convection), or Re$^2 \propto$ Gr. In general, the Nusselt number is a function of the fourth root of the Grashof number (Nu $\propto \sqrt[4]{Gr}$).

The coefficient of thermal expansion appearing in Eq. 3.3.5 can be evaluated simply for gases that approximately obey the ideal gas law. The coefficient of thermal expansion is defined as the relative density change per temperature change

coefficient of thermal expansion

$$\beta = \left(\frac{\rho_\infty - \rho}{\rho}\right)\bigg/(T - T_\infty) \qquad (3.3.7)$$

where ρ_∞ is the density of the bulk fluid (kg/m^3); ρ, the local density of the fluid (kg/m^3); T_∞, the absolute temperature of the bulk fluid (K); and T, the local absolute temperature of the fluid (K). For an ideal gas, where

$$\rho = \frac{m}{v} = \frac{pM}{RT} \qquad (3.3.8)$$

where p is the total pressure (N/m^2); R, the gas constant $[8314.34 \text{ N m}/(\text{kg mol}/K)]$; and M, the molecular weight (kg/kg mol).

$$\beta = \frac{(\rho_\infty/\rho) - 1}{(T - T_\infty)} = \frac{(T/T_\infty) - 1}{T - T_\infty} = \frac{1}{T_\infty} \qquad (3.3.9)$$

bulk fluid
temperature

Thus β may be calculated by inverting the bulk fluid absolute temperature. The bulk fluid temperature is the temperature of the fluid far enough away from the surface to be unaffected by surface temperature.

The other common fluid of interest to biological engineers is water, and the coefficient of thermal expansion for water can be calculated from the following empirical equation:

$$\beta = (0.195 + 0.0712\theta) \times 10^{-4}, \qquad 4 \le \theta < 260°C \qquad (3.3.10)$$

Water at 4°C reaches its maximum density and becomes less dense below 4°C. Thus the value for β at 0°C is $-0.630/K$ (Geankoplis, 1993).

Equations relating Nusselt number to Grashof number will be presented shortly. For now, however, it must be noted that included in the Grashof number is the temperature difference $(\theta_o - \theta_\infty)$. In some cases this difference is not known, and, if it were known, there might be no reason to calculate the Grashof number. This temperature difference must be first assumed, the

iteration
process
necessary

Grashof number calculated, a Nusselt number determined, the convection coefficient obtained, convection heat transfer determined, and then the new temperature difference value calculated. This process is repeated until calculated and assumed values for temperature difference become equal (Figure 3.3.3).

The Prandtl number is the ratio of momentum diffusivity to thermal diffusivity

Prandtl
number

$$\text{Pr} = \frac{\nu}{\alpha} = \left(\frac{\mu}{\rho}\right)\left(\frac{c_p\rho}{k}\right) = \frac{c_p\mu}{k} \qquad (3.3.11)$$

where Pr is the Prandtl number (dimensionless); ν, the kinematic viscosity, or momentum diffusivity (m^2/sec); α, the thermal diffusivity (m^2/sec); μ, the viscosity $[\text{kg}/(\text{m sec})]$; ρ, the density (kg/m^3); c_p, the specific heat at constant pressure $[\text{N m}/(\text{kg °C})]$; and k, the thermal conductivity $[\text{N}/(°\text{C sec})]$.

boundary
layer
thickness

The square root of the Prandtl number determines the ratio of the fluid boundary thickness to thermal boundary layer thickness $(\delta/\delta_t \propto \sqrt{\text{Pr}})$. For Prandtl numbers about 1.0, the fluid boundary layer thickness is nearly the same as the thermal boundary layer thickness. The Prandtl number for air is nearly constant over the temperature range of 0–100°C at 0.7. The Prandtl number for water over the same range varies from 13 to 2.

The Prandtl number for non-Newtonian fluids depends somewhat on geometry. For flow in a tube, the viscosity (μ) in Eq. 3.3.11 is replaced by that

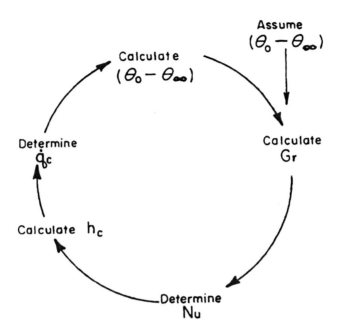

Figure 3.3.3. In the case where surface temperature is not known for a natural convection event, the temperature difference $(\theta_0 - \theta_\infty)$ must be first assumed and then this iterative loop followed. If the amount of convective heat transfer is known, $(\theta_0 - \theta_\infty)$ is not calculated, but adjusted based upon the closeness of calculated and known convective heats.

in Eq. 3.3.6. For fluid between two flat plates moving relative to one another

generalized
Prandtl
number

$$GPr = \frac{c_p K v^{n-1}}{k b^{n-1}} \qquad (3.3.12)$$

where GPr is the generalized Prandtl number (dimensionless); and b, the spacing between plates (m).

Notice that both consistency coefficient (K) and thermal conductivity (k) appear in the same equation and can easily cause confusion or errors if care is not taken to distinguish between the two symbols.

Combinations of the Prandtl number with the Reynolds and Grashof numbers are themselves dimensionless numbers, and are sometimes used in Nusselt number equations. The product of Reynolds and Prandtl numbers is called the Peclet number

Peclet
number

$$Pe = Re \, Pr \qquad (3.3.13)$$

where Pe is the Peclet number (dimensionless). The product of the Grashof number and Prandtl number is called the Rayleigh number

$$Ra = Gr\,Pr \qquad (3.3.14)$$

where Ra = Rayleigh number dimensionless.

The Graetz number is sometimes used for heat transfer of fluid flowing in long pipes. Its definition is:

$$Gz = \frac{\dot{m}c_p}{kL} \qquad (3.3.15)$$

where Gz is the Graetz number (dimensionless); $\dot{m}$, the mass flow rate (kg/sec); c_p, the specific heat of fluid at constant pressure [N m/(kg °C)]; k, the thermal conductivity of fluid [N m/(sec m °C)]; and L, the significant length, usually diameter (m).

EXAMPLE 3.3.1.1-1. Cardiopulmonary Bypass

A cardiopulmonary bypass apparatus is used during major heart or lung surgery to oxygenate and pump blood. A heat exchanger is often used to cool the patient to slow metabolism during the operation. Water at 0–5°C is used to cool the blood and becomes heated to 10–20°C. If the bulk velocity of the water is about 0.05 m/sec, what is its boundary layer thickness 5 cm from the leading edge of the separation between blood and water?

Solution

From Figure 3.3.1, we see that the boundary layer thickness can be calculated from

$$\delta = 5.0\sqrt{\frac{\mu x}{\rho v_\infty}}$$

We obtain viscosity information from Table 2.4.1 and density information from Table 2.5.4, assuming a water temperature of 10°C. Boundary layer thickness is thus

$$\delta = 5.0\sqrt{\frac{(0.00138 \text{ kg/m sec})(0.05 \text{ m})}{(1000 \text{ kg/m}^3)(0.05 \text{ m/sec})}}$$

$$= 0.0059 \text{ m}$$

Remarks

What has just been calculated is the fluid boundary layer. The thickness of the thermal boundary layer is just

$$\delta_t = \delta/\sqrt{Pr} = \frac{0.0059 \text{ m}}{\sqrt{10}} = 0.0019 \text{ m}$$

Both of these boundary layers grow in thickness until either:

1. They completely fill the available flow space (that is, until the boundary layers from opposing walls meet in the center); or
2. Turbulence or other disturbances disrupt the layers.

3.3.1.2 *Forced Convection Equations* A compendium of equations to calculate forced convection Nusselt numbers is given here. The general procedure for using these mathematical correlations of experimental data is diagrammed in Figure 3.3.4. As indicated, one must search through the list of equations for as close a match of geometry, Reynolds number range, and Prandtl number range, as possible. The Nusselt number is then calculated, and the convection coefficient value extracted from the Nusselt number. Finally, any assumptions, especially temperature assumptions, are checked for accuracy.

choice of
equations

Physical properties of the fluid (thermal conductivity, specific heat, viscosity) are often temperature dependent. Values of these properties can be determined at the temperature of the surface, or of the bulk fluid (ambient), or at an intermediate temperature called the film temperature. The film temperature is calculated as the average of surface temperature (θ_0) and ambient temperature (θ_∞).

film
temperature

$$\theta_f = \tfrac{1}{2}\theta_o + \tfrac{1}{2}\theta_\infty \qquad (3.3.16)$$

If physical properties are to be evaluated at the surface temperature, this will be denoted by a subscript "0" (for example, Re_0). If properties are to be evaluated at the temperature of the bulk fluid, this is denoted by a subscript "∞" (for example, Pr_∞). Evaluation at the film temperature is denoted by a subscript "f" (for example, Nu_f). Unless otherwise indicated, all physical properties should be evaluated at the film temperature.

For the common case where heat transfer occurs while a fluid flows inside a pipe, there is usually a difference in bulk fluid temperature between the pipe inlet and outlet (Figure 3.3.5). There is no easily identifiable bulk fluid (or ambient) temperature from which a film temperature can be calculated and physical properties evaluated. In this case, a logarithmic mean temperature is usually used in place of the ambient temperature. Logarithmic mean temperature is calculated as the square root of the product of the inlet and

logarithmic
mean
temperature

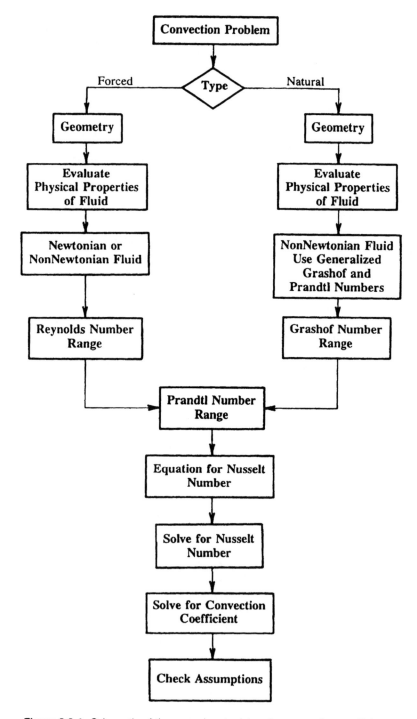

Figure 3.3.4. Schematic of the procedure to determine convection coefficient.

outlet temperatures

$$\theta_{\text{lmt}} = \sqrt{\theta_{\text{in}} \theta_{\text{out}}} \tag{3.3.17}$$

where θ_{lmt} is the logarithmic mean temperature (°C); θ_{in}, the inlet temperature (°C); and θ_{out}, the outlet temperature (°C). (This is called logarithmic mean because $\log \theta_{\text{lmt}} = \frac{1}{2} \log \theta_{\text{in}} + \frac{1}{2} \log \theta_{\text{out}}$.)

corrections for temperature dependence of viscosity

The viscosities of many liquids vary greatly with temperature. It has been suggested that Nusselt numbers calculated from Reynolds and Prandtl numbers based upon viscosity values evaluated at the bulk fluid temperature be corrected by multiplying by a term $(\mu_o/\mu_\infty)^{0.14}$. If the fluid is a gas, then the correction term should be $(T_\infty/T_o)^n$, where T is absolute temperature and $n = 0.25$ for gas heating in a tube and $n = 0.08$ for gas cooling in a tube. These correction terms are only approximate, and are given here as an alternative to evaluation of physical properties at the film temperature.

In the equation listing that follows, there are sometimes several equations given to calculate Nusselt number for a given set of conditions. This is because several different investigators obtained data for nearly the same set of conditions, and presented the correlated data in forms that they deemed appropriate. In these cases, it is not always clear which is the most accurate equation.

1. Flat plate, flow parallel to plate, laminar flow $0.6 < \text{Pr} < 1000$ (Gebhart, 1971; Kreith, 1958)

$$\text{Nu} = 0.664 \sqrt{\text{Re}} \sqrt[3]{\text{Pr}} \tag{3.3.18}$$

2. Flat plate, flow parallel to plate, turbulent flow $0.7 \le \text{Pr}$ (von Kármán) (Gebhart, 1971; Kreith, 1958)

$$\text{Nu} = 0.037 \, \text{Re}^{0.8} \, \text{Pr}^{1/3} \tag{3.3.19}$$

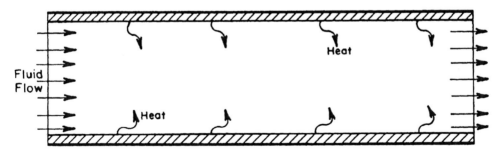

Figure 3.3.5. When fluid flowing inside a pipe is heated by the surrounding pipe walls, the bulk temperature of the fluid increases constantly.

3. Smooth-walled tubes, flow inside, turbulent flow, small temperature differences (Dittus–Boelter) (Gebhart, 1971; Kreith, 1958; Henderson and Perry, 1966). $0.6 < \mathrm{Pr} < 100$; $L/d > 60$

$$\mathrm{Nu}_\infty = 0.023\, \mathrm{Re}_\infty^{0.8}\, \mathrm{Pr}_\infty^m \tag{3.3.20}$$

$m = 0.4$ for fluid being heated; $m = 0.3$ for fluid being cooled; different numbers to account for temperature-dependent properties.

4. Smooth-walled tubes, flow inside, turbulent flow, larger temperature differences (but not large relative to temperature dependence of fluid properties). $L/d > 60$ (Colburn) (Gebhart, 1971)

$$\mathrm{Nu}_\infty = 0.023\, \mathrm{Re}_f^{-0.2}\, \mathrm{Re}_\infty\, \mathrm{Pr}_\infty \tag{3.3.21}$$

5. Smooth-walled tubes, flow inside, turbulent flow, moderate heat-transfer temperature differences, high-viscosity fluids. $L/d > 60$ (Sieder–Tate) (Gebhart, 1971; Henderson and Perry, 1966)

$$\mathrm{Nu}_\infty = 0.023\, \mathrm{Re}_\infty^{0.8}\, \mathrm{Pr}_\infty^{1/3} \left(\frac{\mu_\infty}{\mu_o}\right)^{0.14} \tag{3.3.22}$$

6. Flow inside tubes, turbulent flow, very low Prandtl numbers: liquid metals (Lyon) (Gebhart, 1971; Kreith, 1958)

$$\mathrm{Nu} = 5.0 + 0.025(\mathrm{Re}\, \mathrm{Pr})^{0.8} \tag{3.3.23a}$$

$100 < \mathrm{Re}\, \mathrm{Pr} < 10{,}000$ (Lubarsky–Kaufman) (Gebhart, 1971)

$$\mathrm{Nu} = 0.625(\mathrm{Re}\, \mathrm{Pr})^{0.4} \tag{3.3.23b}$$

7. Flow inside tubes, laminar flow, high Prandtl numbers: oils, heavy hydrocarbons, sugar solutions. $1 < \mathrm{Pr} < 500$ (Friend–Metzner)

$$\mathrm{Nu}_\infty = \frac{\mathrm{Re}_\infty\, \mathrm{Pr}_\infty^{4/3}}{0.150\sqrt{\mathrm{Re}} + 4.17(\mathrm{Pr} - 1)} \tag{3.3.24}$$

8. Flow inside tubes, laminar flow (Sieder–Tate) (Gebhart, 1971; Kreith, 1958)

$$\mathrm{Nu}_\infty = 1.86\, \mathrm{Re}_\infty^{1/3}\, \mathrm{Pr}^{1/3} \left(\frac{d}{L}\right)^{1/3} \left(\frac{\mu_\infty}{\mu_f}\right)^{0.14} \tag{3.3.25}$$

For $\mathrm{Re}\, \mathrm{Pr}(d/L) > 10$, mean temperature is the average between inlet and outlet. $\mathrm{Re}\, \mathrm{Pr}(d/L) < 10$, long pipes: mean temperature is the logarithmic mean $= \sqrt{\theta_1 \theta_2}$.

9. Flow inside tubes, isothermal tube, fluid properties independent of temperature, plug flow: pseudoplastics. Re $\Pr(d/L) < 400$ (Graetz) (Heldman, 1975).

$$\text{Nu} = 0.5 \text{ Re Pr}\left(\frac{d}{L}\right)\left(\frac{1 - 4E'(4/\text{Re Pr}(d/L))}{1 + 4E'(4/\text{Re Pr}(d/L))}\right) \tag{3.3.26a}$$

where

$$E'(4/\text{Re Pr}(d/L)) = \sum_{j=1}^{\infty}\frac{1}{a_j^2}\exp\left(-a_j^2\frac{4}{\text{Re Pr}(d/L)}\right)$$

a_j is the jth root of $J_0(a_j) = 0$ and J_0 is the zeroth-order Bessel function. Re $\Pr(d/L) > 400$ (Heldman, 1975).

$$\text{Nu} = \frac{8}{\pi} + \frac{4}{\pi}\sqrt{\frac{\pi}{4} \text{ Re Pr}\left(\frac{d}{L}\right)} \tag{3.3.26b}$$

10. Non-Newtonian flow in tubes, isothermal tube, (Pigford) (Heldman, 1975)

$$\text{Nu} = 1.75\left(\frac{3n + 1}{4n}\right)^{1/3}\left[\frac{\pi}{4}(\text{GRe})(\text{GPr})\left(\frac{d}{L}\right)\right]^{1/3} \tag{3.3.27}$$

n is the flow behavior index.

11. Pseudoplastic fluids through tubes $0.4 < n < 1.0$ (Charm–Merrill) (Heldman, 1975). n does not vary greatly over wide ranges of temperature.

$$\text{Nu} = 2\left[\frac{\pi}{4}(\text{GRe})(\text{GPr})\left(\frac{d}{L}\right)\right]^{1/3}\left(\frac{K_\infty}{K_0}\frac{3n + 1}{2(3n - 1)}\right)^{0.14} \tag{3.3.28a}$$

K is the consistency coefficient. Convection on surfaces of food particles in non-Newtonian fluid (Bhamidipati and Singh, 1994).

$$\text{Nu} = 0.27 \text{ GRe}^{0.20} \text{ GPr}^{0.33} \tag{3.3.28b}$$

Dimensions are based on particle size.

12. Laminar flow of non-Newtonian fluids in tubes (with natural convection correction) (Mahalingam et al., 1975)

$$\begin{aligned}\text{Nu}_\infty = 1.46&\left(\frac{3n + 1}{4n}\right)^{1/3}\left(\frac{K_\infty}{K_0}\right)^{0.14}\\&\times\left\{\left[\frac{\pi}{4}(\text{GRe})_\infty(\text{GPr})_\infty\left(\frac{d}{L}\right)\right] + 0.083(\text{GGr}_0\ \text{GPr}_0)^{0.75}\right\}^{1/3}\end{aligned} \tag{3.3.29}$$

13. Flow in tubes, noncircular cross section. Use hydraulic diameter (Gebhart, 1971). $L/d > 30$.

14. Flow between concentric cylinders (Gebhart, 1971). $L/d > 15$, $Re > 10^4$, $d = d_2 - d_1$, the annular space.

$$Nu_\infty = 0.023\ Re_\infty^{0.8}\ Pr_\infty^{1/3} \left(\frac{\mu_\infty}{\mu_o}\right)^{0.14} \tag{3.3.30}$$

where mean temperature is the logarithmic mean of inlet and outlet.

15. Flow across and normal to cylinders $Pr \approx 1.0$ (Hilpert) (Gebhart, 1971; Kreith, 1958).

$$Nu_f = B\ Re_f^m\ Pr^{1/3} \tag{3.3.31a}$$

Re	m	B
1–4	0.330	0.998
4–40	0.385	0.920
40–4,000	0.466	0.689
4,000–40,000	0.618	1.315
40,000–250,000	0.805	0.0268

$3 < Re < 100$ (Kreith, 1958).

$$Nu = 0.82\ Re^{0.4}\ Pr^{0.3} \tag{3.3.31b}$$

(Kreith, 1958)

$$Nu = C\sqrt{Re} \tag{3.3.31c}$$

Pr	0.7	0.8	1.0	5.0	10.0
C	1.0	1.05	1.14	2.1	1.7

$0.1 < Re < 1,000$ (Henderson and Perry, 1966).

$$Nu = 0.35 + 0.47\ Re^{0.52}\ Pr^{0.3} \tag{3.3.31d}$$

$1000 < Re < 50,000$ (Henderson and Perry, 1966).

$$Nu = 0.26\ Re^{0.6}\ Pr^{0.3} \tag{3.3.31e}$$

16. Flow across bodies of noncircular cross section (use Eq. 3.3.31a).

Configuration	Re	m	B
$\rightarrow \diamond$	5,000–10,000	0.588	0.249
$\rightarrow \square$	5,000–10,000	0.675	0.103
$\rightarrow \bigcirc$	5,000–100,000	0.638	0.155
$\rightarrow \bigcirc$	5,000–19,500	0.638	0.161
	19,500–100,000	0.782	0.0389
$\rightarrow$ |	4,000–15,000	0.731	0.230

17. Flow across a sphere (Pr not extreme) $1 < Re < 70,000$ (Gebhart, 1971).

$$\mathrm{Nu}_f = 2.0 + 6.0\, \mathrm{Re}_f^{1/2}\, \mathrm{Pr}_f^{1/3} \tag{3.3.32a}$$

$25 < Re < 10^5$ (Kreith, 1958; Henderson and Perry, 1966).

$$\mathrm{Nu} = 0.37\, \mathrm{Re}^{0.6} \tag{3.3.32b}$$

$2000 < Re < 7000$ (Tang et al., 1991).

$$\mathrm{Nu}_f = 0.671\, \mathrm{Re}_f^{0.515} \tag{3.3.32c}$$

$$\mathrm{Nu} = 0.41\, \mathrm{Re}^{0.6}\, \mathrm{Pr}^{0.3} \tag{3.3.32d}$$

18. Turbulent flow of non-Newtonian fluids in tubes (Skelland, 1967).

$$\mathrm{Nu} = 0.041(\mathrm{GRe})(\mathrm{GPr})^{0.4} \tag{3.3.33}$$

A special convection case occurs when two surfaces are separated by a stagnant fluid medium. In this case, heat actually transfers by conduction, but, because the fluid is involved, the problem is treated as if it were convection.

If convection from the surface is equated to conduction in the fluid, the following is obtained

$$h_c A(\theta_0 - \theta_\infty) = \frac{kA}{L}(\theta_0 - \theta_\infty) \tag{3.3.34}$$

$$\frac{h_c L}{k} = 1.0 \tag{3.3.35}$$

where A is the area of the surface (m²); L, the thickness of the boundary layer equal to the separation between the surfaces (m); and k, the thermal conductivity of the fluid [N/(°C sec)].

The term on the left is the Nusselt number, and if the fluid is stagnant so that convection does not occur, but all heat is transferred by conduction, then the Nusselt number equals 1.0.

Although heat is being transferred, in this case by conduction, we are calling it convection. Contrarily, if the Nusselt number just happens to have a value of 1.0, it does not mean that heat is being transferred by conduction.

EXAMPLE 3.3.1.2-1. Convection Coefficient in a Hot Tub

A person stands at the edge of a hot tub, where the water velocity is 0.02 m/sec. The person weighs 608 N (mass = 62 kg) and is 165 cm tall. The temperature in the hot tub is 43°C. What is the value of the convection coefficient at the surface of the skin?

Solution

Human body shapes are irregular, and are often approximated as cylinders. As long as the arms are kept close to the trunk and legs are kept together, this approximation is reasonable. Because convection heat loss depends on skin surface area, the equivalent cylinder is determined to have the same height as the person, and the cylinder diameter is chosen to give the same surface area. From Eq. 3.3.11

$$A = 0.2025(62)^{0.425}(165)^{0.725}$$
$$= 1.68 \text{ m}^2$$

The surface area of a cylinder is

$$A = 2\left(\frac{\pi D^2}{4}\right) + \pi D H_t$$

From which, using the quadratic formula, we obtain

$$D = -H_t \pm \sqrt{H_t^2 + \frac{2A}{\pi}}$$
$$= -1.65 \pm \sqrt{1.65^2 + \frac{(2)(1.68)}{\pi}}$$
$$= 0.30 \text{ m}$$

The Reynolds number must now be calculated. The most appropriate temperature at which to evaluate physical properties is at the film temperature, given as the average between the surface temperature, assumed to be 37°C, and the water temperature

$$\theta_f = \frac{37°C + 43°C}{2} = 40°C$$

From Tables 2.4.2, 2.5.4, 3.2.1, and 3.6.1, we obtain

$$\mu = 6.61 \times 10^{-4} \text{ kg/(m sec)}$$
$$\rho = 1000 \text{ kg/m}^3$$
$$k = 0.637 \text{ N m/(m sec °C)}$$
$$c_p = 4180 \text{ N m/(kg °C)}$$

The Reynolds number is thus

$$\text{Re} = \frac{Dv\rho}{\mu} = \frac{(0.30)(0.02 \text{ m/sec})(1000 \text{ kg/m}^3)}{6.61 \times 10^{-4} \text{ kg/(m sec)}}$$
$$= 9077$$

The Prandtl number is

$$\text{Pr} = \frac{c_p \mu}{k} = \frac{(4180 \text{ N m/(kg °C)})[6.61 \times 10^{-4} \text{ kg/(m sec)}]}{0.637 \text{ N m/(m sec °C)}}$$
$$= 4.34$$

We now search among the forced convection equations to find a set of conditions closely matching the flow of water across the circumference of a cylinder. If the Prandtl number can be considered to be close to 1.0, then Eq. 3.3.31a will suffice. Equations 3.3.31c and 3.3.31e also match the conditions. Equations 3.3.31b and 3.3.31d are inappropriate because our Reynolds number is out of range for those equations.

Choosing first to use Eq. 3.3.31a

$$\text{Nu}_f = (1.315)(9077)^{0.618}(4.34)^{1/3}$$
$$= 599$$

Using Eq. 3.3.31c

$$\text{Nu} = 2\sqrt{9077} = 191$$

Using Eq. 3.3.31e

$$\text{Nu} = 0.26(9077)^{0.6}(4.34)^{0.3} = 96$$

This is a large range of Nusselt numbers, and there is little to guide the selection about which to choose. We chose the first, however, for two reasons: First, it gives the highest value for convection coefficient, and that is usually the most conservative approach; and, second, the Hilpert equation is time tested and found to be satisfactory, if not reliable.

$$h_c = \frac{\text{Nu } k}{D} = \frac{(599)(0.637 \text{ N m}/(\sec \text{ m } °C))}{0.30 \text{ m}}$$

$$= 1270 \text{ N m}/(\sec \text{ m}^2 \text{ } °C)$$

Remarks

This is a very high convection coefficient value, and indicates why spending long times is hot tubes can lead to hyperthermia.

This exercise should illustrate that inordinate precision in calculations of convection coefficient is usually not warranted.

EXAMPLE 3.3.1.2-2. Heat Loss in a Tube

Tomato sauce is being pumped through a 20-m-long stainless steel pipe, 102 mm inside diameter, with wall thickness of 6.02 mm. The outside of the pipe is open to the 27°C air, flowing at 2.7 m/sec across the pipe. If the tomato sauce flows at a rate of 13 kg/sec, enters the pipe at a temperature of 82°C, and exits at 80°C, how much heat was transmitted through the pipe to the ambient environment?

Solution

The total thermal resistance offered by the pipe consists of three parts in series: convection resistance inside the pipe, $1/h_iA_i$; conduction resistance of the pipe wall, $\ln(r_o/r_i)/2\pi kL$; and convection resistance outside the pipe, $1/h_oA_o$ (see Figure 3.7.4). We begin by calculating the outside convection resistance. We begin by assuming the temperature at the outer surface of the pipe is 27°C; thus the film temperature is also 27°C. The Reynolds number of air flowing outside the pipe is

$$\text{Re} = \frac{dv\rho}{\mu} = \frac{(0.114 \text{ m})(2.7 \text{ m/sec})(1.18 \text{ kg/m}^3)}{1.70 \times 10^{-5} \text{ kg}/(\text{m sec})} = 21,280$$

The Prandtl number for air is 0.7. Using Eq. 3.3.31a

$$Nu_f = 0.618 \, Re_f^{1.315} \, Pr^{1/3}$$

$$= 0.618(21,280)^{1.315}(0.7)^{1.3}$$

$$h_o = \frac{Nu \, k}{d} = \frac{(552)(0.026 \text{ N m/(sec m °C)})}{0.114 \text{ m}}$$

$$= 126 \text{ N m/(sec m}^2 \text{ °C)}$$

Outside convection resistance is thus

$$R_o = \frac{1}{h_o A_o} = \frac{1}{h_o \pi \, dL}$$

$$= \frac{1}{\pi[126 \text{ N m/(sec m}^2 \text{ °C)}](0.114 \text{ m})(20 \text{ m})}$$

$$= 1.11 \times 10^{-3} \text{ sec °C/(N m)}$$

Resistance of the pipe wall is

$$R_p = \frac{\ln(r_o/r_1)}{2\pi k_{ss} L} = \frac{\ln(0.057 \text{ m}/0.051 \text{ m})}{2\pi(20 \text{ N m/(sec m °C)})(20 \text{ m})}$$

$$= 4.43 \times 10^{-5} \text{ sec °C/(N m)}$$

This value is very small because of the high thermal conductivity of the stainless steel, and because of the thin pipe walls.

Tomato sauce is a non-Newtonian fluid. From Tables 2.5.4, 2.6.2, 3.2.2, and 3.6.1, we obtain the following physical properties

$$k = 0.8 \text{ N m/(sec m °C)}$$
$$c_p = 3980 \text{ N m/(kg °C)}$$
$$\rho = 1080 \text{ kg/m}^3$$
$$K = 18.7 \text{ N sec}^n/\text{m}^2 \quad \text{(at 32°C)}$$
$$n = 0.4$$

The logarithmic mean temperature in the pipe is

$$\theta_{lmt} = \sqrt{\theta_{in}\theta_{out}} = \sqrt{(82°C)(80°C)} = 81°C$$

If the inside surface of the pipe is assumed to nearly equal the ambient temperature, let us say 29°C, then the film temperature is

$$\theta_f = \tfrac{1}{2}\theta_{lmt} + \tfrac{1}{2}\theta_o = \frac{29°C + 81°C}{2} = 55°C$$

Depending on the equation used to calculate Nusselt number, the value of consistency coefficient (K) must be corrected for temperature, as in Eq. 2.6.2

$$K_{55} = K_{32} \left(\frac{\mu_{H_2O,55}}{\mu_{H_2O,32}} \right)$$

$$= 18.7 \text{ N sec}^n/\text{m}^2 \left(\frac{5.08 \times 10^{-4} \text{ kg/(m sec)}}{7.71 \times 10^{-4} \text{ kg/(m sec)}} \right)$$

$$= 12.3 \text{ N sec}^n/\text{m}^2$$

Velocity of the tomato sauce in the pipe is

$$v = \frac{\dot{V}}{A} = \frac{\dot{m}}{\rho A} = \frac{13 \text{ kg/sec}}{(1080 \text{ kg/m}^3)\pi(0.051 \text{ m})^2}$$

$$= 1.47 \text{ m/sec}$$

The generalized Reynolds number is

$$\text{GRe} = \frac{8d_i^n v^{2-n}\rho}{\left(\frac{1}{n} + 3 \right)^n 2^n K} = \frac{8(0.102 \text{ m})^{0.4}(1.47 \text{ m/sec})^{1.6}(1080 \text{ kg/m}^3)}{\left(\frac{1}{0.4} + 3 \right)^{0.4} 2^{0.4}(12.3 \text{ N sec}^{0.4}/\text{m}^2)}$$

$$= 202$$

The generalized Prandtl number is

$$\text{GPr} = \frac{c_p}{k} \left(\frac{v}{d} \right)^{n-1} \frac{(1/n + 3)^n 2^n K}{8}$$

$$= \frac{(3980 \text{ N m/(kg }^\circ\text{C))}}{0.8 \text{ N m/(sec m }^\circ\text{C)}} \left(\frac{1.47 \text{ m/sec}}{0.102 \text{ m}} \right)^{-0.6}$$

$$\times \frac{\left(\frac{1}{0.4} + 3 \right)^{0.4} 2^{0.4}(12.3 \text{ N sec}^{0.4}/\text{m}^2)}{8}$$

$$= 4026$$

To calculate the Nusselt number in the tube, any one of several equations can be used. We choose Eq. 3.3.27

$$\text{Nu} = 1.75 \left(\frac{3n+1}{4n} \right)^{1/3} \left[\frac{\pi}{4} \text{ GRe GPr} \left(\frac{d}{L} \right) \right]^{1/3}$$

$$= 1.75 \left(\frac{1.2+1}{1.6} \right)^{1/3} \left[\frac{\pi}{4}(202)(4026) \left(\frac{0.102 \text{ m}}{20 \text{ m}} \right) \right]^{1/3}$$

$$= 28.8$$

The inside convection coefficient is then

$$h_i = \frac{Nu\,k}{d} = \frac{(28.8)[0.8\text{ N m}/(\text{sec m }°\text{C})]}{0.102\text{ m}} = 202\text{ N m}/(\text{sec m}^2\ °\text{C})$$

Inside convection resistance is

$$R_i = \frac{1}{h_i \pi d_i L} = \frac{1}{[202\text{ N m}/(\text{sec m}^2\ °\text{C})]\pi(0.102\text{ m})(20\text{ m})}$$
$$= 7.71 \times 10^{-4}\text{ sec }°\text{C}/(\text{N m})$$

Total thermal resistance is thus

$$R_{tot} = 7.71 \times 10^{-4} + 4.43 \times 10^{-5} + 1.11 \times 10^{-3}$$
$$= 1.92 \times 10^{-3}\text{ sec }°\text{C}/(\text{N m})$$

The amount of heat transferred from the tomato sauce to the ambient air is (if all assumptions were correct)

$$\dot{q} = \frac{\theta_{lmt} - \theta_a}{R_{tot}} = \frac{81°\text{C} - 27°\text{C}}{1.9 \times 10^{-3}\text{ sec }°\text{C}/(\text{N m})}$$
$$= 28{,}060\text{ N m}/\text{sec}$$

Now we check temperature assumptions

$$\theta_{\text{inside pipe}} = \theta_{lmt} - \dot{q}R_i$$
$$= 81°\text{C} - \left(28{,}060\ \frac{\text{N m}}{\text{sec}}\right)[7.71 \times 10^{-4}\text{ sec }°\text{C}/(\text{N m})]$$
$$= 59.4°\text{C}$$

$$\theta_{\text{outside pipe}} = \theta_{\text{inside pipe}} - \dot{q}R_p$$
$$= 59.4°\text{C} - \left(28{,}060\ \frac{\text{N m}}{\text{sec}}\right)[4.43 \times 10^{-5}\text{ sec }°\text{C}/(\text{N m})]$$
$$= 58.1°\text{C}$$

$$\theta_{\text{ambient}} = \theta_{\text{outside pipe}} - \dot{q}R_D$$
$$= 58.1°\text{C} - \left(28{,}060\ \frac{\text{N m}}{\text{sec}}\right)[1.11 \times 10^{-3}\text{ sec }°\text{C}/(\text{N m})]$$
$$= 27°\text{C}$$

This last value is as it should be, but the assumptions for the temperature on the inside and outside surfaces of the pipe (29°C, 27°C) are not the same as the calculated temperatures (59.4°C, 58.1°C). These calculations can be repeated using these calculated temperatures as the new assumed temperatures. After three iterations, these results were obtained

$$\theta_{\text{inside pipe}} = 60.4°C$$
$$\theta_{\text{outside pipe}} = 59.1°C$$
$$R_i = 7.18 \times 10^{-4} \text{ sec } °C/(N \text{ m})$$
$$R_o = 1.12 \times 10^{-3} \text{ sec } °C/(N \text{ m})$$
$$\dot{q} = 28,714 \text{ N m/sec}$$

Remarks

Programming this procedure on a computer makes it much easier to iterate to a final solution. However, even with grossly incorrect temperature assumptions, there is little sensitivity of resistance and heat flow values. Even more important is the fact that this problem is overspecified. The temperature of the tomato sauce as it leaves the pipe should reflect loss of stored heat in the tomato sauce material. See Example 3.6.2-2.

---

directional
convection

3.3.1.3 Natural Convection Equations Natural convection is directional; that is, buoyant fluid moves in a direction counter to gravity or to the inertial force. Inclined surfaces, therefore, require a correction to give an equivalent vertical distance over which natural convection is developed. The Grashof number should be calculated using $g \cos \phi$ instead of g, where g is the acceleration due to gravity and ϕ is the angle relative to the vertical. This correction fails for horizontal surfaces, and is not applicable to curved surfaces.

1. Vertical slab, laminar flow, isothermal plate. $Pr > 0.7$; $10^9 > Gr > 10^4$ (Lorenz) (Gebhart, 1971).

$$Nu = 0.548\sqrt[4]{Gr \text{ Pr}} \tag{3.3.36a}$$

(Kreith, 1958).

$$Nu = 0.480\sqrt[4]{Gr} \tag{3.3.36b}$$

(Ostrach) (Gebhart, 1971).

$$Nu = \tfrac{4}{3}F(Pr)\sqrt[4]{Gr} \tag{3.3.36c}$$

Pr	$\frac{4}{3}F(\text{Pr})$	$F_1\,(\text{Pr})$
0.01	0.0765	
0.1		0.237
0.72	0.475	
1.0	0.535	0.573
2.0	0.675	
10.0	1.10	1.17
100	2.06	2.18
1000	3.74	

2. Vertical slab, laminar flow, uniform flux. $10^4 < \text{Gr} < 10^9$ (Gebhart, 1971).

$$\text{Nu} = F_1(\text{Pr})\sqrt[4]{\text{Gr}} \tag{3.3.37}$$

3. Flow parallel to vertical cylinders. Radius of curvature not large compared to thickness of convection layer; temperature difference varies linearly from zero at the leading edge. $10^4 < \text{Gr} < 10^9$ (Gebhart, 1971).

$$\text{Nu} = 1.058\sqrt[4]{\text{Gr}}\sqrt[4]{\frac{\text{Pr}^2}{4 + 7\,\text{Pr}}}$$
$$\text{Gr} = \frac{g\beta d^4}{\nu^2}\left(\frac{d\theta}{dx}\right) \tag{3.3.38}$$

4. Flow over vertical plates and large-diameter cylinders. $10^4 < \text{Gr}\,\text{Pr} < 10^9$ (Gebhart, 1971); $\text{Pr} \approx 1.0$.

$$\text{Nu}_f = 0.59\sqrt[4]{\text{Gr}_f\,\text{Pr}_f} \tag{3.3.39a}$$

$10^9 < \text{Gr}\,\text{Pr}$ (Gebhart, 1971).

$$\text{Nu}_f = 0.13\sqrt[4]{\text{Gr}_f\,\text{Pr}_f} \tag{3.3.39b}$$

$\text{Gr} > 10^{10}$ (Eckert–Jackson) (Gebhart, 1971).

$$\text{Nu}_f = 0.0246\,\text{Gr}_f^{2/5}\,\text{Pr}_f^{7/15}(1 + 0.494\,\text{Pr}_f^{2/3})^{-2/5} \tag{3.3.39c}$$

$\text{Gr} > 10^9$ (Kreith, 1958).

$$\text{Nu} = 0.024\left(\frac{\text{Pr}^{1.17}\,\text{Gr}}{1 + 0.494\,\text{Pr}^{2/3}}\right)^{2/5} \tag{3.3.39d}$$

Vertical cylinders of small diameter (Gebhart, 1971).

$$\frac{d}{L} \geq \frac{35}{\sqrt[4]{\text{Gr}}} \qquad (3.3.39e)$$

use vertical plate solution.

5. Flow over horizontal cylinders and spheres (Nusselt, McAdams) (Gebhart, 1971; Kreith, 1958). $\text{Pr Gr} > 10^4$; $\text{Pr} > 0.5$; $10^3 < \text{Gr} < 10^9$.

$$\text{Nu}_f = 0.52\sqrt[4]{\text{Gr}_f \, \text{Pr}_f} \qquad (3.3.40a)$$

$\text{Gr Pr} > 10^9$ (Henderson and Perry, 1966).

$$\text{Nu} = 0.12\sqrt[3]{\text{Gr Pr}} \qquad (3.3.40b)$$

6. Flow over spheres (Yule) (Gebhart, 1971). $1 < \text{Gr} < 10^5$; $\text{Pr} \approx 1.0$.

$$\text{Nu}_f = 2 + 0.45\sqrt[4]{\text{Gr}_f}\sqrt[3]{\text{Pr}_f} \qquad (3.3.41a)$$

$2.5 \times 10^5 < \text{Gr} < 1.1 \times 10^6$ (Tang et al., 1991).

$$\text{Nu}_f = 0.509 \, \text{Gr}_f^{0.255} \qquad (3.3.41b)$$

7. Flow over miscellaneous shapes (King) (Gebhart, 1971). $10^4 < \text{Gr Pr} < 10^9$.

$$\text{Nu} = 0.60\sqrt[4]{\text{Gr Pr}} \qquad (3.3.42a)$$

where the characteristic length, $L = L_h L_v/(L_h + L_v)$, h, horizontal; v, vertical. Mushroom being heated (Alhamdan et al., 1988)

$$\text{Nu} = 2.2 \, \text{Ra}^{0.21} \qquad (3.3.42b)$$

Mushroom being cooled (Alhamdan et al., 1988)

$$\text{Nu} = 0.42 \, \text{Ra}^{0.27} \qquad (3.3.42c)$$

d, the major axis diameter of mushroom cap.

8. Horizontal square plates, heated surfaces facing upward or cooled surfaces facing downward. $10^5 < \text{Gr Pr} < 10^8$ (Fishenden–Saunders).

$$\text{Nu} = 0.54\sqrt[4]{\text{Gr Pr}} \qquad (3.3.43a)$$

$10^8 < \text{Gr Pr} < 3 \times 10^{10}$ (Gebhart, 1971; Kreith, 1958).

$$\text{Nu} = 0.14\sqrt[3]{\text{Gr Pr}} \qquad (3.3.43\text{b})$$

9. Horizontal square plates, heated surfaces facing downward or cooled surfaces facing upward. $3 \times 10^5 < \text{Gr Pr} < 3 \times 10^{10}$ (McAdams) (Gebhart, 1971; Kreith, 1958).

$$\text{Nu} = 0.27\sqrt[4]{\text{Gr Pr}} \qquad (3.3.44)$$

Horizontal circular disks—use with $/0.9d$ in place of L.

10. Horizontal air space, upper plate at higher temperature (Gebhart, 1971).

$$\text{Nu} = 1.0 = \frac{hS}{k} \qquad (3.3.45)$$

S, the air space.

11. Horizontal air space, lower plate warmer. $\text{Gr} < 1700$ (Gebhart, 1971).

$$\text{Nu} = 1.0 \qquad (3.3.46\text{a})$$

$10^4 < \text{Gr} < 4 \times 10^5$ (Gebhart, 1971).

$$\text{Nu} = 0.195\sqrt[4]{\text{Gr}} \qquad (3.3.46\text{b})$$

$4 \times 10^5 < \text{Gr}$ (Gebhart, 1971).

$$\text{Nu} = 0.069\sqrt[3]{\text{Gr}}\, \text{Pr}^{0.407} \qquad (3.3.46\text{c})$$

where $\text{Gr} = g\beta S^3(\theta_1 - \theta_2)/v^2$ and S is the spacing between plates. Properties evaluated at an average of two surface temperatures.

12. Vertical air spaces (Jakob) (Gebhart, 1971). $\text{Gr} < 2000$; $L/S > 3$.

$$\text{Nu} = 1.0 \qquad (3.3.47\text{a})$$

L, the plate height; $2 \times 10^4 < \text{Gr} < 2 \times 10^5$; $L/S > 3$.

$$\text{Nu} = 0.18\sqrt[4]{\text{Gr}}\left(\frac{L}{S}\right)^{-1/9} \qquad (3.3.47\text{b})$$

$2 \times 10^5 < \text{Gr} < 11 \times 10^6$ (Gebhart, 1971); $L/S > 3$.

$$\text{Nu} = 0.065\sqrt[3]{\text{Gr}}\left(\frac{L}{S}\right)^{-1/9} \qquad (3.3.47\text{c})$$

13. Human body (Clark et al., 1981)

$$Nu = 0.63 \ Gr^{0.25} \ Pr^{0.25} \tag{3.3.48}$$

EXAMPLE 3.3.1.4-1. Convection from a Cow

Cena (1974) gives the surface area of an animal as $A = 0.09 \ m^{2/3}$, where A is the area (m^2), and m, the body mass (kg). Kleiber (1975) gives the metabolic rate of animals as $M = 3.39 \ m^{0.75}$, where M is the basal metabolic rate (N m/sec). Estimate the thermal conductivity of the 5-mm-thick hair coat of a 600 kg cow if the skin surface is at 32°C and the ambient temperature is 15°C. The cow is standing quietly in a room with still air.

Solution

Total metabolic heat produced by the cow is

$$M = 3.39(600 \ kg)^{0.75} = 411 \ N \ m/sec$$

About 15% of this heat is lost as evaporative heat, mostly from the respiratory system. Another 50% is lost from the legs, tail, and ears. This leaves about 35%, or 144 N m/sec, to be lost through the body.

The total surface area of a cow is about

$$A = 0.09(600 \ kg)^{2/3} = 6.40 \ m^2$$

We estimate that about 20% of this surface area is in the legs and tail. This leaves about 5.12 m^2 as the surface area of the body.

The irregularly shaped body of the cow is commonly replaced by a cylinder with equivalent area. If the radius of the cylinder is taken to be 40 cm, then the area of the cylinder, including both ends, is

$$A = 2\pi r L + 2\pi r^2 = 2\pi(0.40)L + 2\pi(0.40)^2 = 5.12 \ m^2$$
$$L = 1.64 \ m$$

For purposes of this example we assume totally still air and natural convection. Normally we would assume air moving at 0.15 m/sec. To calculate the Grashof number, we must assume a temperature for the outer surface of the hair coat. A temperature of 19°C will be assumed.

$$\theta_{film} = \frac{15°C + 19°C}{2} = 17°C$$

$$T_{film} = 17°C + 273°C = 290\ K$$

$$\rho_{air} = \frac{pM}{RT} = \frac{(101325\ N/m^2)(29\ kg/kg\ mol)}{(8314.34\ N\ m/(kg\ mol\ K))(290\ K)}$$

$$= 1.22\ kg/m^3$$

$$T_{amb} = 15°C + 273°C = 292\ K$$

$$\beta = \frac{1}{T_{amb}} = \frac{1}{292\ K} = 3.42 \times 10^{-3}/K$$

$$\mu = 1.81 \times 10^{-5}\ kg/(m\ sec) \quad at \quad \theta_{film} = 17°C$$

$$Gr = \frac{(1.22\ kg/m^3)^2(9.81\ m/sec^2)(3.42 \times 10^{-3}/K)(0.8\ m)^3(19 - 15\ K)}{(1.81 \times 10^{-5}\ kg/(m\ sec))^2}$$

$$= 312 \times 10^6$$

$$Pr = 0.7$$

We choose to use Eq. 3.3.40a

$$Nu_f = 0.52\sqrt[4]{Gr_f\ Pr_f} = 0.52\sqrt[4]{(312 \times 10^6)(0.7)}$$

$$= 63.2$$

The thermal conductivity for air at the film temperature can be computed

$$k = 0.0001477(290\ K)^{0.907} = 0.0252\ N\ m/(sec\ m\ °C)$$

Then

$$h_c = \frac{Nu\ k}{d} = \frac{(63.2)[0.0253\ N\ m/(sec\ m\ °C)]}{(0.80\ m)} = 2.00\frac{N\ m}{sec\ m^2\ °C}$$

The same convection coefficient value is assumed to apply to the ends of the cylinder as well as to the sides.

Total thermal resistance of the body comes from conduction and convection resistances

$$R_{tot} = \frac{(\theta_{sk} - \theta_{amb})}{\dot{q}} = \frac{L_{hair}}{k_{hair}A} + \frac{1}{h_cA}$$

$$= \frac{(32 - 15°C)}{144\ N\ m/sec} = \frac{0.0005\ m}{k_{hair}(5.12\ m^2)} + \frac{1}{[2.00\ N\ m/(sec\ m^2\ °C)](5.12\ m^2)}$$

$$0.118 = \frac{0.005\ m}{k_{hair}(5.12\ m^2)} + 0.0977\ sec\ °C/(N\ m)$$

$$k_{hair} = 0.048\ N\ m/(sec\ m\ °C)$$

We now check the assumed surface temperature

$$
\theta_o = \theta_{sk} - \dot{q}\left(\frac{L}{k_{hair}A}\right) = 32°C - 144\frac{N\ m}{sec}\left(\frac{0.005\ m}{0.048\dfrac{N\ m}{(sec\ m\ °C)}(5.12\ m^2)}\right)
$$

$$
\theta_o = 29°C
$$

This value is now used as the new assumed hair coat surface temperature. Three additional iterations are required before no futher change in θ_0 occurs. Results of these iterations are

θ_0 (assumed)	29	25	26
θ_{film}	22	20	21.5
T_{film}	295	293	293.5
ρ_{air}	1.20	1.21	1.22
μ_{air}	1.71×10^{-5}	1.71×10^{-5}	1.71×10^{-5}
Gr	1,184,304,505	860,089,398	961,800,946
Nu	88.24	81.5	83.8
k_{air}	2.57×10^{-2}	2.53×10^{-2}	2.56×10^{-2}
h_c	2.83	2.57	2.68
k_{hair}	0.0199	0.0232	0.0217
θ_0 (calc)	25	26	26

The correct value for the hair coat of this cow is $k_{hair} = 0.022$ N m/(sec m °C).

Remark

It would also normally be necessary to estimate radiation. In a laboratory, where walls are kept at ambient air temperature, radiation heat loss is nearly indistinguishable from convection.

Comparing the above value to the value found in Table 3.2.2 shows a large discrepancy. Not all biological samples conform to tabulated values.

3.3.1.4 Mixed Convection Not all convection is exclusively forced or
some forced, exclusively natural convection. There is a region where both modes are
some natural important (Figure 3.3.6). An example is the cooling of warm produce in the
convection field where sufficient power is not available to produce strong forced
convection. In this case, weak forced convection combines with strong natural
convection. The convection coefficient for mixed convection should be greater
than that determined for either forced or natural convection alone.

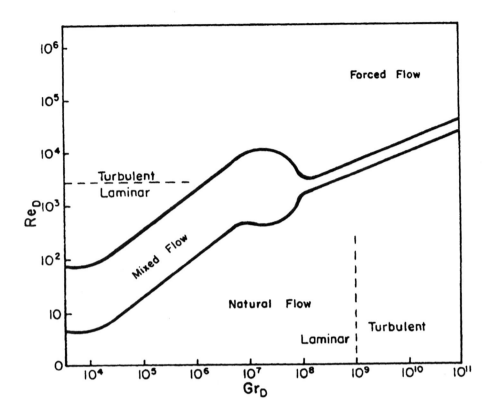

Figure 3.3.6. Diagram showing the regions of forced convection, natural convection, and mixed convection. Laminar and turbulent flow regimes exist for both forced and natural convection.

test for mixed convection

To determine if the region of mixed convection has been entered, check the ratio (Gr/Re^2). If $(Gr/Re^2) \gg 1$, then natural convection predominates; if $(Gr/Re^2) \ll 1$, then forced convection predominates; if (Gr/Re^2) is of the order of 1, then mixed convection prevails.

extension of convection methods

Efforts have been made to determine a mixed convection Nusselt number from forced and natural convection equations. Such a procedure would be an extension of methods thus far covered. Two such procedures have been developed: one for cylinders and one for spheres. Both procedures assume the natural convection Nusselt number is determined by an equation of the form

$$Nu = B(Gr\ Pr)_{d,f}^{m} \qquad (3.3.49)$$

where B is a coefficient (dimensionless); m, an exponent (dimensionless); and d, f denote bases of Gr and Pr calculations are diameter and film temperature.

The forced convection Nusselt number is assumed to be

$$Nu = A(Re)^n_{d,f}(Pr)^p_f \tag{3.3.50}$$

where A is a coefficient (dimensionless); and n, p are exponents (dimensionless).

method of Hatten

Cylinders   The method of Hatten for mixed convection from cylinders first equates the two Nusselt numbers above

$$B(Gr\ Pr)^m_{d,f} = A(Re)^n_{d,f}(Pr)^p_f \tag{3.3.51}$$

and solves for an equivalent Reynolds number

$$Re_{equiv} = [(B/A)(Pr)^{m-p}(Gr)^m]^{1/n} \tag{3.3.52}$$

vectorial addition

If natural convection is vertical and ϕ is the angle between directions of forced and natural convection, then a vectorial addition is used

$$Re^2_{eff} = (Re_{equiv} + Re\cos\phi)^2 + (Re\sin\phi)^2 \tag{3.3.53}$$

where Re_{eff} is the effective Reynolds number (dimensionless); and ϕ, the angle between forced and natural convection directions (rad).

The effective Reynolds number is used in the original forced convection equation

$$Nu = A(Re)^n_{eff}(Pr)^p \tag{3.3.54}$$

where Nu is the mixed convection Nusselt number (dimensionless).

Kirk method

Spheres   The effective diameter scalar addition (EDSA) method devised by Kirk and Johnson (1986) must be used for spheres. In this method, equation forms are assumed to be the same as in Eqs. 3.3.49 and 3.3.50. An effective Grashof number is used instead of the Grashof number normally calculated. The effective Grashof number uses an effective diameter instead of the sphere diameter

$$Gr_{eff} = \frac{g\beta d^3_{eff}(\theta_o - \theta_\infty)}{v^2} \tag{3.3.55}$$

where Gr_{eff} is the effective Grashof number (dimensionless); g, the acceleration due to gravity (9.81 m/sec²); β, the coefficient of thermal expansion (1/°C); d_{eff}, the effective sphere diameter (m); θ_0, the surface temperature (°C); θ_∞, the ambient temperature (°C); and v, the kinematic viscosity (m²/sec).

The effective diameter depends on the angle between forced and natural convective flow directions (Figure 3.3.7). It can be calculated from (Tang et al., 1991)

effective
diameter

$$d_{\text{eff}} = d[\alpha - (2/\pi)\cos\phi \tan^{-1}(\text{Re}^2/\text{Gr})] \qquad (3.3.56)$$

where d is the sphere actual diameter (m); α, a parameter that depends on Grashof number (dimensionless); ϕ, the angle between forced and natural convection directions (rad); Re, the actual Reynolds number (dimensionless); and Gr, the actual Grashof number (dimensionless).

The value of α varies between 0.9 and 1.0, but can be assumed to be 1.0, for practical purposes.

Similar to the method of Hatten, an equivalent Reynolds number is obtained from the effective Grashof number

$$\text{Re}_{\text{equiv}} = [(B/A)(\text{Pr})^{m-p}(\text{Gr}_{\text{eff}})^m]^{1/n} \qquad (3.3.57)$$

The actual Reynolds number is combined with the equivalent Reynolds number by scalar addition

scalar
addition

$$\text{Re}_{\text{eff}} = \text{Re}_{\text{equiv}} + \text{Re} \qquad (3.3.58)$$

Finally, a mixed convection Nusselt number is obtained using Eq. 3.3.54.

The foregoing method is applicable for spheres in parallel flow to cross flow (the angle between forced and natural convection flows is 0 to $\pi/2$ rad). For counterflow (the angle is $\pi/2$ to π rad) no suitable method has been devised (Figure 3.3.8).

Fruits The EDSA method may be used for spheroidal fruits and vegetables
Tang method with a simple modification. Nusselt numbers for forced and natural convection can be calculated as before with the addition of a constant to each equation

$$\text{Nu} = A\,\text{Re}^n\,\text{Pr}^p + M \qquad (3.3.59)$$
$$\text{Nu} = B(\text{Gr}\,\text{Pr})^m + N \qquad (3.3.60)$$

where N, M are constants dependent on the shape of the fruit (dimensionless). The equivalent Reynolds number is thus

$$\text{Re}_{\text{equiv}} = \{[B(\text{Gr}_{\text{eff}}\,\text{Pr})^m - M + N]/A\,\text{Pr}^p\}^{1/n} \qquad (3.3.61)$$

Tang et al. (1991) evaluated physical properties in this manner: Air density and the coefficient of thermal expansion were evaluated at the free stream temperature of the air; air thermal conductivity and viscosity were evaluated at

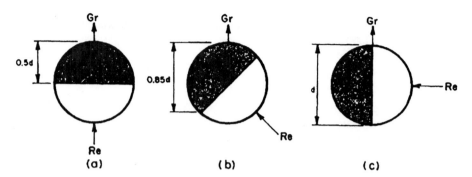

Figure 3.3.7. Flow patterns around the sphere: (a) aiding flow; (b) aiding flow (45°); (c) cross flow (90°). Re represents forced flow and Gr represents natural flow. Natural convective flow develops on the leeward side (shaded area) of the sphere and does not seem to depend on area, but on projected vertical distance.

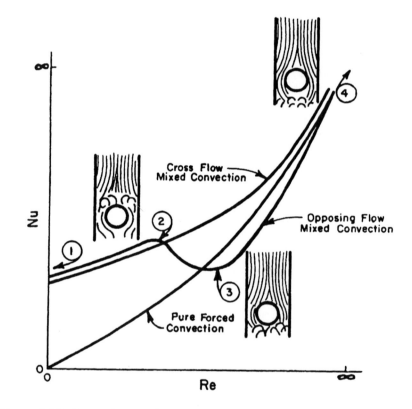

Figure 3.3.8. Curves for pure forced convection cross-flow mixed convection, and opposing-flow mixed convection compared. In opposing-flow mixed convection the forced flow is downward and natural flow is upward. In the region from (1) to (2), natural convection predominates and the Nusselt number approaches that for cross flow. In the region near (4), forced convection predominates and the Nusselt number reflects this. At (3), there is a balance between forced and natural flows that reduces convection heat loss to a minimum.

Table 3.3.2 *M* and *N* Values for Different Fruits

	Apple	Peach	Plum	Strawberry
M	-4	-2	1	5
N	-2	-1	1	3.5

the mean film temperature. The Prandtl number for air is nearly constant at a value of 0.7 over a range greater than -40 to $300°C$.

Values for the constants M and N can be found in Table 3.3.2 (Tang and Johnson, 1992).

EXAMPLE 3.3.1.4-1. Mixed Convection from an Apple

Determine the convection coefficient for a just-picked apple that is immediately being cooled. Temperature of the apple is $21°C$ and the cooling air is $2°C$. Free stream velocity of the air in the field cooler is vertical at 0.22 m/sec and the apple is 10 cm in diameter.

Solution

Because the air velocity is so small, the strength of forced convection will not likely be so great. Hence, the ratio of Re^2/Gr will be checked to see if convection is mixed forced and natural. The following physical properties of air at $2°C$ are determined with the help of Table 2.4.1

$$\text{density } (\rho) = 1.284 \text{ kg/m}^3$$
$$\text{viscosity } (\mu) = 1.72 \times 10^{-5} \text{ kg/(m sec)}$$

For air at $21°C$

$$\text{density } (\rho) = 1.201 \text{ kg/m}^3$$
$$\text{viscosity } (\mu) = 1.81 \times 10^{-5} \text{ kg/(m sec)}$$

The film temperature is $(2°C + 21°C)/2 = 11.5°C$. At this temperature

$$\rho_{11.5} = \rho_2 T_2/T_{11.5} = (1.284 \text{ kg/m}^3)(275.15 \text{ K}/284.65 \text{ K})$$
$$= 1.241 \text{ kg/m}^3$$

$$\mu_{11.5} = \left[\frac{\dfrac{1}{T_{11.5}} - \dfrac{1}{T_2}}{\dfrac{1}{T_{21}} - \dfrac{1}{T_2}} \right] (\mu_{21} - \mu_2) + \mu_2$$

$$= \left[\frac{\dfrac{1}{284.65} - \dfrac{1}{275.15}}{\dfrac{1}{294.15} - \dfrac{1}{275.15}} \right] (1.81 \times 10^{-5} - 1.72 \times 10^{-5}) + 1.72 \times 10^{-5}$$

$$= 1.77 \times 10^{-5} \text{ kg/(m sec)}$$

The coefficient of thermal expansion is calculated as

$$\beta = \frac{1}{T_{21}} = \frac{1}{294.15} = 3.40 \times 10^{-3}/\text{K}$$

The Reynolds number is

$$\text{Re} = \frac{Dv\rho}{\mu} = \frac{(0.10 \text{ m})(0.22 \text{ m/sec})(1.241 \text{ kg/m}^3)}{1.77 \times 10^{-5} \text{ kg/(m sec)}}$$

$$= 1542$$

The Grashof number is

$$\text{Gr} = \frac{\rho^2 g \beta D^3 (\theta_o - \theta_\infty)}{\mu^2}$$

$$= \frac{(1.241 \text{ kg/m}^3)^2 (9.8 \text{ m/sec}^2)(3.40 \times 10^{-3}/\text{K})(0.10 \text{ m})^3 (21°\text{C} - 2°\text{C})}{[1.77 \times 10^{-5} \text{ kg/(m sec)}]^2}$$

$$= 3.11 \times 10^6$$

The ratio $\text{Re}^2/\text{Gr} = 0.77$. Since this ratio shows that Re^2 and Gr are of the same order of magnitude, mixed convection prevails. We must now calculate the Nusselt number based on the EDSA method.

Since the free stream air velocity is vertical and natural convection is vertical as well, the angle between forced and natural convection directions is $0°$. Thus, from Eq. 3.3.56:

$$d_{\text{eff}} = (0.1 \text{ m})\left(1 - \frac{2}{\pi}\cos 0° \tan^{-1}(0.77)\right)$$

$$= 0.0582 \text{ m}$$

The effective Grashof number is

$$\text{Gr}_{\text{eff}} = \text{Gr}\frac{(0.0582 \text{ m})^3}{(0.100 \text{ m})^3} = 6.13 \times 10^5$$

Now we must find equations for Nusselt number for forced and natural convection. Equation 3.3.32b seems to fit our geometric and Reynolds number forced convection conditions

$$\text{Nu} = 0.37 \text{ Re}^{0.6}$$

Equation 3.3.41b seems to fit our geometric and Grashof number conditions

$$Nu = 0.509 \ Gr^{0.255}$$

From Table 3.3.2, the values for M and N are found to be -4 and -2, respectively. The equivalent Reynolds number is thus determined from Eq. 3.3.61 to be

$$Re_{equiv} = [(0.509 \ Gr_{eff}^{0.255} + 4 - 2)/0.37]^{1/0.6}$$
$$= 602$$

We calculate using Eq. 3.3.58

$$Re_{eff} = Re_{equiv} + Re = 602 + 1542 = 2144$$

The Nusselt number is found from Eq. 3.3.32b

$$Nu = 0.37 \ Re_{eff}^{0.6} = (0.37)(2144)^{0.6} = 36.9$$

In Table 3.2.1 are found these values for thermal conductivity of the air:

$$10°C \qquad 0.0247 \ N/(°C \ sec)$$
$$20°C \qquad 0.0255 \ N/(°C \ sec)$$

The thermal conductivity value at the film temperature, 11.5°C,

$$k_{11.5} = k_{10} + \left(\frac{T_{11.5}^{0.85} - T_{10}^{0.85}}{T_{20}^{0.85} - T_{10}^{0.85}} \right)(k_{20} - k_{10})$$
$$= 0.0247 + \left(\frac{284.65^{0.85} - 283.15^{0.85}}{293.15^{0.85} - 283.15^{0.85}} \right)(0.0255 - 0.0247)$$
$$= 0.0248 \ N/(°C \ sec)$$

Thus, the convection coefficient can be determined:

$$h_c = \frac{Nu \ k_f}{D} = \frac{(36.9)(0.0248 \ N/(°C \ sec))}{(0.10 \ m)}$$
$$= 9.16 \ N/(°C \ sec \ m)$$

Remark

Once this value is determined, the rate of cooling of the apple can be found using the Heisler charts in Section 3.7.2.2.

3.3.1.5 Inaccuracies

convection
errors are
high

As explained earlier, convection is a complicated phenomenon, and, as such, calculated values for convection coefficients may be in error by significant amounts. Calculated values within ±3–5% of the true values should be considered excellent; errors of ±10–20% are probably typical; errors of ±100% are not uncommon. Nonetheless, calculated values are much better than guesses (but numerical precision greater than two or three significant figures is not justified).

differences
over various
surfaces

There may be very large differences among convection coefficient values over different portions of a surface. For instance, convection coefficient values on the top and bottom of a cube are likely to be considerably different from convection coefficient values for the sides (Figure 3.3.9). Different equations could be used for different surfaces of the cube, but, given the admonition about error rates, convection coefficient values are often assumed to be the same for all surfaces.

local
differences

Local convection coefficient values are often different from average convection coefficient values, but these differences are not usually accounted for unless local convection conditions are somehow important. Such cases can arise especially in detailed numerical heat transfer simulations.

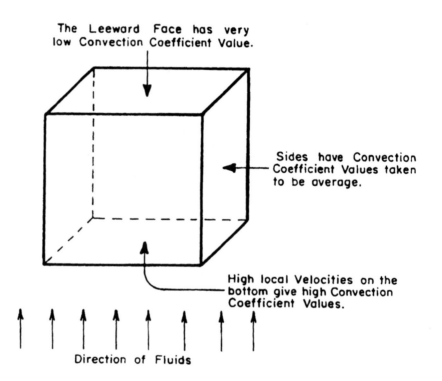

Figure 3.3.9. Convection coefficient values vary over the surfaces of an object.

The highest convection coefficient values occur for forced convection on the side of an object facing the wind, and for natural convection on the side of a hot object facing upward or a cold side facing downward (Figure 3.3.10). Very low convection coefficient values for forced convection occur on leeward sides, and for natural convection on hot sides facing downward or cold sides facing upward (especially in a pocket where cooled fluid cannot spill over the sides). For a horizontal space between a hot upper surface and a cold lower surface, the Nusselt number has a value of 1.0, as indicated by Eq. 3.3.35.

turbulent convection greatest

The convection coefficient value for turbulent flow is almost always greater than that for laminar flow. For the case of forced convection, the transition from laminar to turbulent flow occurs (as discussed in Section 2.5.1) at a Reynolds number of about 2000. For external flow over a bluff body, such as a sphere, vortex shedding, with effects similar to turbulent flow, begins at Reynolds numbers much lower than 2000.

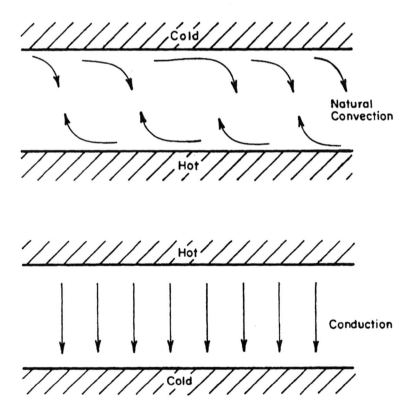

Figure 3.3.10. When the cold surface faces downward and the hot surface faces upward (top), a natural convection fluid circulation arises. When the cold surface faces upward and the hot surface faces downward, buoyant air at the top has nowhere to go; heat can only be transferred by conduction and radiation.

turbulent
natural flow

Turbulent flow can occur in natural convection as well (Figure 3.3.6). The transition from laminar to turbulent flow occurs at a Grashof number of about 10^9.

There is a broad range of transitional flow from laminar to turbulent flow. The uncertainties associated with this range of flows adds to lack of precision for calculated convection coefficient values.

still air

There are times when not enough information is available to calculate convection coefficient values. Some guidance in this case can be obtained from measured and tabulated values. For instance, air movement inside a "still" room is usually considered to be 0.15 m/sec. A convection coefficient value for the outside surfaces of buildings exposed to wind is often assumed to be 34 N/(m sec °C), and for the inside surfaces of buildings a value of 9.3 N/(m sec °C) can be used. Other nominal values are found in Tables 3.3.3 through 3.3.5. Note that these values should not be used unless other calculation procedures fail.

rules of
thumb

3.3.1.6 Convection with Viscous Dissipation

thick fluids

With significant viscous dissipation in the boundary layer next to a surface, enough heat is generated to reduce and perhaps even reverse surface heat flow (Figure 3.3.11). Viscous heating can cause a temperature maximum to appear between a warm surface and a cool fluid. In such a case there is actually a reversal of heat flow from the wall. The temperature difference for convective heat transfer is not really $(\theta_o - \theta_\infty)$, but instead is $(\theta_o - \theta_a)$, where θ_a is an adiabatic temperature that appears within the boundary layer.

This adiabatic temperature is calculated for flow over a flat surface from

adiabatic
temperature

$$\theta_a = \theta_\infty + \sqrt{Pr}\,\frac{v_\infty^2}{2c_p} \qquad (3.3.62)$$

where θ_a is the fluid adiabatic temperature (°C); θ_∞, the bulk fluid temperature (°C); v_∞, the bulk fluid velocity (m/sec); c_p, the fluid specific heat [N m/(kg °C)]; and Pr, the Prandtl number (dimensionless).

Table 3.3.3 Variation in Convection Coefficient with Room Air Motion

Air Velocity (m/sec)	Convection Coefficient [N m/(°C sec m^2)]
0.1–0.18	3.1
0.5	6.2
1.0	9.0
2.0	12.6
4.0	17.7

Source: Reprinted with permission from ASHRAE (1977).

Table 3.3.4 Equations Relating Convection Coefficient h_c to Velocity v

Equation for h_c	Condition	Remarks
$8.3v^{0.6}$	Seated	v is room air movement
$2.7 + 8.7v^{0.67}$	Reclining	v is lengthwise air movement
$8.6v^{0.53}$	Free walking	v is speed of walking
$6.5v^{0.39}$	Treadmill	v is speed of treadmill
$105 + 612v$	Immersed in water	v is relative water speed
580	Swimming	h_c independent of speed

Sources: Holmer and Bergh (1981); ASHRAE (1985).

Table 3.3.5 More Equations Relating Convection Coefficient to Air Velocity

Veloticy Range (m/sec)	Convection Coefficient [N m/(°C sec m²)]
$v < 0.2$	$5.4v^{0.466}$
$0.2 < v < 2$	$6.8v^{0.618}$
$v > 2$	$5.9v^{0.805}$

Source: Nishi and Gagge (1970).

The convection coefficient is calculated in the same way as before for forced convection. Convective heat transfer from the surface thus becomes

$$\dot{q} = h_c A(\theta_o - \theta_a) \tag{3.3.63}$$

This heat transfer is not the same as the heat transferred to the fluid.

EXAMPLE 3.3.1.6-1. Adiabatic Stagnation Temperature

A peregrine falcon can dive at a speed of 90 m/sec. What is the adiabatic stagnation temperature at this speed? Ambient temperature is 20°C.

Solution

The specific heat of air is 1004.8 N m/(kg °C) and the Prandlt number is 0.7. From Eq. 3.3.62

$$\theta_a = 20°C + \frac{\sqrt{0.7}(90 \text{ m/sec})^2}{(2)[1004.8 \text{ N m/(kg °C)}]}$$

$$= 23.4°C$$

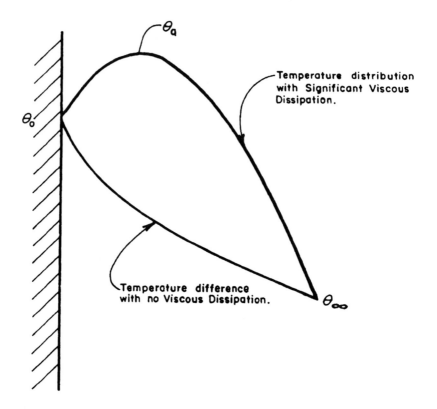

Figure 3.3.11. Viscous dissipation within the fluid significantly alters heat transfer from the surface.

3.3.1.7 Boiling and Condensation

very high convection coefficients

3.3.1.7 Boiling and Condensation   These two processes occur in very complex fashion. Because they do, they are often treated as convective processes, with a convection coefficient similar to the one previously considered. There is a great deal of heat that can be transferred by boiling or condensation, and convection coefficients for these processes are typically 1000 or more times higher than forced convection discussed earlier [typical values are 10,000 to 60,000 $N/(m\ sec\ °C)$].

The highest rate of heat transfer happens when fresh liquid (in boiling) or fresh vapor (in condensation) constantly impinges upon the surface. That means that when vapor bubbles form in boiling they immediately rise from the surface, to be replaced by fresh liquid. Heat is removed from the surface as rapidly as the liquid vaporizes and is carried away. The trick during condensation is to remove the condensed liquid as rapidly as possible, thus exposing the surface to fresh vapor. Heat is added to the surface as rapidly as the vapor condenses.

Heat is transferred during a change of state of the fluid with very little temperature difference. The process is nearly isothermal.

cold spots receive most heat

One useful feature of this type of heat transfer is that cold spots on the surface during condensation attract the most vapor and receive the most heat transfer (Figure 3.3.12). This tends to equalize temperatures at all points on the surface. Similarly, hot spots on a surface are the ones most likely to lose heat during boiling. Cooking griddles have been made that make use of the tendency to deliver more heat to regions that contain cold raw food.

3.3.2 Convection Thermal Resistance

The thermal resistance for convection is just the ratio of heat transfer to temperature difference

$$R_{\text{th}} = \frac{(\theta_o - \theta_\infty)}{\dot{q}} = \frac{1}{h_c A} \qquad (3.3.64)$$

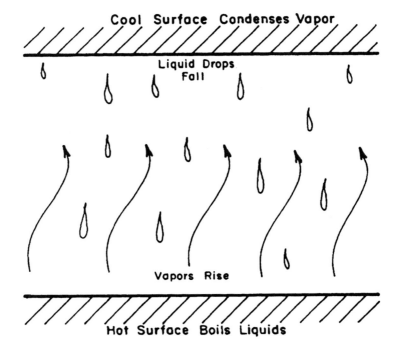

Figure 3.3.12. Heat transfer from a hot surface to a cold surface can occur by boiling and condensation. Gravity is used here to return liquid to the warm surface. In other cases liquid wicking can be used. The vapor will condense on the cool accessible surface no matter where it is located.

where R_{th} is the convective thermal resistance [°C sec/(N m)]. This resistance appears between a surface and the convecting fluid.

convection
considered
steady state

Although there can be transient convective phenomena that influence the value of the convection coefficient over time, convection heat transfer is usually considered to be steady state. Thus the convection resistance previously given is usually considered to be constant. Transient heat transfer processes involving convection are reflected in surface or fluid temperature changes but not in convection coefficient value.

convection
resistar ces

Figure 3.3.13 shows the same composite wall as Figure 1.6.2, except that the equivalent thermal resistance diagram now includes a convection resistance inside the wall and a convection resistance outside the wall. These convection resistances often offer considerable limitations to the amount of heat that flows through the wall.

wall example

There are two thermal resistance diagrams given in the figure. Both could be considered candidates to describe thermal resistances for the wall. In the upper diagram, brick and window areas are used together with the convection coefficient to form one convection resistance on the inside and one convection resistance on the outside of the wall. In the lower diagram, the same convection coefficient value is typically used with separate window and brick areas to form two different convection resistances calculated for the inside and for the outside surfaces. Is there a difference between these two representations, and which is the correct one?

The upper diagram is incorrect because the connections between $R_{concrete}$ and R_{window} on the inside and between R_{brick} and R_{window} on the outside denote equal temperatures on the inside surfaces of the window and concrete wall, and equal temperatures on the outside surfaces of the window and brick surfaces.

connections
denote same
temperatures

Connections on the thermal resistance diagram indicate identical temperatures.

Anyone who has felt the inside surface of a window on a cold day outside will have experienced that the window surface feels colder than the wall surface. These two temperatures are not the same and should not be forced to be the same in a faulty thermal resistance diagram. Hence the lower diagram is correct and the upper diagram is not.

EXAMPLE 3.3.2-1. Convection Coefficient for Freely Swimming Fish

Fish provide interesting examples of animals whose metabolism generally decreases as the temperatures of their surroundings decrease. Vigor of their movements also decreases as temperature declines. Thus heat loss by convection from their bodies should also show a marked temperature dependence. Speculate on the relationship between environmental temperature and the magnitude of the convection coefficient of freely swimming fish.

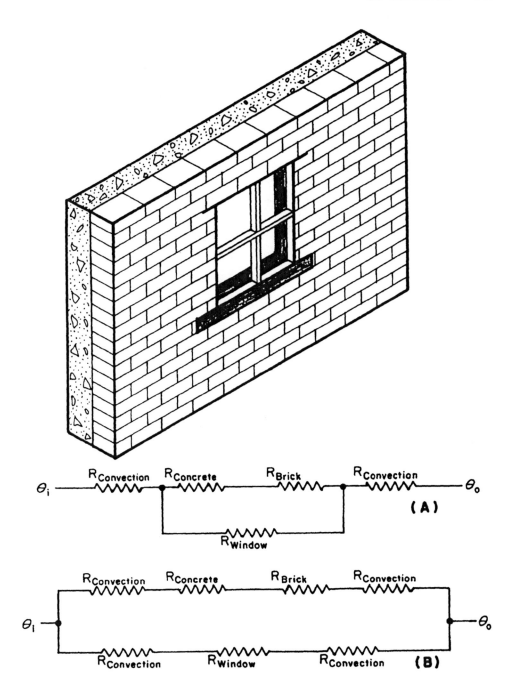

Figure 3.3.13. Equivalent thermal resistance diagram for a composite wall that includes convection resistances. The upper diagram (A) is incorrect because it describes inner surface temperatures for the wall and window as the same value.

Solution

A plot of convection coefficient value as a function of temperature might appear as in Figure 3.3.14. The convection coefficient is highest in regions 3 and 4 due to the activity levels of the fish. The fish are actively swimming and searching for food. Convection across the fish surface is largely forced convection. At temperatures above the peak, activity levels decrease as the fish become less comfortable in the heat and attempt to remain as cool as possible.

In region 2, temperatures have dropped to levels where the fish become sluggish or even still. Convection from the fish scale surface becomes almost predominantly natural convection. Some forced convection is maintained in the gills, but low temperatures do not require large gill flows.

At 4°C, water reaches its greatest density. At temperatures in this vicinity, natural convection virtually ceases because water both cooler and warmer tends to rise. The convection coefficient probably does not actually become zero, because a very low level of heat production and gill activity is maintained by the fish, but convection is not significant compared to conduction.

Below 4°C, in region 1, natural convection can again occur, but upside-down, with cooler water rising and warmer water falling. Convection coefficients are very small.

Below 0°C, fish can freeze unless they are protected by naturally occurring antifreeze compounds (glycoproteins) present in their tissues. These compounds allow the survival of low-temperature species despite below-

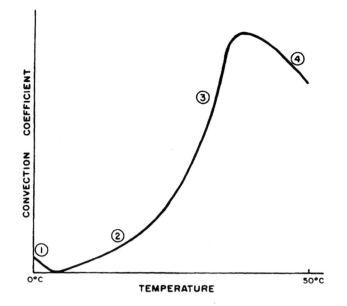

Figure 3.3.14. Convection coefficient (h_c) for freely swimming fish. Peaks and valleys are caused by both biological and physical reasons.

freezing environments. Sea water freezes at a temperature of $-1.9°C$; fish blood containing glycoproteins freezes at $-2.75°C$. Thus fish can survive as long as the water is not frozen around them.

___

3.3.3 Theoretical Relationships among Parameters

In Section 2.4.1 it was revealed that viscosity for gases is dependent upon the square root of the absolute temperature ($\sqrt{T}$). A similar relationship has been found for thermal conductivities of gases of simple molecular structure. Thus it might be expected that there might be some connection between thermal conductivity and viscosity in gases.

thermal conductivity related to viscosity

The kinetic theory of gases can give insight into this relationship. Thermal conductivity represents the value of kinetic energy being transferred from molecule to molecule; viscosity represents the value of momentum diffusion. As a result, a theoretical expression has been derived (Gebhart, 1971) relating some common parameters

$$k = \phi c_v \mu \tag{3.3.65}$$

where k is the thermal conductivity [N m/(sec m °C)]; c_v, the specific heat at constant volume [N m/(kg °C)]; μ, the viscosity [kg/(m sec)]; and ϕ, a dimensionless parameter ($=2.5$ for monatomic gases, or 1.9 for diatomic gases without appreciable vibration; namely, at temperatures below about 100°C).

Prandtl number theoretically determined

Because the Prandtl number is defined in terms of specific heat, viscosity, and thermal conductivity, its value is completely determined by the above relationship

$$\text{Pr} = \frac{c_p \mu}{k} = \frac{c_p}{c_v} \frac{c_v \mu}{k} = \left(\frac{c_p}{c_v}\right) \frac{1}{\phi} \tag{3.3.66}$$

The ratio of specific heat values is 1.67 for a monatomic gas and 1.4 for a diatomic gas (see Section 2.5.7.1). Thus $\text{Pr} = 0.67$ for monatomic gases and $\text{Pr} = 0.74$ for diatomic gases. It should be noted that the value for Prandtl number does not differ by more than 5% from the theoretical value of 0.67 for helium until a temperature of about 200°C is reached. Prandtl numbers for common diatomic gases (air, nitrogen, oxygen, and carbon monoxide) also do not differ from the theoretical value 0.74 by more than 5% for temperatures less than about 200°C. Thus the relationships above are expected to be correct for the normal physiological range of 0 to 100°C.

Energy transfer in materials other than gases has also been investigated, with resulting useful insights (Gebhart, 1971). In general analogous terms, we consider an energy gradient in one single direction, x. Energy can be exchanged from a more energetic entity to a less energetic entity, and is

energy
exchange
over a
distance

expected to be completely exchanged in some distance λ, which is considered to be an extremely small distance. The distance λ represents the mean free path in a gas (see Section 4.3.2.5), the absorption length for radiation, or other appropriate distance concept for the energy-exchange mechanism being considered.

Energy values per unit volume for the two entities located a distance λ apart are e_1 and e_2. The velocity of energy transport is taken as v. The net rate of energy exchange between any two points in the x direction is the difference between transport to the right and transport to the left (see Section 1.7.3.2, and especially Eq. 1.7.18)

$$\dot{q} = A(e_1 v - e_2 v) = -Av(e_2 - e_1)$$

$$= -v \frac{de}{dx} \lambda A = -\lambda v A \frac{de}{d\theta} \frac{d\theta}{dx}$$

$$= c\lambda v A \frac{d\theta}{dx} \qquad (3.3.67)$$

where $\dot{q}$ is the energy transferred (N m/sec); e, the energy levels per unit volume in the material (N m/m^3); v, the velocity of energy exchange (m/sec); A, the area perpendicular to energy flow (m^2); λ, the energy exchange distance (m); x, the distance along direction of energy exchange (m); θ, the temperature (°C); and c, the specific heat for the material [N m/(kg °C)].

Comparing this result to the general equation for conduction heat transfer shows that

$$k = c\lambda v \qquad (3.3.68)$$

thermal
conductivity
determined
by first
principles

where k is the thermal conductivity [N m/(sec m °C). Thus the thermal conductivity is related to other parameters the values for which can be determined from theory.

Gebhart (1971) nicely summarizes the results from theoretical studies for heat transfer in various types of materials. Some of his results appear in Table 3.3.6. By multiplying relationships for c, λ, and v, the temperature dependence of thermal conductivity may be inferred, and this appears in the rightmost column. The results can be used to anticipate expected thermal conductivity changes with temperature, or to interpolate tabulated values.

summary of
temperature
depen-
dencies

3.4 RADIATION

mass not
required for
radiation

Unlike conduction and convection heat transfer, radiation heat exchange does not require that the transmitting body and the receiving body be connected by an intervening mass medium. Conduction heat transfer occurs by molecule-to-molecule heat exchange; convection heat transfer requires a moving fluid. Radiation requires neither of these, and, as a matter of fact, occurs best in a

Table 3.3.6 Summary of Theoretical Relationships[a]

Condition	Specific Heat (c)	Energy Exchange Distance (λ)	Energy Exchange Velocity (v)	Temperature Dependence of Thermal Conductivity
Classical gas, molecular diffusion (gases, porous solids, electrons in semiconductors)	$\dfrac{3nk}{2}$	$\dfrac{1}{ns}$	$\left(\dfrac{kT}{m}\right)^{1/2}$	$\sqrt{T}$
Degenerate electron gas diffusion (metals)	CT	$f(T)$	v_F	$Tf(T)$
Elastic phonon diffusion (vibration in all solids)	$f\left(\dfrac{kT^3}{v_{son}h}\right)$	$e^{\theta_D/T}$	v_{son}	$T^3 e^{\theta_D/T}$
Radiation, electromagnetic photons (internal emission and reabsorption)	$f\left(\dfrac{kT^3}{v_{lgt}h}\right)$	$\dfrac{1}{\alpha}$	v_{lgt}	T^3
Common liquids	$3nk$	$\dfrac{1}{n\lambda^2}$	v_{son}	none

[a] k, the Boltzmann constant; n, the particle number density; m, the particle mass; s, the scattering cross section; T, the absolute temperature; θ_D, the Debye temperature; h, Planck's constant; l, the length; α, the absorption characteristic; v_F, the Fermi level velocity; v_{son}, the velocity of sound; v_{lgt}, the velocity of light.

vacuum. This is one major difference between radiation and the other two modes.

Electromagnetic waves are the means by which heat is transferred by radiation. The energy contained in an electromagnetic wave of a specific frequency is

electro-
magnetic
energy

$$E = hf = h\frac{c}{\lambda} \tag{3.4.1}$$

where E is the electromagnetic energy (N m); h, Planck's constant ($= 6.62554 \times 10^{-34}$ N m sec); f, the frequency of radiation (cycles/sec); c, the speed of light ($= 2.997925 \times 10^8$ m/sec); and λ, the wavelength (m/cycle).

It can be seen that higher energies accompany higher frequencies. Most radiant heat exchange occurs in the range of 10^{12}–10^{15} cycles/sec (see Figure 3.4.8). At these frequencies, radiation occurs along the line of sight. That is,

two objects
must face
each other

one surface must be directly facing another for radiant heat exchange to occur between them. Heat may be transferred around corners and behind objects by conduction and convection. You do not have to be facing a room heater to feel its warmth, but you must face a campfire on a cold night to receive its heat. One may even be extremely hot on the side facing the campfire and cold on the side not facing it because the facing side is receiving radiant heat but the opposite side is losing radiant heat to cold surroundings.

Of the radiant heat received, there are three possible dispositions to the total: the heat may be (1) absorbed, (2) reflected, or (3) transmitted. This is expressed by:

$$1 = \alpha + \rho + \tau \tag{3.4.2}$$

where α is the fraction of incident radiation absorbed, also called absorptivity (dimensionless); ρ, the fraction of incident radiation reflected, also called reflectivity (dimensionless); and τ, the fraction of incident radiation transmitted, also called transmissivity (dimensionless).

For an opaque object, $\tau = 0$, and $\alpha + \rho = 1$. For an opaque object that does not reflect anything, $\tau = \rho = 0$, and $\alpha = 1$. Notice that the absorptivity and reflectivity are surface properties. Only the transmissivity depends on the bulk properties of the material.

3.4.1 Black Body Radiation

An ideal radiator is called a black body. This is because the black body absorbs all incident radiation (absorptivity, $\alpha = 1$) and the lack of reflected light would appear black in the visible portion (400 to 700 nm) of the electromagnetic spectrum (Figure 3.4.1). A black body absorbs heat according to

$$\dot{q} = AF\alpha\sigma T^4 \tag{3.4.3}$$

where q is the radiant energy absorbed (N m/sec); A, the surface area of black body (m^2); σ, the Stefan–Boltzmann constant [$= 5.676 \times 10^{-8}$ N m/(sec m^2 K^4)]; α, absorptivity (dimensionless); T, the absolute temperature (K); and F, the shape factor (dimensionless).

Unlike conduction and convection heat transfer, which are transmitted because of an effort variable of temperature θ, radiation heat transfer depends on T^4. In fact, the effort variable for radiation heat transfer is usually taken to be σT^4, and the radiation thermal resistance is $(1/AF\alpha)$. Of course, $\alpha = 1$ for black bodies.

3.4.1.1 Shape Factors The shape factor, F, is the fraction of the surface area, A, that takes part in radiant heat exchange. Because radiation exchange occurs by line of sight, those parts of the surface area not able to be viewed directly from some part of the other body with which the black body is in radiant contact cannot be used to calculate radiant heat exchange. Thus a body with one-half of its total surface area in line-of-sight contact with another body has a shape factor to that body of 0.5 (Figure 3.4.2).

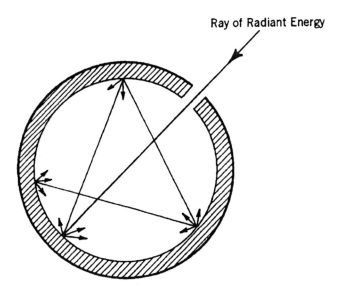

Figure 3.4.1. Reflection of radiation in a cavity. As long as the radiation does not return to the hole, the cavity is a perfect absorber and acts like a black body.

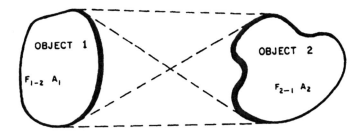

Figure 3.4.2. Two-dimensional representation of line-of-sight areas.

geometrically determined   Shape factor values depend totally on geometry and are calculated for bodies in pairs. Shape factors (also called view factors) are calculated from

$$F_{ij} = \frac{1}{A_i} \int_{A_j} \int_{A_i} \frac{\cos \theta_i \cos \theta_j \, dA_i \, dA_j}{\pi r^2} \tag{3.4.4}$$

where F_{ij} is the shape factor from body i to body j (dimensionless); A_i, the surface area of body i (m^2); A_j, the surface area of body j (m^2); r, the distance between bodies i and j (m); θ_i, the view angle for body i toward body j, taken from a line drawn normal to the surface of body i (rad); and θ_j, the view angle for body j toward body i, taken from a line drawn normal to the surface of body j (rad).

Values for θ_i, θ_j, and r vary across the surfaces A_i and A_j (Figure 3.4.3).

Because they depend only on geometry, shape factors do not require recalculation each time a radiation problem is solved. Tabulated values are available (Howell, 1982) and some are found in Figures 3.4.4–3.4.7 and in Table 3.4.1.

For shape factors that are not given, but are related to tabulated values, there are a number of relationships among them that may help to determine unknown values. First is the principle of reciprocity (the same parts of body i that view body j can be viewed from the other body)

reciprocity

$$A_i F_{ij} = A_j F_{ji} \qquad (3.4.5)$$

sum equals one

The second is the principle of conservation (all parts of a surface can view something)

$$\sum_{j=1}^{N} F_{ij} = 1 \qquad (3.4.6)$$

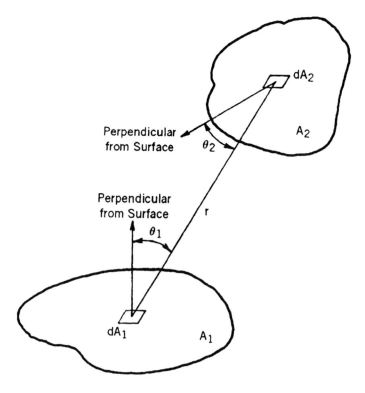

Figure 3.4.3. Geometrical shape factor notation.

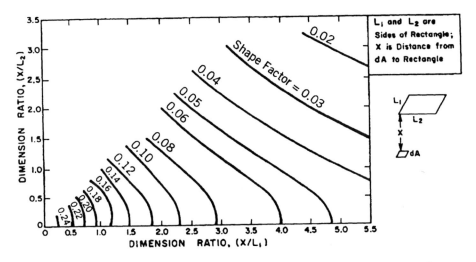

Figure 3.4.4. Shape factor for a surface element and a parallel rectangular surface. Courtesy of *Mechanical Engineering* magazine vol 52, 1930, pp. 699–704; copyright *Mechanical Engineering* magazine (the American Society of Mechanical Engineers International).

There are other techniques that can be used to extend tabulated shape factor values to unique circumstances (Hottel and Sarofim, 1967).

EXAMPLE 3.4.1.1-1. Shape Factor Calculation

Data for mixed convection from a sphere to air, the basis for much of the material in Section 3.3.1.4, was obtained from a wind tunnel 1.9 m long and 30.5 cm diameter (Kirk, 1984). A 5.75-cm-diameter sphere was placed in the center of the wind tunnel test section, or 0.95 m from each end. Determine the shape factors to calculate heat transfer from the heated sphere to the downstream walls of the wind tunnel. Assume thermocouples measured wall temperatures at distances of 22.5 and 75 cm from the end of the tunnel.

Solution

Wind tunnel wall sections appear to the sphere as if they were equivalent to disks, with thermocouples measuring temperatures in the centers of the wall areas corresponding to the disks (see Figure 3.4.8). The shape factor for radiation out the end of the tunnel will be determined as that from a sphere to a disk of diameter equal to 30.5 cm appearing at the end of the tunnel 95 cm

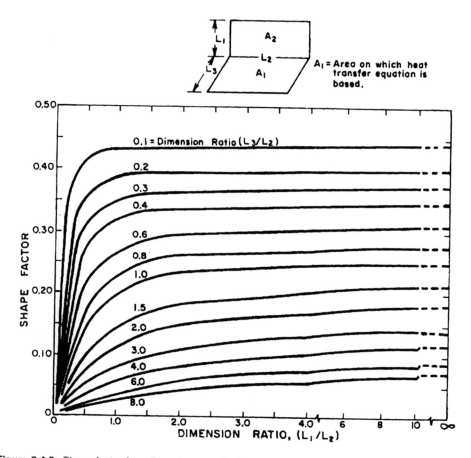

Figure 3.4.5. Shape factor for adjacent perpendicular rectangles. Shape factor must be applied to the heat transfer from area A_1, else a shape factor value of $F_{1-2}A_1/A_2$ must be used. Courtesy of *Mechanical Engineering* magazine vol 52, 1930, pp. 699–704; copyright *Mechanical Engineering* magazine (the American Society of Mechanical Engineers International).

away from the surface of the sphere. Using formula (9) (Howell, 1982) in Table 3.4.1

$$F_{S-1} = \left\{ 1 - \left[1 + \left(\frac{15.25 \text{ cm}}{95 \text{ cm}} \right)^2 \right]^{-0.5} \right\}$$

$$= 0.00632$$

The next equivalent disk will be placed at a position of twice the distance of the first thermocouple from the end of the tunnel, or 55 cm from the tunnel end. In this way, the thermocouple will be measuring the temperature in the center of its section. The radius of this disk is the same as those of the tunnel

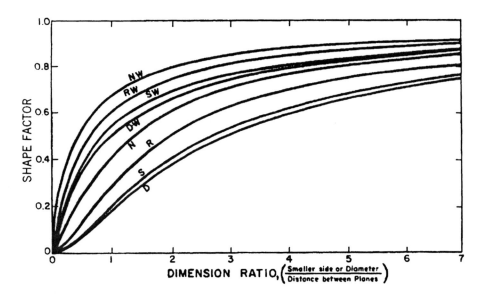

Figure 3.4.6. Radiation between directly opposed parallel planes of equal size and shaped like squares (S), 2 : 1 rectangles (R), narrow rectangles (N), and disks (D). Curves marked with a W relate to opposing planes connected by nonconducting, but reradiating, walls that vary in temperature from one opposing plane to the other. Courtesy of *Mechanical Engineering* magazine vol 52, 1930, pp. 699–704; copyright *Mechanical Engineering* magazine (the American Society of Mechanical Engineers International).

and the first disk (15.25 cm); the distance of this disk from the sphere surface is (95 cm − 55 cm = 40 cm). Thus

$$F_{S-2} = 0.5 \left\{ 1 - \left[1 + \left(\frac{15.25 \text{ cm}}{40 \text{ cm}} \right)^2 \right]^{-0.5} \right\}$$

$$= 0.0328$$

The shape factor from the sphere to this disk, however, also includes the shape factor to the disk at the end of the tunnel. Thus the shape factor value for the ring represented by this portion of the wind tunnel is

$$F_{sr} = F_{S-2} - F_{S-1} = 0.0328 - 0.00632 = 0.0265$$

The shape factor from the sphere to the center of the wind tunnel test section must equal 0.5, since half the sphere surface area is visible from this part of the tunnel. Thus the shape factor from the sphere to the section represented by the second thermocouple is

$$F_{s-w} = 0.5 - F_{S-2} - F_{S-1} = 0.5 - 0.0265 = 0.474$$

Shape factors for the upstream side of the tunnel can be found in a similar fashion.

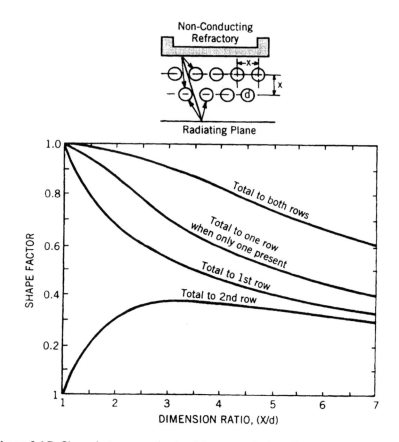

Figure 3.4.7. Shape factor correction involving rows of tubes. This correction must be multiplied by the shape factor between two parallel planes of equal size, one of which is placed in the plane of the tubes. Tubes must be spaced equally within and between rows. Shape factor must be applied to the heat transfer from the radiating plane. Courtesy of *Mechanical Engineering* magazine vol 52, 1930, pp. 699–704; copyright *Mechanical Engineering* magazine (the American Society of Mechanical Engineers International).

3.4.1.2 Spectral Distribution

Black body radiation is not monochromatic. That is, it does not occur at only one frequency. There are simultaneous transmissions of radiant energy over a wide band of frequencies, a fact that complicates radiation heat exchange calculations.

mono-
chromatic
emissive
power

Monochromatic emissive power is the radiant energy emitted per unit area at one particular electromagnetic wave frequency. Thus units of monochromatic emissive power are power (N m/sec) per unit area (m^2) per wavelength (m). Planck's equation relates monochromatic emissive power to radiant wavelength, and is based on quantum theory for a black body in a vacuum

Planck's
equation

$$E_{b\lambda} = \frac{2\pi h c^2}{\lambda^5 [\exp(ch/k\lambda T) - 1]} \qquad (3.4.7)$$

Table 3.4.1 Geometric Shape Factor Values[a]

Surfaces between Which Radiation is Being Interchanged	Shape Factor, F_{1-2}
1. Infinite parallel planes (only facing surface areas used)	1
2. Body A_1 completely enclosed by another body, A_2. Neither of the bodies can see any part of itself	1
3. Surface element dA (A_1) and rectangular surface (A_2) above and parallel to it, with one corner of rectangle contained in normal to dA	See Figure 3.4.4
4. Element dA (A_1) and parallel circular disk (A_2) with its center directly above dA. a, radius of disk; L, separation distance	$a^2/(a^2 + L^2)$
5. Two parallel and equal squares, rectangles, or disks of width or diameter D, a distance L apart	See Figure 3.4.6
6. Two parallel disks of unequal diameter, distance L apart with centers on the same normal to their planes, smaller disk A_1 of radius a, larger disk of radius b	$\dfrac{1}{2a^2}[L^2 + a^2 + b^2 - \sqrt{(L^2 + a^2 + b^2) - 4a^2b^2}]$
7. Two rectangles in perpendicular planes with a common side	See Figure 3.4.5
8. Radiation between an infinite plane A_1 and one or two rows of infinite parallel tubes in a parallel plane A_2 if the only other surface is a nonreflective surface behind the tubes	See Figure 3.4.7
9. Radiation from a sphere of radius a to a perpendicular disk of radius b separated by a perpendicular distance L measured from the center of the sphere to the disk	$0.5\{1 - [1 + (b/L)^2]^{-0.5}\}$
10. Concentric cylinders with the inside cylinder (A_1) of radius a and the outside cylinder (A_2) of radius b	$F_{2-2} = 1 - \dfrac{a}{b}$ $F_{1-2} = 1$ $F_{2-1} = a/b$

[a] For additional shape factor formulas, see Howell (1982).

where $E_{b\lambda}$ is the monochromatic emissive power for a black body [N m/(sec m^3)]; h, Planck's constant ($=6.62554 \times 10^{-34}$ N m sec); c, the speed of light ($=2.9979 \times 10^8$ m/sec); k, Boltzmann's constant ($=1.3802 \times 10^{-23}$ N m/K); λ, the wavelength (m); T, the absolute temperature (K); and $\exp(x)$ denotes e^x.

This relationship is plotted in Figure 3.4.9 and for several surface temperatures, T (Figure 3.4.10). The emission spectrum is rather broad toward the tail at the longer wavelengths, and the curve has a maximum value at a wavelength of

maximum
emission

$$\lambda = 2.898 \times 10^{-3}/T \tag{3.4.8}$$

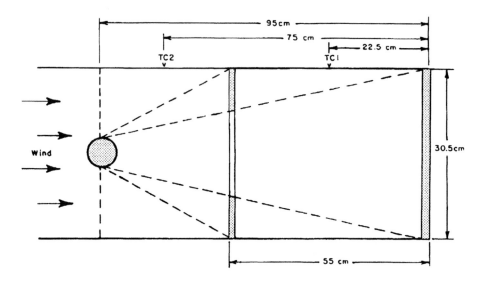

Figure 3.4.8. Shape factor diagram for the experiment to apply a radiation correction to heat lost from a sphere to air in a wind tunnel. Radiation to the walls of the tunnel is equivalent to radiation to disks and rings with diameter of the tunnel and at various distances from the sphere.

As a radiating object becomes hotter, more energy is emitted, and the position of maximum energy falls at shorter wavelengths. Some of the radiant thermal energy falls in the visible spectrum, and the hotter the object, the brighter it glows (until, of course, the maximum moves into the ultraviolet band—but this would require a body to be so energetic that it might no longer be defined as thermal radiation). The sun has an effective surface temperature about 5800 K, which puts the maximum of the sun's thermal radiation in the visible range.

Integrating the monochromatic emissive power over all wavelengths (area under the curve in Figure 3.4.9) and over the entire surface area of a black body gives the total amount of thermally emitted radiation as

relationship among constants

$$\dot{q} = A \int_0^\infty E_{b\lambda} \, d\lambda = \frac{2\pi^5 k^4 A}{15 c^2 h^3} T^4 = \sigma A T^4 \tag{3.4.9}$$

The value of the Stefan–Boltzmann constant, σ, is thus obtainable from the Planck, Boltzmann, and speed of light constants.

3.4.2 Real Surfaces

real surfaces Real surfaces do not have smooth monochromatic emissive power curves as shown in Figure 3.4.9. There are notches and bumps that can sometimes be explained by radiation absorption at particular frequencies (see Section 3.4.5).

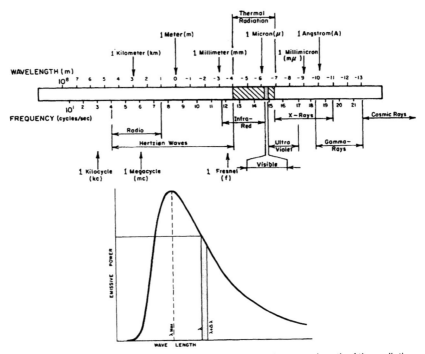

Figure 3.4.9. Monochromatic emissive power plotted against wavelength of the radiation. There is a peak that shows at what wavelength the maximum amount of radiation is emitted. The emitting body is at a constant absolute temperature.

In Figure 3.4.11 is shown the sun's spectral distribution of radiant energy at the Earth's surface and above the atmosphere. Various atmospheric gases modify the shape of the curve of the sun's radiation quite a bit.

gray bodies
A surface that absorbs and radiates a fraction of black body radiation is called a gray body. Gray bodies have the same shape of monochromatic emissive power curve as do black bodies but everywhere the value of the curve is diminished by a constant fraction. This fraction is called the surface emissivity, and radiative heat transfer to a gray body is given as

surface
emissivity

$$\dot{q} = AF\varepsilon\sigma T^4 \qquad (3.4.10)$$

where $\dot{q}$ is the radiant energy absorbed (N m/sec); A, the surface area of grey body (m^2); ε, the surface emissivity (dimensionless); σ, the Stefan–Boltzmann constant [5.676×10^{-8} N m/(sec m^2 K^4)]; T, the absolute temperature (K); and F, the shape factor (dimensionless).

Values for surface emissivity are presented in Table 3.4.2.

emissivity not
same as
absorptivity
Comparing Eq. 3.4.10 with Eq. 3.4.3 should suggest that the emissivity and absorptivity are the same. Normally, the two values are taken to be the same, but there is a key difference: Emissivity is defined for a surface emitting

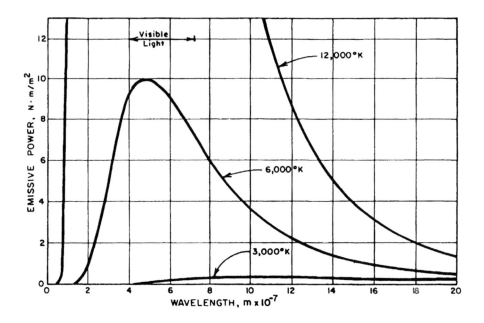

Figure 3.4.10. Plots of emissive power over different wavelengths for emitting bodies at different absolute temperatures. When there is significant radiation in the visible spectrum, we see that the object glows.

radiation, but absorptivity is defined for an absorbing surface. At thermal equilibrium there is no difference between the two because both bodies are at emissivity sometimes taken to equal absorptivity surfaces the measured emissivity varies with temperature, and it is also possible for a surface to absorb radiation from a surface at one temperature and emit radiation at a different temperature. In this case, both bodies are not in thermal equilibrium, and, for body 1 at temperature T_1 absorbing radiation from another body 2 at temperature T_2, the absorptivity of body 1 should be evaluated as the emissivity value of body 1 at T_2, whereas the emissivity of body 1 should be evaluated as the emissivity value at T_1 (Figure 3.4.12). If measured emissivity values do not change with temperature, this procedure is not needed.

From Eq. 3.4.2, radiation that is not absorbed by opaque gray bodies must be reflected

$$\rho = 1 - \alpha \approx 1 - \varepsilon \tag{3.4.11}$$

where ρ is the reflectivity (dimensionless); ε, the emissivity (dimensionless); and α, the absorptivity (dimensionless).

Emissivity of a plant canopy is given in Table 3.4.2 as nearly equal to that plant canopy and fur of a black body. This high value is based mostly on the texture of the plants' surface. Emissivity of an individual leaf, contrarily, is not the same at all

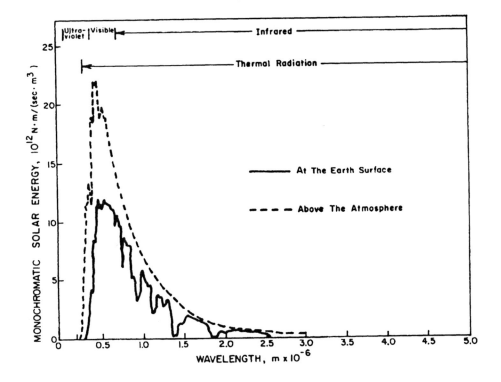

Figure 3.4.11. Characteristics of solar radiation, as given by Gates (1966), and wavelength ranges of the various bands of the radiation spectrum.

wavelengths. The green color radiation is reflected more than blue or red radiation. A similar condition holds for fur, where the overall high-emissivity value is based on surface texture and not on the values for individual hairs.

3.4.3 Radiation Exchange Among Gray Bodies

consider pair
by pair

Because radiation heat exchange occurs between pairs of bodies, a multibody radiant heat exchange system must account for each pair of bodies in the system. Thermal resistance diagrams appear in Figure 3.4.12. Radiant heat exchange between two black bodies is given. There is one resistor between the two effort variables of σT_1^4 and σT_2^4. The value for the resistor is $1/A_1 F_{12} = 1/A_2 F_{21}$. Heat flow between the two is given as

$$\dot{q} = \sigma\,\Delta T^4/R = \sigma A_1 F_{12}(T_1^4 - T_2^4) \tag{3.4.12}$$

two gray
bodies

Radiant heat exchange between two gray bodies must include a correction from the black bodies described above. In Figure 3.4.13 are shown two additional resistances, each with value $R = \rho/A\varepsilon$. These resistances account for the fact that not all heat is absorbed at the surface, and that reflected heat returns to the source. The nodes J_1 and J_2 are equivalent black-body positions,

Table 3.4.2 Emissivities of Various Surfaces

Material	Wavelength (μm) and Average Temperature (°C)				
	9.3 μm 38°C	5.4 μm 260°C	3.0 μm 538°C	1.8 μm 1371°C	0.6 μm Solar
Metals					
Aluminum					
Polished	0.04	0.05	0.08	0.10	~0.30
Oxidized	0.11	0.12	0.18		
Brass					
Polished	0.10	0.10			
Oxidized	0.61				
Chromium, polished	0.08	0.17	0.26	0.40	0.49
Iron					
Polished	0.06	0.08	0.13	0.25	0.45
Cast, oxidized	0.63	0.66	0.76		
Galvanized, new	0.23			0.42	0.66
Galvanized, dirty	0.28			0.90	0.89
Steel plate, rough	0.94	0.97	0.98		
Oxide	0.96		0.85		0.74
Molten				0.3–0.40	
Silver, polished	0.01	0.02	0.03		0.11
Stainless steel					
18-8, polished	0.15	0.18	0.22		
18-8, weathered	0.85	0.85	0.85		
Steel tube, oxidized		0.80			
Building and insulating materials					
Asphalt	0.93		0.90		0.93
Brick					
Red	0.93				0.70
Fire clay	0.90		~0.70	~0.75	
Enamel, white	0.90				
Marble, white	0.95		0.93		0.47
Paper, white	0.95		0.82	0.25	0.28
Paints					
Aluminized lacquer	0.65	0.65			
Cream paints	0.95	0.88	0.70	0.42	0.35
Dull black paint	0.96	0.97		0.97	0.97
Red paint	0.96				0.74
Yellow paint	0.95		0.50		0.30
Oil paints (all colors)	~0.94	~0.0			
White (ZnO)	0.95		0.91		0.18
Miscellaneous					
Ice at 0°C	~0.97				
Water	~0.96				
Carbon					

(*continued*)

Table 3.4.2 (Continued)

Material	Wavelength (µm) and Average Temperature (°C)				
	9.3 µm 38°C	5.4 µm 260°C	3.0 µm 538°C	1.8 µm 1371°C	0.6 µm Solar
T-carbon, 0.9 percent ash	0.82	0.80	0.79		
Filament	~0.72			0.53	
Wood	~0.93				
Glass	0.90				(low)
Skin of white human at 35°C	0.58				
Skin of black human at 35°C	0.80				
Cloth	0.80–0.98				
Bird feathers	0.90–0.98				
Calf skin	0.44–0.64				
Calf fur	0.67–0.80				
Plant canopy (visible light)	0.96–0.98				

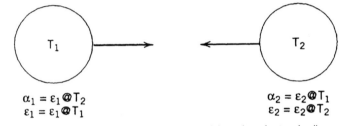

$$\alpha_1 = \varepsilon_1 @ T_2$$
$$\varepsilon_1 = \varepsilon_1 @ T_1$$

$$\alpha_2 = \varepsilon_2 @ T_1$$
$$\varepsilon_2 = \varepsilon_2 @ T_2$$

Figure 3.4.12. Evaluation of absorptivity and emissivity values for two bodies engaged in radiation heat exchange depends on the respective surface temperatures.

and the resistor between J_1 and J_2 is of the same value as the black-body case above. Heat flowing between these two gray bodies is

$$\dot{q} = \frac{\sigma \, \Delta T^4}{R} = \frac{\sigma(T_1^4 - T_2^4)}{\dfrac{\rho_1}{A_1 \varepsilon_1} + \dfrac{1}{A_1 F_{12}} + \dfrac{\rho_2}{A_2 \varepsilon_2}}$$

$$= \frac{\sigma(T_1^4 - T_2^4)}{\dfrac{1 - \varepsilon_1}{A_1 \varepsilon_1} + \dfrac{1}{A_1 F_{12}} + \dfrac{1 - \varepsilon_2}{A_2 \varepsilon_2}} \tag{3.4.13}$$

three surfaces

Heat exchange among three surfaces is also given in Figure 3.4.13. There are now three nodes J_1, J_2, and J_3, and there are resistors between each pair of radiating nodes. Heat exchange between any two bodies is given by an equation similar to Eq. 3.4.13.

Other combinations are also given in the figure. It should be apparent that, when many radiant bodies are involved, the diagrams could become very complicated. This is because each pair of bodies must be included.

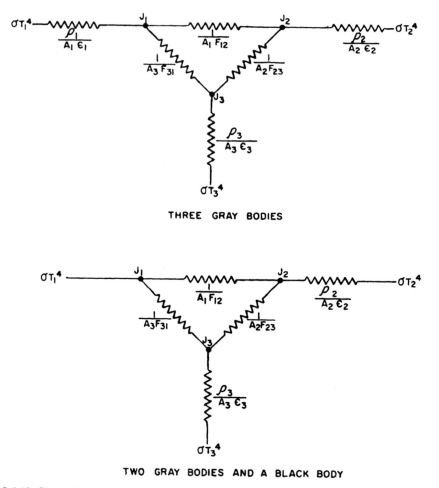

THREE GRAY BODIES

TWO GRAY BODIES AND A BLACK BODY

Figure 3.4.13. Thermal resistance diagrams for radiation heat exchange among combinations of black bodies and grey bodies. When gray bodies are involved, extra resistors of value $(\rho/A\varepsilon)$ are necessary to convert to equivalent black bodies.

Sometimes there is a passively reradiating surface in the system that accepts heat from other sources but that is not maintained at any particular temperature. In this case, the J node is allowed to float without being required that it hold a specific effort variable value. The temperature T_3 in Figure 3.4.14 can be considered to be unknown in this case and varies depending on a balance between T_1 and T_2.

3.4.4 One Body Completely Enclosed in Another

This case occurs commonly enough that it is worthy of separate consideration. One body completely enclosed within another describes the condition of a

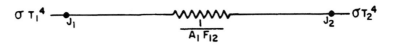

TWO BLACK BODIES

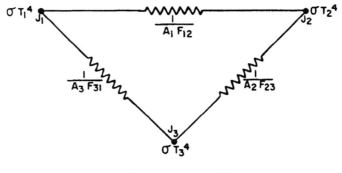

THREE BLACK BODIES

Figure 3.4.13. (*continued*)

single human, animal, or plant within a room, or a living organ within a container, or a pallet of food within a cooler. If the enclosed body (body 1) cannot see any part of itself,

$$F_{12} = 1 \qquad (3.4.14)$$

Furthermore, if the surface area of the enclosing body (body 2) is much greater than the surface area of the enclosed body ($A_1/A_2 \ll 1$), Eq. 3.4.13 becomes

$$\dot{q} = \left(\frac{\rho_1}{\varepsilon_1} + \frac{1}{F_{12}} + \frac{A_1 \rho_2}{A_2 \varepsilon_2}\right)^{-1} \sigma A_1 (T_1^4 - T_2^4)$$

$$= \left(\frac{\rho_1}{\varepsilon_1} + 1 + 0\right)^{-1} \sigma A_1 (T_1^4 - T_2^4)$$

$$= \frac{\varepsilon_1 \sigma A_1 (T_1^4 - T_2^4)}{(\rho_1 + \varepsilon_1)} \qquad (3.4.15)$$

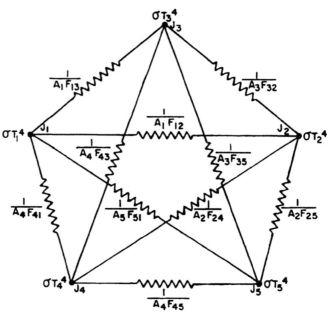

FIVE   BLACK   BODIES

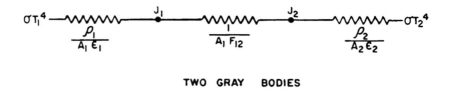

TWO   GRAY   BODIES

Figure 3.4.13. (*continued*)

But, for an opaque body, $\rho_1 + \varepsilon_1 = 1$, and

radiant heat
exchange

$$\dot{q} = \varepsilon_1 \sigma A_1 (T_1^4 - T_2^4) \tag{3.4.16}$$

where $\dot{q}$ is the radiation heat exchange between a body and an enclosure (N m/sec); ε_1, the emissivity of the enclosed body (dimensionless); σ, the Stefan–Boltzmann constant [N m/(sec m^2 K^4)]; A_1, the area of enclosed body (m^2); and T, the absolute temperature (K).

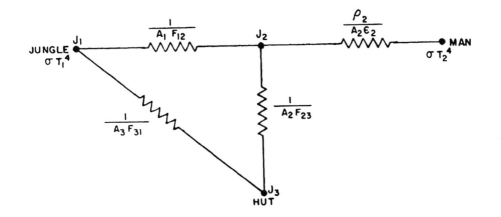

Figure 3.4.14. Radiation thermal resistance diagram for Example 3.4.4-1, where a man stands in the doorway of his hut.

EXAMPLE 3.4.4-1.   Radiation Exchange between Surfaces

A man stands at the door of his hut in the middle of the jungle. Calculate radiation heat exchange.

Solution

There are three distinct radiation thermal resistance components to determine (Figure 3.4.14). There is a resistance from the man to the surrounding jungle. There is also a resistance from the man to the inside of the hut. The third resistance is from the hut to the jungle. We will assume that no part of the man, hut, or facing jungle has radiant contact with the sky.

The jungle will act almost as a black body, according to the emissivity value for a plant canopy in Table 3.4.2. The hut will also be assumed to act as a black body (1) because of the analogy with the cavity in Figure 3.4.1 and (2) because the rough surface of the hut will have a surface emissivity value close to 1.0. The man will radiate heat as a gray body.

The jungle temperature will be that of the ambient environment, perhaps 30°C. Temperature of the man's skin will be slightly less than human body temperature, perhaps 35°C. The temperature of the hut will probably be close to ambient, but because of convection, not radiation. From a radiant standpoint, hut surface temperature floats, between 30 and 35°C, depending on heat gains and losses. For purposes of this example only, we will not specify hut surface temperature; normally its temperature would be specified as ambient.

If the man is an average 70-kg man, then his surface area is 1.8 m^2 (by Eq. 3.2.11). If he stands at the doorway, half inside the hut and half out, then the shape factor from his skin to the jungle is about 0.5. Thus the radiation thermal resistance between points J_1 and J_2 in Figure 3.4.14 is $1/A_2F_{21} = 1.11$ m^{-2}.

The man in the jungle is likely to have dark skin with an emissivity value of 0.80 (Table 3.4.2). That makes the reflectivity value equal $1 - 0.8 = 0.2$. Thus the resistance value between man and point J_2 in Figure 3.4.14 is

$$\frac{\rho_2}{A_2\varepsilon_2} = \frac{0.2}{(1.8 \text{ m}^2)(0.8)} = 0.139 \text{ m}^{-2}$$

The value of resistance from the man to the hut is

$$\frac{1}{A_2F_{23}} = \frac{1}{(1.8 \text{ m}^2)(0.5)} = 1.11 \text{ m}^{-2}$$

To calculate the value of resistance between the hut and jungle requires knowledge about the size of the hut (or about the surface area of the jungle). Except for the small part of the hut represented by the doorway with the man inside, we can assume that the hut surrounded by the jungle acts as one body completely enclosed in another. If, as is likely, the surface area of the hut is much smaller than the surface area of the surrounding jungle, then Eq. 3.4.16 applies, and the resistance between points J_3 and J_1 becomes

$$\frac{1}{A_3F_{31}} = \frac{1}{\varepsilon_1 A_3} = \frac{1}{A_3}$$

Just to put a number on this, let us assume $A_3 = 50$ m^2, making $1/A_3 = 0.02$ m^{-2}.

Radiant thermal effort variable values are σT^4, or [5.676$\times$ 10^{-8} N m/(sec m^2 K^4)] (303 K)^4 = 478.4 N m/(sec m^2) for the jungle and 510.8 N m/(sec m^2) for the human skin.

Referring to Figure 3.4.14, total thermal resistance between man and jungle consists of the series combination of (0.02 m^{-2} and 1.11 m^{-2} = 1.13 m^{-2}), in parallel with the 1.11 m^{-2}, and the 0.139 m^{-2} in series with the previous resistances

$$R_{\text{tot}} = \frac{(1.13 \text{ m}^{-2})(1.11 \text{ m}^{-2})}{1.13 \text{ m}^{-2} + 1.11 \text{ m}^{-2}} + 0.139 \text{ m}^{-2}$$
$$= 0.560 \text{ m}^{-2}$$

Total heat transfer is

$$\dot{q} = \frac{(510.8 - 478.4) \text{ N m/(sec m}^2)}{0.560 \text{ m}^{-2}} = 46.3 \text{ N m/sec}$$

The effort variable value at point J_2 is

$$510.8 \; \frac{\text{N m}}{\text{sec m}^2} - (0.139 \; \text{m}^{-2})\left(46.3 \; \frac{\text{N m}}{\text{sec}}\right) = 504.4 \; \frac{\text{N m}}{\text{sec m}^2}$$

which corresponds to a temperature of 307 K = 34.0°C.

The path for heat transfer splits at point J_2, some heat going directly to the jungle from the man, and some going to the intermediate surface of the hut. To find the hut wall temperature in the absence of convection, we consider only the heat taking the second path. There are several ways to approach this problem, including Kirchhoff's laws (Eqs. 1.7.76 and 1.7.77). The heat flowing through this path is

$$\dot{q} = \frac{(504.4 - 478.4) \; \text{N m}/(\text{sec m}^2)}{1.13 \; \text{m}^{-2}} = 23.0 \; \text{N m/sec}$$

The effort variable value at point J_3 is thus

$$504.4 \; \frac{\text{N m}}{\text{sec m}^2} - \left(23.0 \; \frac{\text{N m}}{\text{sec}}\right)(1.11 \; \text{m}^{-2}) = 478.9 \; \frac{\text{N m}}{\text{sec m}^2}$$

which corresponds to an equilibrium temperature of 30.1°C.

Remark

Because radiation in the absence of convection would cause the hut surface temperature to be only 0.1°C higher than ambient, the assumption of no convection is substantiated. A temperature difference of 0.1°C is well within the error band of all assumptions made in this example.

3.4.5 Radiation Through Absorbing Gases

Many common gases, for example O_2, N_2, H_2, and dry air, have symmetrical molecules and are nearly transparent to thermal radiation. Other gases and vapors important in biological systems, for example, CO_2, H_2O, SO_2, CO, NH_3, hydrocarbons, and alcohols, are heteropolar and either emit or absorb significant amounts of radiant energy at temperatures of practical interest. Two of the most important of these are carbon dioxide and water vapor.

It has already been shown in Figure 3.4.11 that the atmosphere significantly absorbs portions of the solar spectrum. Much of this is due to water vapor in the air. Because there are higher levels of water vapor in the air during the warm days of summer, this absorption is greater during that season.

some
important
gases absorb
energy

Gases absorb and emit radiation only in relatively narrow wavelength bands (Figure 3.4.15).

absorption
not surface
process

Absorption within a gas is not a surface process, as it is with a solid body. The thicker the layer of gas, the more energy that is absorbed. Also of importance are the pressure, shape, and surface area of the gas. The intensity (flux) of monochromatic radiation through a gas varies exponentially with thickness

Beer's law

$$I_{\lambda L} = I_{\lambda 0} e^{-\alpha_\lambda L} \qquad (3.4.17)$$

where $I_{\lambda L}$ is the monochromatic energy intensity at distance L [N m/(sec m^2)]; $I_{\lambda 0}$, the monochromatic energy intensity impinging on the gas surface [N m/(sec m^2)]; α_λ, the monochromatic absorption coefficient (m^{-1}); and L, the distance through gas (m). This equation is known as Beer's law.

The total amount of energy passing through a gas is the integral over all applicable wavelengths of the monochromatic energy intensity. This becomes a very complicated problem, however, depending on pressure, temperature, shape, and direction of the radiation.

From a different standpoint, however, the problem becomes somewhat simpler. If gas is interposed between two black bodies, the gas absorbs radiant energy at the temperature of one of the black bodies, and radiates energy to both bodies at the temperature of the gas

radiant
energy
exchange

$$\dot{q} = \sigma A F (\varepsilon_g T_g^4 - \alpha_g T_b^4) \qquad (3.4.18)$$

where $\dot{q}$ is the net radiant energy exchange between gas and one black body (N m/sec); σ, the Stefan–Boltzmann constant [5.676 × 10^{-8} N m/ (sec m^2 K^4)]; A, the area of gas in contact with black body (m^2); F, the shape factor from gas to black body (often equal to 1.0); ε_g, the emissivity of gas evaluated at the temperature (T_g) of the gas (dimensionless); α_g, the

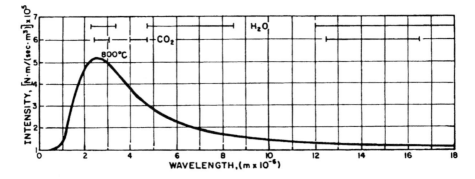

Figure 3.4.15. Emission and absorption bands of water vapor and carbon dioxide. These bands are located at constant wavelengths, but the emissive power spectrum changes with temperature. Thus the effect of gas absorption can be relatively more or less, depending on temperature.

emissivity of gas evaluated at the temperature (T_b) of the black body (dimensionless); and T, the absolute temperature (K).

This equation is true because gas absorption or emission is nearly the same as for a black body as long as the radiant path through the gas is long enough.

In addition to the heat energy coming directly from the gas, there is also radiant energy passing through the gas that is unaffected by the absorption of the gas. As the temperature of the emitting solid body increases, the peak of its energy occurs at a shorter wavelength (Figure 3.4.10), and the relative amount of energy absorbed by the gas at fixed wavelengths (Figure 3.4.15) becomes less consequential.

One additional heat flow path must be mentioned. If the gas is in contact with two bodies at different temperatures, then a convection heat path exists between the bodies. All these paths are diagrammed in Figure 3.4.16.

There are means to determine gas absorptivities and emissivities (Hottel and Egbert, 1942) in order to calculate some of these heat exchanges. The methods are approximate, very empirical, and are limited to black bodies. Significant reflection poses a computational difficulty.

less effect at higher temperatures (margin note)

convection path, also (margin note)

EXAMPLE 3.4.5-1. Radiation through the Atmosphere

Nighttime radiation can cool the Earth and damage crops. The average Earth temperature is about 290 K and the average space temperature is about 3 K. Determine the nighttime temperature of the atmosphere. The radius of the Earth is about 6378 km, and the thickness of the atmosphere is about 80 km.

Solution

We assume that both the Earth and space act as black bodies with the atmosphere interposed. The atmosphere has a long enough radiation path length that the atmosphere acts as a black body to absorb and emit radiation (Kreith, 1958). Thus the temperature of the atmosphere is a result of equilibrium between radiation exchange from atmosphere to Earth and atmosphere to space. Making use of these facts and Eq. 3.4.18 gives a heat balance of

$$\dot{q}_{in} - \dot{q}_{out} = 0 = \sigma A_{earth} F(\varepsilon_g T_g^4 - \alpha_g T_{earth}^4) + \sigma A_{atm} F(\varepsilon_g T_g^4 - \alpha_g T_{space}^4)$$

which becomes ($\varepsilon_g = 1$, $\alpha_g = 1$, $F = 1$)

$$T_g^4 = \frac{\left(\dfrac{A_{atm}}{A_{earth}}\right) T_{space}^4 + T_{earth}^4}{\left[\left(\dfrac{A_{atm}}{A_{earth}}\right) + 1\right]}$$

where T_g is the temperature of the atmosphere.

Convection Through Gas

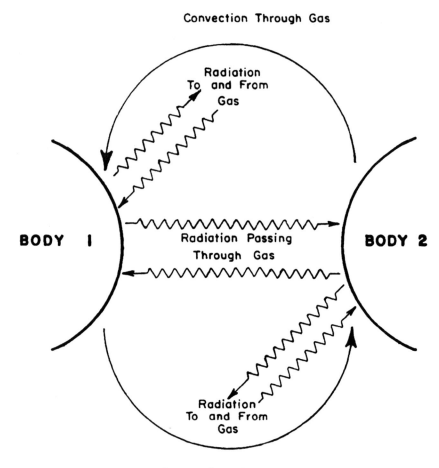

Convection Through Gas

Figure 3.4.16. Heat exchange paths between two bodies separated by a gas-filled gap.

Assuming that one-half the surface of the Earth is radiating to the nighttime sky

$$\frac{A_{\text{atm}}}{A_{\text{earth}}} = \frac{4\pi(80,000 \text{ m} + 6,378,000 \text{ m})^2}{4\pi(6,378,000 \text{ m})^2}$$

$$= 1.025$$

Thus

$$T_g^4 = \frac{1.025(3 \text{ K})^4 + (290 \text{ K})^4}{2.025}$$

$$T_g = 243 \text{ K}$$

Remarks

There are a great number of simplifications that have been made in order to analyze the problem this way. The atmosphere absorbs and emits in specific wave bands, otherwise no sunlight would be seen from Earth; also, the varying gas composition and pressure throughout the atmosphere has not been considered. Nonetheless, the calculated value of $T_g = 243$ K compares reasonably well with the average equivalent radiation temperature of the atmosphere, which is 255 K.

I originally tried to formulate this problem using sunlight. The sun has a radius of 696,000 km and is separated from the Earth by 1.495×10^8 km. It was impossible for me to calculate the shape factor from the Sun to the Earth because the number was so small that it got lost in calculation inaccuracies. It was no better trying to calculate the shape factor for Mercury. Considering that the heat flux from the Sun to the Earth has a mean value of 1390 N m/(sec m^2), and that this is but a very small fraction of emitted power, an immense amount of power is being radiated from the Sun.

3.4.6 Radiation Coefficient

We have thus far seen that calculation of radiant heat exchange can be very complicated. Heat transfer by radiation depends on the fourth power of the absolute temperature, must be calculated separately for each pair of bodies involved in radiation heat exchange, and involves approximations to surface properties that may not be constant over the entire radiation spectrum. It is no wonder, then, that an alternate procedure is sometimes used.

It is convenient to define a radiant heat transfer coefficient, h_r, to be used to calculate radiation heat exchange by a simple temperature difference only

$$\dot{q} = h_r A(T_o - T_e) = h_r A(\theta_o - \theta_e) \tag{3.4.19}$$

an approximation simplifies calculations

where h_r is the radiation coefficient [N m/(sec m^2 °C)]; A, the radiating body surface area (m^2); θ_0, the surface temperature (°C); and θ_e, the mean environmental temperature (°C).

Since calculation of radiant heat exchange by Eq. 3.4.19 is so much different than the correct means of calculation that depends on absolute temperatures to the fourth power, the radiation coefficient must be used with extreme caution and only for very specific circumstances. Nonetheless, there are compelling reasons to use it: (1) It simplifies calculation of radiant heat exchange; (2) it allows the combination of radiation and convection heat transfer calculations insofar as the environmental temperatures for convection and radiation are the same; and (3) it collects radiation heat exchange with all surrounding objects into one equation.

convection and radiation can be combined

One of the specific conditions for which the radiation coefficient is sometimes used is for a human inside a room. Referring to Eq. 3.4.16, which applies to one body completely enclosed in another, the radiation coefficient can be determined from

$$h_r = \frac{\dot{q}}{A(T_o - T_e)} = \frac{\sigma \varepsilon A(T_o^4 - T_e^4)}{A(T_o - T_o)}$$
$$= \sigma \varepsilon (T_o^3 + T_o^2 T_e + T_o T_e^2 + T_e^3) \tag{3.4.20}$$

representative values

Representative values of h_r for clothed humans range from 4.1 to 5.8 N m/(m² sec °C), with an average value for normal environments of 4.7 N m/(m sec °C). Goldman (1978) estimates the radiation coefficient value to be 6.11 N m/(m² sec °C) for a person wearing shorts and tennis shoes and 4.27 N m/(m² sec °C) when the person wears a long-sleeved shirt and trousers.

a sitting person radiates to himself

The surface area used in previous equations cannot automatically be considered to be the actual surface area of the radiating body. One restriction in the development of Eq. 3.4.16 for one body completely enclosed in another is that no part of the enclosed body radiates to another part of itself. Whereas a person standing may, for the most part, be considered to meet this restriction, the same person sitting or crouching cannot. The effective radiation surface area would need to be reduced for the sitting person to about 90% of the standing surface area. Heavy clothing can increase surface area up to about 20% above the unclothed surface area (Section 3.2.2.1).

3.4.7 Solar Flux

Sun is primary energy source

The sun is the primary source for much of the energy used by biological systems occurring naturally on the Earth. The sun sends about 1395 W m/(sec m²) to the Earth (called the solar flux). Because carbon dioxide, water vapor, and other gases absorb solar energy, and atmospheric dust reflects it, only about one-half of this flux reaches the Earth's surface, and the actual amount varies with latitude, cloud cover, air pollution, time of the year, and other variables. Solar intensities for "hazy" clear sky at 1400 hours on representative summer days are given as maximum solar flux values in Table 3.4.3 (Johnson, 1991a). Much work has been expended to calculate accurately solar flux at any given location at various months of the year and times of day (Moon et al., 1981). These calculations are often formidable. Nonetheless, approximate methods can be used to correct values in Table 3.4.3 for time of year and time of day.

surface energy depends on many things

adjustment for time of year

The time of year adjustment can be made by

$$\dot{q}''_{sol,m} = 0.5(\dot{q}''_{sol,max} - \dot{q}''_{sol,min})\{\sin[\pi(d - d_{ve})/183] + 1\}$$
$$+ \dot{q}''_{sol,min} \tag{3.4.21}$$

Table 3.4.3 Direct Solar Fluxes at Different Locations

Region	Representative Elevation (m)	Maximum Solar Flux [N m/(sec m^2)]	Minimum Solar Flux [N m/(sec m^2)]
Tropical rain forest	400	845	794
Tropical savanna	200	890	631
Tropical–subtropical steppe	1600	1013	621
Tropical–subtropical desert	1600	1010	422
Humid subtropical	100	850	292
Humid continental	300	835	237
Subarctic	400	955	253
Tundra	200	985	126
Ice cap	100	1095	0

Sources: Johnson (1991a); Meinel and Meinel (1976).

where $\dot{q}''_{sol,m}$ is the midday solar flux [N m/(sec m^2)]; $\dot{q}''_{sol,max}$, the maximum solar flux [N m/(sec m^2)]; $\dot{q}''_{sol,min}$, the minimum solar flux [N m/(sec m^2)]; and $(d - d_{ve})$, the days past the vernal equinox (days).

adjustment for time of day

The value of solar flux at any time of day can be obtained similarly

$$\dot{q}''_{sol} = \dot{q}''_{sol,m} \sin[\pi(t - t_{sr})/(t_{ss} - t_r)] \qquad (3.4.22)$$

where $\dot{q}_{sol}$ is the solar flux during the day [N m/(sec m^2)]; $\dot{q}''_{sol,m}$, the midday solar flux [N m/(sec m^2)]; t, the time of day (h); t_{sr}, the time of sunrise (h); and t_{ss}, the time of sunset (h).

These adjustments are approximate, but can be used to estimate clear sky solar radiation. The presence of clouds and atmospheric pollution will modify the actual solar flux.

In addition to the direct solar flux, there is diffuse radiation and reflected radiation to consider. These are beyond the scope of this text. See Johnson (1991a) for details.

Solar flux can be considered to be a heat source impinging on a flat surface.

heat absorbed

The actual rate of heat absorbed will be

$$\dot{q} = \alpha A \dot{q}''_{sol} \qquad (3.4.23)$$

where $\dot{q}$ is the absorbed heat (N m/sec); α, the absorptivity (dimensionless); A, the surface area facing the sun (m^2); and $\dot{q}''_{sol}$, the solar flux [N m/(sec m^2)].

We have pointed to the fact that absorptivity and surface emissivity are not the same, yet tabulated emissivity values are often used in place of unknown absorptivity values. Many times this will be sufficient. There are cases, however, where more exact methods are required. One of these is in the

greenhouses are different

analysis of greenhouses, where short-wavelength solar radiation can be transmitted through the glass, but longer-wavelength heat given off by plants is blocked from escape from the greenhouse interior by the same glass.

3.5 HEAT GENERATION

One of the features that distinguishes biological from inanimate systems is the inevitable heat generated by living things. The biological engineer should never forget about generated heat, because designs involving, especially, cooling cannot be sized correctly without accounting for it. Because it is so pervasive, we will investigate the nature of generated heat.

3.5.1 Diffuse Heat Production

Biological heat is usually produced intracellularly. On a small enough scale this could not be considered uniform, but this scale is rarely encountered.

Biological heat is therefore considered to be produced uniformly over some specific volume. Sometimes the rates of heat production are not the same from one site to another. Human and animal viscera differ in heat production rates from one organ to another; the liver has a particularly high rate of heat production. Although not a visceral organ, the brain is another above average heat producer. In both of these organs the blood circulation serves not to bring heat to them, but to cool them.

Since it is diffuse, heat production density is considered to be a uniform $\dot{q}_{gen}'''$, and the total rate of heat production is just $\dot{q}_{gen}''' V$, where V is the volume.

3.5.2 Temperature Dependence

Biological heat production occurs as a result of biochemical processes. There are two general types of these: Anabolism is the formation of complex substances from much simpler chemical building blocks, and catabolism is the reduction of complex molecules into simpler ones. The former process occurs whenever one of the many specialized substances must be made from the relatively small number of stored precursors. As a means of economizing, the body can make most of its essential substances from a very limited number of substrates.

The latter process (catabolism) is essential; otherwise, the intestines would need to be permeable to a very large number of digested ingredients. Breaking these complex substances into smaller parts enables a much more manageable digestion process than would otherwise be the case.

Each of these processes is controlled by enzymatic reactions that are characterized by a high degree of specificity and by a high degree of controllability. The process by which an end product is produced from several basic ingredients often occurs, in biological systems, in many small steps. This allows exquisite control of the process by the biological system.

ATP

The basic energy source for most biological processes is the energy-rich phosphate bond in adenosine triphosphate (ATP). When ATP is hydrolyzed, it forms phosphate plus free energy plus adenosine diphosphate (ADP)

$$ATP \rightleftharpoons ADP + P + \text{free energy} \qquad (3.5.1)$$

The amount of energy released depends on the efficiency of the specific biochemical process (perhaps 40–70%), but can be up to 29 kN m per mole of ATP.

The formation of ATP from glucose and ADP

$$C_6H_{12}O_6 + 38PO_4 + 38ADP + 6O_2 \rightarrow 6CO_2 + 38ATP + 44H_2O \quad (3.5.2)$$

and the reverse process given in Eq. 3.5.1 are both enzyme mediated and, as such, proceed at rates determined by temperature.

The rate of heat generated per unit volume is often temperature dependent. The Van't Hoff equation has been used to describe this effect mathematically

Van't Hoff equation

$$\dot{q}_\theta''' = \dot{q}_o''' Q_{10}^{(\theta-\theta_o)/10} \qquad (3.5.3)$$

where $\dot{q}_\theta'''$ is the heat generation at temperature θ [N/(sec m^3)]; $\dot{q}_o'''$, the heat generation at some reference temperature θ_0 [N/(sec m^3)]; and Q_{10}, the Van't Hoff quotient (dimensionless).

Typical values of Q_{10} are in the range of 2 to 4. The above is similar to the Arrhenius equation, which is sometimes used to describe temperature dependence

Arrhenius equation

$$\dot{q}_\theta''' = \dot{q}_o''' e^{-\mu/T} \qquad (3.5.4)$$

where μ is a constant (K); and T, the absolute temperature (K).

temperature dependence

The increase of reaction rates, and thus of heat production, as temperature increases does not occur indefinitely. When temperature climbs too high, the enzymes used to facilitate biochemical reactions fail to retain their actions. They denature, meaning that they change irreversibly, and the organism dies. This lethal temperature is reached precipitously, not being too different from a somewhat lower temperature where reaction rates are very high. The human body temperature of 37°C is just a few degrees from lethality, but offers the advantage of a high rate of metabolism and a high degree of alertness.

EXAMPLE 3.5.2-1. Composting

Composting is often used to turn unsanitary biological waste into useful product. Thermophilic bacteria produce heat from energy contained in metabolic substrates in the waste material, and the temperature of the compost pile rises. The regulatory criterion for compost pathogen control (largely *E. coli*) is that it must be held for 36 h at 55°C or higher. To reach that temperature usually requires an insulated container, but temperatures higher than 70°C can result in microbial demise. Thus there must be a means to remove excess heat. If a commercial compost pile 7.6 m wide × 46 m long × 4.3 m deep is filled with sewage sludge and wood chips (density about 730 kg/m^3) that produce about 1.4×10^6 N m/sec of heat at 55°C, estimate the amount of heat to be removed at 70°C.

Solution

A modified Eq. 3.5.3 can be used to estimate the heat produced at 70°C. The modification involves estimating total heat ($\dot{q}_\theta$), rather than heat density ($\dot{q}_\theta'''$). We assume a Q_{10} value of 2, a low value because of the high temperatures involved. A more conservative choice for cooling system design would be a Q_{10} value of 4. Using Eq. 3.5.3

$$\dot{q}_{70} = \dot{q}_{55} Q_{10}^{(70-55)/10} = \left(1.4 \times 10^6 \ \frac{\text{N m}}{\text{sec}}\right) 2^{(70-55)/10}$$

$$= 4.0 \times 10^6 \text{ N m/sec}$$

Since any heat produced at 70°C is excess heat, it all must be removed.

3.5.3 Biological Heat Production

Heat generated by different types of biological systems each has the preceding characteristics of diffuse nature and temperature dependence in common. However, depending on the specific biological system of interest, different aspects of generated heat assume relative importance. We consider five such systems: microbial, human and animal, stored produce, living plants, and an ecological system.

3.5.3.1 Microbial Systems Microbes are important in waste treatment, biochemical reactors, alcohol production, fermented food production, and other such systems. Microbes are useful organisms to the biological engineer because they are compact biochemical factories that can produce useful products as long as their basic input, output, and environmental conditions are met. Input requirements include sufficient nutrients to produce the desired product; output requirements often involve removal of waste products; and environmental conditions include maintenance of optimal (or at least tolerable) temperature, concentration, acidity, and others.

microbial requirements

Typical stages of microbial growth are shown in Figure 3.5.1 for a batch culture. If a few cells are introduced into a container of nutrient medium, there
lag stage   is first a short lag time during which the cells store energy and acclimatize to the new surroundings. If the cells were in a vegetative state when they were introduced, then the lag time might be longer than if they originally came from an actively growing culture.

Values of lag time often fall in the range of 10–12 h, unless the inoculant is in a physiologically poor state (Sheppard, 1995). In that case, the lag time might stretch to as long as 1–2 days. This is extremely poor practice, however, and should be avoided if at all possible.

To obtain useful product from the cell culture in as short a time as possible, it is important to minimize the lag time. Thus inoculum cells should be in good health and actively growing when introduced. They should have come from the same kind of nutrient medium that they will be put into, else they will have to adapt to the new medium by producing new enzymes and exercising different metabolic pathways. The medium should be optimized to provide all

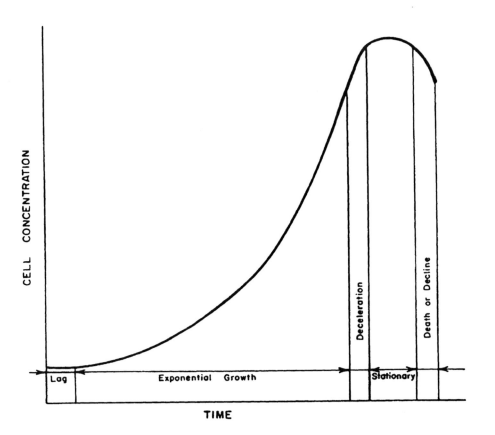

Figure 3.5.1. Typical stages for microbial growth.

necessary nutrients for rapid growth, and the volume of introduced cell culture should be large, perhaps 5–10% by volume of the resulting mix.

logarithmic
growth stage   Once past the lag stage, then the exponential growth stage is reached. Here the rate of increase in the number of cells depends on the number of cells present, an example of an autocatalytic reaction.

The exponential growth is first order

$$\frac{dn}{dt} = \mu n \qquad (3.5.5)$$

where n is the number of cells in the culture (units); t, the time (sec); and μ, specific growth rate (sec^{-1}).

Integration of this equation gives

$$\ln(n/n_o) = \mu t \qquad (3.5.6a)$$

or

$$n = n_o e^{\mu t} \qquad (3.5.6b)$$

where n_0 is the cell number at time t_0 (units). Since the batch volume is constant, cell concentration follows the same equation.

This phase is characterized by a very rapid explosion of cellular numbers. Nutrient concentration does not limit either the cell quantity or composition. Reproduction time depends solely on the metabolic rate of the cells. This rate is usually temperature dependent, but not in a manner similar to the Van't Hoff Eq. 3.5.3. Different cells have different optimal growth temperatures, and varying from the optimum, either higher or lower, can decrease the rate of growth.

The growth phase is typically exponential in character with a time constant of about 30 min for bacteria (*E. coli*), stretching sometimes to a time constant value of 9 to 12 h. Yeasts may have growth phase time constants of 2 to 3–6 h, which may become 14 to 17 h under nonoptimum conditions (Sheppard, 1995).

Nearly the only useful product during the growth phase is the cells themselves. If either the biomass is sought (as in the case of food yeast production), or a cellular component is sought (as in the case of plasmid production), this phase is the most useful one. If a by-product of cellular metabolism is sought (as in the case of amylase or humalin production), it will not usually be produced in useful quantities in this stage.

deceleration
growth phase   The deceleration growth phase follows quite suddenly after the exponential growth phase due to an essential nutrient limitation or the beginning of toxic waste accumulation. The cells adjust themselves to the new limiting environment and do not reproduce unchecked.

stationary
phase

In the stationary phase there is no net growth. This is achieved in one of several ways, either through no new growth, or through equal reproduction and death rates, of either numbers or mass. During this phase the cell produces nongrowth-related metabolites that may be of value. It is during the stationary phase that useful cellular products can be harvested. The composition of the nutrient medium must often be manipulated to induce the culture to produce extraordinary amounts of useful product. This manipulation might involve introduction of a toxin or substitution of one substrate for another. Removal of the toxin or supplying the preferred substrate could induce a second growth period, with the resulting reduction in useful product.

The death phase is caused either by nutrient depletion or toxic product accumulation. For example, alcohol accumulation kills yeast cells. In some early cases of nutrient depletion, dead cells may lyse (decompose), and intracellular nutrients released into the medium are used by the remaining living cells. Toxin accumulation cannot be balanced in this way.

The rate of death is often described similar to growth by a first-order kinetic equation:

$$\frac{dn}{dt} = -\dot{k}_d n \tag{3.5.7}$$

where $\dot{k}_d$ is the first-order death rate constant ($\sec^{-1}$).

Kinetics of the death phase are highly dependent on the type of organism present. Many bacteria form lifeless spores that are not completely dead. These may be capable of becoming active again even after thousands of years, and, therefore, spore formers are not considered to enter the death phase. If death is defined as cell lysis, death may begin 12–24 h after starvation begins and proceed in negative exponential fashion. The cell death phase has not received much attention from those dealing with bioreactors intended to produce biochemical products, but is important in the kinetics of mammalian cell tissue culture survival (Sheppard, 1995) and of sterilization of food and medical materials.

net growth

The effect of temperature on growth and death combines Eqs. 3.5.5 and 3.5.7 to give a net growth rate

$$\frac{dn}{dt} = (\dot{\mu} - \dot{k}_d)n \tag{3.5.8}$$

At high temperatures the thermal death rate exceeds the growth rate, and at low temperatures the growth rate is larger (Shuler and Kargi, 1992). Both growth rate and death rate vary with temperature according to the Arrhenius equation

$$\dot{\mu} = A_\mu e^{E_\mu/RT} \tag{3.5.9a}$$

$$\dot{k}_d = A_k e^{-E_k/RT} \tag{3.5.9b}$$

where E_μ and E_k are the activation energies for growth and death (N m/kg mol); R, the universal gas constant [8314.34 N m/(kg mol K)]; T, the absolute temperature (K); and A_μ and A_k are constants (sec^{-1}).

The activation energy for growth is typically 42 to 84 kN m/kg mol and for death is 250 to 330 kN m/kg mol, showing that thermal death is more sensitive to temperature than is growth.

Other influences on cell growth and metabolism are pH, oxygen availability, solute concentration, carbon dioxide concentration, and glucose availability. There are differences, too, when the culture is composed of many different kinds of coexisting cells. Sometimes different types of bacteria inhibit the growth of other types; sometimes they facilitate growth.

Continuous cultures differ from batch cultures in that new nutrient medium is introduced periodically, and old medium is removed. It is possible, therefore, to maintain the cell culture in either the growth or stationary phase.

Heat generated by these cultures is proportional to the number of cells present times the metabolic rate of each cell. About 40–50% of the energy stored in a carbon-based energy source is converted to ATP energy (Shular and Kargi, 1992). The rest of the energy forms heat. For actively growing cells, metabolism is directed toward growth and not toward maintenance. Thus heat evolution in the exponential growth stage is itself exponentially growing with time.

To estimate the rate of heat production in a reactor requires information about the substrate being used. Heat produced by microbes in bioreactors must be removed for proper bioreactor operation. For aerobic fermentations the rate of metabolic heat generation is empirically given as (Shular and Kargi, 1992)

$$\dot{q}'''_{gen} = 5.02 \times 10^8 \, \dot{Q}_{O_2} \qquad (3.5.10)$$

where $\dot{q}'''_{gen}$ is the rate of heat generation [N m/(sec m^2)]; and $\dot{Q}_{O_2}$, the molar rate of oxygen uptake [kg mol O$_2$/(sec m^3)].

This relationship between heat and oxygen consumption is similar to the value obtained from the oxidation of glucose, Eq. 1.7.3.

EXAMPLE 3.5.3.1-1. Microbial Growth

Growth of microbes is usually expressed in terms of doubling time, defined as the time for a microbial population to double in number. If *Eschericia coli* has an exponential growth rate characterized by a time constant of 30 min, what is the doubling time?

Solution

From Eq. 3.5.6b, the exponential growth is given by

$$n = n_0 e^{\mu t}$$

The parameter μ is equivalent to the inverse of a time constant τ. If we want to find the time difference, $t_2 - t_1$, during which the number n is doubled,

$$\frac{n_2}{n_1} = \frac{n_0 e^{t_2/\tau}}{n_0 e^{t_1/\tau}} = 2$$
$$e^{t_2/\tau} = 2e^{t_1/\tau}$$

Taking natural logarithms

$$t_2/\tau = \ln 2 + t_1/\tau$$
$$(t_2 - t_1) = \tau \ln 2$$
$$= (30 \text{ min})(0.693) = 20.8 \text{ min}$$

EXAMPLE 3.5.3.1-2. Oxygen for Composting

Assuming that the composting operation in Example 3.5.2-1 is aerobic, estimate the rate of oxygen required to sustain microbial activity.

Solution

We can estimate the rate of required oxygen from Eq. 3.5.10

$$\dot{Q}_{O_2} = \frac{\dot{q}_{gen}}{5.02 \times 10^8} = \frac{4.0 \times 10^6 \text{ N m/sec}}{5.02 \times 10^8 \text{ N m/kg mol O}_2}$$
$$= 7.97 \times 10^{-3} \text{ kg mol O}_2/\text{sec}$$

Using the ideal gas law, this amount of oxygen can be translated into volume flow rate of oxygen at 20°C

$$\dot{V}_{O_2} = \frac{\dot{Q}_{O_2} RT}{p}$$
$$= \frac{(7.97 \times 10^{-3} \text{ kg mol/sec})(8314.34 \text{ N m/kg mol K})(293 \text{ K})}{101,300 \text{ N/m}^2}$$
$$= 0.192 \text{ m}^3/\text{sec}$$

Remark

Since air is 20% oxygen, a minimum volumetric flow rate of 0.96 m³/sec must be delivered to satisfy the oxygen requirements of the microbes. More likely, however, is that no more than 5% of the oxygen in the delivered air will be used as it moves through the compost. This is because aerobic organisms are not efficient at removing oxygen from the air at less than 10–12% oxygen. Below this level, composting will turn anaerobic and produce foul-smelling odors. Thus the most likely minimum airflow rate is about 2.0 m³/sec.

3.5.3.2 Human and Animal Heat Production There are three sources of heat production by humans and animals (Figure 3.5.2). The first is basal metabolism, or the summation of heats from all chemical and mechanical processes that must occur to sustain life at a very low level. The second source of heat is due to additional biochemical processes associated with food ingestion. The third source is due to thermodynamic inefficiency of muscular movement.

Basal Metabolic Rate Basal metabolic rate (BMR) varies with body mass (Kleiber, 1975; King and Farner, 1961), as seen in Figure 3.5.3

variation with
body mass
$$BMR = 3.39m^{0.75} \tag{3.5.11}$$

where BMR is the basal metabolic rate (N m/sec); and m, the body mass (kg). Smaller animals with less body mass thus have lower BMRs (Table 3.5.1).

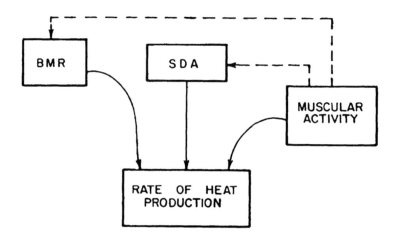

Figure 3.5.2. Factors contributing to the rate of heat production. (SDA is specific dynamic action.)

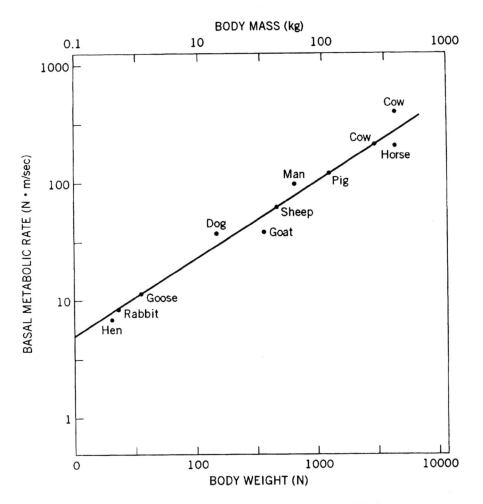

Figure 3.5.3. Logarithmic relation between metabolic rate and body mass.

Table 3.5.1 Basal Metabolic Rates of Various Animals[a]

Animal	Average Weight (N)	Basal Metabolic Rate (N m/sec)
Horse	4325	241
Pig	1255	118
Man	631	105
Dog	149	37.9
Goose	34	11.3
Hen	20	6.9
Rabbit	23	8.4

[a] Also see Altman and Dittmer (1968).

homeo-
therms and
poikilotherms

Some animals are homeothermic (constant body temperature), whereas others are poikilothermic (variable body temperature). BMR, as inferred by the Van't Hoff equation 3.5.3, is strongly influenced by body temperature; thus BMRs of poikilothermic animals must be compared at the same temperature.

temperature
effects

Environmental temperature also influences BMR for homeothermic animals such as humans. At temperatures warmer than thermoneutral (about 20°C), BMR decreases to a minimum amount of 84–91 N m/sec and barely changes with environmental temperature (Table 3.5.2). Below thermoneutral temperatures, BMR increases as environmental temperature decreases.

other
influences

BMR is also affected by age (Table 3.5.3), sex (Table 3.5.3), race [Inuit (Eskimos) > Caucasians > Indians and Chinese], emotional state (anxiety increases heart rate, respiration rate, and circulating epinephrine, all of which increase BMR), climate (increasing amounts of blood thyroxine in cold adaptation increase BMR), body temperature (for each degree Celsius of fever, BMR increases about 9%).

Table 3.5.2 Effect of Environmental Temperature on Resting Metabolic Rate of Clothed Subjects

Room Temperature (°C)	Metabolic Rate (N m/sec)
0	115
10	101
20	84
30	87
40	89
45	91

Table 3.5.3 Age and Sex Differences in BMR Divided by Surface Area

Age (yr)	Basal Metabolic Rate [N m/(m² sec)]	
	Male	Female
2	66.3	61.0
6	61.6	58.8
8	60.2	54.7
10	56.4	53.4
16	53.1	45.1
20	48.1	42.0
30	45.7	41.5
40	44.2	41.5
50	42.7	39.5
60	41.3	37.9

Adult humans eat about 6.3 to 14.7×10^6 N m of food energy per day. Given that there is negligible energy storage as fat or negligible growth as muscle or bone, human daily energy intake must be nearly equal to daily energy production. This gives a heat loss rate of about 120 N m/sec.

digestion produces heat

Food Ingestion   There is an increase of heat production for up to 6 h after ingestion of food that is caused mainly by catabolic processes associated with digestion. This increase is called the specific dynamic action (SDA) of food. Only food that is digestible contributes to SDA, but food, such as cellulose, that is undigestible to humans may be digestible to ruminants. Hence an SDA would be expected in cows and sheep when they eat grass and hay.

For humans, 30% of the digestible protein, 6% of carbohydrate, and 4% of fat is transformed into SDA (Johnson, 1991a). The remainder of the food energy is available for useful bodily purposes.

greatest heat production

Muscular Activity   By far the largest contribution to heat production is muscle activity. It is so important that it is used by poikilotherms to warm themselves when they are cold. Honey bees, for instance, tense their flight muscles on a cold morning to warm the muscles and prepare for flight; the muscles must be warm before they can contract rapidly enough to sustain wing beating frequencies necessary for flight.

muscular efficiency

Muscle contraction is at most 20–25% efficient, and this efficiency is associated with the larger muscles of the legs and arms. Smaller muscles are usually used for actions requiring finer control, and these actions are often performed by antagonistic muscles arranged to act in opposite directions. The efficiency of these muscles approaches 0%. However, the smaller muscles do not contribute greatly to overall heat production.

muscular inefficiency

If muscular action is 20% efficient, then muscular action is 80% inefficient, and the 80% portion of energy shows up as heat. In Table 3.5.4 are shown energy expenditures of various activities (Astrand and Rodahl, 1970). To calculate heat production for each of these, subtract the 105 N m/sec normally taken to be the energy expenditure at rest, take 80% of the difference, and add 105 N m/sec to the result

$$\dot{q}_{\text{gen}} = 0.8(\dot{W} - 105) + 105 \tag{3.5.12}$$

where $\dot{q}_{\text{gen}}$ is the total heat production (N m/sec); and $\dot{W}$, the energy expenditure for the activity (N m/sec).

Aerobic work is more efficient than anaerobic work, and the effects of anaerobic work last for a longer period of time following the cessation of exercise. The excess postexercise oxygen consumption (EPOC) is the oxygen consumption in addition to that required to sustain the BMR following a bout of exercise. EPOC is particularly large following anaerobic exercise.

EPOC

Table 3.5.4 Energy Expenditures for Various Activities

Activity	Energy Expenditure (N m/sec)
Sleeping	49–98
Personal necessities	
Dressing and undressing	160–272
Washing, showering, brushing hair	174–195
Locomotion	
Walking on the level	
0.89 m/sec	133–265
1.34 m/sec	185–370
1.79 m/sec	223–488
Running on the level	
4.47 m/sec	1320–1400
Recreation	
Lying	91–112
Sitting	105–140
Standing	126–174
Playing with children	237–698
Driving a car	63–223
Horseback riding (walk–gallop)	209–488
Cycling 5.81 m/sec	314–774
Dancing	328–886
Gardening	300–698
Gymnastics	174–453
Volleyball	244–698
Golf	342–356
Archery	356–363
Tennis	495–698
Football	614–621
Swimming: crawl	802–977
Cross-country running	747–747
Climbing	698–851
Skiing	698–1400
Domestic work	
Sweeping floors	70–112
Polishing	105–195
Scrubbing	202–488
Cleaning windows	209–251
Washing clothes	160–349
Making beds	265–370
Mopping	293–405
Light industry	
Light engineering work (drafting, drilling watch repair, etc.)	112–167

(*continued*)

Table 3.5.4   (Continued)

Activity	Energy Expenditure (N m/sec)
Medium engineering work (tool room, sheet metal, plastic molding, machinist, etc.)	147–272
Heavy engineering work (machine idling, loading chemical into mixer, etc.)	251–412
Printing industry	147–174
Manual labor	
Shoveling	377–726
Pushing wheelbarrow	349–488

This additional oxygen consumption is accompanied by additional heat production. It is due to

1. Sustained, increased heart rate following exercise;
2. Sustained, increased respiration following exercise;
3. Replenishment of ATP stores, glucose, glycogen, and other compounds from lactic acid in the muscle and blood;
4. Elevated body temperature following exercise.

EPOC may add 50% to the BMR following exercise and can last up to a day after exercise ceases.

EXAMPLE 3.5.3.2-1. Ventilation Requirements for Office and Gym

Compare amounts of heat to be removed from an office containing five human adults and from a gymnasium containing ten human adults.

Solution

Heat produced by five people working in an office, each expending about 120 N m/sec (Table 3.5.4, light engineering work) is

$$\dot{q}_{gen} = 5[0.8(120 - 105) + 105]$$
$$= 585 \text{ N m/sec}$$

Heat produced by ten people in the gym, each expending 350 N m/sec (Table 3.5.4, gymnastics) is

$$\dot{q}_{gen} = 10[0.8(350 - 105) + 105]$$
$$= 3010 \text{ N m/sec}$$

For each instance, generated heat must be removed by the ventilation system.

Remark

Heat produced by lights is often a larger contribution to the ventilation heat load than are the activities of the human occupants.

3.5.3.3 *Living Plants* Plants must sustain themselves with biochemical reactions, many similar or identical to those of animals. Thus plants produce some heat, and this heat production is temperature dependent according to the Van't Hoff equation 3.5.3.

plants cannot perambulate

Poikilothermic animals may have variable body temperatures, but they do have means to regulate their thermal states by moving to other environments or by muscular actions to generate heat. Plants are much less able to regulate their own temperatures, and thus they are more likely to be at the same temperatures as their environments. Plants do lose significant amounts of heat through diffusion of water from the stomata on the undersides of their leaves (some water plants have the stomata on the upper side). Water-stressed plants close their stomata to conserve water, and the temperature of the leaf canopy increases as a result. Well-watered plants have a surface temperature 0.5–3°C lower than ambient.

no significant metabolic heat

Plants produce no significant metabolic heat. Thus heat produced by living plants is not often of great interest to biological engineers. Those plants produced in close confinement, for instance, in glass houses, do not produce significant amounts of heat compared to the solar or supplemental light required for photosynthesis. Since photosynthesis is only 0.2–5% efficient, nearly all the impinging energy results in heat load for cooling systems.

3.5.3.4 *Stored Fruits and Vegetables* Fruits and vegetables are stored at reduced temperatures to lower their rates of ripening and quality deterioration. As given by the Van't Hoff equation, enzymatic activity is reduced by one-half to one-quarter for each 10°C reduction of environmental temperature. Microbial activity is also reduced at lowered temperatures, and thus spoilage is retarded.

reduced temperature important

The biological engineer cannot forget that these are still living, respiring tissues. As such, they produce heat and give off moisture. Both of these are important considerations for cooling system design.

Table 3.5.5 presents measured rates of heat liberation for various fruits and vegetables stored at different temperatures. Estimates of Q_{10} values can be obtained from these tabulated values, and thus heat production rates at other temperatures can be estimated.

It is common to store many of these fruits and vegetables in atmospheres containing elevated carbon dioxide and reduced oxygen levels (Section

Table 3.5.5 Approximate Heat Evolution Rates of Fresh Fruits and Vegetables Stored at Temperatures Shown [N m/(sec Mg)]

Commodity	Temperature (°C)			
	0	5	10	15
Apples[a]	10–12	15–21	41–61	41–92
Apricots	15–17	19–27	33–56	63–101
Artichokes, globe	67–133	94–177	161–291	229–429
Asparagus	81–237	161–403	269–902	471–970
Avocados	—	59–89	—	184–464
Bananas, ripening	—	—	65–116	87–164
Beans, green or snap	—	101–103	161–172	251–276
Beans, lima (unshelled)	31–89	58–106	—	296–369
Beets, red (roots)	16–21	27–28	35–40	50–69
Blackberries	46–68	85–135	154–280	208–431
Blueberries	7–31	27–36	69–104	101–183
Broccoli, sprouting	55–63	102–474	—	514–1000
Brussels sprouts	46–71	95–143	186–250	282–316
Cabbage	12–40	28–63	36–86	66–169
Cantaloupes	15–17	26–30	46	100–114
Carrots, topped	46	58	93	117
Cauliflower	53–71	61–81	100–144	136–242
Celery	21	32	58–81	110
Cherries, sour	17–39	38–39	—	81–148
Cherries, sweet	12–16	28–42	—	74–133
Corn, sweet	125	230	331	482
Cranberries	—	12–14	—	—
Cucumbers	—	—	68–86	71–98
Figs, Mission	—	32–39	65–68	145–187
Garlic	9–32	17–29	27–29	32–81
Gooseberries	20–26	36–40	—	64–95
Grapefruit	—	—	20–27	35–38
Grapes, American	9	16	23	47
Grapes, European	4–7	9–17	24	30–35
Honeydew melons	—	9–15	24	35–47
Horseradish	24	32	78	97
Kohlrabi	30	48	93	145
Leeks	28–48	58–86	158–201	245–346
Lemons	9	15	33	47
Lettuce head	27–50	39–59	64–118	114–121
Lettuce, leaf	68	87	116	186
Mushrooms	83–129	210	297	—
Nuts, kind not specified	2	5	10	10
Okra	—	163	258	431
Onions	7–9	10–20	21	33
Onions, green	31–66	51–201	107–174	195–288
Olives	—	—	—	64–115

(*continued*)

Table 3.5.5 (Continued)

Commodity	Temperature (°C)			
	0	5	10	15
Oranges	9	14–19	35–40	38–67
Peaches	11–19	19–27	46	98–125
Pears	8–20	15–46	23–63	45–159
Peas, green (in pod)	90–138	163–226	—	529–599
Peppers, sweet	—	—	43	68
Plums, Wickson	6–9	12–27	27–34	35–37
Potatoes, immature	—	35	42–62	42–92
Potatoes, mature	—	17–20	20–30	20–35
Radishes, with tops	43–51	57–62	92–108	207–230
Radishes, topped	16–17	23–24	45–47	82–97
Raspberries	52–74	92–114	82–164	243–300
Rhubarb, topped	24–39	32–54	—	92–134
Spinach	—	136	327	529
Squash, yellow	35–38	42–55	103–108	222–269
Strawberries	36–52	48–98	145–280	210–273
Sweet potatoes	—	—	39–95	47–85
Tomatoes, mature green	—	21	45	61
Tomatoes, ripening	—	—	42	79
Turnips, roots	26	28–30	—	63–71
Watermelons	—	9–12	22	—

[a] As an example, 1 kg of apples stored at 10°C will produce about 0.050 N m/sec of heat.

modified atmospheres

4.3.3.2). Such modified atmospheres reduce respiration many fold compared to normal atmospheric storage. Measurements of heat production for specific environmental conditions must be consulted for quantitative information necessary to design cooling systems for these cases.

specific needs

As a last issue, another important consideration is the requirements of the fruits and vegetables themselves. Some tropical fruits cannot be stored at very low temperatures; some vegetables cannot tolerate high CO_2 levels. As living tissues, the needs of these fruits and vegetables should be respected, and some biological response to changing environments should always be expected.

EXAMPLE 3.5.3.4-1. Heat Produced by Warm Apples

Estimate the heat produced by apples at 20°C.

Solution

In Table 3.5.5 are values for heat produced by apples at 0, 5, 10, and 15°C. We need to estimate evolved heat at 20°C. Extrapolation of data is always dangerous, but this is ameliorated somewhat by the use of a sufficiently accurate model. Our model in this case is the Van't Hoff equation 3.5.3. First, we must visually check to see that the data could conceivably fit the Van't Hoff equation. Transforming this equation by taking logarithms

$$\log \dot{q} = \log \dot{q}_o + \left(\frac{\theta - \theta_o}{10}\right) \log Q_{10}$$

or, by assigning θ_0 to $0°C$

$$\log \dot{q} = \log \dot{q}_o + \frac{\theta}{10} \log Q_{10}$$

We recognize this as being of linear form

$$y = a + bx$$

where $y = \log \dot{q}$, $a = \log \dot{q}_o$, $b = (\log Q_{10}/10)$, and $x = \theta$.

We thus obtain $\dot{q}$ and θ from Table 3.5.5 and plot them on semilog graph paper (Figure 3.5.4). Since there is a range of values for heat production at each temperature, we plot an average value. From the graph we can see that,

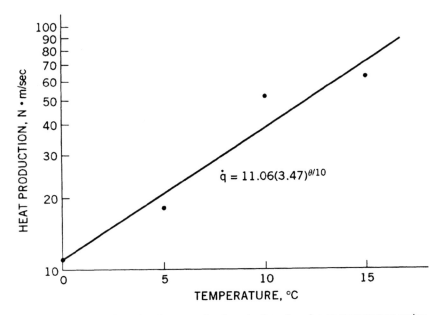

Figure 3.5.4. A semilog plot of average heat production of apples as temperature varies. The appearance of the data does not preclude its description by the Van't Hoff equation. The least-squares best-fit line is drawn on the graph.

although the data do not fall exactly on a straight line, there is no reason to reject the Van't Hoff equation as an adequate model for the data.

Next we must obtain values for $\dot{q}_o$ and Q_{10}. A very common technique with which to accomplish this is least squares (Johnson, 1991b). We form the following table of values:

θ	θ^2	$\dot{q}$	$\log \dot{q}$	$\theta \log \dot{q}$
0	0	11	1.041	0
5	25	18	1.255	6.276
10	100	51	1.708	17.076
15	225	62	1.792	26.886
Sums 30	350	—	5.797	50.238

The intercept a is found as

$$a = \frac{\Sigma y \Sigma x^2 - \Sigma x \Sigma xy}{n \Sigma x^2 - (\Sigma x)^2}$$
$$= \frac{(5.797)(350) - (30)(50.238)}{4(350) - (30)^2}$$
$$= 1.044 = \log \dot{q}_o$$

The coefficient b is found from

$$b = \frac{n \Sigma xy - \Sigma x \Sigma x}{n \Sigma x^2 - (\Sigma x)^2}$$
$$= \frac{(4(50.238) - (30)(5.797)}{4(350) - (30)^2}$$
$$= 0.0541 = (\log Q_{10})/10$$

Thus

$$\dot{q}_o = \text{antilog}(1.044) = 11.06$$
$$Q_{10} = \text{antilog}(0.541) = 3.47$$

Our estimate of heat produced at 20°C by apples is

$$\dot{q} = 11.06(3.47)^{(20/10)} = 133 \text{ N m/sec}$$

Remark

The heat demand on an apple storage system will be greatest at the highest temperature, not only because the apples produce the most heat at this temperature, but also because heat stored in the apples must be removed.

whole Earth modelling

3.5.3.5 Ecological Scale Ecological modelling can be an important application area for biological engineering, especially when one is considering the effects of changes resulting from human activities of astrophysical perturbations. Ecological modelling is the ultimate systems level modelling for the Earth, and results can only be as valid as the model is accurate.

ecological inter-connections complex

A schematic diagram for an ecological model is given in Figure 3.5.5. As with all biological systems, the model has many possible components and extremely complex interconnections.

Earth's heat generation

Two major inputs will be discussed here: Earth's heat generation and solar radiation. Allen (1973) gives the Earth's heat generation as 0.058604 N m/(m^2 sec). The average radius of the Earth at the equator is 6378.164 km and in the polar direction is 6356.779 km. This gives an average radius of about 6368 km. The area of a sphere is $4\pi r^2$, giving a surface area of 9.096×10^{14} m^2, and total heat generation of the Earth as 3.222×10^{13} N m/sec.

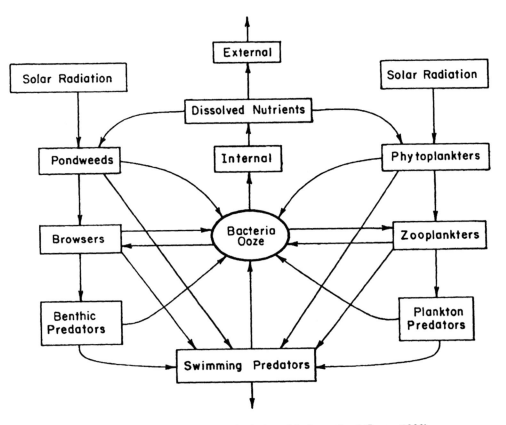

Figure 3.5.5. Diagram for an ecological model of a wetland (Burns, 1989).

Using the expression for thermal resistance of a sphere from Eq. 1.7.47 as

$$R = \frac{1/r_i - 1/r_o}{4\pi k}$$

(1.7.39)

where R is the thermal resistance [°C sec/(N m)]; r_i, the inner radius (m); r_0, the outer radius (m); and k, the thermal conductivity [N/(°C sec)]. The thermal resistance for a 10 km depth of the Earth is 7.86×10^{-12}, using a thermal conductivity value of 2.5 N/(°C sec).

The temperature difference expected over this 10 km difference in depth is $\dot{q}R$, which is calculated as 253°C. This value compares to 20°C/km given by Allen (1973) and 32°C/km given by Clark (1962).

Solar radiation reaching the Earth has been averaged over many years and is called the solar constant (see Section 3.4.7). Its value is 1395 N m/(sec m²). Thus the Earth received about 7.1×10^{17} N m/sec from the Sun. Of course, not all this power reaches the Earth, as shown by Figure 3.4.11. And of the power that is not absorbed by atmospheric gases, some is scattered to reach the ground as diffuse radiation, and some is received as direct radiation. The relative amounts of these depends on the angle of the Sun.

3.5.4 Nonbiological Heat Production

There are many nonbiological sources of heat production to be accounted for in designs of environmental control systems. Motors driving fans and pumps, lights, chemical processes, engines, and mechanical components all generate heat. Even concentrations of electrical wiring or fluid-carrying pipes in a room can be significant sources of heat.

Since there are so many possibilities and so few generalities, there will be no detailed discussion of other sources of heat production inside a closed space. Some further information is presented in Section 4.8.1.6 (ventilation). The best that can be done is to provide a checklist, with the constant question: "What other sources of heat are present?"

3.5.4.1 *Microwaves and Other Electromagnetically Induced Heat* Microwave energy can be used to produce heat inside a material with polar molecules (Figure 3.5.6). Microwaves are electromagnetic waves generated inside special vacuum tubes such as klystrons or magnetrons. Klystrons are usually used at 2450 million cycles per second, and magnetrons usually operate at 915 million cycles per second. These are packaged in microwave generator equipment capable of delivering from 1000 N m/sec (used in the home) to 75 kN m/sec.

Although microwaves are generated as electromagnetic radiation, and thus are subject to the same line of sight that governs other radiation, microwave enclosures are built to reflect radiation so that it reaches the enclosed object from many different directions. Microwaves striking the object can then be

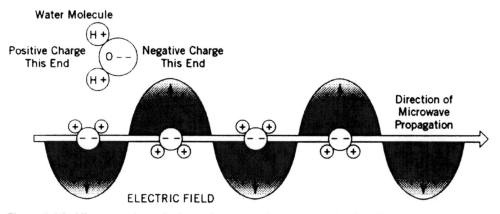

Figure 3.5.6. Microwaves heat dipolar molecules, such as water molecules. These molecules rapidly align themselves with the microwave electric field that changes direction millions of times per second, thus generating heat.

absorbed and transformed into heat distributed throughout the object (although less heat is generated the farther into the object away from the surface). The amount of heat is limited only by the ability of the material to absorb microwave energy and the amount of energy applied. Thus microwave energy can be a very efficient means to dry products, for instance (Figure 3.5.7).

efficiency   Although microwave heating can efficiently generate heat inside an object, microwave systems have overall energy efficiencies as low as 50%. Conventional fuel-fired heating processes are generally 10–60% efficient. Therefore, costs for the heating process must be determined for the particular situation at hand.

As a very rough first approximation, microwave heating can be considered to generate heat uniformly internally in the object, and is therefore similar to metabolically generated heat. Diathermally generated heat produced by a focused high-frequency alternating electric field is similar and is often used in biomedical applications.

3.5.5 Conduction with Heat Generation

A great many biological materials, tissues, organisms, and systems generate internal heat. Whether a microbial reactor or an animal appendage is being considered, heat generation must often be taken into account. There are several possible types of solutions required, and some of these will be examined in the next sections.

3.5.5.1 *Constant Rate of Heat Production* One of the simplest possible problems is to determine the amount of heat being conducted through an object while it is generating heat. In this case, the problem can be solved by

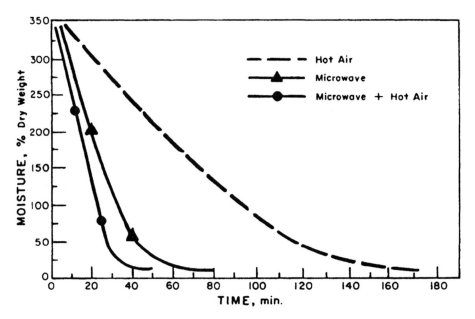

Figure 3.5.7. Comparison of drying rates with hot air, microwave-generated heat, and a combination of the two. Microwaves alone produced drying much faster than hot air alone in this food sample.

add both
components

relying on superposition: Heat conducted in response to external temperature gradients is added to heat generated in the object. The total amount of conducted heat is just the sum of the two components (Figure 3.5.8). (See also Section 1.7.3.2 and Figure 1.7.7.)

The total amount of heat generated in a planar slab of thickness $2L$ is

$$\dot{q}_{gen} = \dot{q}'''_{gen}V = 2\dot{q}'''_{gen}LA \qquad (3.5.13)$$

where $\dot{q}_{gen}$ is the generated heat (N m/sec); $\dot{q}'''_{gen}$, the uniform heat density [N/(sec m^3)]; V, the volume (m^3); L, the half thickness of the slab (m); and A, the surface area of the slab (m^2).

total rate of
heat
generation

The total amount of heat generated in a circular cylinder is

$$\dot{q}_{gen} = \dot{q}'''_{gen}V = \pi\dot{q}'''_{gen}r_o^2L \qquad (3.5.14)$$

where r_o is the outside radius (m); and L, the cylinder length (m).

And for a sphere it is

$$\dot{q}_{gen} = \tfrac{4}{3}\pi\dot{q}'''_{gen}r_o^3 \qquad (3.5.15)$$

no net heat
transfer in
slab center

An odd effect happens when considering the slab. If there is no external temperature difference, then uniform internal heat generation causes a

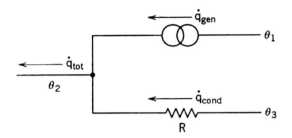

Figure 3.5.8. To determine the amount of heat coming from the object with generated heat, just add heat conducted ($\dot{q}_{cond}$) in response to external temperature gradients ($\theta_3 - \theta_2$) to the total generated heat ($\dot{q}_{gen}$).

temperature maximum to appear along the center line of the slab. Accompanying the maximum temperature is a temperature gradient of zero. Since heat is conducted along a temperature gradient, a zero gradient results in no heat conducted from the center of the slab (Figure 3.5.9).

The situation becomes even more odd when considering both faces of the slab. The temperature gradient at one face is exactly the opposite in sign to the temperature gradient at the other face. Thus the net amount of heat transferred is zero (Figure 3.5.10).

heat transfer is directional

The amount of heat emanating from the slab is really not zero; it must equal $2\dot{q}'''_{gen}LA$, but transferred heat is directional; so the $\dot{q}'''_{gen}LA$ removed from the one face exactly balances with the $-\dot{q}'''_{gen}LA$ removed from the other. Imposition of an external temperature difference moves the position of the temperature maximum and causes an imbalance between the fractions of total heat flowing from the two surfaces. The fraction of the generated heat flowing from either surface is always one-half, but generated heat can either add or subtract from heat conducted as a result of the temperature difference between the faces.

3.5.5.2 Temperature-Dependent Heat Production

superposition does not work

Addition of temperature dependence greatly complicates the solution to the heat transfer problem because it links heat generation and temperature distribution. Superposition can no longer be used.

As an illustration of the means by which problems like this can be solved, one of the simplest of cases is considered. More complicated problems are probably better treated numerically on computer.

The basic differential equation (see Section 1.7.32) for a planar slab of thickness $2L$ is

$$\frac{d^2\theta}{dx^2} = -\frac{\dot{q}'''_{gen}}{k} = -\frac{\dot{q}'''_o Q_{10}^{(\theta-\theta_o)/10}}{k} \tag{3.5.16}$$

Since this is a second-order differential equation, two boundary conditions must be specified to determine the two unknown constants of integration that

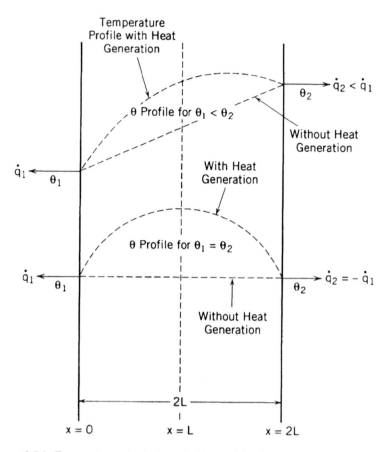

Figure 3.5.9. Temperature distributions inside a slab with and without an imposed temperature difference at the surfaces. The point where $d\theta/dx = 0$ moves from $x = L$ toward $2L$ as θ_2 becomes greater than θ_1.

boundary
conditions

arise when the equation is integrated twice. These boundary conditions can be of three forms: (1) specification of temperature (θ) at the two faces ($x = 0$ and $2L$), or (2) specification of the amount of heat transfer ($\dot{q} = -kA\, d\theta/dx$) at the two faces, or (3) a combination of temperature specification at one face and heat transferred at the other face. As you can see, specification of the amount of heat transferred is equivalent to the definition of the first derivative of temperature with distance x.

To solve Eq. 3.5.16, first substitute

substitution

$$p = \frac{d\theta}{dx} \qquad (3.5.17)$$

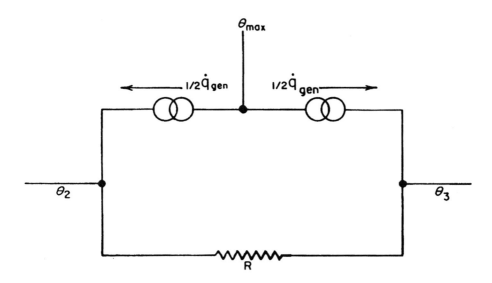

Figure 3.5.10. The thermal situation in a planar slab of thickness $2L$. Heat flows through the slab in both directions and may balance out to zero.

Thus, using the chain rule,

chain rule
used

$$\frac{d^2\theta}{dx^2} = \frac{d}{dx}\left(\frac{d\theta}{dx}\right) = \frac{dp}{dx} = \frac{dp}{d\theta}\frac{d\theta}{dx}$$

$$= p\,\frac{dp}{d\theta} \tag{3.5.18}$$

So

$$p\,\frac{dp}{d\theta} = \frac{-\dot{q}_o''' Q_{10}^{(\theta-\theta_o)/10}}{k} = AQ_{10}^{\theta/10} \tag{3.5.19}$$

where $A = -\dot{q}_o''' Q_{10}^{-\theta_o/10}/k$. Manipulating the equation to separate like variables to different sides,

separating
values

$$p\,dp = AQ_{10}^{\theta/10}\,d\theta \tag{3.5.20}$$

Integrating

integration

$$\frac{p^2}{2} = \frac{10AQ_{10}^{\theta/10} + C_1}{\ln Q_{10}} \tag{3.5.21}$$

where C_1 is an unknown constant of integration ($°C^2/m^2$). Logarithms are taken to the base e.

Thus

$$p = \frac{d\theta}{dx} = \sqrt{\frac{20AQ_{10}^{\theta/10} + 2C_1}{\ln Q_{10}}} \qquad (3.5.22)$$

Again, separating variables,

$$\sqrt{\ln Q_{10}} \int \frac{d\theta}{\sqrt{20AQ_{10}^{\theta/10} + 2C_1}} = \int dx \qquad (3.5.23)$$

Substituting,

$$y = Q_{10}^{\theta/10} \qquad (3.5.24)$$

$$dy = \frac{\ln Q_{10}}{10} Q_{10}^{\theta/10} \, d\theta = \frac{\ln Q_{10}}{10} y \, d\theta \qquad (3.5.25)$$

gives

$$\frac{10}{\sqrt{\ln Q_{10}}} \int \frac{dy}{y\sqrt{20Ay + 2C_1}} = \int dx = x \qquad (3.5.26)$$

which yields

solution

$$x = \frac{10}{\sqrt{2C_1 \ln Q_{10}}}$$

$$\times \ln\left(\frac{\sqrt{20AQ_{10}^{\theta/10} + 2C_1} - \sqrt{2C_1}}{\sqrt{20AQ_{10}^{\theta/10} + 2C_1} + \sqrt{2C_1}}\right) + C_2 \qquad (3.5.27)$$

constant
evaluation
not easy

At this point, it is not difficult to see that evaluating the constants C_1 and C_2 by substituting the boundary values is not going to be easy, and probably the best method is to proceed with a numerical solution. However, by proceeding this far, the numerical solution can be obtained more rapidly and with more accuracy than if the derivation had stopped before this point.

Notice several things about the solution. First, it involves a logarithmic relationship of the variable x, and, second, once past the logarithm, the solution for θ is quadratic with x.

There are several ways to check the accuracy of this solution:

checking
solution
accuracy

1. The final solution, when differentiated twice, should satisfy the original second-order differential equation 3.2.16.

2. Each term at each stage in the solution should have units identical to all other terms. The units of A are those of $\dot{q}_0'''/k$, or $°C/m^2$; the units of $Q_{10}^{\theta/10}$ are

null; the units of the constant 10 in the exponent of $Q_{10}^{\theta/10}$ are °C; the units of $d\theta/dx$ are °C/m, and the units of $d^2\theta/dx^2$ are °C/m^2; Q_{10} is unitless, as is $\log Q_{10}$.

3. If temperatures at the two faces are both zero and $Q_{10} = 1$, then the total amount of heat transferred from the two surfaces should be zero if sign is taken into account or the total should be $2\dot{q}_0''' LA$ if absolute values are summed.

All solutions should be checked with all these criteria before they are taken seriously.

3.6 HEAT STORAGE

Like thermal aspects of change of state, heat storage in a mass is a means for heat to accumulate. However, unlike change of state, heat storage is accompanied by a change in temperature. Thus the heat balance, Eq. 3.1.1, contains a term for heat storage that results as the sum of all the heat gains and all the heat losses experienced by the system.

temperature change

The amount of heat stored depends on the temperature rise, the mass present, and a physical property called the specific heat:

heat storage

$$q = mc_p \, \Delta\theta \tag{3.6.1}$$

where q is the stored heat (N m); m, the mass (kg); c_p, the specific heat [N m/(kg °C)]; $\Delta\theta$, the temperature change (°C).

3.6.1 Specific Heats

Specific heat, also known as heat capacity, of a system is the ratio of the heat absorbed by a unit mass to the concomitant rise in temperature. There are two specific heats common to gaseous systems: those at constant pressure and those at constant volume.

The specific heat at constant volume is defined as

specific heat at constant volume

$$c_v = \frac{dq_v}{d\theta} = \left(\frac{\partial u}{\partial \theta}\right)_v \tag{3.6.2}$$

where c_v is the specific heat at constant volume [N m/(kg °C)]; u, the specific internal energy (N m/kg); and q_v, the heat stored at constant volume (N m/kg).

The specific heat at constant pressure is defined as

specific heat at constant pressure

$$c_p = \frac{dq_p}{d\theta} = \left(\frac{\partial h}{\partial \theta}\right)_p \tag{3.6.3}$$

where c_p is the specific heat at constant pressure [N m/(kg °C)]; h, the specific enthalpy (N m/kg); and q_p, the heat stored at constant pressure (N m/kg).

Because the definition of specific enthalpy is

$$h = u + pv \tag{3.6.4}$$

it follows that

$$c_p = \left(\frac{\partial h}{\partial \theta}\right)_p = \left(\frac{\partial u}{\partial \theta}\right)_p + p\left(\frac{\partial v}{\partial \theta}\right)_p \tag{3.6.5}$$

relationships between the two

where v is the specific volume (m³/kg); and p, the pressure (N/m²).

Note that $(\partial u/\partial \theta)_p$ generally does not equal $(\partial u/\partial \theta)_v$; so the first term on the right-hand side of Eq. 3.6.5 is not c_v. There is, however, a predictable difference between c_p and c_v that can be derived. For gases

$$c_p - c_v = R/M \tag{3.6.6}$$

where R is the universal gas constant [8314.34 N m/kg mol K)]; and M, the molecular weight (kg/kg mol).

The molecular weight for air is 28.97 kg/kg mol; so the difference between specific heats for air over the range of 0 to 200°C is about 0.0687 N m/(kg °C). For monatomic gases

$$c_v = 3R/2M \tag{3.6.7}$$

whereas for diatomic gases

$$c_v = 5R/2M \tag{3.6.8}$$

Equation 3.6.1 includes specific heat at constant pressure for two reasons: (1) Many gaseous processes occur at atmospheric pressure, which is for most practical purposes isobaric; and (2) for liquids and solids (that have very little change in volume with pressure or temperature), there is essentially no difference between c_p and c_v.

temperature dependence

Specific heat values are very temperature dependent, but only slightly pressure dependent. It has usually been found to be sufficient to express the temperature dependence of specific heat with a second-order polynomial, the coefficient values for which are found by experiment. For instance, specific heat at constant volume for air is (Stuart and Kiefer, 1975)

$$c_v = 711.3 + 0.0249T + 0.2030448T^2 \tag{3.6.9}$$

where T is the absolute temperature (K).

Specific heat values for monatomic gases do not vary with temperature, and the ratio c_p/c_v is 1.66 for these gases. Specific heat values for diatomic gases vary with temperature, but are often considered to be constant. The ratio of specific heats for diatomic gas is approximately constant at 1.40 over a wide range of temperature. For more complex gases, generalizations are not possible. Many gases of physiological interest (carbon dioxide, water vapor, ammonia, sulfur dioxide, and others) are in this category. Specific heat generally increases with molecular complexity, and the range of c_p/c_v decreases.

For those situations where specific heat cannot be considered constant, Eq. 3.6.1 must be modified to

$$q = m \int_{\theta_1}^{\theta_2} c_p \, d\theta \qquad (3.6.10)$$

Specific heat is nearly always assumed to be constant, making the use of Eq. 3.6.10 unnecessary.

mixtures   Specific heat values for mixtures may be obtained from specific heats of the pure constituents by

$$c_{pm} = X_1 c_{p_1} + X_2 c_{p_2} + \cdots \qquad (3.6.11)$$

where X_i is the mass fraction of constituent i, (kg i)/(kg total), $= m_i / \sum_j m_j$; and c_{pi}, the specific heat of constituent i [N m/(kg $°C$)].

Thus an estimate of specific heat for a biological or food substance can be obtained from

$$c_p = 4200(X_{H_2O} + 0.5X_{fat} + 0.3X_{solids}) \qquad (3.6.12)$$

Specific heat values for aqueous solutions of salts may be approximated as the mass fraction of water times the specific heat of water alone [about 4200 N m/(kg $°C$)]. Some specific heat values are found in Table 3.6.1.

Specific heat values often change during a change of state. Water, for instance, has a specific heat value of about 2000 N m/(kg $°C$) as ice, 4200 as liquid, and 1900 as steam. The specific heats of foods and biological materials change of may also change greatly between frozen and unfrozen states. To calculate state specific heat values, Eq. 3.6.11 can be used with mass fractions of the substance in different states multiplied by their respective specific heat values.

3.6.2 Flow Systems

Although the mass storing heat can be static, as in a teakettle of water on the static case stove, the mass may also be moving, as in a fluid through a tube. In the static case, heat storage is easily monitored through a difference in temperature, and the amount of stored heat is given by Eq. 3.6.1. The *rate* of heat storage can be calculated from

$$\dot{q} = mc_p \frac{\Delta\theta}{\Delta t} \tag{3.6.13}$$

where $\dot{q}$ is the rate of heat storage (N m/sec); m, the mass (kg); c_p, the specific heat [N m/(kg °C)]; $\Delta\theta$, the change of temperature (°C); and Δt, the time interval (sec).

Fluid flowing in a pipe, when it reaches thermal steady state, does not always exhibit an obvious change of temperature. Take, for example, fluid being heated as it flows through the pipe. Such a situation could exist for milk being pasteurized while it is flowing through a heated pipe. The milk enters the pipe at one temperature and exits at a temperature of 72°C or greater at the other end (after being held at that temperature for 15 sec to kill the *Mycobacterium tuberculosis* organisms). Monitoring temperatures at both ends of the pipe would not show any differences of temperature at those ends, but there is a temperature difference, and thus heat storage, between the ends. Thus, instead of calculating the rate of heat storage from Eq. 3.6.13, the rate of heat storage for a flow system is calculated from

flowing fluid

$$\dot{q} = \dot{m}c_p \, \Delta\theta \tag{3.6.14}$$

where $\dot{m}$ is the mass flow rate (kg/sec); and $\Delta\theta$, the temperature difference between inlet and outlet (°C).

EXAMPLE 3.6.2-1. Addition of Generated Heat

The heat of fermentation in a mixture of grape juice (must) and yeast is 1220 N m/sec. If the vat contains 38,000 L of must, what is the rate of rise of temperature in the vat if there is no cooling system to remove excess heat?

Solution

The specific heat of grape juice is nearly the same as the specific heat of water, or about 4000 N m/(kg °C). The density of water is about 1000 kg/m³, or 1 kg/L. Thus the heat capacity of the vat of must is

$$mc_p = (38,000 \text{ L})(1 \text{ kg/L})[4000 \text{ N m/(kg °C)}]$$
$$= 1.52 \times 10^8 \text{ N m/°C}$$

Table 3.6.1 Specific Heats of Different Materials

Material	Specific Heat Value [N m/(kg °C)]
Metals	
Aluminum	900
Brass	380
Bronze	340
Copper	380
Iron	450
Lead	130
Magnesium	1010
Nickel	450
Silver	230
Steel	460–490
Tin	230
Tungsten	130
Zinc	380
Nonmetals	
Brick	840
Glass	840
Glass wool	700
Grains	
Corn, yellow dent	$1460 + 35.5 \times$ moisture content
Oats	$1280 + 32.6 \times$ moisture content
Rape seed	$1360 + 32.0 \times$ moisture content
Rice, short-grain	$1270 + 34.9 \times$ moisture content
Flour, wheat	1720
Starch, corn	1880
Food materials	
Applesauce	4020
Asparagus, frozen	2010
Bacon, lean	3430
Banana purée	3660
Beef, ground	3520
Beef, lean	3430
Bread, white	2720–2850
Butter	2050–2140
Cheese, Swiss	2680
Cheese, cottage	3260
Corn, frozen	1770
Cream	3060–3270
Eggs, fresh	3180
Eggs, frozen	1680
Fish, canned	3430
Fish, fresh	3600
Flour	1800–1880
Ice cream	1880

(*continued*)

Table 3.6.1 (Continued)

Material	Specific Heat Value [N m/(kg °C)]
Lamb	3180
Macaroni	1840–1880
Meat emulsion	3600
Milk, whole	3860
Milk, dry	1760
Milk, skim	3980–4020
Olive oil	2010
Peas, frozen	1760
Pea soup	4100
Pork, fresh	2850
Pork, frozen	1340
Poultry, fresh	3310
Poultry, frozen	1550
Sausage, fresh	3600
Sausage, frozen	2350
String beans, frozen	1970
Veal	3220
Fruits and vegetables	
Apple	3730–4020
Asparagus	3940
Cantaloupe	3940
Carrot	3810–3940
Corn, sweet	3320
Cucumber	4100
Oranges	3770
Peas	3310
Plums	3520
Potato	3520
String beans	3810
Tomatoes	3980
Ecological materials	
Clay	940
Granite	820
Limestone	900
Marble	800
Sandstone	710
Fir wood	2720
Maple wood	2400
Oak wood	2400
Pine wood	2800
Water	4187
Biological materials	
Beef cattle manure, slurry (85% total solids)	2050
Compost (70% moisture)	4190

(*continued*)

Dentine	1170
Enamel	750
Lactose	1200
Oxalic acid	1610
Paraffin	2890
Porcelain	1090
Urea	1340
Wool	1360
Skeleton	2090
Tissue fat	2520
Blood	3740
Other human tissues	3780
Plastics	
Acrylic	1460
Cellulose acetate	1260–2090
Nylon	1590–2430
Phenolic	1260–1670
Polycarbonate	1260
Polyethylene	
Low density	2300
High density	2300
Polypropylene	1920
Polystyrene	1340
Polysulfone	1300
Polyvinyl chloride	837–1170
Building and insulating materials	
Asbestos	1050
Asphalt	920
Brick, fireclay	830
Cement, portland	780
Concrete	630
Rubber, vulcanized	2010
Liquids	
Acetic acid	2240
Acetone	2210
Benzene	1700
Ethyl alcohol	2430
Formic acid	2130
Hydrochloric acid	2470
Glycerol	2410
Mercury	13.9
Methyl alcohol	2580
Salt brine	3430
Sulfuric acid	1400
Toluene	1760
Water	4180
Gases	
Air	1005

(*continued*)

Table 3.6.1 (Continued)

Material	Specific Heat Value [N m/(kg °C)]
Carbon dioxide	854
Nitrogen	1042
Oxygen	921

From Eq. 3.6.13 we see that

$$\frac{\Delta\theta}{\Delta t} = \frac{\dot{q}}{mc_p} = \frac{1220 \text{ N m/sec}}{1.52 \times 10^8 \text{ N m/}^\circ\text{C}}$$

$$= 8.03 \times 10^{-6} \, ^\circ\text{C/sec} = 0.69^\circ\text{C/day}$$

EXAMPLE 3.6.2-2. Heat Storage in a Pipe

In Example 3.3.1.2-2, tomato sauce was flowing through a stainless steel pipe. Both inlet and outlet temperatures were specified. Determine the correct outlet temperature and heat loss.

Solution

The heat loss calculated in the previous example from overall temperature difference and thermal resistance between pipe and ambient air was $\dot{q} = 28,714$ N m/sec. However, the heat balance equation indicates that this much heat must come from stored heat in the tomato sauce.

Stored heat is given by Eq. 3.6.14. The outlet temperature can be calculated from

$$\theta_o = \theta_i - \frac{\dot{q}}{mc_p}$$

because $\dot{q}$ can be calculated as in the previous example. When θ_o is computed in this way, it changes other variables in the problem, beginning with logarithmic mean temperature. The solution can be obtained by iteration using the approach in Example 3.3.1.2-2. After five iterations, we obtain

$$\dot{q} = 29,128 \text{ N m/sec}$$

$$R_i = 7.16 \times 10^{-4} \text{ sec } ^\circ\text{C/(N m)}$$

$$R_o = 1.12 \times 10^{-3} \text{ sec } ^\circ\text{C/(N m)}$$

$$\theta_{\text{inside pipe}} = 60.9^\circ\text{C}$$

$$\theta_{\text{outside pipe}} = 59.6^\circ\text{C}$$

Remark

These values are not too different from those obtained before, but can be very different depending on the particular problem to be solved.

3.6.3 Convection Determination

capillary
blood flow

Sometimes convection systems are so complicated that there is no practical means to calculate convection coefficients and heat transfer. An example of this is blood flowing through a capillary bed. If the blood enters a limb at deep body temperature and is cooled by the limb as it flows through the vascular system, it may be extremely difficult to calculate the rate of heat loss of the flowing blood. A means around this problem is to calculate convection heat loss as a heat storage problem in a flow system, as in the previous section. Therefore, if entering and exiting temperatures are known, the rate of convective heat transfer can be obtained as the amount of heat storage given by Eq. 3.6.14. The tissue through which the blood flows is usually assumed to be isothermal, and the convective process is usually assumed to be effective enough that the exiting temperature of the blood equals the tissue temperature (Johnson, 1991a). Since blood is composed nearly entirely of water, a great deal of heat can be transferred by this process.

EXAMPLE 3.6.3-1. Touching a Hot Object

Why does a hot metal object feel warmer than a nonmetallic object of the same temperature? The reason involves the higher thermal conductivity possessed by the metallic object. When you touch the object, heat is transferred from the object to your skin, and away from the skin surface by convection and conduction. Temperature is sensed by small, unencapsulated nerve endings called thermoreceptors, located just beneath the surface of the skin. Compare touching aluminum and rubber.

Solution

The preferred skin temperature of the hands is 28.6°C, and we assume that temperature to exist 3 mm from the point of contact between the finger and the object (Figure 3.6.1). We must determine the temperature at the point of contact by means of a heat balance

(heat conducted to point through the object)
− (heat conducted from point through finger)
− (heat convected from point by blood) = 0

Ingersoll et al. (1954) give the temperature distribution (Eq. 1.7.56) for spherical steady-state conduction where heat flows from a point source inside an infinite solid as

$$\dot{q} = 4\pi k r (\theta_s - \theta_o)$$

where $\dot{q}$ is the heat from source (N m/sec); k, the thermal conductivity [N/(°C sec)]; r, the radial distance from point heat source to assumed isothermal surface (m); θ_s, the source temperature at center of sphere (°C); and θ_o, the isothermal temperature on surface of sphere (°C).

In this case, the point of contact between the flat surface and finger becomes a heat sink rather than a source. Also, the spherical area of the isothermal surface in the object is only one-half the area of an entire sphere. Therefore, the first term in the heat balance becomes

$$\dot{q} = 2\pi k_o r_o (\theta_s - \theta_o)$$

where k_o is the thermal conductivity of the object [N/(°C sec)]; r_o, the radial distance in the object (assumed to be 1 cm); θ_s, the temperature at point of contact (°C); θ_o, the isothermal surface temperature (assumed to be 40°C).

Heat conducted from the point of contact through the finger is

$$\dot{q}_k = \frac{-k_f A_f}{L_f} (\theta_s - 28.6)$$

where k_f is the thermal conductivity for finger tissue, [assumed to be 0.252 N/(sec °C)]; A_f, the finger cross-sectional area (m^2); and L_f, the distance

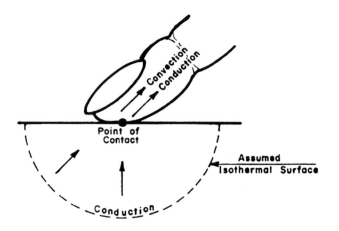

Figure 3.6.1. Touching a hot object involves thermal processes of conduction, convection, and heat storage.

to the point of contact to the deep temperature of 28.6°C (assumed to be 3 mm).

The finger is assumed to be a cylinder 13 mm in diameter.

Convection heat loss occurs as the blood removes heat from the skin and shallow tissues. Blood heat transfer is calculated as the amount of heat required to raise the temperature of the blood to its new level from its entering level. Blood temperature is assumed to equilibrate first with the deep temperature of 28.6°C and then with the temperature at the point of contact

$$\dot{q}_c = -\dot{m}c_p(\theta_s - 28.6)$$

where $\dot{m}$ is the blood mass flow rate (assumed to be 0.0350 kg/sec); and c_p, the specific heat of the blood [assumed to be 3.99×10^{-3} N m/(kg °C)].

Adding these terms,

$$2\pi k_o r_o(\theta_s - \theta_o) + \frac{k_f A_f}{L_f}(\theta_s - 28.6) + \dot{m}c_p(\theta_s - 28.6) = 0$$

Solving for θ_s

$$\theta_s = \frac{2\pi k_o r_o \theta_o + 28.6\left(\dfrac{k_f A_f}{L_f} + \dot{m}c_p\right)}{2\pi k_o r_o + \dfrac{k_f A_f}{L_f} + \dot{m}c_p}$$

$$\frac{k_f A_f}{L_f} = \frac{[0.252 \text{ N}/(\text{sec °C})]\pi(0.013 \text{ m})^2/4}{0.003 \text{ m}}$$

$$= 11.1 \times 10^{-3} \text{ N m}/(\text{sec °C})$$

$$\dot{m}c_p = (0.0350 \text{ kg/sec})[3.99 \times 10^{-3} \text{ N m}/(\text{kg °C})]$$

$$= 0.140 \times 10^{-3} \text{ N m}/(\text{sec °C})$$

$$2\pi r_o = 2\pi(0.01 \text{ m}) = 62.8 \times 10^{-3} \text{ m}$$

Thermal conductivity for aluminum is 206 N/(sec °C). Thus touching an aluminum object results in a surface temperature of

$$\theta_s = \frac{\begin{array}{c}(62.8 \times 10^{-3} \text{ m})[206 \text{ N}/(\text{sec °C})](40°C) \\ + (28.6°C)(11.1 \times 10^{-3} + 0.140 \times 10^{-3}) \text{ N m}/(\text{sec °C})\end{array}}{(62.8 \times 10^{-3} \text{ m})[206 \text{ N}/(\text{sec °C})] + (11.1 \times 10^{-3}) \text{ N m}/(\text{sec °C})}$$

$$= 39.99°C$$

Touching a rubber object with a thermal conductivity of 0.15 N/(sec °C) gives

$$\theta_s = \frac{\begin{array}{l}(62.8 \times 10^{-3}\text{ m})[0.15\text{ N/(sec °C)}](40°C)\\ + (28.6°C)(11.1 \times 10^{-3} + 0.140 \times 10^{-3})\text{ N m/(sec °C)}\end{array}}{\begin{array}{l}(62.8 \times 10^{-3}\text{ m})[0.15\text{ N/(sec °C)}]\\ + (11.1 \times 10^{-3} + 0.140 \times 10^{-3})\text{ N m/(sec °C)}\end{array}}$$

$$= 33.8°C$$

The aluminum surface is perceived as being much hotter despite the fact that both aluminum and rubber objects are at the same temperature. The hotter the surface, the bigger the difference: At 100°C object temperature, the perceived temperatures are 99.9 and 61.1°C. Determining temperature by feel is often not very accurate.

3.6.4 Heat Storage in Biological Systems

Storage of heat seems to be a simple operation: Fill thermal capacity with heat, and temperature rises. However, just to underscore the fact that heat storage does not happen unless heat gain plus heat generated is greater than heat loss, biological systems often have means to change heat loss, heat generated, or heat gained in order that storage of heat does not happen if unneeded.

biological accommodations

Nearly all animals have locomotive responses to high or low temperatures. They can move to positions that are more comfortable thermally. Plants can lose heat by evaporation, although they cannot change locations. Some can also fold their leaves in the heat to reduce heat gain.

Warm-blooded animals can regulate their body temperatures through vasodilation or vasoconstriction of skin blood vessels. This has the effect of changing skin temperature to lose either more or less heat. When that does not completely answer the challenge, sweating or panting helps to remove excess heat. Shivering and thermogenesis can be used to generate extra heat, if necessary.

thermoregulation

The result of this thermoregulation is a long delay between the time of entry into a hostile thermal environment and the beginning of noticeable heat storage. There is at least a 10 min delay between the onset of exercise and a measurable rise in deep body temperature of humans. Clearly, biological systems do not act the same as inanimate objects when it comes to heat storage.

3.6.5 Thermal Capacity

Capacity is a storage element; thermal capacity stores heat. Heat storage can be visualized as a capacity element with a value of mc_p. Larger values of heat capacity exist for larger masses or for larger specific heats. The presence of a heat capacity element adds memory to a heat exchange system. That is, the amount of heat stored depends not only on the present balance between heat

oceans

gains and losses, but also on their variations over time. The oceans, for instance, with their larger specific heat values than those of land masses, tend to store (absorb) more heat energy than land during hot periods and release more heat energy during cool periods. Thus continental land masses are often subject to more extreme temperature differences than are coastal areas.

EXAMPLE 3.6.5-1. Thermal Capacity of an Orange

Calculate the thermal capacity for an orange 6 cm in diameter. Also calculate the thermal time constant if the convection coefficient is 1.5 N m/(sec m^2 °C).

Solution

An orange is shaped like a sphere, with a volume of

$$V = \tfrac{4}{3}\pi r^3 = \tfrac{4}{3}\pi (0.03 \text{ m})^3 = 1.13 \times 10^{-4} \text{ m}^3$$

From Table 2.5.4 we see that the density of water is 1000 kg/m^3. Oranges float in water; so they must be less dense. We assume a density of 850 kg/m^3. Also, from Table 3.6.1, the specific heat of an orange is 3770 N m/(kg °C). Thus

$$
\begin{aligned}
\text{thermal capacity} &= (1.13 \times 10^{-4} \text{ m}^3)(850 \text{ kg/m}^3)[3770 \text{ N m/(kg °C)}] \\
&= 362 \text{ N m/°C}
\end{aligned}
$$

The thermal time constant of this system can be calculated simply only if the convection thermal resistance is much larger than the conduction thermal resistance inside the orange, and is diagrammed in Figure 3.2.3. If conduction appreciably limits the rate of heat storage in the orange, then we must treat this as a distributed-parameter system instead of a lumped-parameter system (see Section 3.7.2.2). As a rough check, we calculate external convection thermal resistance from Eq. 3.3.64

$$
\begin{aligned}
R_c &= \frac{1}{h_c A} = \frac{1}{1.5 \text{ N m/(sec °C m}^2)4\pi(0.03 \text{ m}^2)} \\
&= 58.9 \ \frac{\text{sec °C}}{\text{N m}}
\end{aligned}
$$

with internal conduction resistance from Eq. 1.7.57 (the value for thermal conductivity is from Table 3.2.2)

$$R_k = \frac{1}{4\pi k r_o} = \frac{1}{4\pi [0.43 \text{ N}/(^\circ\text{C sec})](0.03 \text{ m})}$$

$$= 6.17 \frac{\text{sec } ^\circ\text{C}}{\text{N m}}$$

Since convection limits heat flow by about ten times as much as conduction, the lumped-parameter representation can be used. Thus the thermal time constant is

$$\tau = RC = \left(58.9 \frac{\text{sec } ^\circ\text{C}}{\text{N m}}\right)(362 \text{ N m}/^\circ\text{C})$$

$$= 21,400 \text{ sec} = 5.9 \text{ h}$$

EXAMPLE 3.6.5-2. Thermal Capacity of the Human Body

Stolwijk and Hardy (1977) have modeled the human body and its appendages as a series of cylinders. The trunk has the following composition: total mass, 38.50 kg; skeleton, 2.83 kg; viscera, 9.35 kg; muscle, 17.9 kg; fat, 7.07 kg; and skin, 1.35 kg. Calculate the thermal capacity of the trunk.

Solution

Thermal capacity elements due to different types of tissues are in parallel with each other. Thus the total thermal capacity is the sum of the individual capacity elements. The following table is prepared

Tissue	Mass (kg)	Specific Heat (N m/kg °C)	Capacity (N m/°C)
Skeleton	2.83	2090	5,915
Viscera	9.35	3780	35,343
Muscle	17.9	3780	67,662
Fat	7.07	2520	17,816
Skin	1.35	3780	5,103

The total thermal capacity is thus 132,000 N m/°C.

Remark

In order for body temperature to rise 1°C, 132,000 N m of energy must be stored. From Table 3.5.4 we can see that running at 4.47 m/sec for almost

2 min, totally covered by heavy clothes for almost perfect insulation, would generate this much heat.

3.7 MIXED-MODE HEAT TRANSFER

Several very important heat transfer applications depend upon simultaneous conduction through solid material and convection to flowing fluids. The two applications to be discussed here are heat exchangers and transient heat transfer to a surrounding fluid.

3.7.1 Heat Exchangers

no fluid
mixing

There are very many cases where it is desired to transfer heat from one fluid to another without mixing the fluids. Several of these come easily to mind: removal of heat from the circulating engine coolant to air through the car's radiator; conservation of heat from outlet ventilation air to inlet ventilation air in a building; removal of heat from the blood prior to hypothermic surgery. In each of these cases, mixing of the fluids must be prohibited for reasons of safety, efficiency, or operability. Fortunately, if the fluids are separated by a good solid conductor of heat with enough surface area, resistance to heat transfer between the two fluids can be made small while resistance to flow from one fluid to the other is maintained extremely high. The function of heat exchangers is thus to transfer heat efficiently while preventing fluid mixing.

3.7.1.1 Heat Exchanger Types   There are four basic types of heat exchangers to be described. The first involves a change of state and is, strictly speaking, not a heat exchanger in the same fashion as the other three types.

Change of State   The change of state heat exchanger allows the fluid that is to be heated or cooled to flow across a substance that is changing state. There may or may not be a physical barrier between the change of state substance and the flowing fluid. The change of state substance is often stationary, as in the case of air flowing across melting ice. For this example, the air cools as it flows across the ice, and the heat transferred by convection to the ice melts the ice, but does not increase the temperature of either the ice or the melted liquid until there is little or no more ice remaining to be melted. Given a long enough time in contact with the ice, the air eventually reaches the melting temperature of ice, no higher and no lower. This type of heat exchanger can be used when it is important to know very definitely what the final temperature of the flowing fluid will be. Other substances besides water and other changes of state besides melting can be used to design a heat transfer system capable of performing as desired.

melting ice

final
temperatures
known

Referring to Figure 3.7.1, fluid *a* is the flowing fluid, and substance *b* is the substance changing state. Substance *b* may or may not be stationary. In the accompanying temperature diagram is shown that the temperature of *a* falls to the temperature of *b*, until all of *b* is melted. The amount of heat transferred depends on the mass flow rate, specific heat, and initial temperature of *a*, or, equivalently, the mass of *b* melted and the latent heat of melting.

Naturally, the change of state heat exchanger will not operate correctly if the change of state does not occur. It cannot be used for heating fluid *a* if melting or evaporation is the change of state of substance *b*, and it cannot be used for cooling fluid *a* if freezing or condensation is the change of state of substance *b*.

Parallel Flow   The parallel-flow heat exchanger uses two flowing fluids, separated by a solid. The two fluids enter at the same end and flow in the same direction. The accompanying temperature diagram in Figure 3.7.1 shows that the final temperatures of both fluids fall somewhere between the entrance temperature of fluid *a* and the entrance temperature of fluid *b*. The actual final temperatures will depend on the amounts of heat able to be given or accepted by the two fluids. Because the temperature difference between the two fluids is at first large and then small, the rate of heat transfer is also initially high and falls to a very low value at the end. Indeed, a very long flow path is required for the exit temperatures of both fluids to be considered to be the same, because of the decreasing heat transfer rate along the flow path. This type of heat exchanger can be used in unusual circumstances where, for some reason, a gentle final temperature conditioning process is necessary or desired; they are also used to achieve very rapid temperature changes upon entry where the temperature difference between the two fluids is very great.

gentle
conditioning

Counter Flow   The counter-flow heat exchanger is similar to the parallel-flow heat exchanger except that the fluids flow in opposite directions. Fluid *a* enters at the location where fluid *b* exits, and vice versa. As seen in Figure 3.7.1, the counter-directional flow causes the temperature difference between the two fluids to be large and nearly the same along the entire flow path. Heat transfer in this type of heat exchanger occurs at a high rate at all locations, and this type of exchanger can be made relatively small for any amount of heat to be exchanged. The rate of heat transfer per unit volume of heat exchanger is highest with the counter-flow heat exchanger, making it the most efficient type. The exit temperature of fluid *a* becomes the entrance temperature of fluid *b*, and the exit temperature of fluid *b* becomes the entrance temperature of fluid *a*. This is not true, of course, if the heat exchanger is made too small to allow a long enough flow path for all heat to be exchanged between the fluids.

most efficient
type

There is a counter-flow heat exchanger in each limb of most warm-blooded animals. Arms and legs have high ratios of surface area to volume; they thus lose heat very easily, but do not generate much heat. In cool environments, a great deal of body heat can be lost from the limbs unless some method of heat conservation is used.

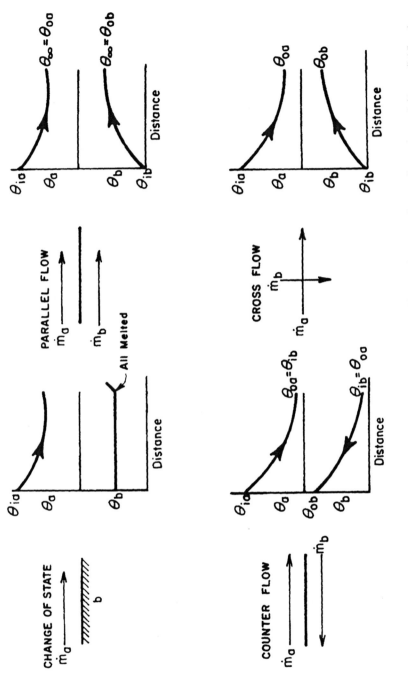

Figure 3.7.1. Variation of temperature with distance for fluids flowing in four different types of heat exchangers: change of state (upper left), parallel flow (upper right), counter flow (lower left), and cross flow (lower right). Subscripts a and b refer to the two different fluids, and subscripts o and i refer to the outlet and inlet.

artery–vein heat exchange

There is in the center of the arm an artery and a vein in close proximity (Figure 3.7.2). These two act as a counterflow heat exchanger. Heat from the warm arterial blood entering the arm becomes lost to the cooler venous blood returning from the arm. Blood that reaches the hand has thus been cooled, the hand temperature reduced, and heat transferred from the hand and arm to the environment has been made smaller. At the same time, blood returning to the torso has been heated to the point where it does not cause reduction of deep body temperature. For some arctic animals, there are much more efficient counter-flow heat exchange structures called *retes* that are able to reduce limb temperature to nearly the temperature of the surrounding air or water while conserving body heat.

vasodilation

Temperate environment animals, such as humans, do not always need to conserve body heat. At times the body must lose heat to maintain a stable body temperature. In that case, returning blood is shunted from the deep vein pictured in Figure 3.7.2 to surface veins. You may have noticed more prominent hand and arm blood vessels in hot weather. This surface return path shuts off the counter-flow heat exchange mechanism, allows the hand and arm surface temperature to rise, and removes a great deal of heat from the body.

ventilation systems

Counter-flow heat exchangers are also commonly used in building ventilation systems. Exhaust air may be humid, or contain dust and other contaminants, or have various amounts of noxious gases, depending on the purpose of the building and the type of occupants (human, animal, or plant). Chemically purer fresh air is important to use, but ambient air is often at a temperature quite different from that maintained inside the building. A counter-flow heat exchanger, where one fluid is the entering air, and the other fluid is the exhaust air, allows fresh air to be supplied without losing the energy equivalent of the warm or cool air inside the building.

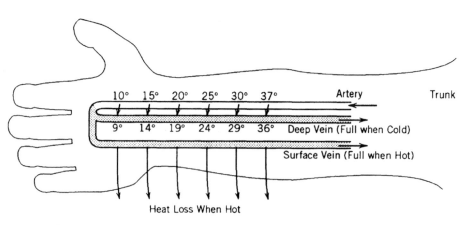

Figure 3.7.2. Schematic of the heat maintenance/heat loss system of the arm and hand. When heat requires conserving, parallel deep artery and vein act as a counter-flow heat exchanger to maintain blood heat. When heat must be lost, the return path is through surface veins.

Cross Flow   The last type of heat exchanger to be described is the cross-flow type. Fluids *a* and *b* flow in directions perpendicular to one another. This is usually done because of some geometrical constraints. For instance, the radiator in your car (called a radiator, but really functions as a convector) allows heat exchange between engine coolant and outside air in a compact space. The temperature diagram in Figure 3.7.1 shows that the final temperatures for fluids *a* and *b* are not predictable in general, but can be determined by experimental testing.

There are many different configurations for cross-flow heat exchangers that use multiple passes and serpentine flow paths. One of these is shown in Figure 3.7.3. For further information on the design or use of these, see Kreith (1958).

3.7.1.2 Heat Transferred
As with all thermal problems, the rate of heat transferred in a heat exchanger is dependent on temperature difference and thermal resistance

heat
exchanged

$$\dot{q} = \Delta\theta_{avg}/R \tag{3.7.1}$$

where $\dot{q}$ is the rate of heat transferred from one fluid to another (N m/sec); $\Delta\theta_{avg}$, the average temperature difference between the fluids (°C); and R, the thermal resistance [°C sec/(N m)].

There is a difference of approach in calculating heat transfer depending on whether all inlet and outlet temperatures are known or not. We consider first the case where all fluid temperatures are known.

All Inlet and Outlet Temperatures Known   There is some problem in determining both the average temperature difference and the effective thermal resistance in Eq. 3.6.1. The temperature difference between the two fluids may be nearly constant for the counter-flow type, but the temperature difference varies greatly with location for the other types. Thus heat flows from the hotter fluid to the colder fluid in varying amounts depending on location. It is desired to be able to calculate the total amount of heat transferred without summing the incremental heat transfer amounts along the length of the heat exchanger. An effective average temperature difference is sought that could be used to calculate this overall heat transfer rate.

Taking a cue from the heat transferred to a fluid flowing in a pipe (Section 3.3.1.2), the average temperature difference has been found to be related to a

logarithmic
mean
temperature
difference

logarithmic mean temperature difference (LMTD). The average temperature difference is thus

$$\Delta\theta_{avg} = \text{LMTD} = \frac{\Delta\theta_1 - \Delta\theta_2}{\ln(\Delta\theta_1/\Delta\theta_2)}$$

$$= \frac{(\theta_{a_1} - \theta_{b_1}) - (\theta_{a_2} - \theta_{b_2})}{\ln[(\theta_{a_1} - \theta_{b_1})/(\theta_{a_2} - \theta_{b_2})]} \tag{3.7.2}$$

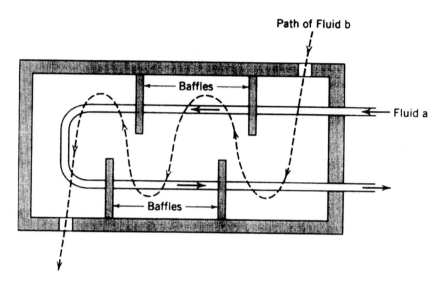

Figure 3.7.3. Cross-flow heat exchanger incorporating multiple passes and serpentine flow pattern.

where θ is the temperature (°C); a, b refer to the two fluids; and 1, 2 refer to the two ends of the heat exchanger.

Note that, for the sake of generality, the two ends, noted as 1 and 2, of the heat exchanger have not been identified as entrance or exit. For the parallel-flow type, end 1 can correspond to the entrance for both fluids. For the counter-flow type, end 1 corresponds to the entrance for one fluid and the outlet for the other. The LMTD can be extended to the cross-flow type, but the meaning of end 1 and end 2 must be defined. Thus, for a parallel flow heat exchanger, the temperature difference between fluid a and fluid b at the entrance, and the temperature difference between fluid a and fluid b at the exit are used in the equation to determine LMTD.

Effective resistance is also to be determined. Heat transferred occurs by convection inside the tube, conduction through the tube, and convection outside the tube (Figure 3.7.4). Since the same amounts of heat are transferred by each of these steps, thermal resistances are in series, and the total thermal resistance is (Eqs. 1.7.44 and 3.3.64)

thermal resistance

$$R = \frac{1}{h_{ci}A_i} + \frac{\ln(r_o/r_i)}{2\pi kL} + \frac{1}{h_{co}A_o} \qquad (3.7.3)$$

where h_{ci} is the inside convection coefficient [N/(sec m °C)]; h_{co}, the outside convection coefficient [N/(sec m °C)]; r_i, the inside radius of the tube (m); r_o, the outside radius of the tube (m); k, the thermal conductivity [N/(sec °C)]; and L, the length of tube (m).

Various components of the above thermal resistance may vary with location, and heat exchangers may be built with much more complicated configurations. Extended surfaces (Section 3.7.3) in the form of fins are often used on the tubes, for instance. Hence, the most reliable means of determining thermal resistance is to measure it. Manufacturers provide this information for their products in the form of $R = 1/AU$, where A is some average effective area (m^2) for heat transfer and U is the overall heat transfer coefficient [N/(sec m °C)]. Actually, the AU product is obtained by dividing LMTD by measured heat transferred.

Thus, knowing LMTD, and thermal resistance obtained from manufacturers' specifications gives the amount of heat transfer, insofar as all inlet and outlet temperatures are known.

All Inlet and Outlet Temperatures Not Known The LMTD can be calculated if all four temperatures are known. It is more likely that they are all not known. A heat exchanger to cool milk and heat water in a dairy barn, for instance, is often used to perform two useful functions in a very efficient way. The milk needs to be cooled to inhibit microbial growth and preserve quality; there is also a great deal of use of hot water for washing purposes. Rather than expend energy to perform both functions, waste heat from the milk is used to preheat well water that later is to be heated further.

dairy application

milk temperatures

The inlet and outlet temperatures for the milk are likely to be known. Inlet temperature is close to body temperature of the cows from which it came. Outlet temperature should be as low as possible to reduce refrigeration costs. Outlet temperature can be specified as reasonably close to inlet water temperature, assuming a countercurrent heat exchanger is used.

groundwater temperatures

Likewise, inlet water temperature is often known. The water is most likely to come from a deep well, and groundwater temperature varies very little with season at different geographical locations. Groundwater temperatures are often given in isothermocline maps. In the middle United States, for instance, groundwater temperature is about 13°C.

What is not known in this example is the outlet temperature of the water. This depends largely on the flow rate of the water.

While it is possible to modify the procedures used when all temperatures are known to fit the case when all temperatures are not known, a different approach can be taken. For this approach, we can use the fact that heat lost by one fluid must equal heat gain by the other fluid.

heat-limiting fluid

Of the two heat exchanger fluids, one is likely to limit the amount of heat that can be transferred. For instance, if the two fluids are air and water, both flowing at the same rate, then the water, by virtue of its larger specific heat, would be capable of accepting much more heat than the air can give. The rate of heat transfer is limited by the air; air temperatures between inlet and outlet would differ by a great deal more than would the water temperatures.

If both fluids are water, but one is flowing at ten times the rate of the other, then the slower fluid limits the amount of heat that can be transferred (Figure

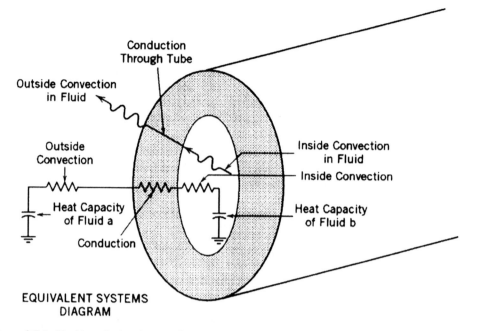

Figure 3.7.4. Heat transfer in a heat exchanger consists of the three processes of convection in fluid a, conduction through the solid tube, and convection outside the tube in fluid b. Heat removed from one fluid becomes stored in the other.

3.7.5). The difference between inlet and outlet temperatures of the slower fluid would be greater by far than the temperature difference of the other fluid. This indicates that the slower fluid has come closer than the faster fluid to accepting or losing the maximum possible amount of heat given the temperature differences available.

It is the product of mass flow rate and specific heat ($\dot{m}c_p$) that determines which fluid limits the amount of heat transferred. Whichever fluid has the smaller $\dot{m}c_p$ product is called the limiting fluid. From a heat balance viewpoint, for the same amount of heat transferred to or from the limiting fluid to the nonlimiting fluid, smaller $\dot{m}c_p$ terms must be accompanied by larger temperature differences. Thus the temperature change for the limiting fluid is greater than the temperature change for the nonlimiting fluid. As shown in Figure 3.7.6, the exiting temperature for the limiting fluid in a counter flow heat exchanger of finite length is, for all practical purposes, the entrance temperature of the nonlimiting fluid. The exit temperature of the nonlimiting fluid is not the entrance temperature of the limiting fluid.

The maximum rate of heat transfer for the counter-flow heat exchanger is

$$\dot{q}_{max} = [\dot{m}c_p(\theta_i - \theta_o)]_{lim}$$

$$= (\dot{m}c_p)_{lim}(\theta_{i\,lim} - \theta_{i\,nonlim}) \qquad (3.7.4)$$

where $\dot{q}_{max}$ is the maximum heat transfer rate (N m/sec); $\dot{m}$, the mass flow rate (kg/sec); c_p, the specific heat [N m/(kg °C)]; θ, the temperature (°C); i, o denote inlet and outlet; and lim, nonlim denote limiting and nonlimiting fluids.

The temperature difference $(\theta_{i\,lim} - \theta_{i\,nonlim})$ represents the maximum temperature difference available for heat to flow.

effectiveness   Heat exchanger effectiveness is the actual rate of heat transfer relative to the maximum rate of heat transfer, as given in Eq. 3.7.4. Thus

$$E = \frac{\text{actual heat transfer rate}}{\text{maximum heat transfer rate}}$$

$$= \frac{[(\dot{m}c_p)(\theta_i - \theta_o)]_{lim}}{(\dot{m}c_p)_{lim}(\theta_{i\,lim} - \theta_{i\,nonlim})}$$

$$= \frac{(\theta_i - \theta_o)_{lim}}{(\theta_{i\,lim} - \theta_{i\,nonlim})} \tag{3.7.5}$$

where E is the heat exchanger effectiveness (dimensionless).

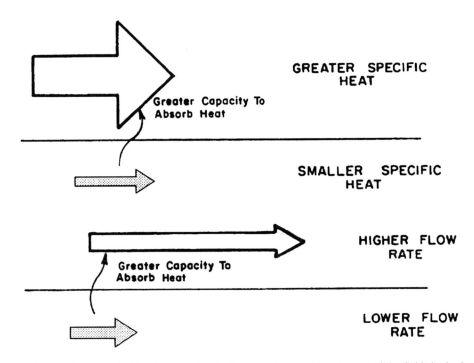

Figure 3.7.5. The amount of heat either absorbed or released is limited by one of the fluids in the heat exchanger. The larger capacity belongs to the fluid with higher mass flow rate and specific heat.

Henderson and Perry (1966) give the heat exchanger effectiveness for counter-flow heat exchangers as

counter-flow effectiveness

$$E = \frac{1 - \exp\left[\dfrac{-UA}{(\dot{m}c_p)_{\text{lim}}}\left(1 - \dfrac{1}{\Re}\right)\right]}{1 - \dfrac{1}{\Re}\exp\left[\dfrac{-UA}{(\dot{m}c_p)_{\text{lim}}}\left(1 - \dfrac{1}{\Re}\right)\right]}$$

(3.7.6)

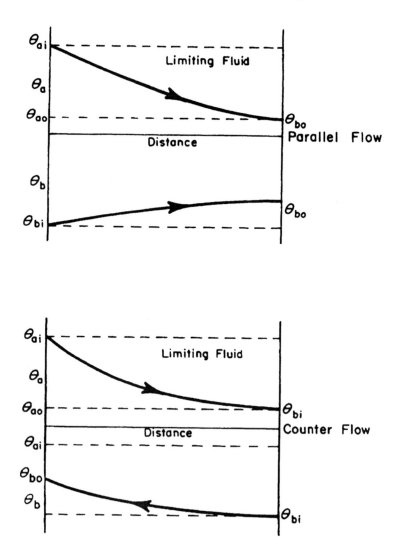

Figure 3.7.6. The temperature change of the limiting fluid is much greater in the heat exchanger than the temperature change of the nonlimiting fluid.

where $\Re = (\dot{m}c_p)_{\text{nonlim}}/(\dot{m}c_p)_{\text{lim}}$; and exp denotes the base of the natural logarithm taken to the exponent in the parentheses, except for the case where $\Re = 1$, for which

$$E = \left(1 + \frac{(\dot{m}c_p)_{\text{lim}}}{UA}\right)^{-1} \qquad (3.7.7)$$

A graph of counter-flow heat exchanger effectiveness appears in Figure 3.7.7.

parallel-flow effectiveness
    For the parallel-flow heat exchanger, effectiveness is given as

$$E = \frac{1 - \exp\left[\dfrac{-UA}{(\dot{m}c_p)_{\text{lim}}}\left(1 - \dfrac{1}{\Re}\right)\right]}{1 + 1/\Re} \qquad (3.7.8)$$

Curves for this effectiveness are found in Figure 3.7.8.

cross-flow effectiveness
    A similar analysis can be developed for cross-flow heat exchangers. The results are more complicated than the previous equations because of the flow and temperature gradient in two directions. The results of this analysis is presented graphically in Figure 3.7.9.

    Heat exchanger effectiveness actually equals a temperature difference ratio, $(\theta_{i\,\text{lim}} - \theta_{o\,\text{lim}})/(\theta_{i\,\text{lim}} - \theta_{i\,\text{nonlim}})$, and this appears on the ordinates of the graphs. On the abscissa is another dimensionless ratio $(R\dot{m}c_p)$, where R is the thermal resistance $(R = 1/AU)$ of the heat exchanger. The ratio of the mass flow rates and specific heat products for the limiting and nonlimiting fluids, $(\dot{m}c_p)_{\text{nonlim}}/(\dot{m}c_p)_{\text{lim}}$, appears as the parameter distinguishing the family of curves. If calculated correctly, this parameter should always be greater than or equal to 1.0. In case a mistake has been made, one curve for a parameter value of 0.5 is included in the graphs.

graphical approach
    These graphs can be used to determine actual rates of heat transfer in heat exchangers. Once effectiveness is known, the rate of heat transfer can be calculated from:

heat transferred

$$\dot{q} = E(\dot{m}c_p)_{\text{lim}}(\theta_{i\,\text{lim}} - \theta_{i\,\text{nonlim}}) \qquad (3.7.9)$$

for parallel-, counter-, or cross-flow types.

    With the rate of heat transfer known, temperature in the nonlimiting fluid may be determined by equating (negative) heat transferred to the nonlimiting fluid to the (positive) heat lost by the limiting fluid. Thus

$$\dot{q}_{\text{lim}} = \dot{q}_{\text{nonlim}}$$

temperature calculations
$$E(\dot{m}c_p)_{\text{lim}}(\theta_{i\,\text{lim}} - \theta_{i\,\text{nonlim}}) = -(\dot{m}c_p)_{\text{nonlim}}(\theta_i - \theta_o)_{\text{nonlim}} \qquad (3.7.10)$$

where $\dot{q}$ is the rate of heat transfer (N m/sec); $\dot{m}$, the mass flow rate (kg/sec); c_p, the specific heat [N m/(kg °C)]; θ, the temperature (°C); i, o denote inlet and outlet; and lim, nonlim refer to limiting and nonlimiting fluids.

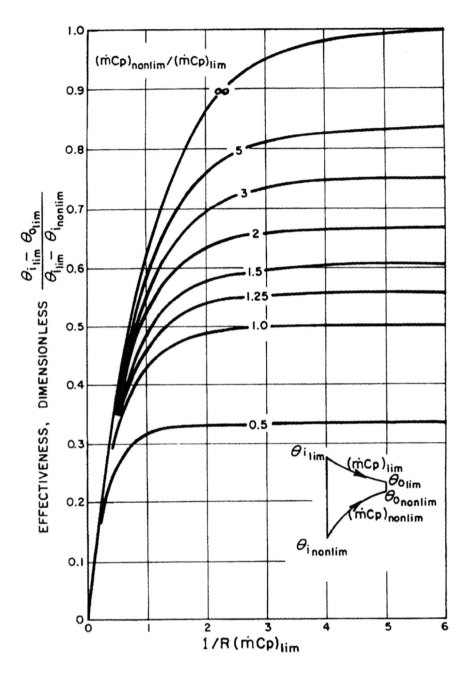

Figure 3.7.7. Effectiveness curves for parallel-flow heat exchangers.

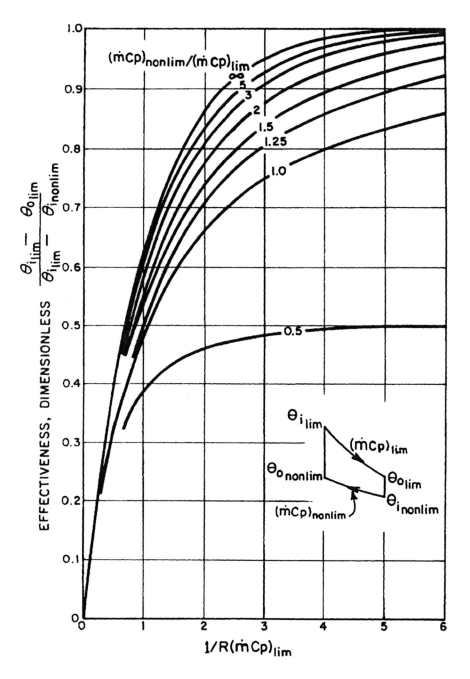

Figure 3.7.8. Effectiveness curves for counter-flow heat exchangers.

Unknown temperature values can hence be determined.

simple pipe

For the simple case of a fluid flowing inside a pipe and a fluid flowing outside the same pipe, there have now been presented at least two means to analyze the heat transfer problem. One of these approaches is outlined in Section 3.6.2, and involves calculation of convection coefficients and logarithmic mean temperature. The other approach presented here is to determine the heat transfer effectiveness for the pipe and to use the methods presented in this section.

3.7.1.3 Fouling Factors

Heat exchanger surfaces do not always remain clean. Surface deposits of sediment often form from suspended particles carried by liquids. These surface deposits add thermal resistance to the heat transfer surfaces.

surface
deposits

To the resistances included in Figure 3.7.4 and Eq. 3.7.3 must be added a resistance on the inside surface of the tubing and another resistance on the outside surface of the tubing. Each resistance is given a mathematical form similar to convection resistance, except that a fouling factor substitutes for the convection coefficient

$$R = [(\text{area})(\text{fouling factor})]^{-1} \tag{3.7.11}$$

Some typical values for fouling factors appear in Table 3.7.1. Notice that the greater the fouling factor value, the lower the additional resistance.

3.7.1.4 Heat Exchanger Specification

The heat exchanger to be chosen for a particular design application is likely going to be a standard manufactured item chosen from a manufacturer's specification sheet. One of the first decisions to be made concerns the general type of heat exchanger: parallel flow, counter flow, or cross flow. This will not always be a very critical decision, and any of these types could suffice. In certain circumstances, geometrical or size constraints may help to narrow the number of choices. In

manu-
facturers'
specifications

Table 3.7.1 Typical Fouling Factor Values

Fluid	Fouling Factor [N m/(m^2 sec °C)]
Distilled and seawater	11,350
City water	5,680
Muddy water	1,990–2,840
Gases	2,840
Vaporizing liquids	2,840
Vegetable and gas oils	1,990

Source: Transport Processes and Unit Operations, 3/E, by Geankoplis, C. J., 1978. Reprinted by permission of Prentice-Hall, Inc., Upper Saddle River, N.J.

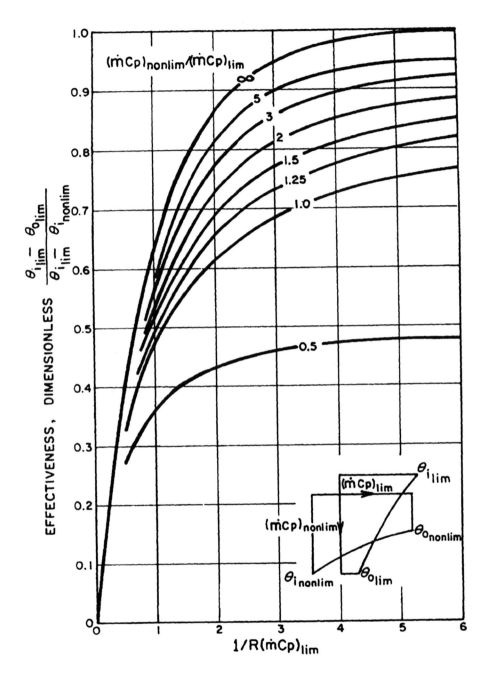

Figure 3.7.9. Effectiveness curves for cross-flow heat exchangers.

other cases, there may be a particular premium to be paid for heat transfer efficiency, leading to the choice of a counter-flow heat exchanger, or gentleness to the fluids, leading to the choice of a parallel-flow exchanger. Cost is also sometimes an important determinant of the choice of type.

thermal resistance

Once the type is determined, the target value for *UA* is next. This is likely to be ascertained in a trial-and-error means, unless something is known about the *UA* values available to be chosen.

Either all inlet and outlet temperatures of both fluids must be known or *UA* must be known. For this discussion, we assume that neither of these is known, and that some initial value of *UA* must be determined to start the design process. As given in Figure 3.7.10, the procedure is to assume all unknown inlet and outlet temperatures, combine the assumed values with the known values, and determine an initial value for *UA* using the general procedure diagrammed in Figure 3.7.11. With this trial *UA* value, manufacturers' specifications are consulted and a heat exchanger tentatively chosen. Taking the manufacturer's *UA* value, the procedure diagrammed in Figure 3.7.11 is reversed to determine values for the unknown inlet and outlet temperatures. If these are acceptable, then a correct choice has been made.

special consider-ations

There are additional important considerations before a final choice is made. If the heat exchanger is to be used with food materials, then it must be sanitary and able to be cleaned; materials must not add flavors or impurities to the food. If the heat exchanger is to be used for blood or other bodily fluids, then it must be sterilizable, cannot contain pockets where blood can accumulate, and cannot promote thrombus formation (clotting); materials must be extremely inert. If the heat exchanger is to be used with waste liquids, then it must be capable of passing large suspended particles without clogging; materials must often be as inexpensive as possible. If the heat exchanger is to cool air, then it must be capable of removing condensate water. There are many practical considerations contributing to heat exchanger choice.

easily opened

Heat exchangers containing food materials must be able to be cleaned in place on a daily basis. Not only must the materials of which the exchanger is made be inert to the food materials (often highly acidic), but they must also be immune to damage by caustic cleaning materials (often highly alkaline). Heat exchangers for use with food or bodily fluids must be able to be easily opened and inspected to assure that contaminants, old material, or microbial growth are not lodged within. Heat exchangers to be used with waste materials must also be able to be opened for cleaning.

one pressurized fluid

Of the two fluids situated on both sides of the common heat exchanger surface, one will usually be at a higher pressure than the other. Usually, it is more critical that one fluid maintain its purity. For instance, a food material and water could be the two fluids; the food material must remain pure. Thus the food material should be more highly pressurized than the water; in case a leak develops in the heat exchanger, the food will leak into the water rather than vice versa.

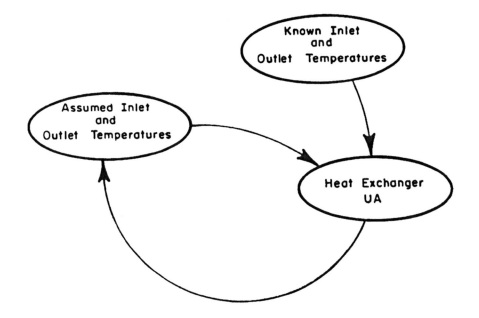

Figure 3.7.10. Schematic of overall means to determine initial and final heat exchanger *UA* specification.

finned tubes

To increase heat transfer efficiency, that is, to maximize heat transfer rate in the smallest possible space, interfacial surface areas between the two fluids must be large compared to the volume of fluids contained in the passages. Sometimes extended surfaces called fins are included in heat exchanger tubes. These may be of many different configurations, and are the object of active research to provide more efficient heat transfer. Heat transfer efficiency is promoted by turbulence and mixing in the fluid, but the penalty for this is the high fluid pressure drops from one side of the exchanger to the other.

jacketed kettle

As presented by Pelosi (1972), there are several embodiments of heat exchangers normally used in food processing. Perhaps the simplest is a jacketed kettle, a tank with an outer jacket through which flows heating or cooling fluid. This type of heat exchanger can be used for all types of food products. It is limited to batch operations, but there is a strong trend to replacement of batch operations with more efficient flow systems.

plate type

Plate heat exchangers consist of numerous metal heat transfer plates clamped together in a frame (Figure 3.7.12). Gasket spacers hold adjacent plates apart and form narrow passages through which the two fluids flow. The same fluid flows through alternate channels, allowing a large amount of heat transfer contact area. Fluids are collected in tubes running the length of the exchanger.

concentric tubes

Concentric tube heat exchangers have tubes within tubes to contain the heating or cooling medium in one tube or annulus and the liquid food in others

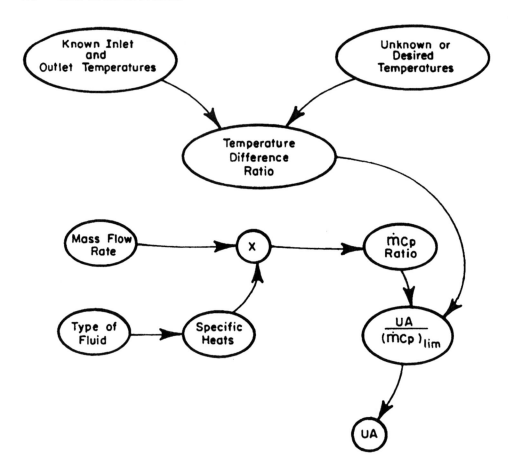

Figure 3.7.11. Schematic of the procedure to determine *UA* of a heat exchanger starting with known and assumed temperatures.

(Figure 3.7.13). The shell and tube exchanger shown in Figure 3.7.14 is a special case of this where there are many tubes within a large tube acting as a shell.

spiral type   A spiral heat exchanger is shown in Figure 3.7.15. Channels for food and heating/cooling fluid are bent to form concentric spirals that allow a large amount of heat transfer area interfacing between the two fluids to be contained in a compact volume.

When the fluid either has high viscosity or is subject to fouling or crystallizing, a scraped-surface heat exchanger (Figure 3.7.16) must be used.
scraped   Examples of the type of food requiring a scraped-surface heat exchanger are
surface   peanut butter, gravy, and thick soups. In this exchanger, a blade rotates within the inner chamber to mix and remove product from the walls constantly.

The different types of heat exchangers are compared in Table 3.7.2. Each type has places where it is particularly suited for use. In Table 3.8.1 are typical values for the overall heat transfer coefficient (U) for several types of heat exchangers and a simple jacketed kettle often used for food.

EXAMPLE 3.7.1.4-1. Heat Exchanger for Cardiopulmonary Bypass

Cardiopulmonary bypass, otherwise known as a heart–lung machine, is used to replace the blood pumping function of the heart and the gas exchange functions of the lung to enable open-heart surgery. Cooling the blood after it leaves the body and before it is returned to the body has several advantages, including increased absorptivity of blood gases and induction of body hypothermia to reduce metabolism and oxygen requirements. A heat exchanger is to be designed from stainless steel coated with polymers to minimize blood interactions with the material. Blood is pumped through the heat exchanger at 5 L/min, entering at 37°C body temperature and leaving at 25°C. Ice water enters the heat exchanger at 0°C and leaves at 15°C. What type of heat exchanger is to be used, what is the AU of the heat exchanger, and what flow rate of water is to be used?

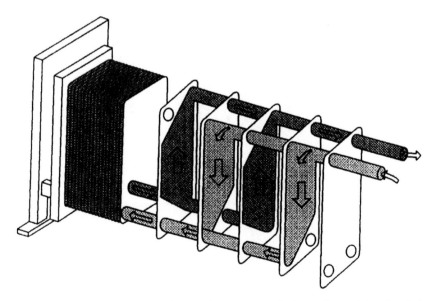

Figure 3.7.12. A plate heat exchanger has corrugated plates clamped together in a frame. Plates are spaced by gaskets to form an uninterrupted space for liquid flow. They are generally used for liquid-to-liquid service.

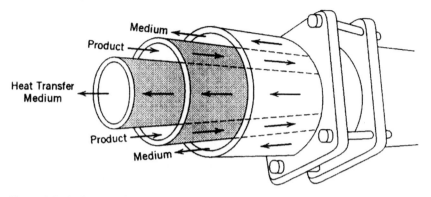

Figure 3.7.13. A tubular heat exchanger (triple tube) has three concentric tubes. Heating or cooling media flow through the inner tube and through the annular space below the outer tube. This unit handles liquids of high viscosity and with solids to 40%.

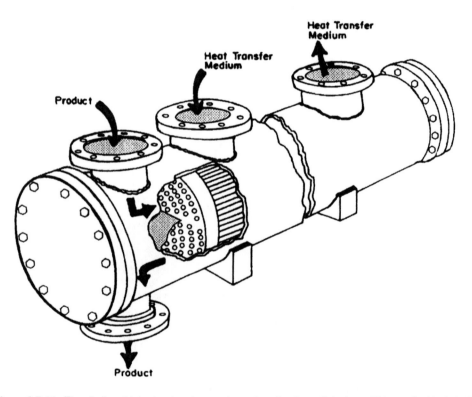

Figure 3.7.14. The shell-and-tube heat exchanger has a bundle of parallel tubes within a cylindrical shell. Its use is restricted to liquids with low solids content and viscosity.

Figure 3.7.15. A spiral heat exchanger employs long strips of plate assembled to form a pair of concentric spiral passages. It is recommended for heating and cooling viscous liquids.

Solution

This is a case where all fluid inlet and outlet temperatures are known. The rate of heat lost by the blood is

$$\dot{q} = \dot{m}c_p(\theta_i - \theta_o)$$

$$= \left(8.33 \times 10^{-5} \; \frac{m^3}{sec}\right)\left(1050 \; \frac{kg}{m^3}\right)\left(3740 \; \frac{N\,m}{kg\,{}^\circ C}\right)(37^\circ C - 25^\circ C)$$

$$= 3927 \; N\,m/sec$$

We will choose a cross-flow heat exchanger for space reasons, and to assure that the blood is not overcooled. Either parallel-flow or counter-flow exchangers could also be used.

To determine heat exchanger specifications, we assume 100% efficiency of heat transfer between the two fluids. This can be assured by adequate insulation around the entire exchanger apparatus. The LMTD of the water is, from Eq. 3.7.2,

$$LMTD = \frac{(37^\circ C - 0^\circ C) - (25^\circ C - 15^\circ C)}{\ln(37 - 0)/(25 - 15)]}$$

$$= 20.6^\circ C$$

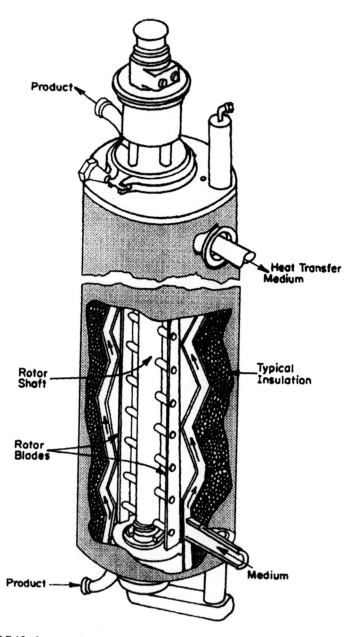

Figure 3.7.16. A scraped-surface heat exchanger consists of a jacketed insulated tube with concentric rotor shaft. Blades on the shaft continually wipe product from the tube wall. It is used for high-viscosity products and for such other operations as crystallization, aeration, and mixing.

Table 3.7.2   Food Applications Guidelines for Different Types of Heat Exchangers

Type of Heat Exchanger	Use
Plate	liquid-to-liquid solids content $<5\%$ nonabrasive solids viscosity <20 N sec/m^2
Concentric tubes	liquids with high viscosities gas-to-liquid, or steam-to-liquid solids content up to 30–40% large-size particulates
Spiral	liquid-to-liquid gas-to-liquid, or steam-to-liquid solids content <5–10% high viscosity sludges, slurries, suspended solids not sanitary
Scraped surface	liquid-to-liquid, gas-to-liquid, or steam-to-liquid high viscosity (up to 100 N sec/m^2) heavy fouling or crystallizing liquids solids content $>75\%$ very large-size particles
Shell and tube	liquid/liquid, gas/liquid, or steam/liquid low solids content viscosity <10 N sec/m^2 usually nonfood use

And, from Eq. 3.7.1,

$$R = 1/AU = \frac{\text{LMTD}}{\dot{q}} = \frac{20.6°\text{C}}{3927 \text{ N m/sec}} = 5.255 \times 10^{-3} \frac{°\text{C sec}}{\text{N m}}$$

$$AU = 190 \frac{\text{N m}}{°\text{C sec}}$$

The water flow rate can be calculated as

$$\dot{m} = \frac{\dot{q}}{c_p \, \Delta\theta} = \frac{3927 \text{ N m/sec}}{[4187 \text{ N m/(kg °C)}](15°\text{C} - 0°\text{C})}$$

$$= 6.25 \times 10^{-2} \text{ kg/sec}$$

$$= 6.25 \times 10^{-5} \text{ m}^2/\text{sec} \quad \text{or} \quad 3.75 \text{ L/min}$$

EXAMPLE 3.7.1.4-2. Water Heating and Milk Cooling

Milk must be cooled and water must be heated in a dairy operation. Milk flows through glass or stainless steel pipes at a rate of about 8×10^{-5} m^3/sec. It enters the pipe at the body temperature of a cow, or 38.6°C. A temperature of 13°C or lower is desired for the milk as it enters the bulk cooling tank. Water comes from the ground at a rate of 2.10×10^{-4} m^3/sec and 10°C. Determine the required AU of a counter-flow heat exchanger and the exit temperature of the water.

Solution

All fluid inlet and outlet temperatures in this system are not known. Therefore, we should use heat exchanger effectiveness graphs to determine the required AU. First we must determine which is the limiting fluid. For the water

$$\dot{m}c_p = (2.10 \times 10^{-4} \text{ m}^3/\text{sec})(1000 \text{ kg/m}^3)[4187 \text{ N m}/(\text{kg °C})]$$

$$= 0.880 \ \frac{\text{N m}}{\text{sec °C}}$$

For the milk

$$\dot{m}c_p = (8 \times 10^{-5} \text{ m}^3/\text{sec})(1030 \text{ kg/m}^3)[3860 \text{ N m}/(\text{kg °C})]$$

$$= 0.309 \ \frac{\text{N m}}{\text{sec °C}}$$

The milk is thus the limiting fluid. This is fortunate, because removing heat from the milk is more critical than adding heat to the water. If the water were to limit heat transfer, we should try to increase the flow rate of the water.

Next we calculate

$$\frac{(\dot{m}c_p)_{\text{nonlim}}}{(\dot{m}c_p)_{\text{lim}}} = \frac{0.880}{0.309} = 2.85$$

Heat exchanger effectiveness, from Eq. 3.7.5, is

$$E = \frac{(\theta_i - \theta_0)_{\text{lim}}}{(\theta_{i \text{ lim}} - \theta_{i \text{ nonlim}})} = \frac{(38.6°C - 13°C)}{(38.6°C - 10°C)} = 0.90$$

From Figure 3.7.8, these values correspond to a value of

$$\frac{1}{R(\dot{m}c_p)_{\text{lim}}} = 3.0$$

Thus

$$\frac{1}{R} = AU = (3.0)\left(0.309 \ \frac{\text{N m}}{\text{sec °C}}\right)$$

A typical value for U is 1100 (N m)/(sec m^2 °C). Thus the required heat exchanger area is 8.4×10^{-4} m^2.

From Eq. 3.7.10,

$$E(\dot{m}c_p)_{\lim}(\theta_{i\,\lim} - \theta_{i\,\text{nonlim}}) = (\dot{m}c_p)_{\text{nonlim}}(\theta_i - \theta_o)_{\text{nonlim}}$$

$$(0.90)\left(0.309 \ \frac{\text{N m}}{\text{sec} \ ^\circ\text{C}}\right)(38.6^\circ\text{C} - 10^\circ\text{C}) = -\left(0.880 \ \frac{\text{N m}}{\text{sec} \ ^\circ\text{C}}\right)(10^\circ\text{C} - \theta_o)$$

$$(\theta_o - 10^\circ\text{C}) = 9.04^\circ\text{C}$$

$$\theta_o = 19^\circ\text{C}$$

The exiting temperature of the water is 19°C.

Remark

The extremely small area required means that a heat exchanger can be designed and built very simply, without fancy enhancements for heat transfer. If an area is used that is larger than the 8.4×10^{-4} m^2, heat exchanger effectiveness will be improved, but since effectiveness is already 90%, the improvement will not be great. The payoff for larger area will not be enough to make extra effort in this regard practical.

3.7.2 Transient Heat Transfer

Cooling and heating objects are important thermal processes, especially in food engineering. In this case the objective is to make the food safe to eat and to maintain quality. In other cases, especially those involving pasteurization or sterilization, objects are heated to kill unwanted microbial organisms. These processes are important in medical or veterinary care, tissue culture, and horticulture.

pasteur-
ization and
sterilization

Whether heating or cooling, the process often involves an object that begins nearly uniform in temperature and that is placed in a convective (liquid or gas) medium. Heat is either gained or lost from the object by conduction internally and by convection externally. While all important transient thermal processes do not fit this description, these conditions are common enough that they can fit a large majority of cases.

object in fluid
medium

In these processes, heat to be removed comes from stored heat, and heat to be added is stored in the object. Thus the thermal capacity of the object is important.

Heisler charts to be introduced shortly cannot be used if there is significant heat generation within the object. All heat removed from or added to the object must be stored heat.

low internal
resistance

A second issue of importance is the location of the major resistance to heat flow. If the largest amount of resistance is outside the object, that is, if the thermal conductivity of the material inside the object is high relative to the convection coefficient outside the object, then the diagram in Figure 3.2.3 is a good representation of the actual system, and the temperature change occurs exponentially with a thermal time constant of $\tau = R_{th} C_{th}$.

high internal resistance

Contrarily, if the greatest resistance to heat flow is internal to the object, then the representation in Figure 3.2.3 is not accurate; resistance and capacity must be distributed throughout the object rather than lumped together as in the diagram. Again, the important ratio to consider involves the thermal conduction resistance inside and the thermal convection resistance outside.

3.7.2.1 Dimensionless Numbers The dimensionless Biot number expresses this ratio

Biot number

$$\mathrm{Bi} = \frac{h_c L}{k} \tag{3.7.12}$$

where Bi is the Biot number (dimensionless); h_c, the convection coefficient [N m/(sec m^2 °C)]; k, the thermal conductivity of material inside the object [N m/(sec m °C)]; and L, the significant length (m).

The Biot number is similar to the Nusselt number in Eq. 3.3.2, except that the thermal conductivity in the Nusselt number is that of the external fluid, whereas the thermal conductivity in the Biot number is that of the internal material.

significant length

The significant length for transient heat transfer is one-half the thickness of the object. It is the radius of a sphere (r_o), or the radius of a cylinder (r_o), or half the thickness of a slab (L).

Another dimensionless number of interest is the Fourier number

Fourier number

$$\mathrm{Fo} = \frac{\alpha t}{L^2} = \frac{kt}{\rho c_p L^2} \tag{3.7.13}$$

where Fo is the Fourier number (dimensionless); α, the thermal diffusivity (see Eq. 3.2.2; m^2/sec); t, the time (sec); L, the significant length (m); ρ, the density (kg/m^3); and c_p, the specific heat [N m/(kg °C)].

The time dimension for transient heat transfer is introduced through the Fourier number. The significant length is the same as for the Biot number.

time constants

The ratio L^2/α equals the exponential time constant, $\tau = RC$. Thus the Fourier number is elapsed time expressed as the number of system time constants it represents.

3.7.2.2 Heisler Charts The Heisler charts can be used to determine transient heat transfer results graphically. There are three charts included here,

sphere, cylinder, and slab

one each for a sphere (Figure 3.7.17), an infinite cylinder (Figure 3.7.18), and an infinite slab (Figure 3.7.19). The cylinder has a finite radius but an infinite length, and the slab has a finite thickness but infinite length and width. Real embodiments of these shapes might be melons or oranges (spheres), capillary tubes containing blood (infinite cylinders), or plant leaves (infinite slabs).

Heisler charts for each shape are divided into three sections. At the top is the center temperature, in the middle is the average temperature, and at the

surface temperature

bottom is the surface temperature. Surface heat treatment of a fresh fruit to kill microbes that cause rotting can be analyzed using the lowest part of the chart. Fruits can be dipped (this is done with some fruits such as papaya) in scalding water for a long enough time to kill surface microbes without significantly heating internal tissues, and this can be checked using the chart.

average temperature

The average temperature is important for processes involving the overall temperature of the object. Cooling a melon to retard spoilage is an example of a process that would involve the average temperature.

center temperature

Sterilization of a medical, biotechnological, or food product would involve the center temperature, because the temperature at the center is the slowest to respond to changes. Sterilization requires that all temperatures meet or exceed a minimum level in order to assure death of internal microbes. Likewise, stopping enzymatic activity requires exceeding some minimum temperature, and thus the center temperature is also of interest.

On the ordinate of the Heisler chart is a temperature difference ratio, $(\theta_{x,t} - \theta_\infty)/(\theta_o - \theta_\infty)$. $\theta_{x,t}$ is the temperature at one of the locations of surface, average, or center, and at some time t. θ_o is the initial internal temperature of the object, assumed to be uniform, and θ_∞ is the unchanging temperature of the surrounding fluid, which is also the temperature to which the object would come if left in the fluid for a sufficiently long time.

The parameter distinguishing the family of curves in each section of the chart is the inverse of the Biot number, Bi^{-1}. For extremely high convection coefficient values, Bi^{-1} approaches zero, and the plot of temperature with time (or Fourier number) reacts most quickly. This is the case on the chart for the leftmost curve. As a matter of fact, the curve for $Bi^{-1} = 0$ traces along the ordinate axis for surface temperature, as should be expected.

Viewed from a distance, all curves on the Heisler charts indicate that the difference between the temperature of interest inside the object (center, average, or surface) and the ultimate temperature of the object (θ_∞) becomes less as time increases.

Chart Use The Heisler charts may be used in two ways, depending on the nature of the problem and the type of information known. In the first type of

time known, temperature unknown

usage, time is known and the temperature of the object is to be determined (Figure 3.7.20). Fluid temperature, object initial temperature, and physical properties of the object material are assumed known. The convection coefficient value can be estimated using one of the equations in Section 3.3.1.2, since, in all likelihood, forced convection predominates. The Biot number can be calculated, the Fourier number value can be calculated from the known amount of time, and the corresponding temperature difference ratio obtained by taking the value corresponding to the intersection of the Fourier number value with the proper Bi^{-1} curve. Temperature is then obtained from the temperature difference ratio. If natural convection predominates, or if there

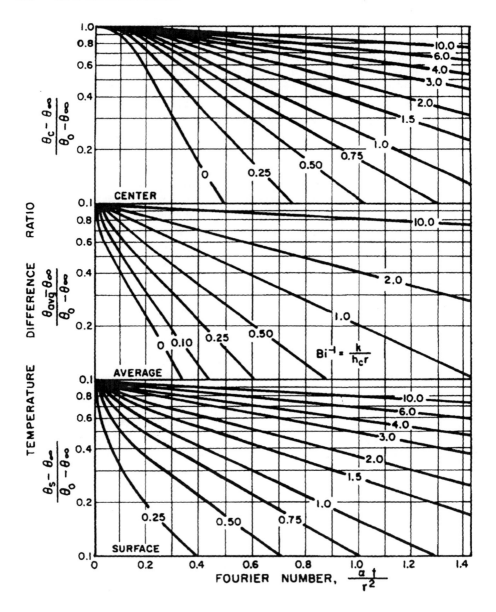

Figure 3.7.17. Transient temperatures in a sphere.

is a significant temperature dependence of the physical properties, then this temperature can be used to check to see if the correct physical property values are used. Because temperature in the object changes with time, an average temperature $(\theta_{x,t} + \theta_o)/2$ can be used to check the values of physical properties.

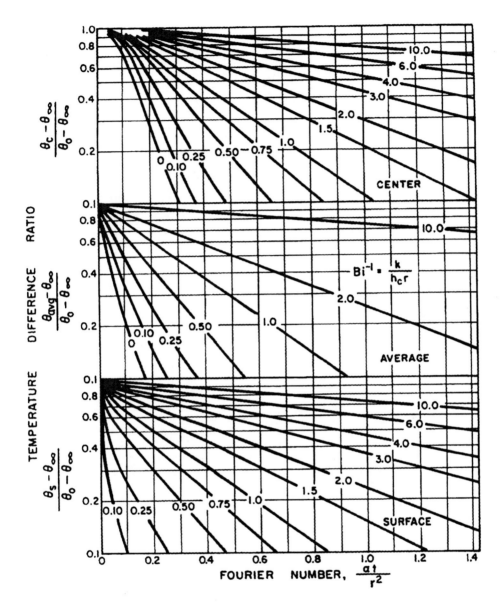

Figure 3.7.18. Transient temperatures in a cylinder.

The inverse problem, which is actually more likely, can also be solved. In this case, the final temperature is known, but the time to reach it is not. The steps are nearly the same as before, but the value of the Fourier number is obtained, and the amount of time is finally discerned.

temperature
known, time
unknown

The case of conduction with heat generation was discussed in Section 3.5.5. It was explained there that there was no net flow of heat along the center line of

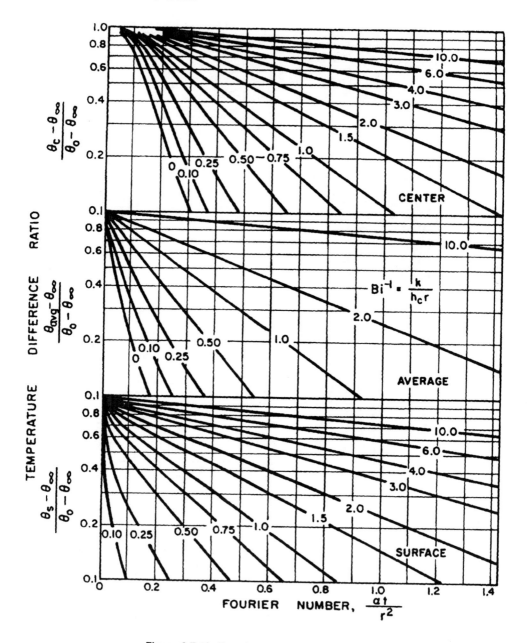

Figure 3.7.19. Transient temperatures in a slab.

a slab, and that one-half of the slab could be equivalently replaced with insulation. In a similar manner, the Heisler chart for a slab of thickness $2L$ can also be correctly applied to a slab of thickness L as long as one side is insulated and the other side is exposed to the convecting fluid.

As one might realize, there is a mathematical basis to the Heisler charts. These are not presented here because they are clearly beyond the scope of this book. The reader is referred to such other heat transfer texts such as Holman (1986) and Schneider (1955) for further information.

Composite Shapes   If the Heisler charts could only be used for spheres, infinite cylinders, and infinite slabs, they would be of little use. Real objects do not reach to infinity. The most useful applications of the Heisler charts are to cans and boxes, shapes that cannot be described as "infinite."

real objects

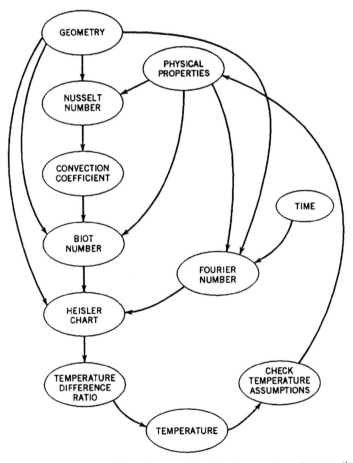

Figure 3.7.20. Using the Heisler chart to determine temperature at some time. To determine time at some temperature, several of these steps must be reversed.

A can is a truncated cylinder, and that shape can be envisioned as the intersection of an infinite cylinder with an infinite slab (Figure 3.7.21). A box is a parallelpiped, and that shape can be envisioned as the intersection of three mutually perpendicular infinite slabs.

Insofar as the temperature scale on the Heisler charts spans the range from 0.0 to 1.0, the charts can be used for real shapes (Carslaw and Jaeger, 1959). The temperature difference ratio for a composite shape is the product of the temperature difference ratios for each of the contributing shapes. Thus the temperature difference ratio for a can is the temperature difference ratio for the cylinder multiplied by the temperature difference ratio for the slab.

product of composite ratios

When using the Heisler charts for this purpose, the significant lengths used in the Biot and Fourier numbers would be the dimensions appropriate for each direction. For the can, they would be the radius of the can (used with the Heisler chart for a cylinder) and half the length of the can (used with the Heisler chart for a slab). A box would have three significant lengths, one for each different direction, and the Heisler chart for a slab would have to be used three different times, each with a different length.

significant lengths

The Heisler charts can be used relatively easily for composite shapes when the time is known and the temperature must be found (Figure 3.7.22). Each Fourier number is calculated, and each Biot number curve is used to find a component temperature difference ratio. The same value for time must be used in each Fourier number calculation, and the same temperature location (center, average, or surface) must be used for each component shape. Multiply the component temperature difference ratios and obtain the unknown temperature from the composite temperature difference ratio.

time known, temperature unknown

Using the charts for the inverse problem when temperature is known but time is not known is not as easy. For this problem, the composite temperature difference ratio can be calculated, and guesses made for component ratio values, the product of which equals the original composite ratio. Each component chart is used in the same way as previously described, and values for Fourier numbers are obtained. From these Fourier numbers, values of time are calculated.

temperature known, time unknown

Because each dimension for the composite shape must coexist, time cannot be different for each component. It is not likely that the initial component temperature difference ratio guesses were correct, and these errors result in times that are not equal. Different guesses must be made for component temperature difference ratios that cause the derived times to become equal. Components with times relatively higher than others must have higher-valued guesses for temperature difference ratios. Those with relatively lower times must have lower initial guesses. This is the process of iteration (Figure 3.7.23).

Because the Fourier number includes the square of the significant dimension, it does not take much of a difference between significant dimensions in different directions before the component temperature difference ratios strongly favor the shorter dimension. Thus, for example, a

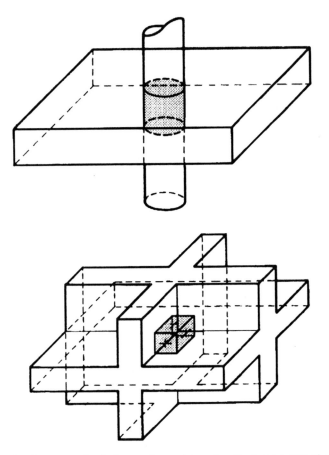

Figure 3.7.21. A truncated cylinder can be envisioned as the intersection of an infinite cylinder with an infinite slab. A parallelpiped can be envisioned as the intersection of three perpendicular slabs.

finite cylinder with length ten times as long as diameter is indistinguishable on the Heisler charts from an infinite cylinder.

Interior Fluid   The Heisler charts are based on the processes of heat transfer by conduction inside the object and convection outside the object. While it is not common that external heat transfer would occur in significant amounts by any other process except convection, the objects may enclose fluids instead of, or with, solids, and fluids enhance heat transfer by convection. Both the Biot number and the Fourier number are calculated using thermal conductivity of the interior material, assuming conduction alone is present.

The effect of convection is somewhat complicated. Neglecting convection underestimates the true value of the Fourier number without time, and causes

enhance heat transfer

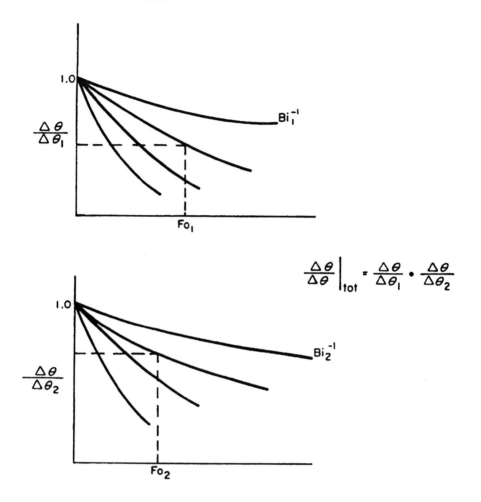

Figure 3.7.22. For a composite shape with two component shapes, each Fourier number and each Biot number is calculated. The overall temperature difference ratio for the composite shape is the product of the two component temperature difference ratios.

time over-estimated   the estimate of time to be higher than it is supposed to be. Neglecting convection reduces the inverse of the Biot number and reduces the estimate of time. The final result is that time to reach a particular temperature is usually overestimated by something like 100%.

Convection thermal resistance would increase heat flow from the interior by acting in parallel with conduction thermal resistance (appearing in simplified form in Figure 3.7.24). However, convection is a very complicated process (Section 3.3), and measurements of convection heat loss would not distinguish conduction heat loss; both would be called convection. Therefore, we do not need to be concerned about parallel conduction and convection thermal resistances; they can be replaced by one convection thermal resistance.

Conduction thermal resistance for a slab is given by L/kA; convection thermal resistance is given by $1/h_cA$. It should be apparent, then, that an

equivalent
conductivity

equivalent thermal conductivity to be used in the Biot number and the Fourier

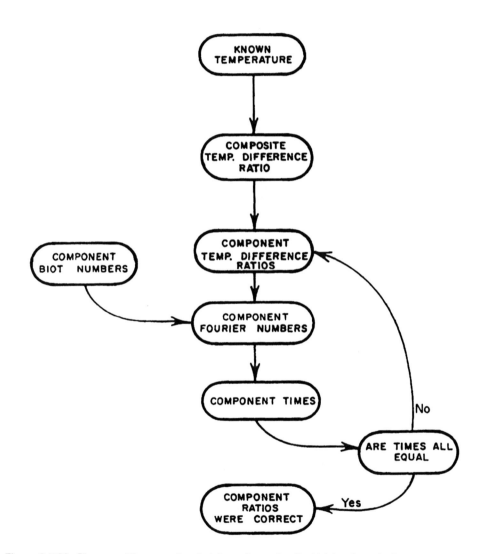

Figure 3.7.23. Diagram of the procedure to follow when using the Heisler charts for known temperatures and unknown times.

number should be $k = h_c L$. Thus, when convection heat transfer dominates the interior of the object,

$$\mathrm{Bi} = \frac{h_{co}}{h_{ci}} \qquad (3.7.14)$$

$$\mathrm{Fo} = \frac{h_{ci} t}{c_p \rho L} \qquad (3.7.15)$$

where Bi is the Biot number (dimensionless); Fo, the Fourier number (dimensionless); h_{co}, the outside convection coefficient [N m/(m^2 sec °C)]; h_{ci}, the inside convection coefficient [N m/(m^2 sec °C)]; t, the time (sec); c_p, the specific heat [N m/(kg °C)]; ρ, the density (kg/m^3); and L, the significant length (m).

The convection coefficient for liquid material inside a can (Ramaswamy and Sablini, 1995) is about 120 N m/(sec m^2 °C). Rotating the can end over end at about 20 rpm enhances the convection coefficient and causes its value to increase to about 180 N m/(sec m^2 °C).

EXAMPLE 3.7.2.2-1. Hydrocooling of Peaches

Peaches are often picked at field temperatures of 24–35°C, but must be cooled rapidly to maintain quality by reducing respiration, ripening, and pathogenic growth. Hydrocooling is popular because it results in rapid cooling of the fruit. In Figure 3.7.25 is shown a conventional flood-type hydrocooler where peaches packed in shipping containers are moved by conveyor through a

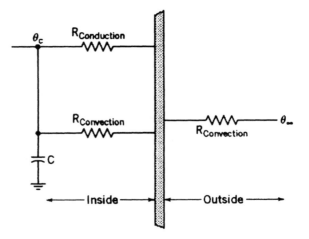

Figure 3.7.24. Liquid inside the container enhances heat transfer by allowing convection to occur. Convection resistance is in parallel with conduction resistance inside the container. The two resistances cannot be easily distinguished by measurements.

cooling tunnel. Chilled chlorinated water flows from overhead flood pans and rains on the fruit in a fine spray that allows complete surface coverage. Peaches have the following physical properties:

$$\text{density } (\rho) = 618 \text{ kg/m}^3 \text{ (as packed) or } 961 \text{ kg/m}^3 \text{ (flesh)}$$
$$\text{specific heat } (c_p) = 3768 \text{ N m/(kg } °\text{C)}$$
$$\text{thermal conductivity } (k) = 0.500 \text{ N m/(sec m } °\text{C)}$$

The convection coefficient for hydrocooled peaches has been determined to be 680–940 N m/(m^2 sec °C). Estimate the time required to cool 64-mm-diameter peaches to 5°C using 2°C water.

Solution

We can use the Heisler chart for spheres for this purpose. The inverse of the Biot number can be calculated from Eq. 3.7.12 as

$$\text{Bi}^{-1} = \frac{k}{h_c r_0} = \frac{[0.500 \text{ N/(sec } °\text{C)}]}{[680 \text{ N/(m sec } °\text{C)}](0.032 \text{ m})} = 0.232$$

If the peaches begin cooling at a uniform temperature of 24°C, and average temperature after cooling is 5°C, then the final temperature difference ratio is

$$\frac{\theta_{\text{avg}} - \theta_{\infty}}{\theta_0 - \theta_{\infty}} = \frac{5°\text{C} - 2°\text{C}}{24°\text{C} - 2°\text{C}} = 0.14$$

From the Heisler chart (Figure 3.7.17), the Fourier number corresponding to the above values is about 0.32. Thermal diffusivity of the peaches is

$$\alpha = \frac{k}{\rho c_p} = \frac{0.500 \text{ N m/(sec m } °\text{C)}}{(961 \text{ kg/m}^3)[3768 \text{ N m/(kg } °\text{C)}]}$$
$$= 1.38 \times 10^{-7} \text{ m}^2/\text{sec}$$

And time is

$$t = \frac{\text{Fo } r^2}{\alpha} = \frac{(0.32)(0.032 \text{ m})^2}{1.38 \times 10^{-7} \text{ m}^2/\text{sec}} = 2373 \text{ sec} = 40 \text{ min}$$

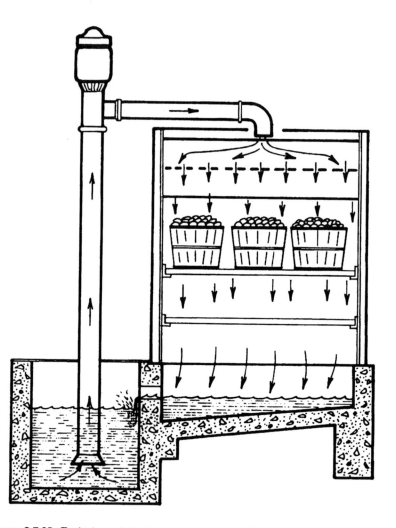

Figure 3.7.25. Typical peach hydrocooling operation. Chilled water is pumped from a sump over the peaches. Chlorine is often added to the water to maintain sterility.

EXAMPLE 3.7.2.2-2. Plant Tissue Culture Medium Sterilization

Plant tissue cultures are prepared to regenerate entire plants from single or multiple plant cells in order to produce pathogen-free plants or to grow transgenic plant samples. Plant tissue cultures are grown in specially prepared media that contain ingredients essential to growth (Sigma, 1996). Tissue culture medium is often used in baby food jars 100 mm in height and 59 mm in diameter (Figure 3.7.26). It may be sterilized before use by autoclaving at 121°C and 103 kN/m^2 steam pressure. Determine the sterilization time required.

Solution

Heisler charts may be used for this purpose, but this is the case of a composite shape and unknown time. Therefore a trial-and-error procedure is required. Sterilization requires that the center temperature reaches the minimum required temperature. Lacking more specific information, we assume that the medium begins at a uniform temperature of 22°C and that the center temperature inside the jar ends somewhat lower than in the autoclave temperature, or about 116°C. Thus the temperature difference ratio for the jar is

$$\frac{\theta_c - \theta_\infty}{\theta_o - \theta_\infty} = \frac{116°C - 121°C}{22°C - 121°C} = 0.0505$$

From information in Geankoplis (1993), condensing steam produces a value of 53,600 N/(M sec °C) for the outside surface convection coefficient (see Section 3.3.1.7).

Physical properties of the culture medium are:

$$\text{thermal conductivity } (k) = 0.62 \text{ N/(sec °C)}$$
$$\text{density } (\rho) = 1073 \text{ kg/m}^3$$
$$\text{specific heat } (c_p) = 4050 \text{ N m/(kg °C)}$$

Biot numbers must be calculated for the two component shapes of cylinder

Figure 3.7.26. Plant tissue culture samples growing in a baby food jar.

and slab. For the cylinder

$$\mathrm{Bi}^{-1} = \frac{k}{h_c r_o} = \frac{0.62 \ \mathrm{N/(sec \ ^\circ C)}}{[53{,}600 \ \mathrm{N/(m \ sec \ ^\circ C)}](0.0295 \ \mathrm{m})}$$
$$= 3.92 \times 10^{-4}$$

For the slab

$$\mathrm{Bi}^{-1} = \frac{k}{h_c L/2} = \frac{0.62 \ \mathrm{N/(sec \ ^\circ C)}}{[53{,}600 \ \mathrm{N/(m \ sec \ ^\circ C)}](0.050 \ \mathrm{m})}$$
$$= 2.91 \times 10^{-4}$$

Since these values are nearly zero, the jar surface temperature becomes 121°C almost immediately. The center temperature, however, is most important for sterilization.

Time must now be found by trial and error by consulting Heisler charts for the slab and cylinder. We begin by calculating the constant part of the Fourier number (Fo/t) for both shapes. For the slab

$$\mathrm{Fo}/t = \frac{k}{\rho c_p (L/2)^2} = \frac{0.62 \ \mathrm{N/(sec \ ^\circ C)}}{(1073 \ \mathrm{kg/m^3})[4050 \ \mathrm{N \ m/(kg \ ^\circ C)}](0.050 \ \mathrm{m})^2}$$
$$= 5.708 \times 10^{-5} \ \mathrm{sec}$$

For the cylinder

$$\mathrm{Fo}/t = \frac{0.62 \ \mathrm{N/(sec \ ^\circ C)}}{(1073 \ \mathrm{kg/m^3})[4050 \ \mathrm{N \ m/(kg \ ^\circ C)}](0.0295 \ \mathrm{m})^2}$$
$$= 1.640 \times 10^{-4} \ \mathrm{sec}$$

Next we assume that the slab center temperature difference ratio is

$$\left. \frac{\Delta\theta}{\Delta\theta} \right|_{\mathrm{slab}} = 0.22$$

Thus

$$\left. \frac{\Delta\theta}{\Delta\theta} \right|_{\mathrm{cyl}} = 0.22$$

because

$$\left. \frac{\Delta\theta}{\Delta\theta} \right|_{\mathrm{slab}} \left. \frac{\Delta\theta}{\Delta\theta} \right|_{\mathrm{cyl}} = 0.05$$

As calculated previously. From the Heisler chart, $Fo_{slab} = 0.74$ and $Fo_{cyl} = 0.23$, leading to time estimates of 12,964 and 1402 sec, respectively.

These times must be equal for the solution to be correct. Thus a new assumption of $(\Delta\theta/\Delta\theta)|_{slab}$ must be made, leading to new values for time. The following is a chart of guesses and results:

$\dfrac{\Delta\theta}{\Delta\theta}\Big|_{slab}$	$\dfrac{\Delta\theta}{\Delta\theta}\Big|_{cyl}$	Fo_{slab}	Fo_{cyl}	t_{slab}	t_{cyl}
0.60	0.084	0.32	0.32	5606	1951
0.70	0.072	0.26	0.34	4555	2073
0.90	0.056	0.13	0.37	2278	2256
0.95	0.053	0.10	0.37	1752	2256

Thus the time required for the center to reach 116°C from an initial value of 22°C is about 2267 sec, or 38 min.

Remarks

It is interesting to compare this time estimate with the recommended autoclaving time. For a container holding a volume of 273 mL $(2.73 \times 10^{-4}$ m$^3)$, the recommended time in the autoclave is 33 min, which includes 15 min held at 121°C to sterilize the medium and 16 min for the liquid volume to reach 121°C. Since our calculated time of 38 is much longer than 16 min, a discrepancy is noted.

The difference in the two times is due to the fact that the jar is filled with liquid, not a solid material. A liquid is free to move, setting up a convective process inside the jar and mixing the cooler and hotter contents. As a result, the jar contents are brought to the desired temperature much more rapidly than calculated.

It can be seen from Eq. 3.7.14 and values already calculated for Bi^{-1}, that increasing the convection coefficient inside the jar will not result in a much lower value for Bi^{-1}. However, manipulating Eq. 3.7.15 and solving for the value of inside convection coefficient that results in a heating time of about one-half of the 2267 sec previously obtained

$$h_{c_i} = \frac{Fo\ c_p \rho r}{t}$$

$$= \frac{(0.37)[4050\ \text{N m}/(\text{kg °C})](1073\ \text{kg/m}^3)(0.0295\ \text{m})}{1133\ \text{sec}}$$

$$= 42\ \text{N m}/(\text{m}^2\ \text{sec °C})$$

This value is about one-third of the value that Ramaswamy and Sablini (1995) found inside cans.

3.7.2.3 Sterilization of Food and Medical Devices

Canned food is a very safe product because the technology of sterilization has been highly developed. It is known that the number of microbes that will survive heat treatments depends on temperature and time of the treatment. As shown in Figure 3.7.27, survival rate is lower for higher temperatures and longer times. Common parameters in food processing (Pflug and Esselen, 1963) are the decimal reduction time D (the number of minutes at a constant temperature required to destroy 90% of the organisms in a population), Z (the number of degrees of temperature required for the thermal death time to change by a factor of 10 along the curve), and F (the number of minutes at a specified temperature, usually 121°C, to destroy a specified number of microorganisms, usually corresponding to $12D$ min, having a specific Z value, usually 10°C).

decimal reduction time

The decimal reduction time D is the time during which the number of viable microbes is reduced to one-tenth of the original number (Figure 3.7.28)

$$n/n_o = 1/10 = e^{-k_d D} \tag{3.7.16}$$

where n is the number of microbes at any time (number); n_o, the initial number of microbes (number); k_d, the first-order death rate constant (sec^{-1}); and D, the decimal reduction time (sec).

Also,

$$\log_{10}(1/10) = -1 = -k_d D \log_{10} e = -k_d D(0.4343) \tag{3.7.17}$$

Thus

$$D = 2.303/k_d \tag{3.7.18}$$

Since

$$\log_{10} x = (\log_{10} e)(\ln x) \tag{3.7.19}$$

$$t = D \log_{10}(n_o/n) \tag{3.7.20}$$

Thus the relationship between microbial kill and time should plot as a straight line on semilog paper. This is usually the case for vegetative cells (actively growing and metabolizing) but may not always hold for bacterial spores (inactive and environmentally resistant forms).

minimum sterilization time

For canned foods it has been decided that the minimum heating process should reduce the number of spores by a factor of 10^{12}. Thus (n/n_o) is 10^{-12}, and $t = 12D$. The time with a value of $12D$ is called the thermal death time, and represents a mathematical concept of microbial reduction that has been found empirically to give effective sterilization.

The decimal reduction time varies with temperature. When $\log_{10} D$ is plotted against a limited range of temperature, a straight line results. This is

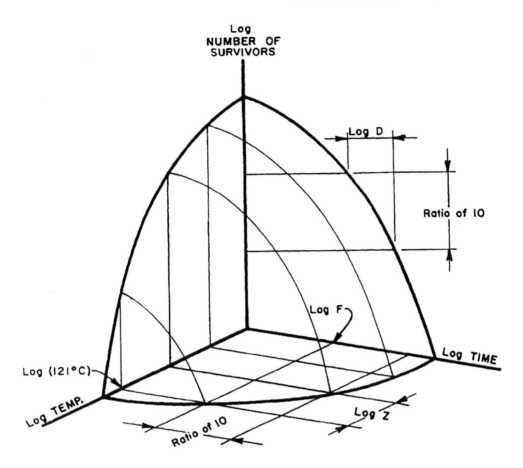

Figure 3.7.27. Schematic relationship among time, temperature, and microbial survivors.

used to relate D at any given temperature to the standard temperature at 121°C. Thus

$$\log_{10} D_\theta - \log_{10} D_{121} = (121 - \theta)/Z \qquad (3.7.21)$$

$$D_\theta = D_{121} 10^{(121-\theta)/Z} \qquad (3.7.22)$$

parameter Z   The parameter Z obtained from this plot is the temperature range over which D changes by a factor of 10. The value of Z depends on the microbe being inactivated. For *Clostridium botulinum* the value of Z is 10°C.

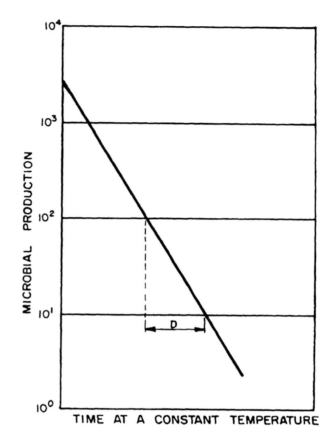

Figure 3.7.28. Semilog plot of the decimal reduction time.

Over a broader range of temperature, the $\log_{10} D$ actually plots as a straight line against the inverse of the absolute temperature. This is because the reaction constant, k_d, is described by an Arrhenius equation

Arrhenius
equation

$$k_d = A_k e^{-E_k/RT} \qquad (3.5.9b)$$

Since $D = 2.303/k_d$,

$$\log_{10} D = \log_{10}(\log_{10} e) + \log 10\, A_k + (\log_{10} e)(-E_k/RT) \qquad (3.7.23)$$

which plots as a straight line with $\log_{10} D$ against $1/T$ (Figure 3.7.29).

The value for F, defined before, is the amount of time at some specified temperature required to destroy a specified ratio of microorganisms. Therefore,

parameter F

$$F_{121} = t_{121} = D_{121} \log_{10} \frac{n_o}{n} = D_\theta 10^{(\theta-121)/Z} \log_{10} \frac{n_o}{n}$$

$$= t_\theta 10^{(\theta-121)/Z} \tag{3.7.24a}$$

where F_{121} is the F value at a standard $121°C$, and t_θ is the time the process remains at temperature θ. This equation can be used to relate sterilization at other temperatures to the standard $121°C$.

Clostridium botulinum

The spore-forming anaerobe *Clostridium botulinum* forms an exotoxin that has a human death rate of 65%. Unless all the spores formed by these bacteria are killed, the food in which the vegetative form of the bacteria grows would be unsafe to eat.

C. botulinum cannot grow in foods more acidic than pH 4.6. These foods, including many canned fruits and pickles, do not require as harsh heat treatment as do higher pH foods, such as meats and many vegetables. Other food constituents can protect bacterial spores and require more time or higher temperature for sterilization if they are present. These include concentrated sugars, fats, and oils. Since time and heat adversely affect food nutrients and flavors, a number of tradeoffs are involved. Certainly, though, the food must be made safe.

Sterilization can take place either in the final container or before packaging. The latter process, called aseptic packaging, occurs when sterilized food is added to previously sterilized containers. The former process can occur in still or agitating retorts. Shaking the containers vastly increases convective heat transfer both outside and inside the containers.

Solid or semisolid foods require longer sterilization times than do less viscous foods because heat transfer inside the can must occur by conduction rather than convection. Sterilization times are determined by temperatures in the center of the can and in the interior of those particles with lowest thermal conductivity.

Since sterilization temperature is not reached instantly, there will be some time when the food temperature is increasing but not yet high enough to achieve acceptable microbial kill. Likewise, cooling does not occur instantly.

food quality affected

Because food quality decreases during heating, sterilization time can be shortened by accounting for the microbial kill that can occur at penultimate temperatures. The equation for the thermal death time curve is

$$F_{121}/t_\theta = 10^{(\theta-121)/Z} \tag{3.7.24b}$$

The quantity F/t_θ is called the lethal rate. Lethal rates can be calculated for temperatures other than the sterilization temperature and then used to reduce

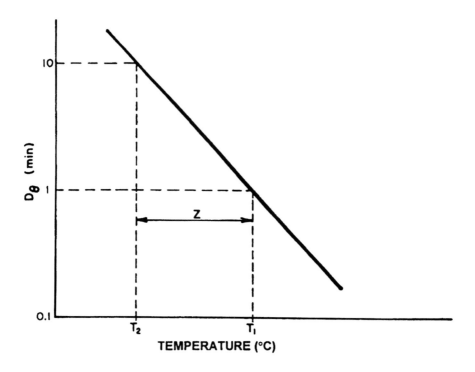

Figure 3.7.29. Semilog plot of the decimal reduction time (*D*) against temperature, showing the parameter *Z*.

the time the food is held at the sterilization temperature. For instance, a temperature of 110°C with a *Z* value of 10°C has a lethal rate of

lethal rate

$$(F_{121}/t_{110}) = 10^{(110-121)/10} = 0.079 \qquad (3.7.25)$$

Thus 1 min at 110°C is equivalent to 0.079 min at 121°C, and the time the food is held at 121°C can be reduced accordingly.

From the Heisler charts we see that final temperatures are reached asymptotically. Once values for convection coefficient, thermal conductivity, specific heat, and density are known, a curve of temperature against time can be constructed from the Heisler charts. When this curve is broken into small time increments, and lethal rates are calculated for rising and falling temperatures, equivalent time at the nominal sterilization temperature can be calculated.

devices and enzymes

These procedures are also used to determine sterilization times and temperatures for medical devices, when sterilization uses heat. These procedures are used as well for inactivation of enzymes in food and biotechnology.

EXAMPLE 3.7.2.3-1. Sterilization of Evaporated Milk

Consider the sterilization of a can of evaporated milk. This product is usually contained in 354 mL cans 7.0 cm in diameter and 9.2 cm long. Evaporated milk has the following physical properties.

$$\text{thermal conductivity } (k) = 0.505 \text{ N/(sec } °C)$$
$$\text{specific heat } (c_p) = 3850 \text{ N m/(kg } °C)$$
$$\text{density } (\rho) = 1050 \text{ kg/m}^3$$

Cans are put into a water-filled retort with steam jacket. We will assume no agitation of the cans or the water in the retort, and that the operating temperature of the water in the retort is 130°C. Physical properties of the water are

$$\text{thermal conductivity } (k) = 0.685 \text{ N/(sec } °C)$$
$$\text{specific heat } (c_p) = 4228 \text{ N m/(kg } °C)$$
$$\text{density } (\rho) = 942 \text{ kg/m}^3$$
$$\text{viscosity } (\mu) = 0.235 \times 10^{-3} \text{ kg/(m sec)}$$
$$\text{Prandtl number (Pr)} = 1.45$$
$$\text{Volume coefficient } (\beta) = 8.64 \times 10^{-4}/°C$$

Calculate the sterilization time.

Solution

Without agitation, heat transfer inside the retort will be solely by convection. An experimental correlation for natural convection in liquids flowing across horizontal cylinders is (Eq. 3.3.40a)

$$\text{Nu} = 0.52\sqrt[4]{\text{Gr Pr}}$$

Since

$$\text{Gr} = \frac{g\beta d^3(\theta_o - \theta_\infty)}{\nu^2}$$

the Grashof number changes magnitude as the can heats. With the rate at which the can heats depending on the convection coefficient, which in turn depends on the Grashof number, an iterative process should be used to determine the exact value of Grashof number to be used during each time interval. Practically speaking, however, it is sufficient to use an average value of can temperature.

If the can begins at a uniform temperature of 100°C, then the logarithmic mean temperature is

$$\theta_{lmt} = \text{antilog}[\tfrac{1}{2}\log(100) + \tfrac{1}{2}\log(130)] = 114°C$$

The logarithmic mean temperature is used for the average can temperature because the temperature rise in the can is close to exponential in time. Thus

$$Gr = \frac{g\beta d^3 \rho^2 (\theta_o - \theta_\infty)}{\mu^2}$$

$$= \frac{(9.8 \text{ m/sec}^2)(8.64 \times 10^{-4}/°C)(9.2 \times 10^{-2} \text{ m})^3 (942 \text{ kg/m}^3)(130 - 114)}{(0.235 \times 10^{-3} \text{ kg/(m sec)})^2}$$

$$= 1.80 \times 10^6$$

$$Nu = 0.52\sqrt[4]{Gr\ Pr} = 0.52\sqrt[4]{(1.80 \times 10^6)(1.45)} = 20.9$$

$$h_c = \frac{Nu\ k}{d} = \frac{(20.9)[0.685 \text{ N/(sec °C)}]}{(9.2 \times 10^{-2} \text{ m})} = 156 \text{ N/(sec m °C)}$$

We assume the same convection coefficient value over the ends of the can as well as over the sides. Since two Heisler charts must be used, those for cylinder and slab, a separate convection coefficient could be calculated for the ends of the cans.

The inverse of the Biot number is calculated for the cylinder as:

$$Bi^{-1} = \frac{k}{h_c r} = \frac{[0.505 \text{ N/(sec °C)}]}{[156 \text{ N/(sec m °C)}](0.046 \text{ m})} = 0.070$$

That for the slab is (remembering to use half the length of the can)

$$Bi^{-1} = \frac{k}{h_c L} = \frac{[0.505 \text{ N/(sec °C)}]}{[156 \text{ N/(sec m °C)}](0.035 \text{ m})} = 0.092$$

Time in the retort can be considered in increments and temperatures calculated. The smaller the time increment, the more accurate can be the results. For purposes of this example, we use 1000-sec increments and construct the following table.

Time	$\dfrac{kt}{c_p \rho r^2}$	$\dfrac{kt}{c_p \rho L^2}$	$\dfrac{\Delta\theta}{\Delta\theta}\bigg|_{cyl}$	$\dfrac{\Delta\theta}{\Delta\theta}\bigg|_{slab}$	$\dfrac{\Delta\theta}{\Delta\theta}\bigg|_{tot}$	θ_{final} (°C)	θ_{avg} (°C)
0	0	0	1.0	1.0	1.0	100	
							101
1000	0.059	0.102	0.99	0.97	0.96	101	
							104
2000	0.118	0.204	0.89	0.84	0.75	108	
							112
3000	0.177	0.306	0.63	0.69	0.43	117	
							120
4000	0.236	0.408	0.45	0.55	0.25	122	
							124
5000	0.295	0.510	0.33	0.45	0.15	126	

The Fourier number for the cylinder was calculated from

$$\text{Fo} = \frac{kt}{c_p \rho r^2} = \frac{[0.505 \text{ N}/(\text{sec } °\text{C})]t}{[3850 \text{ N m}/(\text{kg } °\text{C})](1050 \text{ kg/m}^3)(0.046 \text{ m})^2}$$

$$= 5.90 \times 10^{-5}t$$

The Fourier number for the slab is $\text{Fo} = 1.02 \times 10^{-4}t$.

Temperature difference ratios for slab and cylinder were determined from the Heisler charts. Center temperature was used because the middle of the can is expected to be the coolest location. The overall temperature difference ratio for the composite can shape is the product of the cylinder and slab temperature difference ratios. Final temperature was calculated from the definition of the temperature difference ratio

$$\left.\frac{\Delta\theta}{\Delta\theta}\right|_{\text{tot}} = \frac{\theta_c - \theta_\infty}{\theta_o - \theta_\infty} = \frac{\theta_c - 130}{100 - 130}$$

$$\theta_c = 130 - 30 \left.\frac{\Delta\theta}{\Delta\theta}\right|_{\text{tot}}$$

The overall F value for the can was calculated from

$$F = \sum_{i=1}^{N} t_i 10^{(\theta_i - 121)/Z}$$

Using average temperatures during the 1000-sec intervals in the table, and a Z value (usually determined from experimental data) of 10

$$F = 1000 \times 10^{(101-121)/10} + 1000 \times 10^{(104-121)/10}$$

$$+ 1000 \times 10^{(112-121)/10} + 1000 \times 10^{(120-121)/10}$$

$$+ 1000 \times 10^{(124-121)/10}$$

$$= 10.0 + 20.0 + 126 + 794 + 1995 = 2945 \text{ sec} = 49 \text{ min}$$

Remarks

Notice that the logarithmic relationship between temperature and F value results in the greatest contribution to the total F from the highest temperature. If the required F value for these cans of evaporated milk is not greater than 49 min, retort heating can be shut off after 5000 sec (83 min) and the cans allowed to cool. Note that during cooling there will continue to be some sterilization occurring, but this contribution to the F value is not added in order to be conservative.

The slower the cooling, the more conservative will be the sterilization treatment of the food. However, other factors must also be considered: Many nutrients (for example, thiamine) are degraded by heat and slow cooling results in more degradation than does fast cooling. Also, flavor ingredients are often adversely affected by heat, and slow cooling is to be avoided. Thus the retort in this example would likely be filled with cooling water after heating has ceased.

There were many assumptions and approximations made in this example, and these were not checked for validity. A few words of explanation must be given about this procedure.

First, the Heisler charts were designed to be used when heat transfer inside the can occurs by conduction. Since evaporated milk is a liquid, convection is expected greatly to enhance heat transfer inside the can. Thus the can should heat much faster than we calculated, and, in addition, convective movement of liquid inside the can should mix the contents to reduce any temperature differences between the center of the can and the surface.

Next, heat transfer in the retort was assumed to arise mainly from natural convection. Several approximations were made in the calculation of the convection coefficient, including assumptions of can surface temperature and convection coefficient over the ends of the cans. Normally, more care should be taken that assumptions are justified and checked, if possible. However, in this case, overall accuracy was compromised by failing to account for convection inside the can. Good food quality practice would also probably require some stirring inside the retort, thus enhancing forced convection.

Food safety is too important to be left to unchecked assumptions and faulty calculations. Experimental measurements of temperatures inside the cans would be required before the sterilization process could be approved. Any modification of the process would require experimental verification of the adequacy of sterilization before approval. Thus the food industry must completely test and establish beyond any reasonable doubt that its procedures are adequate. The time and expense involved in this testing leads to a very conservative attitude toward processing that avoids changes in procedures.

Adequate testing should demonstrate that bacterial spores are incapable of growing and producing toxin. Since *Clostridium botulinum* is so dangerous, spores from the nonpathogenic *Bacillus stearothermophilus*, which are of similar heat resistance to *C. botulinum*, are often used to test heat treatments.

If experimental verification is required, what is the value of calculations similar to the ones in this example? The values are several: (1) Used for design purposes, these calculations lead to results not far from the actual outcomes; (2) they can demonstrate unproductive avenues without the expense of actually building and testing the system; and (3) they aid conceptual understanding of the process. Thus, for food engineers, these calculations are of great importance to making good design engineers who are rarely, if ever, wrong and who have the judgment to know the difference between practical and impractical solutions to problems they encounter.

3.7.3 Extended Surfaces

fins

You may have noticed the thin metal pieces that extend from the tubes of your car radiator or from the condenser on your home refrigerator. These solid pieces of metal of relatively small cross section protruding into a fluid of different temperature are called fins, and are meant to increase the rate of heat transfer.

    The thermal resistance from a planar surface is $R = L/kA$. Although this planar surface resistance hardly describes the thermal resistance of a fin, some insight can still be gained by considering it. To increase heat transfer from a surface requires that the resistance decreases. One way to do this is to increase

multiply area

the surface area greatly, and fins can thus be thought of as surface area multipliers.

    Fins can be of many shapes and sizes. There are rectangular fins, tapered fins, and pin fins, among others. Thin, slender, and closely spaced fins offer superior heat transfer compared to fewer and thicker fins. Almost all intentional manmade fins are made of metal because of its high thermal conductivity. Protruding areas of animals or plants can also act as fins: Birds, for instance, extend their wings and feet in hot environments to enhance heat loss, but fold their wings close to their bodies and squat down on their feet to

natural fins

conserve heat. Hands and feet in people become cold first because of the high ratio of surface area to volume in these appendages. Arctic animals that usually must conserve heat can be seen to have body shapes much more compact and stocky than do equatorial animals that must often lose heat.

    This is not to say that fins are always advantageous for heat transfer. If the

limitations

surrounding fluid is turbulent or moving rapidly, sufficient heat transfer may occur from the surface without fins. In that case, the effect of adding fins may be to create pockets of still or slowly moving fluid that are less, not more, effective in transferring heat. Increased heat transfer from the addition of pin fins of uniform cross section can only be expected if (Kreith, 1958)

$$(h_c A/Pk) \leq 1 \qquad (3.7.26)$$

where h_c is the average convection coefficient [N/(m sec °C)]; A, the cross-sectional area of the fin (m^2); P, the pin cross-sectional perimeter (m); and k, the thermal conductivity of the fin material [N/(°C sec)].

    For pin fins of circular cross section, this reduces to

$$(h_c d/4k) \leq 1 \qquad (3.7.27a)$$

where d is the pin diameter (m).

For pin fins of ectangular cross section, this becomes

$$[h_c ab/2(a + b)k] \leq 1 \tag{3.7.27b}$$

where a, b are pin cross-sectional dimensions (m).

Kreith (1958) has presented an analysis of the temperature distribution and heat flow from a pin fin of uniform cross section. Starting with the pin fin diagrammed in Figure 3.7.30, and with the assumption that heat is transferred by conduction in one direction along the length of the fin only, he began with a steady-state heat balance

| Rate of heat flow by conduction into an element at x | = | Rate of heat flow by conduction from the element at $(x + dx)$ | + | Rate of heat flow by convection from the surface between x and $(x + dx)$ | (3.7.28) |

Or, symbolically

$$-kA \frac{d\theta}{dx} = \left[kA \frac{d\theta}{dx} + \frac{d}{dx}\left(-kA \frac{d\theta}{dx}\right) dx \right] + h_c P \, dx(\theta - \theta_\infty) \tag{3.7.29}$$

where k is the thermal conductivity of the pin material [N/(°C sec)]; A, the cross-sectional area of the pin (m^2); θ, the temperature (°C); x, the direction along the pin (m); h_c, the average convection coefficient on the surface of the pin [N/(m sec °C); P, the perimeter of the pin at a cross section (m).

The term in square brackets in Eq. 3.7.29 has two components: The second is the difference between the heat flowing into and out of the element, and the first is the heat flowing in; the sum of these two is the heat escaping from the element by conduction at $(x + dx)$. The rightmost term is the standard convection equation (Eq. 3.3.1) with the surface area replaced by its equivalent $P \, dx$.

After algebraic manipulation, this equation becomes

$$\frac{d^2\theta}{dx^2} = \left(\frac{h_c P}{kA}\right)(\theta - \theta_\infty) \tag{3.7.30}$$

The general solution to this ordinary second-order linear differential equation is

$$(\theta - \theta_\infty) = C_1 e^{mx} + C_2 e^{-mx} \tag{3.7.31}$$

where θ_∞ is the ambient temperature (°C); C_1, C_2, the constants of integration (°C); and m, the $\sqrt{h_c P/kA}$ (dimensionless).

The values for C_1 and C_2 will depend on the boundary conditions chosen for this problem. One boundary condition is that $\theta = \theta_s$ at $x = 0$, where θ_s is the surface temperature (°C). The other boundary condition depends on whether the pin fin is assumed to be (for all practical purposes) infinitely long, or of finite length and insulated at the end, or of finite length with end convection.

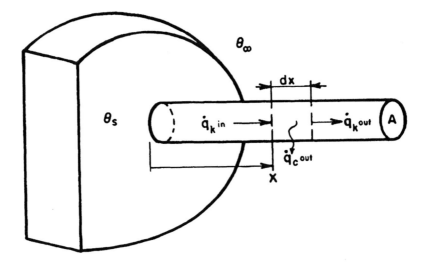

Figure 3.7.30. Schematic diagram of a pin fin attached to a surface. For each elemental volume ($A\, dx$), heat is conducted in ($\dot{q}_{k\,\text{in}}$), conducted out ($\dot{q}_{k\,\text{out}}$), and lost by convection ($\dot{q}_{c\,\text{out}}$).

For the infinite case, the temperature distribution becomes

$$\frac{(\theta - \theta_\infty)}{(\theta_s - \theta_\infty)} = e^{-mx} \tag{3.7.32}$$

and the heat lost from the fin is

$$\dot{q}_{\text{fin}} = -kA\left.\frac{d\theta}{dx}\right|_{x=0} = mkA(\theta_x - \theta_\infty)e^{-mx}\Big|_{x=0}$$

$$= \sqrt{h_c P k A}(\theta_s - \theta_\infty) \tag{3.7.33}$$

For the finite, but insulated case, the temperature distribution becomes

$$\frac{(\theta - \theta_\infty)}{(\theta_s - \theta_\infty)} = \frac{\cosh m(L - x)}{\cosh(mL)} \tag{3.7.34}$$

with heat loss

$$\dot{q}_{\text{fin}} = \sqrt{h_c P k A}(\theta_s - \theta_\infty)\tanh(mL) \tag{3.7.35}$$

For the finite, but uninsulated, case the temperature distribution becomes

$$\frac{(\theta - \theta_\infty)}{(\theta_s - \theta_\infty)} = \frac{\cosh m(L - x) + (h_c/mk)\sinh m(L - x)}{\cosh mL (h_c/mk)\sinh mL} \qquad (3.7.36)$$

with heat flow from the fin

$$\dot{q}_{\text{fin}} = \sqrt{Ph_c Ak}(\theta_s - \theta_\infty)\frac{\sinh mL + (h_c/mk)\cosh mL}{\cosh mL + (h_c/mk)\sinh mL} \qquad (3.7.37)$$

The temperature distribution for this last case is diagrammed in Figure 3.7.31. From the nonlinear decrease of temperature with pin length, it should be clear that there is a point where increasing the length of the pin fin does little to increase the rate of heat loss.

EXAMPLE 3.7.3-1. Heat Loss from Honey Bee Legs

The metabolic power delivered by the wing muscles of freely foraging honey bees has been measured at about 0.025 N m/sec at 25–35°C ambient temperature. Flying bees can easily overheat; they thus use heat loss by convection caused by their beating wings, and when this is not sufficient, they extend their legs to act as pin fins to remove additional heat. A flying bee (Figure 3.7.32), with body mass of 75–80 mg, has a thorax temperature about 3°C higher than ambient (down from 7°C preflight), and 2°C higher when dangling the legs. Calculate the heat lost from the legs.

Solution

First, we shall calculate the apparent convection coefficient, then we will use Eq. 3.7.37 to calculate heat flow from the leg. The total power produced is 0.025 N m/sec. Assuming the muscles are about 20% efficient, about 0.020 N m/sec must be removed as heat. The convection coefficient can be approximated from the information given, and

$$h_c = \dot{q}/(A\ \Delta\theta)$$

The surface area of a honey bee is assumed to be about three times the thorax surface area, and the thorax is assumed to be a sphere 4 mm in diameter. Thus

$$A = 3\pi d^2/4 = 3\pi(0.004)^2/4 = 3.77 \times 10^{-5}\ \text{m}^2$$
$$h_c = (0.020\ \text{N m/sec})/(3.77 \times 10^{-5}\ \text{m}^2)(3°\text{C})$$
$$= 177\ \text{N m/(sec m}^2\ °\text{C})$$

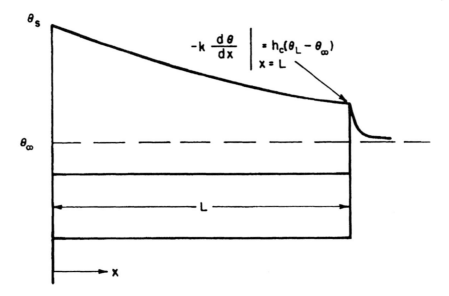

Figure 3.7.31. Diagram of the temperature distribution along a pin fin with a base temperature of θ_s losing heat by convection to the surrounding fluid at θ_∞. The end of the fin has been assumed to be subject to convective heat loss.

This is a very high value for convection coefficient, but it is reasonable considering the forced airflow due to wing movement.

Leg length is assumed to be 5 mm when extended, and leg diameter is assumed to be 0.5 mm. Thus leg perimeter is about

$$P = \pi d = \pi(0.005 \text{ m}) = 1.57 \times 10^{-3} \text{ m}$$

Cross-sectional area is

$$A = \pi d^2/4 = \pi(0.0005 \text{ m})^2/4 = 1.96 \times 10^{-7} \text{ m}^2$$

Thermal conductivity of the leg, composed of blood-perfused muscle tissue covered with a thin layer of epidermis, is assumed to be the same as for human muscle tissue. From Table 3.2.2, $k = 0.50 \text{ N}/(^\circ\text{C sec})$. Thus

$$m = \sqrt{h_c P/kA}$$
$$= \sqrt{(177 \text{ N}/(\text{sec m }^\circ\text{C})(1.57 \times 10^{-3} \text{ m})/[0.50 \text{ N}/(^\circ\text{C sec})]}$$
$$\times (1.96 \times 10^{-7} \text{ m}^2)$$
$$= 1684 \text{ m}^{-1}$$
$$mL = (1684 \text{ m}^{-1})(0.005 \text{ m}) = 8.42$$
$$\sinh mL = \tfrac{1}{2}(e^{mL} - e^{-mL}) = 2268$$
$$\cosh mL = \tfrac{1}{2}(e^{mL} + e^{-mL}) = 2268$$

Figure 3.7.32. A flying honey bee can lose heat from her legs when flying. The legs acting as fins are efficient at heat loss.

From Eq. 3.7.37

$$\dot{q} = \sqrt{Ph_c Ak}(\theta_s - \theta_\infty)\frac{\sinh mL + (h_c/mk)\cosh mL}{\cosh mL + (h_c/mk)\sinh mL}$$

Because the values for sinh mL and cosh mL are equal in this case, the last term becomes 1.0. Thus

$$\dot{q} = (2°C)\sqrt{(1.57 \times 10^{-3}\ \text{m})(177\ \text{N}/(\text{sec m °C}))}$$
$$\times (1.96 \times 10^{-7}\ \text{m}^2)[0.50\ \text{N}/(°C\ \text{sec})]$$
$$\dot{q} = 3.30 \times 10^{-4}\ \text{N m/sec}$$

There are six legs on a honey bee; so the total heat loss from the legs is (6) $(3.30 \times 10^{-4}\ \text{N m/sec}) = 1.96 \times 10^{-3}\ \text{N m/sec}$. This estimate is about 10% of the total heat loss of the bee.

3.8 CHANGE OF PHASE

requires
energy

Change of phase includes change of state (solid, liquid, gas) and the structure of a particular state (crystal formation, solution). Both are important for food and biological materials. Heat released or absorbed during a change of phase must be added to the net rate of heat transfer appearing as the first two terms of

Eq. 3.1.1. Sometimes, as in the case of condensation inside a potato storage facility, or as in the case of moisture evaporation from the tongue of a panting dog, the change of phase heat term is the largest of any heat transfer term.

3.8.1 Change of State

water

When substances change state there is either a release or absorption of energy. Water, probably the most important change of state substance involved in living systems, releases about 0.334×10^6 N m/kg of heat when ice forms and about 2.44×10^6 N m/kg of heat when steam condenses. The latent heat of sublimation (ice becoming vapor) is 2.83×10^6 N m/kg.

3.8.1.1 Freezing

protecting blossoms

Because of the large amount of heat released when ice forms, an agricultural strategy has evolved that protects fragile blossoms from freezing temperatures by spraying them with water. As the water freezes, at least some of the heat produced satisfies the required amount of heat to be transferred from the blossom to the air (Figure 3.8.1) and the blossom does not freeze.

freezing

Freezing is used to protect food materials from deterioration. Several mechanisms are important here. First, the growth rate of microorganisms is slowed considerably below their optimum growth temperature. Second, the growth of microorganisms depends on available free water, and this is nearly eliminated at low enough freezing temperatures. Third, enzyme activity within the food is reduced by a factor of 2–3 for each 10°C reduction in temperature, and the food thus becomes more biochemically stable at freezing temperatures. Fourth, other important chemical reactions are slowed by the same rate of 2–3 for each 10°C as temperature is lowered (see the Van't Hoff equation 3.5.3). Oxidation is an important reaction that is not completely stopped during frozen storage, but that is slowed considerably.

cellular disruption

Ice crystal formation disrupts cellular structure, and, for that reason, may damage food and biological materials. Large ice crystals require some time to grow, however, and so very rapid cooling can be used to circumvent undesirable effects of slow freezing. Freezing and storage in liquid nitrogen (−195.8°C) is thus used for sperm, some microoganisms, plant and human tissues, and small organs. The challenge to rapid freezing of larger organs is to protect the interior tissues from damage while the outer tissues are being frozen. Some organisms, such as arctic fish that spend much of their time in waters of subfreezing temperature, contain natural antifreeze substances that help protect them from freezing and cellular ice-crystal formation.

Freezing-Point Depression When pure water freezes it maintains a constant temperature until all water is frozen. Food and biological materials, however, are much more complex substances than are pure water: Various substances are in solution, suspension, and/or immiscible relationships with

effect of solutes

the water. In Figure 3.8.2 is shown a diagram of temperature with time for a single solute solution undergoing freezing. As heat is removed, and freezing progresses, there is not one, but a range of freezing temperatures. Most biological materials do not freeze at 0°C, but more likely freeze at -2 to -3°C due to the solutes present within them. Heldman (1992) has given an equation to predict the temperature at which ice crystals begin to form in a solution of complex composition

$$T_s = \left(\frac{1}{T_{H_2O}} - \frac{R}{H_m} \ln \mu_{H_2O} \right)^{-1} \qquad (3.8.1)$$

where T_s is the absolute temperature when freezing begins in the solution (K); T_{H_2O}, the absolute temperature when freezing begins in pure water $(= 273.15 \text{ K})$; R, the universal gas constant [$= 8314.34$ N m/(kg mol K)]; H_m, the molar latent heat of fusion of water $(= 6.0134 \times 10^6$ N m/kg mol); and μ_{H_2O}, the mole fraction of water in the solution, kg mol H_2O/kg mol solution.

The mole fraction of water can be computed from other concentration designations using Table 4.3.12, or it can be computed from

$$\mu_{H_2O} = \frac{m_{H_2O}/M_{H_2O}}{m_{H_2O}/M_{H_2O} + m_{sl}/M_{sl}} \qquad (3.8.2)$$

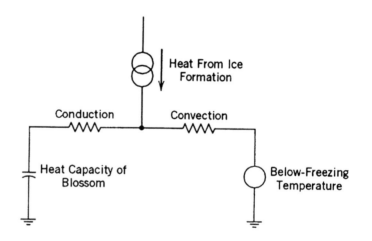

Figure 3.8.1. Schematic of the thermal system when water is sprayed on fragile blossoms to protect them from freezing. Without the heat from ice formation, heat would flow from the thermal storage of the blossom to the air at below-freezing temperature, and the blossom itself would eventually freeze. Depending on relative values of thermal resistances, heat capacities, heat formation, and temperatures, the blossom may warm somewhat or just cool more slowly. A small amount of heat production due to blossom cellular metabolism has been neglected in this diagram.

where m_{H_2O} is the mass of water in the product (kg); m_{sl}, the mass of solids in the product (kg); M_{H_2O}, molecular weight of water ($= 18.0153$ kg/kg mol); and M_{sl}, the molecular weight of solids (kg/kg mol).

solute concen- tration
If there are several solutes present, or if the solutes dissociate into ionic form, then effective solute masses and effective molecular weights are used to determine freezing point depressions.

Equation 3.8.1 can also be used to estimate the amount of unfrozen water fraction in a partially frozen product. To do this, the assumption is made that all solute is concentrated in the unfrozen water fraction. T_s must be known, and the equation is solved for the value of μ_{H_2O}.

Sometimes a simpler means to calculate either freezing point depression or, conversely, the solute concentration in solution is to use the fact that the freezing point of water is reduced by 1.86°C per gram mole of solute particles in 1 kg of water (Sienko and Plane, 1957).

Chilling   Foods that cannot be frozen, fresh fruits and vegetables, organs for transplant, and some laboratory samples, are not frozen, but instead are chilled. Due to the reduction in enzymatic activity with lowered temperature, quality can be maintained for extended lengths of time with chilling.

Chilling is almost exclusively performed by circulating a cool fluid around the outside of the product. In this case, heat is removed from the product by conduction inside the product and convection outside the product (Figure 3.8.3). Sometimes chilling is accomplished by placing slabs of meat or other biological materials to be cooled on a plate contact chiller (or freezer). Heat is transferred here by conduction inside the product and conduction outside the product. Good contact must be assured between the product and the cold plate or else a dead air layer, with relatively high thermal conduction resistance, will exist between them. Usually good contact is assured when the product is both pliable and wet.

air-blast cooling
Product cooling often occurs by means of a blast of cold air. The convection coefficient for air-blast cooling varies from 8.5 to 40 N m/(m² K sec), depending on air velocity and turbulence present. Liquids may also be used for chilling. These include water and brine, and have convection coefficients that vary from 80 to 1700 N m/(m² K sec). Table 3.8.1 includes representative values of equivalent heat transfer coefficients for several different types of chilling systems.

Sometimes the product is covered with a box or a thin plastic film. In most cases, this covering is assumed to fit the product tightly so that there is no dead gas layer between the product and the covering. Also, the covering is usually assumed to have negligible thermal capacity. With these two assumptions, the resistance of the covering (conduction heat transfer) is added to the resistance

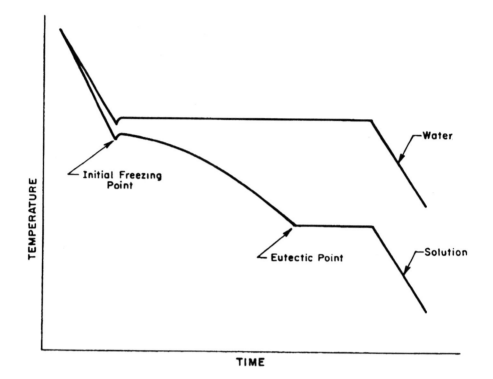

Figure 3.8.2. Schematic freezing curves for pure water and for an aqueous solution with one solute. Solutions with many solutes, found in real foods, have much more complex freezing curves.

equivalent convection coefficient

of the flowing fluid (convective heat transfer) and an equivalent convection coefficient is obtained

$$R_{tot} = R_{conduction} + R_{convection} \tag{3.8.3a}$$

$$R_{tot} = \frac{L}{kA} + \frac{1}{h_c A} = \frac{1}{h_{equiv} A} \tag{3.8.3b}$$

$$h_{equiv} = \frac{1}{R_{tot} A} = \left(\frac{L}{k} + \frac{1}{h_c} \right)^{-1} \tag{3.8.3c}$$

where R_{tot} is the total thermal resistance [N m/(°C sec)]; $R_{conduction}$, the conduction resistance of the covering [N m/(°C sec)]; $R_{convection}$, the convection resistance of the flowing fluid [N m/(°C sec)]; L, the covering thickness (m); k, the thermal conductivity [N m/(°C sec)]; A, the surface area (m^2); h_c, the flowing fluid convection coefficient [N m/(sec m^2 °C)]; and h_{equiv}, the equivalent convection coefficient, including the covering [N m/(sec m^2 °C)].

This equivalent convection coefficient is then used with the Heisler charts (Section 3.7.2.2) to obtain chilling temperatures or chilling times.

physical property uncertainty

There can be sizable errors in time or temperature estimates obtained from the Heisler charts. These result from uncertainties in actual values of the physical properties of density, specific heat, and thermal conductivity; and in the correct characterization of the shapes of the objects being chilled. Heldman (1992) gives a number of techniques for precise prediction of physical properties of food materials, but there will always be an amount of uncertainty involved with measurements of properties of biological materials.

Freezing Time   Knowledge of steady-state heat transfer is sufficient to know that a product will eventually freeze if it is placed in a medium that is

how much time?

cold enough. A very important secondary question is how long it will take to freeze. Maintaining cold temperatures for longer than necessary can be expensive, because the process of freezing a product uses more energy than the process of maintaining a frozen product. Hence, commercial freezing often occurs separately from frozen product storage.

Before products are frozen they must be chilled to freezing temperature. That is one reason why the discussion on chilling preceded this discussion of freezing times. Calculation of freezing times involves at least three steps: (1) chilling to freezing temperatures, (2) time for actual freezing to occur, and (3)

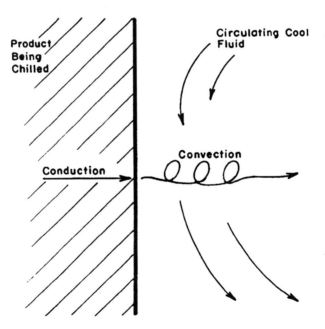

Figure 3.8.3. When a product is being chilled with a flowing cool liquid, the heat flows by conduction inside the product and by convection outside.

Table 3.8.1 Approximate Convection Heat Transfer Coefficients for Different Types of Heat Exchange Systems

Type	Convection Coefficient [N m/(m² K sec)]
Air blast	8.5–40
Agitated vessel	170
Scraped-surface heat exchanger	1700–2270
Shell-and-tube heat exchangers	
Water to water	1140–1700
Water to brine	570–1140
Water to organic liquids	570–1140
Water to vegetable oil	110–230
Water to air (finned tube)	110–230
Water and brine (forced)	80–1700
Hydrocoolers	680–940
Kettle and jacket (steam in jacket)	
Water	850
	1420 (agitated)
Paste	710 (agitated)
Milk	1130
	1700 (agitated)
Tomato purée	170 (agitated)

chilling below freezing to desired storage temperature. The first and third steps are similar in that they both involve use of the Heisler charts. The only difference is that physical properties of the frozen product are generally different from those of the unfrozen product.

There is a method by Cleland and Earle for determining freezing time that is based on sound empirical justification (Heldman, 1992), but that requires a great deal of computation that may or may not be justified depending on the

Plank method

particular application. The approximate Plank method is often sufficient for engineering purposes and is given here. For this method, the entire product is assumed to begin unfrozen at the freezing temperature. Thermal conductivity, density, and latent heat of fusion are all considered to be constant. Heat is removed from the product by conduction inside and by convection at the surface (Figure 3.8.4). The frozen layer thickens, beginning at the surface, and moves inward as heat is removed. The entire process occurs slowly enough to be considered to be nearly at steady state.

Under these conditions, freezing time is given as

freezing time

$$t_{\text{fr}} = \frac{C\rho HL}{\theta_{\text{fr}} - \theta_\infty} \left(\frac{1}{h_\text{c}} + \frac{L}{4k} \right) \qquad (3.8.4)$$

where t_{fr} is the freezing time (sec); ρ, the unfrozen product density (kg/m^3); H, the latent heat of fusion of the product, which can often be assumed to be that of the water in the product (3.34 × 10^5 N m/kg product); θ_{fr}, the freezing temperature (°C); θ_∞, the temperature of surrounding fluid (°C); L, the thickness of a slab, diameter of a sphere, or diameter of a long cylinder (m); h_c, the convection coefficient at the surface of the product [N m/(sec m^2 °C)]; k, the thermal conductivity of the frozen product [N m/(sec m °C)]; and C, a coefficient (dimensionless; $= \frac{1}{2}$ for infinite slab, $\frac{1}{6}$ for sphere, or $\frac{1}{4}$ for infinite cylinder).

For shapes that cannot be approximated as an infinite slab, a sphere, or an infinite cylinder, graphical means have been developed to estimate the coefficient C (Heldman, 1992).

thawing time   Calculation of thawing times can also be made using Eq. 3.8.4 as long as the value of thermal conductivity that is used is of the unfrozen, rather than frozen, product. The value for density must be that for the frozen material.

packages   There are instances commonly encountered where the substance to be frozen is packaged inside another container and thus the surface of the substance is not directly exposed to convection heat loss. In this case, freezing time predicted in Eq. 3.8.4 may be considerably in error. One way to correct Eq. 3.8.4 is to substitute an equivalent convection coefficient (h_{equiv}) for the actual convection coefficient (h_c) appearing in the equation. If the surface of the substance to be frozen is exposed to the surrounding fluid, then the surface resistance to heat transfer is $R = 1/h_c A$. If additional layers of other material are interposed between the substance and the surrounding fluid, then the thermal resistance, R_{tot}, will include resistances of these other layers in series with the convection resistance. The equivalent convection coefficient can then be calculated as $h_{equiv} = 1/R_{tot} A$.

EXAMPLE 3.8.1.1-1. Freezing-Point Temperature

Calculate the freezing point of whey containing 6% total solids by weight. Molecular weight of the solids is about 256.

Solution

We can calculate freezing point depression in two ways. First we use Eq. 3.8.1. Mole fraction of water in 1 kg of whey is

$$\mu_{H_2O} = \frac{0.94 \text{ kg H}_2\text{O}/(18 \text{ kg/kg mol})}{0.94 \text{ kg H}_2\text{O}/(18 \text{ kg/kg mol}) + 0.06 \text{ kg solids}/(256 \text{ kg/kg mol})}$$
$$= 0.996 \text{ kg mol H}_2\text{O/kg mol solution}$$

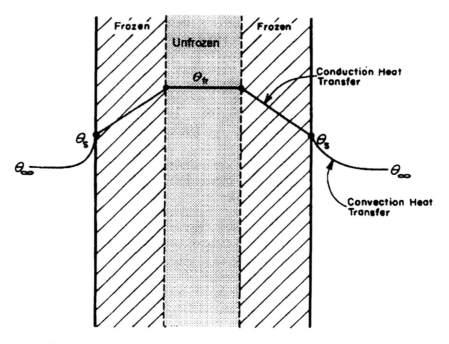

Figure 3.8.4. Condition for the Plank method for calculating freezing time. The frozen layer thickens with time.

From Eq. 3.8.1

$$T_s = \left(\frac{1}{273.15 \text{ K}} - \frac{8314.34 \text{ N m/(kg mol K)}}{6.0134 \times 10^6 \text{ N m/kg mol}} \ln 0.996 \right)^{-1}$$

$$= 272.69 \text{ K} = -0.46°\text{C}$$

As another means to calculate the freezing-point depression, we can determine the number of gram moles of solids in 1 kg of water. Since the whey is 94% water,

$$\frac{\text{grams of solids}}{\text{kg water}} = \frac{(0.06 \text{ kg solids})(1000 \text{ g/kg})}{0.94 \text{ kg water}} = 63.83 \ \frac{\text{g solids}}{\text{kg water}}$$

$$\text{freezing-point depression} = \left(\frac{1.86°\text{C kg water}}{\text{g mol solids}} \right) \left(\frac{63.83 \text{ g solids}}{\text{kg water}} \right)$$

$$\times \left(\frac{\text{g mol solids}}{256 \text{ g}} \right)$$

$$= 0.46°\text{C}$$

Thus the freezing point is $-0.46°\text{C}$ either way we calculate.

EXAMPLE 3.8.1.1-2. Freezing Time for Meat

Hamburger patties for use in schools and restaurants are made individually and frozen in bulk. Patties for one particular operation are 10 cm in diameter and one-third cm in thickness. A thin piece of paper is placed between patties, and a stack of patties 1.5 m tall is wrapped in a layer of 0.5-mm-thick polystyrene and frozen in an air-blast freezer. Estimate the time required to freeze the stack if the meat begins at 10°C and the temperature of the air in the freezer is −29°C.

Solution

This problem involves two processes: (1) chilling of the meat from 10°C to the freezing temperature, and (2) freezing itself. For these calculations we need various physical constants that can be obtained from tables in this book. These values are:

$$
\begin{array}{ll}
\text{unfrozen meat} & \rho = 1060 \text{ kg/m}^3 \\
 & c_p = 3520 \text{ N m/(kg °C)} \\
\text{frozen meat} & k = 1.1 \text{ N m/(m sec °C)} \\
\text{paper} & k = 0.17 \text{ N m/(m sec °C)} \\
\text{polystyrene} & k = 0.10 \text{ N m/(m sec °C)} \\
\text{air-blast freezer} & h_c = 20 \text{ N m/(m}^2 \text{ sec °C)}
\end{array}
$$

As a first step, an equivalent convection coefficient is determined. The layer of polystyrene on the outside of the stack reduces the effectiveness of convective heat loss. Areas of the sides of the stack is

$$A = \pi\, dL = \pi(0.10 \text{ m} + 0.005 \text{ m})(1.5 \text{ m}) = 0.474 \text{ m}^2$$

Total thermal resistance is

$$R_{\text{tot}} = \frac{1}{h_c A} + \frac{\ln(r_o/r_i)}{2\pi k L}$$

$$= \frac{1}{\left(20 \dfrac{\text{N}}{\text{m sec °C}}\right)(0.474 \text{ m}^2)} + \frac{\ln(0.505 \text{ m}/0.050 \text{ m})}{2\pi[0.10 \text{ N/°C sec}](1.5 \text{ m})}$$

$$= 0.1160 \frac{\text{sec °C}}{\text{N m}}$$

And the equivalent convection coefficient is

$$h_{\text{equiv}} = \frac{1}{R_{\text{tot}} A} = \frac{1}{0.1160[\text{sec °C/(N m)}](0.474 \text{ m}^2)} = 18.2 \frac{\text{N m}}{\text{m}^2 \text{ °C sec}}$$

As a second step, the freezing-point temerature can be calculated using Eq. 3.8.1. For this we need an estimate of moisture content and molecular weight of the tissue. Meat products often have a moisture content of about 75%. The solutes in the meat are mostly proteins, and an effective molecular weight of 225 kg/kg mol will be assumed. From Eq. 3.8.2,

$$\mu_{H_2O} = \frac{\dfrac{0.75 \text{ kg H}_2\text{O}}{18 \text{ kg H}_2\text{O/kg mol H}_2\text{O}}}{\dfrac{0.75 \text{ kg H}_2\text{O}}{18 \text{ kg H}_2\text{O/kg mol H}_2\text{O}} + \dfrac{0.25 \text{ kg tissue}}{225 \text{ kg tissue/kg mol tissue}}}$$

$$= 0.974$$

$$T_s = \left(\frac{1}{273.15 \text{ K}} - \frac{8314.34 \text{ N m/(kg mol K)}}{6.0134 \times 10^6 \text{ N m/kg mol}} \ln 0.974 \right)^{-1}$$

$$= 270.46 \text{ K} = -2.69°\text{C}$$

The temperature difference ratio, to be used with the Heisler chart is, therefore,

$$\frac{\theta_{avg} - \theta_\infty}{\theta_0 - \theta_\infty} = \frac{(-2.69°\text{C}) - (-29°\text{C})}{10°\text{C} - (-29°\text{C})} = 0.675$$

The chilling time can now be determined. The Biot number for the cylindrical shape of the stack is

$$B_i = \frac{h_c r}{k} = \frac{[18.2 \text{ N m/(sec m}^2 \text{ °C)}](0.505 \text{ m})}{1.1 \text{ N m/(m}^2 \text{ °C sec)}} = 0.836$$

We use only the Heisler chart for an infinite cylinder for several reasons: (1) The length of the stack is more than 10 times the diameter, and (2) the pieces of paper between the patties, with their thermal conductivity much lower than that of the meat, helps assure that flow is almost completely radial and not longitudinal. From the Heisler chart, Figure 3.7.18, we determine Fo = 0.19. Thus

$$t = \frac{\text{Fo } \rho c_p r^2}{k} = \frac{(0.19)(1060 \text{ kg/m}^3)[3520 \text{ N m/(kg °C)}](0.0505 \text{ m})^2}{1.1 \text{ N m/(m °C sec)}}$$

$$= 1644 \text{ sec} = 27.4 \text{ min}$$

Finally, the freezing time can be calculated. The latent heat of fusion is

$$H = (0.75)(3.34 \times 10^5 \text{ N m/kg H}_2\text{O}) = 250{,}500 \, \frac{\text{N m}}{\text{kg product}}$$

$$t_{tr} = \frac{(1/4)(1060 \text{ kg/m}^3)(250{,}500 \text{ N m/kg})(0.101 \text{ m})}{(-2.69°\text{C}) - (-29°\text{C})}$$
$$\times \left(\frac{1 \text{ m sec } °\text{C}}{18.2 \text{ N}} + \frac{0.101 \text{ m}}{4[1.1 \text{ N/(}°\text{C sec)]}} \right)$$
$$= 19{,}851 \text{ sec} = 5.51 \text{ h}$$

The entire chilling and freezing process thus takes about 6 h.

Remarks

The value for 6 h is, of course, only an estimate because of the number of assumptions and imprecise property values used. However, the value of 6 h is probably reasonably correct. To cool the frozen product to a temperature lower than $-2.69°\text{C}$ requires additional time that can be estimated again from the Heisler chart, but this time using physical property values for the frozen meat.

---

coupled
mass and
heat transfer

respiratory
evaporation

sweating

3.8.1.2 Evaporation   Evaporation is used to concentrate food products and to remove heat from the body, among other things. In order to be effective, vapor must be removed from the evaporation site or else evaporation ceases. Thus the process of evaporation involves mass transfer as well as heat transfer, and the two are intimately connected. The mass transfer relationships associated with evaporation are found in Section 4.6.2.3.

Evaporation from humans occurs both in the respiratory system and on the skin surface. A semiempirical equation attributed to Varène gives respiratory evaporative heat loss as (Johnson, 1991a)

$$\dot{q}_{\text{evap,resp}} = 0.001 \dot{V}(59.34 + 0.53\theta_a - 0.0116 p_{\text{H}_2\text{O}}) \tag{3.8.5}$$

where $\dot{q}_{\text{evap,resp}}$ is the evaporative heat loss from the respiratory system (N m/sec); $\dot{V}$, the respiratory ventilation rate (m^3/sec); θ_a, the ambient temperature (°C); and $p_{\text{H}_2\text{O}}$, the ambient water vapor pressure (N/m^2).

Because the evaporation occurs in the respiratory system, the entire heat loss represented by respiratory evaporation is available to cool the human (or animal) body.

Evaporative heat loss from sweating skin is both regulatory and nonregulatory. Unstressed skin always sweats about 6% of its maximum capacity (Johnson, 1991a), except that skin dehydrated by prolonged exposure to low humidities may cause this percentage to drop as low as 2%. Maximum

heat
equivalents

remote
evaporation
of sweat

cooling capacity depends on skin surface area, but is about 1200 N m/sec for an average man.

Water requires about 2.447×10^6 N m/kg to evaporate. Because sweat contains salts as well as water, its heat equivalent is about 8% higher, or 2.6×10^6 N m/kg.

The calculation of heat loss due to sweating can be very complicated. First, there is the mass transfer process found in Section 4.8.2.3. Once the water vapor has been formed, there is a simple equivalence between mass of water evaporated and heat absorbed. The major complication comes when considering that sweat can fall off the body (in which case it absorbs no body heat when it evaporates), or that sweat can wick through clothes and evaporate at a site removed from the body surface. Conceptually, sweating produces cooling in a manner given in Figure 3.8.5. The conceptual diagram can be more complicated, however, if the slight temperature dependence of evaporation is included, or if evaporation occurs at various levels within the clothing.

EXAMPLE 3.8.1.2-1. Houseplant Evaporation

Energy required to evaporate moisture from the leaf surfaces of indoor house plants comes from household heating systems. Evaporation cools the leaves, and that draws heat from the building interior, which, in turn, is replenished from the heating system. Unless the heat of vaporization is recovered through water condensation somewhere in the building interior, the cost of evapotranspiration is paid by the building owner. Estimate the cost.

Solution

Although there is a great deal of variability, the average house plant evaporates about 100 mL (1×10^{-4} m^3) of water each day. At a density of 1000 kg/m^3, and a latent heat of 2.447×10^6 N m/kg, the amount of heat used for evaporation is

$$\dot{q} = (1 \times 10^{-4} \text{ m}^3/\text{day})(1000 \text{ kg/m}^3)(2.447 \times 10^6 \text{ N m/kg})$$
$$= 2.447 \times 10^5 \text{ N m/day}$$

If the cost of heat is 10¢ per kwh, then

$$\text{cost} = \left(2.447 \times 10^5 \frac{\text{N m}}{\text{d}}\right)\left(\frac{1 \text{ d}}{24 \text{ h}}\right)\left(\frac{\text{h}}{3600 \text{ sec}}\right)\left(\frac{1 \text{ W sec}}{\text{N m}}\right)\left(\frac{1 \text{ kW}}{1000 \text{ W}}\right)$$
$$\times \left(\frac{7 \text{ d}}{\text{wk}}\right)\left(\frac{24 \text{ h}}{\text{d}}\right)\left(\frac{1 \text{ kWh}}{\text{kW h}}\right)\left(\frac{10}{\text{kWh}}\right)$$
$$= 4.76¢/\text{week}$$

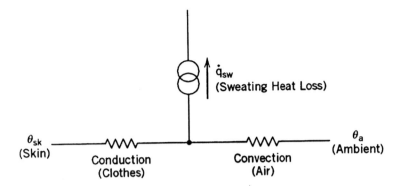

Figure 3.8.5. Sweat that wicks through clothes and evaporates at the clothing surface cools both the skin and the surrounding air. The amount of skin cooling depends on the thermal resistance to conduction through the clothes compared to the thermal resistance to convection in the air. The amount of heat loss due to sweating is somewhat dependent on the temperature at the local site of evaporation because of the dependence of local water vapor pressure on temperature.

3.8.1.3 Sublimation Some biological and food materials are too sensitive to withstand any heat. Drying these products occurs while the water is frozen, and is called freeze drying. Water is removed by sublimation under vacuum. The heat of sublimation is supplied by conduction or radiation.

freeze drying

Freeze-dried foods are of the highest quality because the structure of the food is not severely damaged, and a porous, nonshrunken structure remains after the water is removed. These foods are easily rehydrated for use.

Geankoplis (1993) and Okos et al. (1992) both present analyses of the freeze-drying process with the objective to predict the time required to dry a product from an initial moisture content M_o to a final mixture content M_∞. The approach is similar to that used to predict freezing time. The original material is frozen, and the vapor pressure is lowered to sublime the ice. The plane of sublimation moves from the surface into the interior of the product. Simultaneous heat and mass transfer must occur for the ice to sublimate. As water vapor is transferred by diffusion through the layer of dried material to the outside, heat must be transferred by conduction to the sublimation plane from the outside (Figure 3.8.6). Heat transferred inside the material is supplied to the surface by convection from the surrounding dry air

drying time

heat transfer

$$\dot{q} = h_c A(\theta_a - \theta_s) = \frac{kA}{\Delta L}(\theta_s - \theta_{sub}) \tag{3.8.6}$$

where $\dot{q}$ is the rate of heat transfer at the surface (N m/sec); h_c, the convection coefficient [N m/(m² sec °C)]; k, the thermal conductivity of the dried solid [N m/(m sec °C)]; ΔL, the thickness of the dried product (m); A, the surface area (m^2); θ_a, the ambient temperature (°C); θ_s, the surface temperature (°C); and θ_{sub}, the sublimation temperature (°C).

A similar set of equations must be written for the mass transfer process (details are found in Chapter 4)

mass transfer

$$\dot{m} = \frac{DM_{H_2O}A}{RT\,\Delta L}(p_{sub} - p_s) = k'_G A(p_s - p_a) \tag{3.8.7}$$

where $\dot{m}$ is the mass flow rate (kg/sec); D, the effective water vapor diffusivity in the dry layer (m^2/sec); R, the universal gas constant [8314.34 N m/(kg mol K)]; T, the average absolute temperature in the dry layer (K); M_{H_2O}, the molecular weight of water vapor ($= 18.0153$ kg/kg mol); ΔL, the thickness of dry layer (m); k'_G, the mass transfer coefficient (sec/m); A, the area (m^2); p_{sub}, the partial pressure of water vapor at the sublimation temperature (N/m^2); p_s, the partial pressure of water vapor at the surface (N/m^2); and p_a, the partial pressure of water vapor in the surrounding air (N/m^2).

Of the two processes of heat and mass transfer, the thermal process is considered to be the slower, and limits the rate of drying. In addition, the convection coefficient h_c is considered to be large, meaning that there is negligible thermal resistance at the surface. From this beginning, the sublimation time has been found to be

rate limited
by heat
transfer

sublimation
time

$$t_{sub} = \frac{L^2 H \rho_0}{4k_{sol}} \frac{1}{\theta_a - \theta_{sub}} \left(\frac{M_\infty^2}{2M_0} - M_\infty + \frac{1}{2}\right) \tag{3.8.8}$$

where t_{sub} is the time to dry by sublimation (sec); L, the slab thickness (m); H, the latent heat of sublimation of water in the product ($= 2.83 \times 10^6$ N m/kg H_2O); k_{sol}, the thermal conductivity of dry solids [N m/(m/sec/°C)]; ρ_0, the initial density of solid (kg/m^3); M_0, the initial moisture content (wet basis; kg H_2O/kg wet solid); M_∞, the final moisture content (wet basis; kg H_2O/kg wet solid); θ_a, the ambient temperature (°C); and θ_{sub}, the sublimation temperature (°C).

Because sublimation requires so much energy, extra heat must be added to the solid. If not, then the heat of sublimation comes from the solid itself, with a resulting temperature drop. The lower temperature is accompanied by a lower water vapor pressure and more difficult sublimation (see Section 4.8.2).

EXAMPLE 3.8.1.3-1. Freeze Drying Bacteria

Microorganisms and single cells (such as red blood cells) are sometimes freeze dried to preserve them without refrigeration. Calculate the time required to freeze dry a bacterium 20 μm in diameter if the final moisture content of the cell is no lower than about 1% to ensure survival (Meryman, 1966; Fry, 1966).

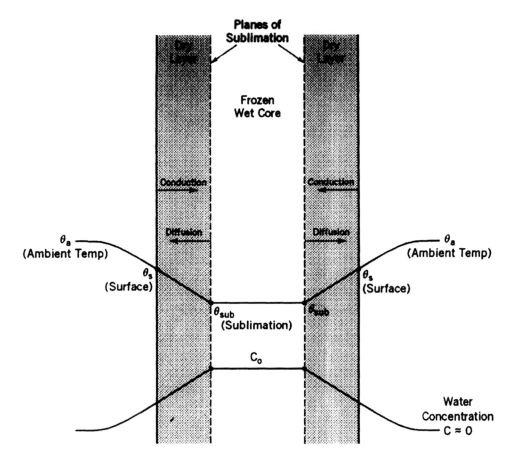

Figure 3.8.6. Model for freeze drying by sublimation. Simultaneous heat and mass transfer are present inside the material. The plane of sublimation moves inward as the material dries.

Solution

Although this application of Eq. 3.8.8 is not entirely valid because the conditions for its derivation were much different from those during freeze drying of bacteria, it is still at least an estimate of the required time. The major difficulty with this application of Eq. 3.8.8 is to estimate required physical properties.

Initial moisture contents of bacteria are probably not much different from moisture contents of animal tissue, which are about 75%. The density of bacteria is probably about 1060 kg/m^3 (from Table 2.5.4), and the thermal conductivity of dried bacteria is likely very low, perhaps as low as 0.04 N/(°C sec) as given in Table 3.2.2 for dried beef. Freeze drying of bacteria is usually accomplished at about -30 to -40°C, and the plane where sublimation occurs is at a lower temperature than ambient. We thus assume -30 and $-35°C$ for θ_a and θ_{sub}, respectively.

$$t_{sub} = \frac{(20 \times 10^{-6} \text{ m})^2(2.83 \times 10^6 \text{ N m/kg})(1060 \text{ kg/m}^3)}{4[0.04 \text{ N/(°C sec})](-30°C + 35°C)}$$
$$\times \left(\frac{0.01^2}{2(0.75)} - 0.01 + 0.5 \right)$$
$$= 0.74 \text{ sec}$$

Remark

This is a very short time, and likely to be considerably longer for large numbers of bacteria. Microorganisms are often freeze dried in a liquid medium containing buffers of glucose, sucrose, sodium glutamate, or other mixture to guard against killing the cells (Meryman, 1966). Fry (1966) states that "the literature of freeze drying is rather like a cookery book, freely sprinkled with recipes for mixtures which different workers have considered best to ensure the survival of the particular organisms which they were trying to preserve." The medium must be stirred vigorously to keep the convection coefficient high and the resistance to transfer of water vapor low.

3.8.2 Heat of Solution

solubilities

Air, carbon dioxide, oxygen, and all other common gases are more soluble in water at lower temperatures (see Table 4.6.1). That is one reason why bubbles are observed to form in a container of cold water as it warms. Sodium bicarbonate, magnesium sulfate (Epsom salts), sodium chloride, and most other common salts are more soluble in water at higher temperatures (two exceptions are calcium hydroxide, hydrated lime, and calcium sulphate, gypsum). When gases dissolve, there is heat that evolves. When salts dissolve that are more soluble at higher temperatures, heat is absorbed. This heat is called the heat of solution.

energy exchanges

The process of solution involves the two steps of (1) breaking up the solid crystal lattice into independent ions and (2) surrounding the ions with weakly bound water molecules (hydration). The first step requires energy, and the second step liberates energy. The relative energy amounts determine the net absorption or release of energy. A substance that requires energy to dissolve releases an identical amount of energy on precipitation

$$\text{heat} + \text{substance} + \text{water} \rightleftharpoons \text{solution} \qquad (3.8.9)$$

LeChatlier's
principle

LeChatlier's principle states that "If a stress is applied to a system at equilibrium, then the system readjusts, if possible, to reduce the stress." Substances that require heat to dissolve will increase dissolved concentration if the temperature is raised. Substances that evolve heat upon dissolving will come out of solution if the temperature is raised.

3.8.3 Phase Changes

glass
transition
temperature

There are many phase transitions that can occur in food and biological materials. For instance, anhydrous sucrose has a glass transition temperature (θ_g) at 62°C. Below this temperature, a frozen, yet unordered, molecular structure, called a glass, is formed. At 103°C, instant crystallization occurs wherein a frozen, but ordered, molecular structure forms. Crystallization can actually occur anywhere between 62 and 103°C. Sucrose melts at 185°C. Every amorphous material has a random structure; it resembles a liquid above its glass transition temperature.

physical
properties
change

The change from a glass to a crystal is accompanied by physical properties changes: There are changes in values of specific heat, density, thermal expansion coefficient, mass diffusivity, viscosity, dielectric constant, and enthalpy. Annealing and aging below the glass transition temperature can also change physical property values. Aging does not occur in crystalline substances.

Cryopreservation of tiny organisms, cells, and tissues depends on rapid cooling to produce amorphous solids, or glasses (this process is called vitrification). If ice crystals form inside cells, the result is often death. Crystal formation can also occur on thawing; so rapid heating is also required.

component
interactions

There are, in addition, interactions between food components that can change physical property values. Water and starch can form a gel. Sugar or salt added to the gel modify the gel. Water can plasticize many compounds. Lipids form complexes with amylose.

fats and oils

Oils and fats that are solid at low temperatures melt, or become softer, over a range of temperatures. The melting temperatures of fatty acids increase with increasing carbon chain length and increase with increasing saturation (hydrogen bonded to suitable sites). Fats deposited at various sites in the body generally have higher melting points than fats deposited in the hands and feet. This allows the hands and feet to operate at cooler temperatures than would be the case with stiffer fats. There is a latent heat of fusion associated with melting of fat.

protein
denaturation

Proteins that have been heated too much have their natural structure disrupted, and are said to be denatured. During the denaturation process, the natural, folded structure is changed to an unfolded structure, and the protein no longer functions in the same way. The presence of water, sugars, or salts affects both temperature and rate of denaturation, as well as the heat of denaturation. Proteins can also exist in amorphous or crystalline states, and exhibit glass transition temperatures.

3.9 HEAT SYSTEM DESIGN

heat balance

There are many opportunities for designing systems for the addition or removal of heat. Some of the more important of these are summarized in Table 3.1.1. Most heat system designs rely fundamentally on a heat balance, as diagrammed in Figure 3.9.1 and given verbally as Eq. 3.1.1.

geometry

Much depends on geometry. Geometrical configuration can influence all modes of heat transfer, including conduction, convection, and radiation. Mathematical expressions for heat transfer coefficients for each of these modes have been seen to depend on configuration.

There are other important matters in heat transfer that do not appear in Figure 3.9.1. Change of state covers thawing, freezing, evaporation, and condensation. These processes can be important for food and biological materials.

Not all heat system designs depend on an explicit heat balance. Freezing times or heat exchanger sizes are two of these. Nonetheless, methods for each of these depend on heat balances, and often the real problems are to provide sufficient cooling to freeze food in a certain amount of time or to determine the total heat system for which the heat exchanger is just a part.

It is easy to forget to include all heat sources when writing a heat balance. Lights add heat; so does electrical or other machinery. Even if the prime mover (a motor or engine) is not in the room, mechanical inefficiency generates heat.

remember all sources

Do not forget to include heat from chemical reactions, heat produced with condensation, or heat coming from room occupants. It is not easy to remember to include all heat sources.

As with fluid flow and mass transfer, practical considerations can be an important determinant of the specific hardware to be used. There are few industry standard sizes and capacities for heat system equipment; so manufacturers' specifications must be consulted before making final choices.

PROBLEMS

3.1-1. **Water in diet.** It has been observed that some desert animals never drink water; they take in less water in their food than they evaporate, and yet they do not decrease the water content of their body. How do they get the extra water?

3.1-2. **Heat balance of a cow.** A cow ingested daily a ration containing 88.6×10^6 N m metabolizable energy. She produced 9 kg of milk, which represented 27.4×10^6 N m of chemical energy; and she lost body substance representing 3.30×10^6 N m metabolizable energy per day. Calculate the daily heat production of the cow.

3.1-3. **Energy of activities.** A man of 686 N body weight climbs 1000 m:

(a) How much work does he perform? (b) How much glucose represents the equivalent of that work expressed in terms of chemical energy? (c) In what form does that energy appear while the man rests on the mountain? (d) To what form is this energy transferred when

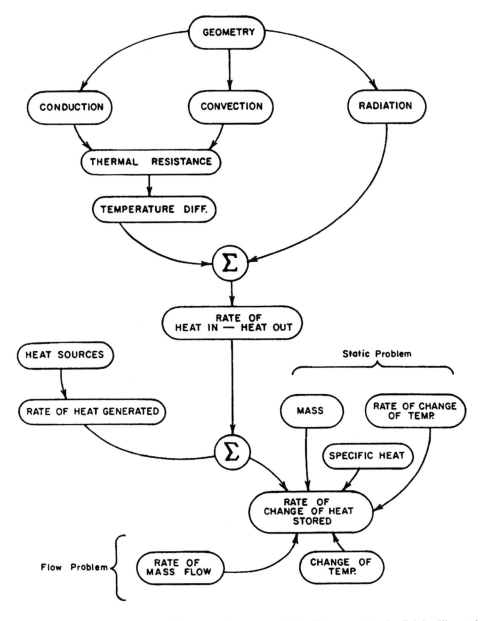

Figure 3.9.1. Schematic diagram of heat transfer system design. The procedure is slightly different for flow systems compared to static systems.

the man walks down a trail? (e) To what form is this energy transferred when he skis down a snow field?

3.1-4. **Coffee in a thermos.** Draw a thermal systems diagram for heat transfer from hot coffee in a Thermos bottle sitting on a picnic table in cold air.

3.1-5. **Organ in a bottle.** Organs for transplant are often transported in vacuum bottles similar to the Thermos bottle of the previous problem, except that the organs are chilled to retard degradation. Draw a thermal systems diagram of a kidney in a vacuum bottle held on someone's lap in an ambulance. How would this diagram differ if the diagram was meant to depict loss of moisture?

3.2-1. **Fleece resistance.** If the thermal resistance of a sheep fleece is 0.3 ($^\circ$C sec)/(N m), estimate the rate of heat lost by the sheep to an environment at 10°C.

3.2-2. **Skin thermal resistance.** Calculate the thermal resistance of a cm^2 of human skin (2 mm thick).

$$\text{thermal conductivity} = 0.627 \text{ N/(}^\circ\text{C sec)}$$
$$\text{specific heat} = 3470 \text{ N m/(kg }^\circ\text{C)}$$
$$\text{density} = 1100 \text{ kg/m}^3$$

3.2-3. **Measuring thermal conductivity.** One common means to measure thermal conductivity of a material is to insert a long metallic probe axially into the center of a long cylindrical container of the material. The probe is maintained at some uniform temperature θ_i, and the outside of the container is maintained at a temperature θ_o. Inside the metallic probe is an electrical heater for which the electrical power is measured. If the diameter of the probe maintained at 60°C is 0.5 cm, the outer diameter of the container maintained at 20°C is 4 cm, the length of the cylindrical container is 60 cm, and the power input is 40.6 N m/sec, what substance appearing in Table 3.2.2 is likely filling the container?

3.2-4. **Thermal conductivity of fur.** Suggest a means to measure the thermal conductivity of fur.

3.2-5. **Body surface area.** Calculate your nude body surface area using the DuBois formula and compare to the surface area of an animal with the same body mass. Which estimate do you think is correct, and why?

3.2-6. **Mean skin temperature.** Estimate mean skin temperature of a clothed human in a room maintained at 22°C.

3.2-7. **Thermal resistance of clothing.** A guy with body mass of 68 kg wears a clothing ensemble of T shirt, briefs, blue jeans, and sandals to his college class on Transport Processes in Biological Systems. If the room he sits in is kept at 20°C, estimate: (a) thermal resistance of his clothing, (b) his mean skin temperature, and (c) his heat loss to his surroundings.

3.2-8. **Polar bears in air and water.** Compare expected heat loss amounts from adult polar bears in water and air.

3.2-9. **Separation of variables.** Explain the method of separation of variables as a means to solve partial differential equations.

3.2-10. **Transient heat transfer.** Suggest a biological example where the thermal capacity is located inside and the thermal resistance is predominantly outside, so the example can be treated as a lumped-parameter model. Justify your choice.

3.3-1. **Comfort in the wind.** The more the wind blows, the colder we feel. Wind removes heat by convection. Assume subjective comfort is related to the rate of heat lost by convection and plot a curve of comfort level (100% comfort is attained for a nude human in still air at 22°C) against wind speed for a constant air temperature of 22°C.

3.3-2. **Film temperature.** The temperature of fluid entering a long pipe is 20°C and the temperature of the same fluid leaving the pipe is 120°C. If the pipe wall temperature is maintained at 150°C, what is the film temperature?

3.3-3. **Heating neonates.** Newborn human babies are dried and often placed on open bassinets, exposed to room air at 22°C. There is less than 0.1 mm of moisture on the skin that evaporates quickly. Neonates generate very little heat; so they are placed under 2 infrared heat lamps at 500 W each. Show by means of a heat balance that this is sufficient heat to maintain adequate body temperature.

3.3-4. **Thermal resistance of Emperor Penguins.** The male Emperor Penguin amazingly incubates its egg in the midst of the Antarctic winter, when temperatures vary between -20 and $-60°C$ and winds blow as fast as 200 km/h. Males exist for 4 months without eating a single meal. By the end of incubation, they have lost about 40% of their normal 40 kg body mass. This loss is mostly fat with an energy density of 37.7×10^6 N m/kg. Keeping their heads down and wings to their sides, they huddle together in slowly moving groups of thousands to conserve about 50% of the heat they would otherwise lose. The eggs are held on their feet and covered with a loose fold of overhanging skin; eggs are maintained at 30°C for a brooding period of 9 weeks. These penguins are 1 m tall and rather cylindrical in shape with diameters of about 40 cm. They maintain body temperatures of 39°C. Calculate the equivalent thermal resistance of their covering of feathers, skin, and fat layer. Compare this to the thermal resistance for a man with the same surface area as a penguin and wearing long underwear, heavy long-sleeved shirt, heavy vest, heavy trousers, heavy jacket, socks, and boots.

3.3-5. **Woman sitting in room.** A young woman weighing 510 N and 162 cm in height wears the following clothing: bra, panties, light blouse, light slacks, light sweater, panty hose, and pumps. her metabolic heat production is 70 N m/sec. The convection coefficient in the 22°C room in which she sits is 9.0 N/(°C sec m). Will she remain comfortable?

3.3-6. **Convection from musk ox.** Arctic animals, such as the musk ox, are much more compact than temperate or tropical animals. Thus they lose less heat from protruding body parts such as ears, legs, and tails. Estimate the convection heat loss for a musk ox standing in still air at −20°C.

3.3-7. **Greenhouse convection.** Air flows inside a greenhouse from one end to the other at a rate of 0.5 m/sec. Wind blows along the outside of the same greenhouse at a rate of 7 m/sec. If the greenhouse is 30 m long and has a semicircular cross-sectional shape with a diameter of 9 m, calculate the convection coefficients inside and outside.

3.3-8. **Cooling of mushrooms.** Mushrooms are grown in environments maintained at about 13°C. Shipping the mushrooms after they are harvested requires cooling to 1–2°C. For mushrooms of 4 cm cap diameter being cooled in chilled water at 1°C, estimate the convection coefficient if natural convection prevails.

3.3-9. **Convection coefficient inside an Artery.** Blood flows at a rate of 0.44 L/min inside an artery 4 mm in diameter and 5 cm long. Estimate the convection coefficient. What assumptions do you have to make in order to solve this problem?

3.3-10. **Mixed convection from a strawberry.** Determine the convection coefficient for a just-picked strawberry that is immediately being cooled. Temperature of the strawberry is 32°C and the cooling air is 4°C. Free stream air velocity in the field cooler is vertical at 0.22 m/sec. The strawberry is 2 cm in diameter.

3.3-11. **Convection coefficient of swimmer.** Someone tells you that the convection coefficient for a swimmer doing the side stroke is 2% less than for the same swimmer doing the free style (crawl). Should you go back and correct your homework problem where you calculated the heat loss from a swimmer? Why or why not?

3.3-12. **Adiabatic temperature in olive oil.** Olive oil is being pumped at a rate of 27 m/sec between two plates 5 cm apart. What is the adiabatic temperature if the oil begins at a temperature of 25°C?

3.3-13. **Immersion heaters.** Immersion heaters are often used to control temperatures of insulated water baths in which medical diagnostic tests are incubated. Estimate the convection resistance within the water outside a vertical cylindrical (1 cm diameter by 20 cm long) immersion heater maintained with a surface temperature of 37°C. The water is not stirred. Will the eventual water bath temperature be higher, lower, or the same if the convection resistance is doubled in value?

3.3-14. **Lower heat loss in women.** Anderson et al. (1995) report that at ambient temperatures as low as 5°C, average skin temperature in women is 1–2°C cooler in women than in men. Estimate the reduction in heat loss per unit body surface area for women compared to men for : (a) completely nude, and (b) normally clothed bodies.

3.4-1. **Radiation thermal resistance.** What is the thermal resistance of black body radiation?

3.4-2. **Measuring air temperature.** A butt-welded thermocouple having an emissivity of 0.8 is used to measure the temperature of air (assume to be transparent) flowing in a long cylindrical duct 5 cm in diameter. The 0.5-mm-diameter thermocouple wire is placed concentrically within the duct. The duct wall is at a temperature of 200°C, and the thermocouple indicates a temperature of 30°C. If the convection coefficient for the thermocouple immersed in air is 60 N m/(m^2 sec °C), estimate the true air temperature.

3.4-3. **Radiation shield.** Repeat Problem 3.4-2, except a cylindrical radiation shield made from thin stainless steel is placed around the thermocouple. The diameter of the shield is 1 cm and the convection coefficient on both inside and outside surfaces is 45 N m/(m^2 sec °C). Draw a thermal resistance diagram of this system and estimate true air temperature.

3.4-4. **Infrared thermometry.** An infrared thermometer is a hand-held device used to measure remote surface temperatures. The device consists of a case with a small opening through which radiation passes and strikes an internal absorbent surface typically 0.2 cm^2 in area. Radiant heat thus warms the absorbent internal surface until thermal equilibrium is reached between internal and measured surfaces. Small convective and conductive losses from the internal surface are compensated for.

In use, the infrared thermometer is pointed toward the surface whose temperature is to be measured, and the temperature is read from a meter. If the measured surface acts as a black body, then the reading is correct as given. If the measured surface acts as a gray body with known emissivity, the indicated temperature can be corrected.

If the infrared thermometer indicates a measured surface temperature of 35°C, how much heat is being transmitted from the measured surface to the infrared thermometer?

3.4-5. **Radiation in a room.** Which of the following statements is true? The rate of heat loss by radiation from a warm body in a room can be decreased by: (a) heating the air around the radiating body; (b) lowering the wall temperature; (c) raising the wall temperature; (d) blackening the walls; (e) covering the walls with aluminum foil.

3.4-6. **Lion lying in the sun.** Estimate the solar heat load on an adult lion reclining in the sun in Africa in midmorning.

3.4-7. **Human in a cold room.** Using the radiation coefficient for a person wearing a long-sleeved shirt and trousers, estimate the radiation exchange between the solitary person and the cold room surface (2°C). The room is 3 m high × 3 m wide × 4 m long.

3.4-8. **Bioremediation site.** A procedure used to clean petroleum products from contaminated soil is to remove the soil to a remote site, supply oxygen, nitrogen compounds, water, and other required nutrients, and allow natural thermophilic bacteria to metabolize the petroleum. When complete, the soil is returned to its original site. Because the bacteria require a closely monitored temperature of $35 \pm 3°C$, heat is usually added to the soil by hot water running through pipes at the bottom of the pile. The pile can be covered by thermally insulated blankets to minimize heat loss. Assume the pile is 500 m × 500 m × 5 m (high) and that the insulating blanket is a series of coarse-woven polypropylene tarpaulins layered to 5 cm thick. Estimate the radiant heat loads on the blankets at noon and at midnight in a humid continental climate. Draw a thermal systems diagram of the blanketed pile and explain why the blankets will minimize effects on the pile of the radiation load fluctuation at the blanket surface.

3.4-9. **Greenhouse thermal curtains.** Greenhouses are subject to large heat losses through the glass. Opaque thermal curtains can be installed inside the glass to reduce heat lost mainly by radiation to the enviroment at night. List the considerations for timing of the opening and closing of these curtains.

3.4-10. **Sunshine on my shoulder makes me happy.** A person sits on the beach in Ocean City, MD, with 0.5 m² of skin exposed to the sun. The sun, with a mean diameter of 700,000 km and a distance of 150 million km from the person, has a shape factor of 2.18×10^{-5}. Water vapor in the atmosphere absorbs a portion of the sun's energy before it reaches the Earth. The following table indicates the effects of water vapor and skin absorption in four wavelength ranges.

Wavelength Range (nm)	Portion of Solar Power	Water Vapor Transmissivity	Human Skin Absorptivity
0–520	0.263	0.90	1.00
520–900	0.378	0.90	0.50
900–1550	0.256	0.40	0.50
> 1550	0.103	0.90	1.00

Find the total amount of power tranferred from the sun to the beachgoer.

3.5-1. **Heat produced by apples.** If apples produce heat at the rate of 120 N m/sec at 30°C, estimate their heat production at 0°C.

3.5-2. **Body heat storage in a horse.** Taylor (1980) gives the relationship between the oxygen cost of terrestrial locomotion and body size as

$$\dot{V}_{O_2} = 5.31 \times 10^{-7} m^{0.70} v + 2.57 \times 10^{-7} m^{0.61}$$

where $\dot{V}_{O_2}$ is the oxygen consumption ($m^3 O_2$/sec); m, the body mass (kg); and v, the running speed (m/sec). Oxygen consumption can be converted into rate of energy production by using the energy equivalence of oxygen of 20.18×10^6 N m/m^3. Calculate the approximate rate of rise of body temperature for a 680 kg racehorse running at 15 m/sec. How would this value change if the horse was running on a treadmill instead of over open ground? What could you do about it?

3.5-3. **Hypothermia.** Humans undergoing open heart surgery are often cooled to 30°C to temporarily reduce their dependency on oxygen. Compare the basal metabolic rates of a 60 kg person at normal body temperature and during hypothermia. Use the energy equivalence of oxygen given in Problem 3.5-2 to estimate oxygen use under each condition.

3.5-4. **Microbes in hamburger.** Some virulent strains of E. coli can grow in ground beef and cause extreme sickness or death if the beef is not cooked properly. Assume beef is contaminated at the time of processing with three E. coli microbes. The beef is refrigerated for a week before it is cooked. Estimate the microbial population at that time.

3.5-5.   Growth of ingested microbes. The hamburger in the previous question is eaten. Unfortunately, the cooking process left 50 *E. coli* microbes unkilled. If the population level must reach 3×10^{16} before it causes illness in an adult human, how long do you estimate before this can occur?

3.5-6.   Pigging out. A man goes on an eating binge and gains 250 N. What effect will this have on his basal metabolic rate?

3.5-7.   Diet pills. A new diet pill is claimed to cause weight loss by doubling basal metabolic rate. How effective will this pill be without making any dietary adjustments? Show how you arrived at your answer.

3.5-8.   Heat production for sports. Choose five sports from Table 3.5.4. Compare heat production for these five. What impact would this have on the ventilation system design for each of these?

3.5-9. Q_{10} of peaches. Estimate the Q_{10} for heat evolution of stored peaches.

3.5-10.   Indirect calorimetry. Gagge and Nishi (1983) gave the relationship between oxygen uptake and metabolic rate in an adult human as

$$\dot{M} = 21.14\dot{V}_{O_2}[(0.23\dot{V}_{CO_2}/\dot{V}_{O_2}) + 0.77]$$

where $\dot{M}$ is the metabolic rate (kN m/sec); $\dot{V}_{O_2}$, the oxygen uptake (L/sec); and $\dot{V}_{CO_2}$, the carbon dioxide production rate (L/sec). The ratio $\dot{V}_{CO_2}/\dot{V}_{O_2}$ depends on the type of nutrient substrate being metabolized, but resting $\dot{V}_{CO_2}/\dot{V}_{O_2}$ is normally about 0.8.

   If the oxygen uptake is measured at 5.3 mL/sec, what is the metabolic rate of the resting subject? Estimate the body weight.

3.5-11.   Heat produced by compost. Compost mixture contains sludge (assumed to be $C_{10}H_{19}O_3N$) and woodchip bulking material ($C_6H_{10}O_5$) that react with oxygen in these ways

$$C_{10}H_{19}O_3N + 12.5O_2 \rightarrow 10CO_2 + 8H_2O + NH_3 + \text{energy}$$
$$C_6H_{10}O_5 + 6O_2 \rightarrow 6CO_2 + 5H_2O + \text{energy}$$

Energy produced is about 13.62×10^6 N m/kg O_2 used. Only 37.5 percent of dry sludge and 19 percent of dry woodchips are the biodegradable volatile solids given in the above chemical equations. In addition, wet sludge contains only about 20 percent solids (dry sludge) and woodchips are only about 52 percent solids (dry woodchips). A common mixture is 1 part wet sludge per 0.3 parts wet woodchips (by weight). For a compost pile of 1000 kg (1 metric ton) wet sludge, determine the total heat production.

3.6-1.   Time constant. If thermal capacity is 100^{-2} N m/°C and thermal resistance is 3.02°C sec/(N m), what value is the thermal time constant?

3.6-2. **Plotting cooling data.** To determine the thermal time constant of the system, you decide to determine experimentally the relationship between temperature and time as the object cools. When you plot the data on semilog graph paper, you discover that it plots as a straight line. What data should you plot on each axis of the graph in order to get a straight line, and what information can you obtain from the slope of the straight line?

3.6-3. **Temperature dependence of properties.** Properties for air are:

Temperature (°C)	0	10
Specific heat [N m/(kg °C)]	1.0048	1.0048
Viscosity (N sec/m^2)	1.72×10^{-5}	1.78×10^{-5}
Thermal conductivity [N m/(sec m °C)]	0.02423	0.02492
Density (kg/m^2)	1.293	1.246

Determine physical property values at 5°C.

3.6-4. **Interior convection.** A liquid with a specific heat of 6.5 N m/(kg °C) flows at a rate of 5 kg/min through a long pipe with 19.5 m^2 inside surface area. The entrance temperature of the fluid is 16°C and the exit temperature is 4°C. The average temperature of the surface of the pipe is 10°C. What is the value of the convection coefficient in the pipe?

3.6-5. **Starving animals.** What do we mean when we say that an exponential function fits the body weight of a starving animal better than does a linear function between body weight and starvation time?

3.6-6. **Survival time when starving.** Assuming that death comes when 50% of the body weight is lost, estimate the likely survival time of a 70 kg individual: (a) assuming a constant rate of weight loss; (b) assuming an exponential function of body weight with starvation time. Assume that adipose tissue, with an energy density of about 27×10^6 N m/kg, is lost while starving.

3.6-7. **Stored heat in a cow.** A cow of 4900 N body weight contains 60% water and produces 465 N m/sec of heat. If this cow were suddenly made into an adiabatic system, but could continue her heat production at a constant rate, how long would it take to raise her mean body temperature 1°C?

3.6-8. **Stored heat in water.** Shaking can raise the temperature of water. Assume that shaking for 5 min increased the temperature of 100 mL of water by 1°C. Assume also that the container was so well insulated that the heat loss by conduction was negligible and that the heat capacity of the container was negligible as compared with that of the 100 mL water. Under these assumptions, calculate the work of the 5-min shaking.

3.6-9. **Water temperature in a fish culture tank.** Van Waversveld et al. (1989) have given the heat production of fish at 20°C as $\dot{q} = 0.194\ m^{0.85}$, where $\dot{q}$ is heat production (N m/sec) and m is body mass (kg). A water-filled aquaculture tank 1.5 m tall and 2.4 m diameter resting on a concrete floor inside a dark room contains 100 kg of fish of 2 kg each. Room air is still and is maintained at 20°C. In order to supply oxygen demand in this recirculating system, air is bubbled through the water at a rate of 560 m³/day. Write a heat balance on this system and estimate water temperature.

3.6-10. **Refrigerator warming.** The cold compartment of a home refrigerator is about 63 cm high × 63 cm wide × 48 cm deep. Compare the rates at which the temperature rises in the center of the cold compartment when the door is open and (a) the refrigerator is empty, and (b) the refrigerator is full of food. State all assumptions.

3.6-11. **Dental drill.** Specifications for the handpiece of a dental drill are:

Speed ranges	8.3 and 333 rev/sec
Optimum air pressure at handpiece	310 kN/m²
Air consumption	9.9×10^{-4} m³/sec at 310 kN/m²
Material	Stainless steel
Cleaning	Clean with soap and water to remove debris prior to autoclaving; do not submerse in water

Estimate the heat production when this drill is used. A 5–6°C rise is lethal for tooth tissue. Recommend a means (including numerical amounts) to limit the temperature rise in the tooth.

3.6-12. **Regenerative thermal oxidation.** Thermal oxidation of volatile organic compound (VOC) vapors in the air is often used to clean exhaust air. Regenerative thermal oxidation, capable of up to 95% heat energy recovery requires the air stream containing the VOCs to be fed through a ceramic heat exchange bed, which heats the air before it enters the central combustion chamber. The hot, purified gases are then exhausted back through another heat exchanger, which absorbs the energy from the air stream and stores it for the next cycle. Contaminated air alternates between heat exchange beds; one is heating while the other is cooling. If 25 m³/sec of contaminated air is to be processed, and if the air begins at 25°C and is to be heated to 1000°C by the ceramic bed, what is your recommendation for the ceramic mass if the switch between heating and cooling does not occur any more quickly than $\frac{1}{2}$ hour? What shape would you make the ceramic material?

3.6-13. **Hot air blower.** Curing adhesives used in biomedical applications uses air heated to 130°C. If a hot air blower uses 8.3 L/sec of room temperature air to be heated and blown on the adhesive, calculate the required amount of heat.

3.7-1. **Limiting fluid.** Two fluids flow through a heat exchanger. Fluid A is hotter than fluid B when it enters. The following is a table of physical parameters. Which fluid limits the heat transferred through the heat exchanger?

	Fluid A	Fluid B	
Density	0.596	0.221	kg/m³
Thermal conductivity	0.0254	0.0484	N m/(sec m K)
Specific heat	1700	1499	N m/(kg K)
Viscosity	0.0228	0.0555	10^{-3} N/(m² sec)
Flow rate	122	136	kg/sec

3.7-2. **Counterflow heat exchange in the arm.** Consider the counterflow heat exchange diagrammed in Figure 3.7.2. Estimate the average convection coefficient values in the brachial artery and the two brachial veins located on both sides of the artery (venae comites). The artery has a lumen diameter of about 0.4 cm, with wall thickness of 1 mm; the veins have lumen diameters of 0.3 cm and wall thicknesses of 0.5 mm. All are about 30 cm long. Assume arterial blood flow is about 50 mL/min and that heat is transferred only between artery and veins. Draw the thermal resistance diagram for this system. Do you think the artery or the veins will have the larger convection coefficient value? Why?

3.7-3. **Cardiopulmonary bypass for dogs.** Repeat Example 3.7.1.4-1, except for dogs. Resting dogs with average body masses of 6.4 kg and 38.9°C body temperature have cardiac outputs of 1.82 L/min.

3.7-4. **Turkey facility ventilation.** A turkey production facility in Minnesota houses 20,000 birds. The ventilation system for this facility supplies about 5 m³/sec of fresh air from the outside. In the winter, ambient temperature can be very low, and excess heat is required to maintain an indoor temperature of 20°C. Specify a heat exchanger to transfer heat from the outlet air to the inlet air so that supplemental heat will not be required until outside air temperature falls below -20°C. How does condensation affect your calculations?

3.7-5.   **Recuperative thermal oxidation.** Thermal oxidation of air contaminated with volatile organic compounds (VOCs) sometimes uses tubular, plate, or shell-type heat exchangers to recover heat from the exhaust stream from the combustion chamber. Such systems are called recuperative thermal oxidation. For the conditions in Problem 3.6-12, specify a heat exchanger that can be used in this application.

3.7-6.   **Culture medium sterilization.** How much time would it take to sterilize plant tissue culture medium in a globe-shaped flask 10 cm in diameter? The autoclave is at 121°C.

3.7-7.   **Heating of a can of food.** A can of food at uniform initial temperature of 20°C is suddenly immersed in boiling water. The Heisler chart for a cylinder with the can radius gives a temperature difference ratio of 0.45, the Heisler chart for a slab with a thickness equal to the length of the can gives a temperature difference ratio of 0.30, the Heisler chart for a sphere with the radius of the can gives a temperature difference ratio of 0.35, and the Heisler chart for a slab with a thickness of one-half the height of the can gives a temperature difference ratio of 0.50. What is the overall temperature difference ratio of the can?

3.7-8.   **Cooling of hot dogs.** Hot dogs come from the cooker with a uniform temperature of 74°C. They are then sprayed with tap water to cool to about 32°C in 10 min. What is the equivalent convection coefficient of the water? If jumbo hot dogs (twice the diameter, 1½ times the length) are to be cooled after cooking, how long will it take?

3.7-9.   **Cooling beer.** Assuming natural convection heat transfer, does it take longer to cool a can of beer (or soda, if you will) in the refrigerator if the can is sitting upright or on its side?

3.7-10. **Caterpillar warming.** On cool spring mornings eastern tent caterpillars will bask together in the sun. In air temperatures of around 10°C, basking caterpillars can raise their body temperatures by about 30°C in 20 min. Write a heat balance on a caterpillar 3.8 cm long by 0.4 cm diameter and determine the effective convection coefficient. State all assumptions.

3.7-11. **Cooking french fries.** French fried potatoes are to be cooked in hot grease at 150°C. If the french fries are 1 cm × 1 cm × 8 cm, how long will it take their center temperatures to reach 66°C, assuming that they began at a uniform temperature of 22°C? Use a convection coefficient value typical for boiling.

3.7-12. **Flamingos.** Flamingos are large water-loving birds that have long (80 cm), thin (2 cm diameter) legs. They have average body temperatures of about 40°C. Fortunately, they are semitropical birds, because otherwise they would lose too much body heat from those legs. However, their legs are much longer than necessary to lose body heat. Assume that he feet are kept at ambient temperature (27°C). How short would their legs have to be to account for 90% of the total amount of heat lost from their legs?

3.7-13. **Fennec ears.** A diminutive fox called the fennec is native to Algeria, Libya, Morocco, and Tunisia. It is equipped with huge triangularly shaped ears 15 cm long by 7.6 cm wide by 3 mm thick. These ears help the fennec to locate prey by amplifying sounds and also help to dissipate the desert's ferocious heat. In ambient temperatures of 35°C, estimate the heat lost through the ears. Assume the natural convection coefficient value is 1.5 N/(°C sec m) and the fennec body temperature is 38°C.

3.7-14. **Thermal inactivation of pectin methyl esterase (PME).** PME is a very heat-resistant enzyme in orange juice that must be inactivated to maintain orange juice quality. From the following decimal reduction time (D) data (Tajchakavit and Ramaswamy, 1994), determine the temperature difference required to result in a tenfold change in D values (Z)

temperature (°C)	50	55	60	65
D value (sec)	53	15.8	10.2	4.2

3.7-15. **Sucrose inversion kinetics.** The same concepts used for sterilization can be applied to biochemical reaction kinetics. The Z value for sucrose inversion (the same as hydrolysis, or the splitting of sucrose into simpler sugars) has been found to be 31.4°C in the temperature range of 120–140°C (Pompei and Rossi, 1994). If sucrose is held at 120°C for 2 min, what is the equivalent amount of time for sucrose inversion at 140°C?

3.8-1. **Sugar concentration.** A sugar solution begins to freeze at a temperature of −1.65°C. What concentration of sugar is present?

3.8-2. **Salt concentration.** A salt solution begins to freeze at a temperature of −1.65°C. What concentration of salt (NaCl) is present?

3.8-3. **Polyethylene coating.** A shell-and-tube heat exchanger is used to transfer heat from water to brine. Because of the caustic nature of the brine, a 0.5 mm layer of high-density polyethylene is used to coat the brine-side surface. Estimate the equivalent convection coefficient.

3.8-4. **Freezing of blood.** It is proposed to freeze blood in cylindrical containers 7.5 cm in diameter and 25 cm long. The containers are to be put in liquid nitrogen at −70°C and agitated. The estimated convection coefficient value is 1500 N m/(sec m^2 °C). Estimate the freezing time.

3.8-5.   **Freezing of cartilage.** Human cartilage for orthopedic repair can be preserved by freezing until needed. If the cartilage is frozen in sheets 2 mm thick by 3 cm long by 2 cm wide, and the cartilage is frozen in liquid nitrogen at −70°C, estimate the freezing time.

3.8-6.   **Chilling and freezing of steaks.** Repeat Example 3.8.1.1-2, except for steaks 3 cm thick and 15 cm in diameter.

3.8-7.   **Sweat rate.** If all the heat produced by a person playing tennis had to be removed by sweating, what would be the required sweat rate? What would be your answer if you account for convective heat loss? Why is the actual sweat rate likely to be higher than this?

3.8-8.   **Evaporative cooling in the ostrich.** Crawford and Schmidt-Nielsen (1967) give data showing that respiratory evaporation in an ostrich exposed to high ambient temperatures can be expressed as

$$\dot{m} = 0.143\theta - 2.43, \qquad 20 \le \theta \le 50°C$$

where $\dot{m}$ is water evaporated (g/min); and θ, the ambient temperature. Estimate the rate of heat lost by respiratory evaporation at a temperature of 45°C.

3.8-9.   **Lyophilization of collagen.** Freeze drying (or lyophilization) is often used to preserve collagen. If this material is lyophilized in sheets 2 mm thick, estimate the sublimation time. If one face of the sheet was in contact with an impermeable surface, how would your estimate change? If the sheets are 1.5 m wide and 3 m long, how much heat must have been required to remove the water?

3.8-10.   **Biological vitrification.** Judge the suitability of the following for preservation by vitrification:

      liver
      kidney
      heart valves
      red blood cells
      bone marrow
      skin
      platelets
      embryos
      islets of Langerhans
      corneal tissue

3.9-1.   **Impossible question.** Under what conditions is heat loss by an animal impossible?

3.9-2. **Fermenting grape juice.** The maximum heat production in a 19,000 L tank of fermenting grape juice occurs soon after the beginning of fermentation. An ideal fermentation activity decreases linearly from maximum to zero in about 30 d. If the total amount of heat to be removed from the fermenting wine over these 30 d is 8.0×10^9 N m, and about 50% of the heat production is lost from the vat by convection, what is the cooling capacity of an internal cooling coil to remove excess heat? If the efficiency of the refrigeration unit is 15%, and the cost of the electricity to run the unit is 15 cents/kWh [4.17×10^{-6} cents/(N m)], how much cost must be added to each 1 L bottle of wine to cover the cost of cooling? Remember that the refrigeration unit will not be working to full capacity except at the beginning of fermentation.

3.9-3. **Skin blood heat loss.** Human skin blood flow can be estimated from

$$\dot{V}_{bl} = [\dot{M}/(\theta_{core} - \theta_{sk}) - 5.3A]/(3.49 \times 10^6)$$

where $\dot{V}_{bl}$ is the skin blood flow (m³/sec); $\dot{M}$, the net metabolic heat production (N m/sec); $(\theta_{core} - \theta_{sk})$, the gradient between core and mean skin temperatures (°C); and A, the body surface area (m²) (Kenny and Havenith, 1993). For a 540 N female sitting in still air at 19°C, write a heat balance for the skin and determine skin temperature.

3.9-4. **Soaking logs.** The wood veneer-making process is a series of steps in which logs are cut in half lengthwise, cooked in hot water, planed, sliced, dried, clipped, bundled, and shipped. After cutting, logs (2.4 m by 0.61 m diameter) are chained in groups of five and soaked in hot water vats at 27°C or higher for one or two days. The vats are 2.4 m × 2.4 m × 3.7 m and are insulated on the sides and bottoms by 15 cm of enclosed fiberglass insulation. Estimate the amount of heat required to maintain water temperature at 27°C.

REFERENCES

Alhamdan, A., S. Sastry, and J. Blaisdell, 1988, Experimental Determination of Free Convective Heat Transfer from a Mushroom-Shaped Particle Immersed in Water, Paper #88-6595, American Society of Agricultural Engineers, St. Joseph, MI 49085-9659.

Allen, C. W., 1973, *Astrophysical Quantites*, The Athlone Press, London.

Altman, P L., and D. S. Dittmer, 1968, *Metabolism*, Federation of American Societies for Experimental Biology, Bethesda, MD.

Anderson, G. S., R. Ward, and I. B. Mekjavić, 1995, Gender Differences in Physiological Reactions to Thermal Stress, *Eur. J. Appl. Physiol.* **71**: 95–101.

ASHRAE, 1977, *Handbook of Fundamentals*, Amer. Soc. Heating, Refrigerating, and Air Conditioning Engineers, New York, NY.

ASHRAE, 1985, Physiological Principles, Comfort and Health, in *Handbook of Fundamentals*, American Society for Heating, Refrigeration, and Air-Conditioning Engineers, Atlanta, GA pp. 8.1–8.36.

Astrand, P.-O., and K. Rodahl, 1970, *Textbook of Work Physiology*, New York, McGraw–Hill.

Bhamidipati, S., and R. K. Singh, 1994, Fluid to Particle Heat Transfer Coefficient Determination in a Continuous System, Paper #94-6542, American Society of Agricultural Engineers, St Joseph, MI.

Blaxter, K. L., N. M., Graham, and F. W. Weinman, 1959, Environmental Temperature, Energy Metabolism, and Heat Regulation in Sheep. III. The Metabolism and Thermal Exchange of Sheep with Fleeces, *J. Agr. Sci.* **52**: 41–49.

Burns, T. P., 1989, Lindeman's contradiction and the trophic structure of ecosystems. *Ecology* **70**: 1355–62.

Carslaw, H. S., and J. C. Jaeger, 1959, *Conduction of Heat in Solids*, Clarendon Press, Oxford.

Cena, K., 1974, Radiative Heat Loss from Animals and Man, in *Heat Loss From Animals and Man: Assessment and Control*, J. L. Monteith and L. E. Mount (Eds.), London, Butterworths, pp. 33–58.

Cena, K., and J. A. Clark, 1979, Transfer of Heat through Animal Coats and Clothing, in *International Review of Physiology, Environmental Physiology III*; Vol. 20, D. Robertshaw (Ed.), University Park Press, Baltimore, MD, pp. 1–42.

Charm, S. E., 1971, *Fundamentals of Food Engineering*, AVI, Westport, CT.

Clark, J. A., K. Cena, and J. L. Monteith, 1973, Measurements of the Local Heat Balance of Animal Coats and Human Clothing, *J. Appl. Physiol.* **35**: 751–54.

Clark, J. A., A. J. McArthur, and J. L. Monteith, 1981, The Physics of the Microclimate, in *Bioengineering, Thermal Physiology, and Comfort*, K. Cena and J. A. Clark (Ed.), Elsevier, Amsterdam, pp. 13–27.

Clark, S. P., Jr., 1962, Temperatures in the Continental Crust, in *Temperature: Its Measurement and Control in Science and Industry*, Part I, F. G. Brickwedde (Ed.), Reinhold, New York, pp. 779–90.

Crawford, E. C., Jr., and K. Schmidt-Nielsen, 1967, Temperature Regulation and Evaporative Cooling in the Ostrich, *Am. J. Physiol.* **212**: 347–53.

Fry, R. M., 1966, Freezing and Drying of Bacteria, in *Cryobiology*, H. T. Meryman (Ed.), Academic Press, New York, pp. 665–96.

Gagge, A. P., and Nishi, Y., 1983, Heat Exchange between Human Skin Surface and Thermal Environment, in *Handbook of Physiology: Reactions to Environmental Agents*, D. H. K. Lee (Ed.), American Physiological Society, Bethesda, MD, pp. 69–92.

Geankoplis, C. J., 1993, *Transport Processes and Unit Operations*, Prentice Hall, Englewood Cliffs, NJ.

Gebhart, B., 1971, *Heat Transfer*, McGraw–Hill, New York.

Gebremedhin, K. G., W. P. Porter, and R. G. Warner, 1984, Heat Flow through Pelage of Calves—A Sensitivity Analysis, *Trans. ASAE* **27**: 1140–43, 1149.

Goldman, R. F., 1967, Systematic Evaluation of Thermal Aspects of Air Crew Protective Systems, in *Behavioral Problems in Aerospace Medicine*, Conference Proceedings No. 25 of the Advisory Group for Aerospace Research and Development (AGARD), Paris, France.

Goldman, R. F., 1978, Prediction of Human Heat Tolerance, in *Environmental Stress*, L. J. Folinsbee, J. A. Wagner, J. F. Borgia, B. L. Drinkwater, J. A. Gliner, and J. G. Bedi (Eds.), Academic Press, New York, pp. 53–69.

Hammel, H. T., 1955, Thermal Properties of Fur, *Am. J. Physiol.* **182**: 369–76.

Heldman, D. R., 1975, *Food Process Engineering*, AVI, Westport, CT.

Heldman, D. R., 1992, Food Freezing, in *Handbook of Food Engineering*, D. R. Heldman and D. B. Lund (Ed.), Marcel Dekker, New York, pp. 277–315.

Henderson, S. M., and R. L. Perry, 1966, *Agricultural Process Engineering*, 2nd ed., Henderson and Perry, Davis, CA.

Holman, J. P., 1986, *Heat Transfer*, 6th ed., McGraw-Hill, New York, NY.

Holmer, I., and U. Bergh, 1981, Thermal Physiology of Man in the Aquatic Environment, in *Bioengineering, Thermal Physiology, and Comfort*, K. Cena and J. A. Clark, (Ed.), Elsevier, Amsterdam, pp. 145–56.

Hottel, H. C., 1930, Radiant Heat Transmission, *Mech. Eng.* **52**: 699–704.

Hottel, H. C., and R. B. Egbert, 1942, Radiant Heat Transmission from Water Vapor, *AIChE Trans.* **38**: 531–68.

Hottel, H. C., and A. F. Sarofim, 1967, *Radiative Transfer*, McGraw-Hill, New York, NY.

Howell, J. R., 1982, A Catalog of Radiation Configuration Factors, McGraw-Hill, New York, NY.

Ingersoll, L. H., O. J. Zobel, and A. C. Ingersoll, 1954, *Heat Conduction with Engineering, Geological, and Other Applications*, University of Wisconsin Press, Madison, WI.

Johnson, A. T., 1991a, *Biomechanics and Exercise Physiology*, John Wiley and Sons, New York.

Johnson, A. T., 1991b, "Curvefitting," in *Digital Biosignal Processing*, R. Weitkunat (Ed.), Elsevier, Amsterdam, pp. 309–35.

Kenney, W. L., and G. Havenith, 1993, Thermal Physiology of the Elderly and Handicapped, *J. Thermal Biol.* **18**: 341–44.

King, J. R., and D. S. Farner, 1961, Energy Metabolism, Thermoregulation, and Body Temperature, in *Biology and Comparative Physiology of Birds*, A. J. Marshall (Ed.), Academic Press, New York, pp. 215–88.

Kirk, G. D., 1984, Empirical Analysis of Mixed Convective Heat Transfer from a Sphere to Air, Unpublished Master of Science Thesis, University of Maryland, College Park, MD.

Kirk, G. D., and A. T. Johnson, 1986, Experimental Determination of Mixed Convective Heat Transfer from a Sphere to Air, *Intern. Comm. Heat Mass Transf.* **13**: 369–87.

Kleiber, M., 1975, *The Fire of Life*, Krieger, Huntington, NY.

Kreith, F., 1958, *Principles of Heat Transfer*, International Textbook Company, Scranton, PA.

Mahalingam, R., L. O. Tilton, and J. M. Coulson, 1975, Heat Transfer in Laminar Flow of Non-Newtonian Fluids, *Chem. Eng. Sci.* **30**: 921–29.

McCullough, E. A., 1993, Factors Affecting the Resistance to Heat Transfer Provided by Clothing, *J. Therm. Biol.* **18**: 405–7.

Meinel, A.B., and M. P. Meinel, 1976, *Applied Solar Energy*, Addison-Wesley, Reading, MA.

Meryman, H. T., 1966, Freeze-Drying, in *Cryobiology*, H. T. Meryman (Ed.), Academic Press, New York, pp. 609–63.

Moon, S. H., K. E. Felton, and A. T. Johnson, 1981, Optimum Tilt Angles of a Solar Collector, *Energy* **6**: 895–99.

Nishi, Y., and A. P. Gagge, 1970, Moisture Permeation of Clothing, A Factor Covering Thermal Equilibrium and Comfort, *ASHRAE Trans.* **76**(1): 137–45.

Okos, M. R., G. Narsimhan, R. K. Singh, and A. C. Weitnauer, 1992, Food Dehydration, in *Handbook of Food Engineering*, D. R. Heldman and D. B. Lund (Eds.), Marcel Dekker, New York, pp. 437–562.

Pelosi, M., 1972, Heat Exchangers, *Food Engineering* (February issue), pp. 79–85.

Pflug, I. J., and W. B. Esselen, 1963, Food Processing by Heat Sterilization, in: Heid, J. L., and M. A. Joslyn, 1963, *Food Processing Operations* Vol. II (AVI: Westport, CT), pp. 410–36.

Polley, S. L., O. P. Snyder, and P. Kotnour, 1980, A Compilation of Thermal Properties of Foods, *Food Technology* (November), pp. 76–94.

Pompei, C., and M. Rossi, 1994, Use of a Model Solution for the Evaluation of Heat Damage in Milk Treated in an Ultrahigh Temperature Heat Exchanger, *J. Agr. Food Chem.* **42**: 360–65.

Ramaswamy, H. S., and S. S. Sablini, 1995, Heat Transfer Coefficients in Cans During End-Over-End Processing as Influenced by System Operating Parameters, Poster #6, Northeast Agricultural/Biological Engineering Conference, Bar Harbor, ME.

Schneider, P. J., 1955, *Conduction Heat Transfer*, Addison-Wesley, Reading, Mass.

Scholander, P. F., V. Walters, R. Hock, and L. Irving, 1950, Body Insulation of Some Arctic and Tropical Mammals and Birds, *Biol. Bull.* **99**: 225–36.

Sheppard, J., 1995, personal communication, Agricultural and Biosystems Engineering Dept., MacDonald College, McGill University, Ste. Anne de Bellevue, Quebec, Canada.

Shuler, M. L., and F. Kargi, 1992, *Bioprocess Engineering, Basic Concepts*, Prentice Hall, Englewood Cliffs, NJ.

Sienko, M. J., and R. A. Plane, 1957, *Chemistry*, McGraw-Hill, New York, NY.

Singh, R. P., and D. R. Heldman, 1993, *Introduction to Food Engineering*, 2nd ed., Academic Press, New York.

Stolwijk, J. A. J., and J. D. Hardy, 1977, Control of Body Temperature, in *Handbook of Physiology, Section 9: Reactions to Environmental Agents*, D. H. K. Lee (Ed.), Williams and Wilkins, Baltimore, pp. 45–68.

Stuart, M. C., and P. J. Kiefer, 1975, Engineering Thermodynamics, in *Handbook of Engineering Fundamentals*, O. W. Eshbach and M. Souders (Eds.), John Wiley and Sons, New York, pp. 824–933.

Tajchakavit, S., and H. S. Ramaswamy, 1994, Microwave Inactivation Kinetics (Batch Mode) of Pectin Methyl Esterase in Orange Juice, Paper #94224, Northeast Agricultural/Biological Engineering Conference, Guelph, ON.

Tang, L., and A. T. Johnson, 1992, Mixed Convection About Fruits, *J. Agr. Engr. Res.* **51**: 15–27.

Tang, L., A. T. Johnson, and R. H. McCuen, 1991, Empirical Study of Mixed Convection about a Sphere, *J. Agr. Engr. Res.* **50**: 197–208.

Taylor, C. R., 1980, Mechanical Efficiency of Terrestrial Locomotion: A Useful Concept?, in *Aspects of Animal Movement*, H. Y. Elder and E. R. Trueman (Eds.), Cambridge University Press, London, pp. 235–44.

Taylor, J. R. E., 1981, Thermal Insulation of the Down and Feathers of Pygoscelid Penguin Chicks and the Unique Properties of Penguin Feathers, *The Auk* **103**: 160–68.

Tregear, R. T., 1965, Hair Density, Wind Speed, and Heat Loss in Mammals, *J. Appl. Physiol.* **20**: 796–801.

van Waversveld, J., A. D. F. Addink, G. Vandenth, and H. Smit, 1989, Heat Production of Fish: A Literature Review, *Comparative Biochemistry and Physiology A* **92**(2): 159–162.

Wathes, C. M., and J. A. Clark, 1981, Sensible Heat Transfer from the Fowl: Thermal Resistance of the Pelt, *Br. Poultry Sci.* **22**: 175–83.

Webster, A. J. F., 1974, Heat Loss from Cattle with Particular Emphasis on the Effects of Cold, in *Heat Loss from Animals and Man: Assessment and Control*, J. L. Monteith and L. E. Mount (Eds.), London, Butterworths, pp. 206–31.

4

MASS TRANSFER

"Everything is incredible, if you can skin off the crust of obviousness our habit put on it."

Aldous Huxley

4.1 INTRODUCTION

three modes
Mass is transferred from one location to another in three different ways. In diffusion mass transfer, a difference in concentration causes spontaneous molecular movement analogous to conduction heat transfer. Convection mass transfer occurs when mass is swept in bulk from one place to another by some physical force such as a pressure difference. Convection mass transfer is usually treated similarly to convection heat transfer. The third mode of mass transfer is not spontaneous, but instead occurs by active physical or biophysical processes and results in mass concentration against diffusion gradients. Examples of this are reverse osmosis used in food processing or concentration of potassium ions inside neural cellular membranes. In this chapter we will consider these processes and present means for analyzing them.

Mass transfer, like heat and fluid flow, obeys many of the same general laws and can be analyzed with many of the same concepts that were developed in Chapter 1 (see Table 4.1.1). Mass transfer usually deals with resistance, and sometimes with capacity; inertia is usually negligible. Unlike heat transfer, mass is physical and cannot normally be considered to be generated (although, sometimes, specific kinds of mass, such as water, are considered to be generated). Transient problems may be important in mass transfer, as in drying a product, but many more mass transfer design problems are steady state. One issue that can be confusing concerns the pertinent effort and flow variables.

comparisons

Diffusion and flow through nonporous membranes usually are treated as processes with effort variable equal to concentration difference (kg/m^3) and flow variable equal to mass flow rate (kg/sec). Convection and flow through porous membranes are usually treated as processes with effort variable equal to pressure difference (N/m^2) and flow variable equal to volume flow rate (m^3/sec). The latter can be treated with methods developed for fluid flow (Chapter 2) and flow through porous media (Section 1.8.1). As might be suspected, the uses of two possible effort variables and two possible flow variables can lead to confusion when describing mass transfer resistance and capacity.

Thus, depending on the type of problem, there may be a good deal of overlap between mass transfer and fluid flow. They can both share the same

Table 4.1.1 Systems Approach to Mass Transfer

Modes of action	Two: diffusion and convection (including dispersion)
Effort variable	There are two: chemical activity, often considered to be equivalent to concentration, and pressure; concentration is usually coupled with mass rate of flow, and pressure is used with volume flow rate
Flow variable	There are two: molecular particle flow, herein considered to be equivalent to rate of mass flow, and volume rate of flow; rate of mass flow is usually used with concentration difference and volume rate of flow with pressure difference
Resistance	Related to material physical properties
Capacity	Depends on solubility of mobile species in surrounding medium
Inertia	There is no significant inertia
Power	Not usually significant
Time	Convection is usually analyzed as steady state, diffusion can be either steady state or transient
Substance generation	Particular molecular species can be generated through chemical processes or added through mechanical or thermal processes
General field equation	Used for diffusion in porous medium, not used for convection
Balances written	Total mass balances; Materials balances for identifiable substances
Typical design problem	1. Determine required permeability for thin plastic film: work from product respiration and desired intrapackage gas concentrations
	2. Determine drying times: use equilibrium moisture curves and either thin layer or thick layer drying equations
	3. Determine dispersion of particulate or gaseous species within lungs or in ambient environment

effort and flow variables. Microscopic diffusion of mass, however, is completely different from fluid flow in that mass flow is not rapid and not in bulk. The power to move this mass is usually much smaller than for either heat or fluids.

In Table 4.1.1 are shown systems concepts applied to mass transfer; these can be compared to those for fluid flow (Table 2.1.1) and heat flow (Table 3.1.1). Design problems are not as extensive as they are for heat transfer; they are usually specific and confined. In this chapter the tools will be introduced to make at least a first cut in solving these problems.

4.2 MASS BALANCE

Basic to mass transfer is the mass balance

$$(\text{rate of mass in}) - (\text{rate of mass out}) + (\text{rate of mass generation})$$
$$= (\text{rate of mass accumulation}) \tag{4.2.1}$$

materials
balance

Mass generation is usually considered to be zero. However, as stated earlier in Section 1.7.3.1, a general mass balance can reduce to a materials balance on a particular species (for instance, water or sugar) as long as the species mantains its identity throughout the process. In the case of species generation, as with water formation or oxygen usage during respiration, the rate of mass generation term in Eq. 4.2.1 is nonzero. For more environmental applications and problems written in a clear and understandable style, see Vesiland (1997).

4.3 MOLECULAR DIFFUSION

random
molecular
movement

We have already had limited discussion of molecular diffusion mass transfer in Section 1.8.3. Diffusion occurs because all molecules at temperatures above absolute zero are subject to random motion. Since the motion is random, the number of molecules traveling in any particular direction at any particular time will depend on the total number of molecules present at the location of interest (Figure 4.3.1). If the total number of molecules is smaller, fewer molecules are expected to move simultaneously in any particular direction.

If there are two adjacent locations, 1 and 2, where there are more M molecules at 1 than 2, then the random movement of molecules determines that there are more Ms moving from 1 to 2 than from 2 to 1 at any time. Thus, over time, the number of M molecules in the two locations should equalize.

Because the net transfer of Ms from 1 to 2 is greatest when the disparity in numbers of Ms between the two locations is greatest, and because the net transfer decreases as the number of Ms in locations 1 and 2 approaches parity,

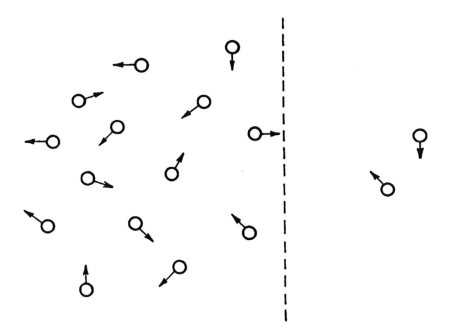

Figure 4.3.1. Volumes where there are larger numbers of molecules at the same temperature and relative freedom of movement tend to lose molecules to volumes where there are smaller numbers of molecules. In the diagram above the circles represent molecules and the arrows represent directions of movement. In the region to the left of the line (the line does not represent a membrane) five molecules have movement directions that have a rightward component, but in the region to the right of the line there is only one molecule with a leftward directional component. Thus there will be a net movement of four molecules to the right.

the time responses of concentrations in locations 1 and 2, and the net mass moving between 1 and 2, appear to be exponentially shaped. In Chapter 1 we saw that most transport processes are described by exponential time responses, and that these responses can be characterized by time constants. In Section 1.8c we saw that the time constant for mass diffusion into a site maintained at zero concentration is calculated by the product of the diffusion resistance and the original mass volume (Figure 4.3.2).

4.3.1 Fick's Laws

Diffusion resistance can be obtained directly from Fick's law and the particular geometrical situation to which it is applied. Fick's first law has been introduced in Eq. 1.8.29 as

Fick's first law

$$\dot{m}_M = -D_{MN} A \, \frac{dc_M}{dz} \tag{4.3.1}$$

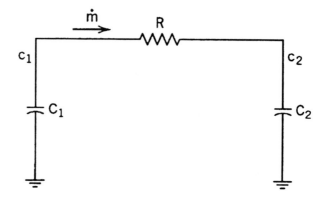

Figure 4.3.2. Movement of mass from one place to another by diffusion can be represented by two capacity elements separated by a resistance. If the concentration in C_1 is greater than the concentration in C_2, there will be a flow of particles through the resistance R. As soon as the concentrations in both capacity elements are equal, mass no longer moves from one to the other. The time constant for this process is given by the product of R with the series combination of C_1 and C_2, or $RC_1C_2/(C_1 + C_2)$.

where $\dot{m}_M$ is the mass flow rate of substance M (kg/sec); D_{MN}, the mass diffusivity of substance M through medium N (m^2/sec); A, the area perpendicular to the direction of diffusion (m^2); c_M, the concentration of substance M (kg/m^3); z, the Cartesian dimensions of length (m); and t, time (sec).

The negative sign indicates that positive mass transfer occurs with a negative concentration gradient.

The material through which diffusion occurs must be homogeneous in order to apply this form of Fick's first law. Concentration discontinuities can exist at boundaries, and diffusion can then occur at these boundaries from regions of lower concentration to regions of higher concentration. More exactly, diffusion occurs along a solute chemical potential gradient that does not change at the boundary. For example, if the affinity of the medium for the solute is greater on one side of a boundary than on the other side, a higher concentration of the solute can be maintained where the solubility is higher despite the solute concentration discontinuity at the boundary (see Section 4.3.3.1).

concentration discontinuities

Relative freedom of movement is thus also important to diffusion occurrence. If a molecule is tightly bound by chemical attraction, or if a molecule cannot move far before it encounters other molecules as obstacles (as in a crystal lattice), it is less likely to diffuse than if it were free to move in any direction for a long distance. It is difficult to measure directly the relative freedom of movement of molecules in any particular region, but movement freedom is a large contributor to a parameter called *chemical activity* that is sometimes applied to diffusion across interfaces (Figure 4.3.3).

freedom of movement

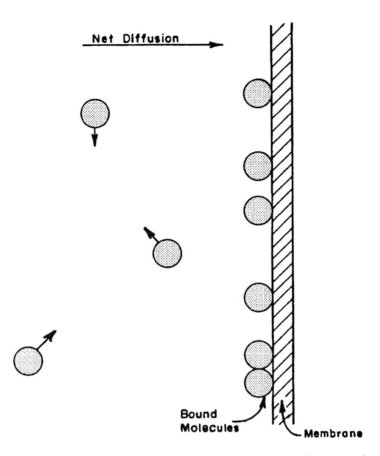

Figure 4.3.3. Diffusion of molecules from a region of lower concentration to a region of higher concentration can occur if the molecules in the higher-concentration region become bound, as to a membrane or to other ions, for instance. In this case the chemical activity of the bound molecules, manifested by vapor pressure, is less than the chemical activity of the mobile molecules. Diffusion takes place from a region of higher to lower activity.

Another formulation of Fick's law is often used for mass in gaseous form. This is

Fick for gases

$$\dot{m}_M = -D'_{MN}A\,\frac{dp_M}{dz} \qquad (4.3.2)$$

where p_M is the partial pressure of gas M (N/m^2); and D'_{MN}, the mass diffusivity [kg m/(sec N), or sec].

Gas partial pressure is the pressure exerted only by that particular gas, and that contributes to the total gas pressure exerted by the total number of gases. By the ideal gas law,

$$p_M = \left(\frac{m_M}{V}\right)\frac{RT}{M_M} = c_M RT/M_M \tag{4.3.3}$$

where m_M is the mass of substance M (kg); c_M, the concentration of M (kg/m^3); V, the volume (m^3); T, the absolute temperature (K); R, the universal gas constant [8314.34 N m/(kg mol K)]; and M_M, the molecular weight for substance M (kg/kg mol).

The two formulations for Fick's law are equivalent as long as appropriate values for diffusivity are also used. Thus

$$D'_{MN} = DM_M/RT \tag{4.3.4}$$

and Eq. 4.3.2 becomes

$$\dot{m}_M = -\frac{DM_M A}{RT}\frac{dp_M}{dz} \tag{4.3.5}$$

Expressing gaseous diffusion as a function of pressure leads to an understanding of diffusion in a binary mixture of gases. In this case, the two gases exist as mixtures within a closed volume. If two of these volumes are connected by an open tube, then gas movement can occur by two means: (1) diffusion, as long as a concentration difference occurs between the two original containers, and (2) convection (also called advection), as long as a pressure difference occurs between the two original containers. If no pressure difference exists, but the concentration of one gas is higher in one of the containers (thus the concentration of the second gas must be lower), then diffusion alone will occur. Molecules of both gases will move between containers from higher to lower concentrations to equalize concentrations of both gases in both containers (Figure 4.3.4). For the total pressure in both containers to remain constant requires that the net movement of molecules ($\dot{m}/M$) of both gases to be equal in magnitude but opposite in sign

$$\frac{\dot{m}_A}{M_A} = -\frac{\dot{m}_B}{M_B} \tag{4.3.6}$$

where $\dot{m}_A$, $\dot{m}_B$ are the net rates of mass movement of gas A or B (kg/sec); and M_A, M_B, the molecular weight of gas A or B (kg/kg mol).

Equating the mass movements as given by Fick's law

$$\frac{\dot{m}_A}{M_A} = -\frac{D_{AB}A}{M_A}\frac{dc_A}{dz} = -\frac{\dot{m}_B}{M_B} = \frac{D_{BA}A}{M_B}\frac{dc_B}{dz} \tag{4.3.7}$$

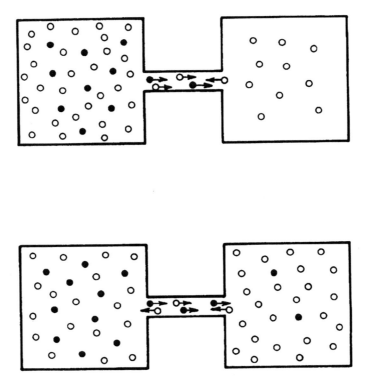

Figure 4.3.4. When total pressure in the left container is higher than total pressure in the right container, the net movement of particles is from left to right by convection (top). When total pressures in the two containers are equal, but concentration is higher in one than the other, molecules move from higher to lower concentration by diffusion (bottom)

But since the total pressure in both containers is constant and equal to the sum of the partial pressures of gases A and B,

$$p = p_A + p_B = c_A R'_A T + c_B R'_B T = \frac{c_A}{M_A} RT + \frac{c_B}{M_B} RT \qquad (4.3.8)$$

Differentiating the above equation,

$$0 = d\left(\frac{c_A}{M_A}\right) + d\left(\frac{c_B}{M_B}\right)$$

$$d\left(\frac{c_A}{M_A}\right) = -d\left(\frac{c_B}{M_B}\right)$$

$$(4.3.9)$$

Inserting this result into Eq. 4.3.7 gives

binary
diffusivities
equal

$$-D_{AB}A\,\frac{d(c_A/M_A)}{dz} = D_{BA}\,\frac{d(c_B/M_B)}{dz}$$

$$= -D_{BA}A\,\frac{d(c_A/M_A)}{dz} \qquad (4.3.10)$$

$$D_{AB} = D_{BA}$$

Thus the mass diffusivity of one binary gas within a second is numerically equal to the mass diffusivity of the second through the first. This result is useful because it means that the diffusing gas and the medium do not have to be identified. Measurement of one mass diffusivity is sufficient to cover two situations.

A point to be emphasized here is that mass diffusion depends directly on the number of *molecules* moving, but not on the amount of mass represented by those molecules. Thus most texts refer to Fick's equations in terms of the

molar flows

number of moles moving per unit time. Because design engineers are often interested in the amount of mass, rather than the number of moles, moved by diffusion, Fick's equation has been presented here in terms of mass flow rate. The difference between the two approaches is just the molecular weight of the diffusing substance.

Fick's first law, Eq. 4.3.1, is related by analogy to the general steady-state rate balance equation 1.7.18. Fick's second law describes non-steady-state diffusion, which we have already derived in Section 1.7.3.2 as the general field equation. Recall that a substance balance was written for a differential element with dimensions of dx, dy, and dz. The result was Eq. 1.7.26.

Two simplifications were made: (1) A constant medium parameter value (k) was assumed, and (2) no distributed source was located within the differential element. The result was termed the Fourier equation 1.7.28.

The Fourier equation for diffusion mass transfer is called Fick's second law

Fick's second
law

$$\frac{\partial^2 c_M}{\partial x^2} + \frac{\partial^2 c_M}{\partial y^2} + \frac{\partial^2 c_M}{\partial z^2} = \frac{1}{D_{MN}}\,\frac{\partial c_M}{\partial t} \qquad (4.3.11)$$

where c_M is the concentration of substance M (kg/m^3); D_{MN}, the mass diffusivity of substance M in medium N (m^2/sec); t, time (sec); and x, y, z, Cartesian dimensions (m).

In one dimension

$$D_{MN}\,\frac{\partial^2 c_M}{\partial x^2} = \frac{\partial c_M}{\partial t} \qquad (4.3.12)$$

If we assume steady-state conditions, $\partial c_M / \partial t = 0$, and

$$D_{MN} \frac{d^2 c_M}{dx^2} = 0 \qquad (4.3.13)$$

Integrating once gives

$$D_{MN} \frac{dC_M}{dx} = \text{const} \qquad (4.3.14)$$

which is the same form as Fick's first law, Eq. 4.3.1.

ecological
application

The previous discussion may have left the impression that diffusion only applies on a microscopic scale. Diffusion concepts have also been used on an ecological scale to analyze population dynamics: the movement of species into areas underpopulated for that species. The same concepts of random movement and concentration gradients important for molecular diffusion also apply there (Aronson, 1985; Nagylaki, 1989).

4.3.2 Mass Diffusivity

The rate of diffusion mass transfer depends very greatly on the value for mass diffusivity. Higher values of mass diffusivity give larger amounts of mass movement for any given time. Gases diffusing within gases have diffusivities of the order of $10^{-5} \, m^2/\text{sec}$. Gases and liquids diffusing within a liquid medium have diffusivities of the order of $10^{-9} \, m^2/\text{sec}$, decreasing, to $10^{-11} \, m^2/\text{sec}$ for larger molecules. Gases, liquids, and solids diffusing within a solid medium have mass diffusivities that are generally smaller yet, ranging from 10^{-10} to $10^{-34} \, m^2/\text{sec}$. This trend reflects the general loss of freedom of movement of molecules as they are surrounded by rare gas, somewhat dense liquid, and very dense solid mediums.

Most biological diffusion involves air or water media. Therefore, special attention will be paid to these in the discussion to follow.

experimentally
determined

4.3.2.1 Gas Diffusivities Values for mass diffusivities for various gases are determined largely by experiment. There are a number of ways to do this, including gas escape from a well, gas diffusion from one bulb to another through a capillary tube, and evaporation from a sphere. Although a number of theoretical and semitheoretical mathematical means have been developed that can be used to predict diffusivity values, they should not be depended upon to determine diffusivities if experimental values are available. Fortunately, there are only a few diffusivity values normally needed with biological systems, and most of these have been measured. A few of these appear in Table 4.3.1.

Of all the mathematical predictions of gaseous mass diffusivity values that have been proposed, the one by Fuller and colleagues (1966) has utility for the

Table 4.3.1   Selected Values of Binary Gas Diffusivities at Atmospheric Pressure

System	Temperature (°C)	Diffusivity (10^{-4} m²/sec)
Air–NH_3	0	0.198
Air–H_2O	0	0.220
	25	0.260
	42	0.288
Air–He	44	0.765
Air–CO_2	0	0.138
	3	0.142
	44	0.177
Air–H_2[a]	0	0.611
Air–O_2	20	0.192[b]
Air–N_2	20	0.193[b]
Air–C_2H_5OH	25	0.135
	42	0.145
Air–CH_3COOH	0	0.106
Air–n-hexane	21	0.080
Air–benzene	25	0.0962
Air–toluene	25.9	0.086
Air–n-butanol	0	0.0703
	25.9	0.087
H_2–N_2	25	0.784
	85	1.052
He–N_2	25	0.687
He–O_2	25	0.729
CO_2–N_2	25	0.167
CO_2–O_2	0	0.139
	20	0.153
H_2O–CO_2	0	0.138
	34.3	0.202
CO–N_2	100	0.318
O_2–N_2	0	0.181
	20	0.201[b]
O_2–H_2O	20	0.256[b]
N_2–H_2O	20	0.256[b]
He–H_2O	20	1.10[b]
He–CO_2	20	0.768[b]

[a] For example, the diffusivity for air–H_2 is 0.611×10^{-4} m²/sec.
[b] Estimated using Eq. 4.3.15.

biological engineer. This is a semiempirical method obtained by correlating many experimental data. Their equation is

calculated
diffusivity

$$D_{AB} = \frac{0.0103 T^{1.75}(1/M_A + 1/M_B)^{1/2}}{p[(V_A)^{1/3} + (V_B)^{1/3}]^2} \qquad (4.3.15)$$

where D_{AB} is the mass diffusivity of gas A within gas B (m²/sec); T, the absolute temperature (K); M_A, M_B, the molecular weights of A and B (kg mass/kg mole); p, the total pressure (N/m²); and V_A, V_B, the atomic diffusion volumes (m³).

atomic
diffusion
volumes

Values for atomic diffusion volumes have been given by Fuller and colleagues (1966) and other authors (Geankoplis, 1993; Johnson, 1991). Their values, however, were determined empirically only for a few substances, and experimental values of mass diffusivity are already available for many of the same substances. Selected atomic diffusion volumes are given in Table 4.3.2.

These diffusion volumes are not closely related to the actual molecular volume for a gas. All gases at STP conditions (0°C, 1 atm pressure) have a molar volume of 22.4×10^{-3} m³/mol (or 22.4 m³/kg mol) and contain 6.0235×10^{23} molecules/mol (Avagadro's number). Thus all gases should have the same molecular volume of 3.72×10^{-26} m³/molecule. Values in Table 4.3.2 are related, however, to the distances of molecules between collisions with other molecules (see Cussler, 1984; Johnson, 1991).

Table 4.3.2 Atomic Diffusion Volumes for Use in Estimating Binary Gaseous Diffusivities

Molecule	Diffusion Volume
Air	20.1
Cl_2	37.7
CO	18.9
CO_2	26.9
H_2	7.07
H_2O	12.7
He	2.88
N_2	17.9
N_2O	35.9
NH_3	14.9
O_2	16.6
SF_6	69.7
SO_2	41.1

Atomic and structural diffusion volume increments to be used to estimate diffusion volumes of additional compounds

C	16.5
Cl	19.5
H	1.98
O	5.48
N	5.69
S	17.0
Atomic ring	−20.2
Heterocycle ring	−20.2

A main value of Eq. 4.3.15 lies not in its ability to calculate primary values of mass diffusivities, but in its ability to correct experimental values for

temperature and pressure variations. From Eq. 4.3.15, we see that mass diffusivities vary directly with $T^{1.75}$ and inversely with pressure. Thus higher temperatures and lower pressures lead to higher mass diffusivity values. Interpolation of mass diffusivities for temperatures not included in a table then uses a linear approach

$$D_2 = D_1 + (D_3 - D_1)(T_2^{1.75} - T_1^{1.75})/(T_3^{1.75} - T_1^{1.75}) \qquad (4.3.16)$$

where D_2 is an unknown mass diffusivity at absolute temperature T_2 (m^2/sec); and $T_3 > T_2 > T_1$, the absolute temperature (K).

Interpolation of mass diffusivity for different pressures uses

$$D_2 = D_1 + (D_3 - D_1)\left(\frac{1}{p_2} - \frac{1}{p_1}\right) \bigg/ \left(\frac{1}{p_3} - \frac{1}{p_1}\right) \qquad (4.3.17)$$

where p is the absolute pressure (N/m^2); and $p_3 < p_2 < p_1$.

When diffusion is not binary, that is, when more than two gas constituents

are present, all at varying concentrations, then other procedures are required. In this case, each individual constituent diffusivity must be calculated before the diffusion of each gas can be determined.

As noted before, air is a multicomponent gas that does not change composition. Air can thus be treated as a separate gas. If the atmospheric composition does change, as it does in the alveolar regions of the lung, in some closed animal-rearing facilities, within plant leaf stomata, in controlled-atmosphere storages, or in fermentation vessels, then air cannot be treated as a binary gaseous constituent.

Proposed methods to estimate individual constituent mass diffusivity values all suffer from inaccuracies for one condition or another: Some do not reduce to self-diffusion when only one constituent is present in the mix; others predict concentration-dependent binary diffusivity values. It can be seen from Table 4.3.1 that mass diffusivity values for gases of normal interest to biological engineers fortunately do not differ from each other by great amounts, and, therefore, errors expected in calculating individual constituent mass diffusivity values probably will be about the same as for estimates of binary diffusion values. One means of calculating individual mass diffusivity values from binary mass diffusivity values is

$$D_A = \left[\sum_{j=1}^{N}\left(\frac{\mu_j}{D_{Aj}}\right)\right]^{-1} \qquad (4.3.18)$$

where D_A is the mass diffusivity of constituent A in the multicomponent system (m^2/sec); μ_j, the mole fractions of gaseous constituents (dimensionless); D_{Aj}, the binary mass diffusivity of constituent A in gas j (m^2/sec).

When calculating mass diffusivity values in multicomponent systems, corrections for temperature and pressure must be first made following the procedure outlined by Eqs. 4.3.16 and 4.3.17 and then multicomponent mass diffusivities calculated by Eq. 4.3.18.

EXAMPLE 4.3.2.1.-1. Estimation of Mass Diffusivity

Estimate the value for mass diffusivity of oxygen in nitrogen at 0°C.

Solution

Using Eq. 4.3.15 and values from Table 4.3.2, we have

$$D_{O_2-N_2} = \frac{0.01013(273.15\ \text{K})^{1.75}\left(\dfrac{1\ \text{kg}}{32\ \text{kg mol}} + \dfrac{1\ \text{kg}}{28\ \text{kg mol}}\right)^{1/2}}{(101,300\ \text{N/m}^2)[(16.6\ \text{m}^3)^{1/3} + (17.9\ \text{m}^3)^{1/3}]^2}$$

$$= 0.178 \times 10^{-4}\ \text{m}^2/\text{sec}$$

Remark

This value differs from the measured value appearing in Table 4.3.1 of 0.181×10^{-4} by less than 2%.

EXAMPLE 4.3.2.1-2. Temperature Correction for Oxygen–Air Diffusivity

Estimate the mass diffusivity for oxygen in air at 37°C.

Solution

From Table 4.3.1, we find the mass diffusivity for oxygen in air to be $0.192 \times 10^{-4}\ \text{m}^2/\text{sec}$ at 20°C. Using Eq. 4.3.15, we obtain

$$D_{37} = D_{20}(T_{37}/T_{20})^{1.75}$$

$$= (0.192 \times 10^{-4}\ \text{m}^2/\text{sec})(310\ \text{K}/293\ \text{K})^{1.75}$$

$$= 0.212 \times 10^{-4}\ \text{m}^2/\text{sec}$$

EXAMPLE 4.3.2.1-3. Temperature Correction for Water–Air Diffusivity

Estimate the mass diffusivity for water vapor in air at 37°C.

Solution

This is a case where values of mass diffusivity can be found for several temperatures, 0, 25, and 42°C. Using Eq. 4.3.16

$$D_{37} = D_{25} + (D_{42} - D_{25})(T_{37}^{1.75} - T_{25}^{1.75})/(T_{42}^{1.75} - T_{25}^{1.75})$$

$$= 0.260 \times 10^{-4} \text{ m}^2/\text{sec} + (0.288 - 0.260 \text{ m}^2/\text{sec})(10^{-4})$$

$$\times \frac{(310^{1.75} - 298^{1.75})}{(315^{1.75} - 298^{1.75})}$$

$$= 0.280 \times 10^{-4} \text{ m}^2/\text{sec}$$

EXAMPLE 4.3.2.1-4. Diffusion in an Air–Oxygen–Water Vapor Mixture

Estimate the mass diffusivities for each constituent in a mixture of 56.8% air, 40% oxygen, and 3.22% water vapor. The percentages are mole fractions, and the temperature is 37°C.

Solution

From the previous two examples, we have seen that the mass diffusivity values at 37°C for $D_{\text{air}-O_2} = 0.212 \times 10^{-4} \text{ m}^2/\text{sec}$ and $D_{\text{air}-H_2O} = 0.280 \times 10^{-4} \text{ m}^2/\text{sec}$. In a similar manner, we find $D_{O_2-H_2O} = 0.283 \times 10^{-4} \text{ m}^2/\text{sec}$.

Likewise, the self-diffusivities must be calculated using Eq. 4.3.15. For instance,

$$D_{O_2-O_2} = \frac{0.01013(310 \text{ K})^{1.75}(2 \text{ kg mol}/32 \text{ kg})^{1/2}}{(101300 \text{ N/m}^2)[2(17.6 \text{ m}^3)^{1.3}]^2}$$

$$= 0.220 \times 10^{-4} \text{ m}^3/\text{sec}$$

Similarly, $D_{H_2-H_2O} = 0.351 \text{ m}^2/\text{sec}$ and $D_{\text{air}-\text{air}} = 0.203 \text{ m}^2/\text{sec}$.
Using Eq. 4.3.18

$$D_{O_2} = \left(\frac{\mu_{\text{air}}}{D_{O_2-\text{air}}} + \frac{\mu_{H_2O}}{D_{O_2-H_2O}} + \frac{\mu_{O_2}}{D_{O_2-O_2}} \right)^{-1}$$

$$= \left(\frac{0.568}{0.212 \times 10^{-4}} + \frac{0.0322}{0.283 \times 10^{-4}} + \frac{0.400}{0.220 \times 10^{-4}} \right)^{-1}$$

$$= 0.217 \times 10^{-4} \text{ m}^2/\text{sec}$$

Also, by the same method,

$$D_{air} = \left(\frac{0.568}{0.203 \times 10^{-4}} + \frac{0.0322}{0.280 \times 10^{-4}} + \frac{0.400}{0.212 \times 10^{-4}} \right)^{-1}$$

$$= 0.208 \times 10^{-4} \ m^2/sec$$

$$D_{H_2O} = \left(\frac{0.568}{0.280 \times 10^{-4}} + \frac{0.0322}{0.351 \times 10^{-4}} + \frac{0.400}{0.283 \times 10^{-4}} \right)^{-1}$$

$$= 0.283 \times 10^{-4} \ m^2/sec$$

4.3.2.2 Liquid Diffusivities

E. L. Cussler (1984) states:

"In chemistry, liquid diffusion limits the rate of acid–base reactions; in physiology, diffusion limits the rate of digestion; in metallurgy, diffusion can control the rate of surface corrosion; in industry, diffusion is responsible for the rates of liquid–liquid extractions. Diffusion in liquids is important because it is slow."

liquid diffusion is slow

Liquid diffusion is much slower than gaseous diffusion because liquid molecules are more closely spaced than are gas molecules. Collisions between molecules occur much more frequently in the liquid phase than in the gaseous phase, thus impeding diffusion more. Mass diffusivities are about 10^4 higher for gases than for liquids as a result. Somewhat compensating for the lower

higher concentrations

diffusivities is the fact that concentrations are much higher in liquids. Take, for example, liquid water and steam at 100°C: liquid water has a density (or concentration) of about 960 kg/m^3; steam has a density of 0.60 kg/m^3. The mass concentration in the liquid phase is about 1500 times greater than in the gaseous phase.

Values for liquid diffusion coefficients appear in Table 4.3.3 (see also Lightfoot, 1995). As with gases, it is best to use experimentally determined diffusivity values when available. When experimental values at desired conditions are not available, the Stokes–Einstein equation may be used to correct tabulated values. Also, it can be seen from Table 4.3.3 that $D_{AB} \neq D_{BA}$ as was true with gases.

The Stokes–Einstein equation was derived assuming a rigid solute sphere diffusing in a continuum of solvent. Hence the solute molecules must be at least five times larger than solvent molecules. The Stokes–Einstein equation is

Stokes–
Einstein
equation

$$D_{AB} = \frac{\sigma T}{2\pi\mu V_A^{1/3}} \tag{4.3.19}$$

where D_{AB} is the mass diffusivity of solute A in solvent B (m^2/sec); σ, the Boltzmann's constant ($=1.38 \times 10^{-23}$ N m/K); T, the absolute temperature

Table 4.3.3 Selected Values for Mass Diffusivities in Dilute Solutions

Solute	Solvent	Temperature	Diffusivity ($10^{-9} m^2/sec$)
NH_3	Water	12	1.64
		15	1.77
		25	2.00
O_2	Water	18	1.98
		25	2.41
		37	3.00
O_2	Blood plasma	37	2.00
O_2	Whole blood	37	1.40
O_2	Red cell interior	37	0.95
CO_2	Water	25	2.00
Cl_2	Water	25	1.44
H_2	Water	25	4.80
Methyl alcohol[a]	Water	15	1.26
		25	1.60
Ethyl alcohol	Water	10	0.84
		25	1.24
n-Propyl alcohol	Water	15	0.87
Formic acid	Water	25	1.52
Acetic acid	Water	9.7	
		25	0.769
			1.26
Caffeine	Water	—	0.63
Propionic acid	Water	25	1.01
HCl (gas)	Water		
(9 kg mol/m^3)		10	3.30
(2.5 kg mol/m^3)		10	2.50
		25	3.10
Benzoic acid	Water	25	1.21
Acetone	Water	25	1.28
Acetic acid	Benzene	25	2.09
Urea	Ethanol	12	0.54
Water	Ethanol	25	1.13
Potassium chloride	Water	25	1.870
Potassium chloride	Ethylene glycol	25	0.119
Glucose	Water	25	0.69
Glycerol	Water	25	0.94
H_2S	Water	25	1.61
Lactose	Water	25	0.49
Maltose	Water	25	0.48
Nicotine	Water	25	0.60
Nitric acid	water	25	2.98
N_2	Water	25	1.90

(*continued*)

Table 4.3.3 (*continued*)

Nitrous oxide	Water	25	1.80
Oxalic acid	Water	25	1.61
Sucrose	Water	25	0.56
	Intracellular fluid	—	0.20
SO_2	Water	25	1.70
Sulfuric acid	Water	25	1.97
Urea	Water	25	1.37
Water	Glycerol	25	0.021
Most organic environmental pollutants	Water	—	≈ 1
Fibrinogen	Blood serum	37	0.020
Ovalbumin	Water	25	0.078
Hemoglobin	Water	25	0.069
Sorbitol	Water	—	0.94
	Intracellular fluid	—	0.50
Methylene blue	Water	—	0.40
	Intracellular fluid	—	0.15
Inulin	Water	—	0.15
	Intracellular fluid	—	0.03
Actin	Water	—	0.053
	Intracellular fluid	—	0.0003
Bovine serum	Water	—	0.069
Albumin	Intracellular fluid	—	0.001

^{a} For example, the diffusivity for methyl alcohol in water is $1.26 \times 10^{-9} \, \text{m}^2/\text{sec}$.

(K); μ, the solvent viscosity [kg/(m sec) or N sec/m^2]; V_A, the effective solute volume (m^3).

hydrated molecules

Macromolecules, such as proteins present in aqueous solution, are often surrounded by water of hydration: Molecules of water attach themselves by weak chemical bonds to local charges on the molecule surface (Figure 4.3.5). These hydrated molecules are much larger than unhydrated molecules and the volume, V_A, in the above equation must be the volume of the entire hydrated complex. Large molecules are often not spherical in shape, either, and thus the cube root of the volume gives only an average molecular radius.

viscosity effect

Equation 4.3.19 predicts an inverse relationship between solvent viscosity and mass diffusivity. However, diffusion becomes independent of viscosity for very viscous materials. The diffusion of sugar in gelatin, for instance, is very nearly equal to the diffusion of sugar in water.

gels

Gels are porous semisolid materials often found in biological systems or formed by technicians to be used with biological systems. Gels are composed of macromolecules in dilute aqueous solution that form a loose structure enclosing water-filled pores. The main effect of this structure is to increase the path length for diffusion (Table 4.3.4). Because the gel structure does not

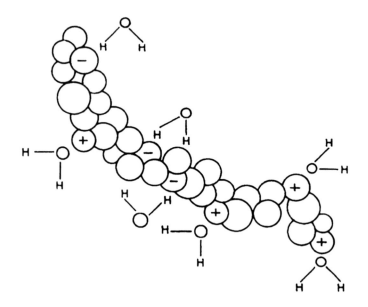

Figure 4.3.5. The water molecule is highly polar, with the oxygen end of the molecule negative relative to the hydrogen end. This polarity makes it easy for water molecules to attach more or less strongly to locally charged regions of large molecules.

move when the solute diffuses, there is a net movement of mass, called mass convection, accompanying the diffusion process. Simultaneous diffusion and convection is treated in Section 4.7.2.

When the diffusing solute is a macromolecule, the Stokes–Einstein equation 4.3.19 cannot be used to determine mass diffusivity because it is

macro-
molecules

inaccurate. Several modifications to Eq. 4.3.19 have been proposed for substances with molecular weights above 1000 or for molar volumes greater than 0.5 m³/kg mol. One of these replaces the $V_A^{1/3}$ with $M_A^{1/3}$, where M_A is the molecular weight of the macromolecular solute. This states that mass diffusivity is inversely related to the cube root of the molecular weight. Nevertheless, this modification is not sufficiently accurate for molecules with nonspherical shapes. Many common macromolecules, including long-chain proteins, polysaccharides, and polypeptides, are more linear than spherical.

biological
solutions

Biological solutions (Table 4.3.5) are ordinarily very complex. There are often many different types of macromolecules present and many different types of smaller solutes present. Some of these are ionic, some are not; some are polar, some are not; some are hydrated, some are not. Some macromolecules can interfere with movement of both other macromolecules and other smaller solutes. Some macromolecules contain binding sites for smaller solutes. In this case the movement of the smaller solute becomes linked strongly to the movement of the macromolecules. An example of this is the binding of atomic oxygen to hemoglobin. About 100 times more oxygen is

Table 4.3.4 Selected Values for Mass Diffusivities for Dilute Solutes in Water-Based Biological Gels

Solute	Gel	Wt% Gel in Solution	Temperature (°C)	Diffusivity $(10^{-9}\,m^2/sec)$
Sucrose	Gelatin	0	5	0.285
		3.8	5	0.209
		10.35	5	0.107
		5.1	20	0.252
Urea	Gelatin	0	5	0.880
		2.9	5	0.644
		5.1	5	0.609
		10.0	5	0.542
		5.1	20	0.859
Methanol	Gelatin	3.8	5	0.626
Urea	Agar	1.05	5	0.727
		3.16	5	0.591
		5.15	5	0.472
Glycerin	Agar	2.06	5	0.297
		6.02	5	0.199
Dextrose	Agar	0.79	5	0.327
Sucrose	Agar	0.79	5	0.247
Ethanol	Agar	5.15	5	0.393
Salt (NaCl)	Agarose	0	25	1.511
(0.05M)		2	25	1.398
Water	Potato starch	20	25	0.024
Water	Gelatinized corn	—	70	0.030

[a] For example, the diffusivity for methanol in gelatin is $0.626 \times 10^{-9}\,m^2/sec$.

transported in the blood as oxygen bound to hemoglobin than as oxygen in solution.

The Stokes–Einstein equation was formulated assuming an infintely dilute solution. When solute concentration increases, the movement of solute particles becomes linked, and the flow of some solute particles enhances the ability of other particles to flow. Thus mass diffusivity tends to increase as concentration increases. The mass diffusivity situation is really not always that simple, however, and liquid mass diffusivity values may vary by several hundred percent and sometimes with a minimum or maximum. For the design engineer who has no better experimental data to work from, a concentration correction may be applied to the infinite-dilution liquid mass diffusivity value

concentrated solutions

concentration correction

$$D'_{AB} = D_{AB}(1 + 1.5\phi) = D_{AB}(1 + 0.015\ \text{vol}\ \%) \tag{4.3.20}$$

where D'_{AB} is the mass diffusivity in concentrated solution (m^2/sec); D_{AB}, the mass diffusivity value in dilute solution (m^2/sec); ϕ, the volume fraction of the solute (m^3/m^3); and vol %, the volume percent $(m^3/100\,m^3$; see Table 4.3.12).

Table 4.3.5 Selected Values for Mass Diffusivities for Dilute Biological Solutes in Aqueous Solution

Solute	Molecular Mass	Temperature (°C)	Diffusivity (m²/sec)
Urea	60.1	20	1.20×10^{-9}
		25	1.378×10^{-9}
Glycerol	92.1	20	0.825×10^{-9}
Glycine	75.1	25	1.055×10^{-9}
Sodium caprylate	166.2	25	8.78×10^{-10}
Bovine serum albumin	67,500	25	6.81×10^{-11}
Urease	482,700	25	4.01×10^{-11}
		20	3.46×10^{-11}
Soybean protein	361,800	20	2.91×10^{-11}
Lipoxidase	97,440	20	5.59×10^{-11}
Fibrinogen, human	339,700	20	1.98×10^{-11}
Human serum albumin	72,300	20	5.93×10^{-11}
γ-Globulin, human	153,100	20	4.00×10^{-11}
Creatinine	113.1	37	1.08×10^{-9}
Sucrose	342.3	37	0.697×10^{-9}
		20	0.460×10^{-9}

Source: From *Transport Processes and Unit Operations* 3/E by Geankoplis, D. J., © 1978, Reprinted by permission of Prentice-Hall, Inc., Upper Saddle River, NJ.

EXAMPLE 4.3.2.2-1. Molecular Volume for Sucrose

Determine the effective volume for a sucrose molecule in dilute solution in water at 25°C.

Solution

We may use the Stokes–Einstein equation 4.3.19 to give us the desired value. From Table 4.3.3 we find the diffusivity value to be $0.56 \times 10^{-9}\,\text{m}^2/\text{sec}$, and from Table 2.4.4 we find the viscosity of water at 20°C to be $1.0050 \times 10^{-3}\,\text{kg/(m sec)}$ and at 30°C the viscosity value is $8.007 \times 10^{-4}\,\text{kg/(m sec)}$. Applying the temperature correction Eq. 2.4.6,

$$
\mu_{25} = \mu_{20} + (\mu_{30} - \mu_{20}) \left(\frac{\dfrac{1}{T_{25}} - \dfrac{1}{T_{20}}}{\dfrac{1}{T_{30}} - \dfrac{1}{T_{20}}} \right)
$$

$$
= 1.0050 \times 10^{-3}\,\frac{\text{kg}}{\text{m sec}} + (8.007 \times 10^{-4} - 1.0050 \times 10^{-3})
$$

$$\times \frac{\text{kg}}{\text{m sec}} \frac{\left(\dfrac{1}{298} - \dfrac{1}{293} \right)}{\left(\dfrac{1}{303} - \dfrac{1}{293} \right)}$$

$$= 9.011 \times 10^{-4} \text{ kg/(m sec)} = 9.011 \times 10^{-4} \text{ N sec/m}^2$$

Rearranging Eq. 4.3.19,

$$V_{\text{sucrose}} = \left(\frac{\sigma T}{2\pi\mu D_{\text{sucrose}-\text{H}_2\text{O}}} \right)^3$$

$$= \left(\frac{(1.38 \times 10^{-23} \text{ N m/K})(298 \text{ K})}{2\pi(9.011 \times 10^{-4} \text{ N sec/m}^2)(0.56 \times 10^{-9} \text{ m}^2/\text{sec})} \right)^3$$

$$= 2.18 \times 10^{-27} \text{ m}^3$$

Remark

This example shows how one physical property can be determined indirectly from other measurements.

EXAMPLE 4.3.2.2-2. Mass Diffusivity of Ethanol in Water

The mixture of ethanol in water is a very historical curative for many common human complaints. Estimate the mass diffusivity of ethanol in water at 25°C using the Stokes–Einstein equation.

Solution

The difficulty in applying the Stokes–Einstein equation is knowing the value for molecular volume to use in the equation. Molecular weights of substances are easy enough to determine, but substituting molecular weights for effective molecular volumes is not successful. So we will use the Stokes–Einstein equation to give a ratio of mass diffusivities for two substances, for one of which the mass diffusivity value is known and the other is unknown. From Table 4.3.3 we select the value for oxygen in water at 25°C as known. Then

$$D_{\text{Eth}} = D_{\text{O}_2}(V_{\text{O}_2}/V_{\text{Eth}})^{1/3}$$

because values for viscosity and temperature are the same for both mass diffusivities.

From Table 4.3.2, we find $V_{\text{O}_2} = 16.6$ and V_{eth} for ethanol (C_2H_5OH) is

$$V_{\text{Eth}} = 2V_c + 6V_H + V_O = 2(16.5) + 6(1.98) + 5.48$$

$$= 50.36$$

Thus

$$D_{Eth} = (2.41 \times 10^{-9} \text{ m}^2/\text{sec})(16.6/50.36)^{1/3}$$

$$= 1.66 \times 10^{-9} \text{ m}^2/\text{sec}$$

Remarks

Comparing the calculated value to the value $(1.24 \times 10^{-9} \text{ m}^2/\text{sec})$ in Table 4.3.3 shows that there is a considerable error when using this technique. However, results are at least somewhat consistent: mass diffusivity for acetic acid (CH_3CO_2H) calculated this way becomes $1.65 \times 10^{-9} \text{ m}^2/\text{sec}$, and mass diffusivity for nitric acid (HNO_3) becomes $2.12 \times 10^{-9} \text{ m}^2/\text{sec}$.

---

small magnitudes

4.3.2.3 Solid Diffusivities   Values for mass diffusivities in solids are smaller than those in liquids by several orders of magnitude. They also span a wide range of magnitudes. There are many types of solid structures, and attempts to use theory to derive equations to predict mass diffusivity values in some of these types (for example, crystalline lattices) have been made (Cussler, 1984). However, there is not a unifying or basic equation for solids similar to the Stokes–Einstein equation in liquids.

lack of theory

Since the purpose for our study of mathematical predictive equations for mass diffusivity values has mainly been to be able to extend experimentally obtained values to conditions other than those for which they were obtained, this lack of equations for solids is not an insurmountable obstacle. It means, however, that mass diffusivity experimental values must be known for conditions close to those for which the values will be used.

temperature effects

Temperature affects mass diffusivity values for some solids diffusing within other solids greatly and nonlinearly. For some cases, an increase in temperature results in a large increase in mass diffusivity. For other cases, the mass diffusivity value varies as the inverse of the absolute temperature. Mass diffusivities for gases dissolving in a solid and moving by diffusion are related to absolute temperature by an Arrhenius equation ($D_{AB} \propto e^{-k/T}$). Values for mass diffusivity appear in Table 4.3.6. Unlike gas diffusivities, D_{AB} does not equal D_{BA} (see also Section 4.3.3.2).

When a gas or liquid diffuses into a solid material, there is usually no equimolar counter diffusion. Thus there is a net mass movement (convection) into the solid. See Section 4.7.2.

4.3.2.4. Porous Solids   Many biological solids are porous rather than completely and uniformly solid. Liquids and gases moving through porous solids can often travel faster through the pores, which are liquid or gas filled, than through the solid particles (Figures 4.3.6 and 4.3.7). Two parameters are used to describe the diffusion path through such a porous solid. The first is the void fraction ε, defined as the ratio of pore area to total cross-sectional area

Table 4.3.6 Diffusivities and Permeabilities in Solids[a]

Solute	Solid	Temperature (°C)	Diffusivity (m²/sec)	Solubility $\left(\dfrac{m^3 \text{ gas } m^2}{m^3 \text{ solid N}}\right)$	Permeability $\left(\dfrac{m^3 \text{ gas } m}{\text{sec N}}\right)$
O_2	Vulcanized rubber	25	2.1×10^{-10}	6.9×10^{-7}	1.50×10^{-16}
N_2	Vulcanized rubber	25	1.5×10^{-10}	3.5×10^{-7}	5.33×10^{-17}
CO_2	Vulcanized rubber	25	1.1×10^{-10}	8.9×10^{-6}	9.97×10^{-16}
O_2	Polyethylene	30			4.12×10^{-17}
N_2	Polyethylene	30			1.50×10^{-17}
O_2	Nylon	30			2.86×10^{-19}
N_2	Nylon	30			1.50×10^{-19}
Air	English leather	25			$1.48 - 6.71 \times 10^{-10}$
H_2O	Wax	33			1.58×10^{-16}
H_2	Cellophane	38			$0.898 - 1.80 \times 10^{-15}$
H_2O	Human skin	37	1.2×10^{-13}		

[a] Gas volumes at standard temperature and pressure (STP: 0°C and 101.3 kN/m²) conditions.

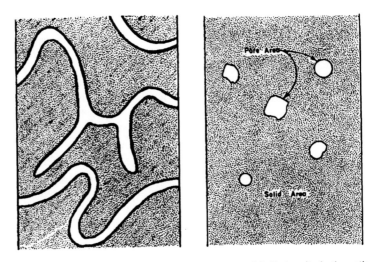

Figure 4.3.6. Tortuosity and void fraction in a porous solid. Tortuosity is the ratio of a path length through pores to the thickness of the solid, representing a straight path from one side to the other. Void fraction is the ratio of pore area to total surface area of the solid.

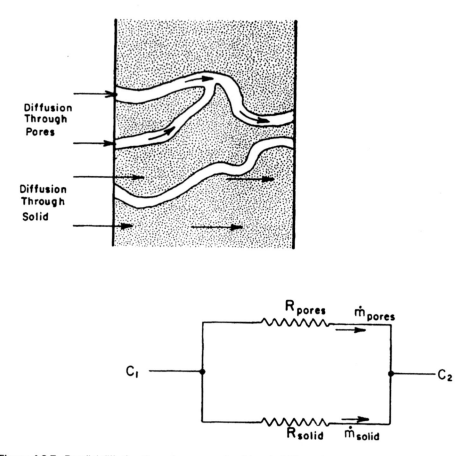

Figure 4.3.7. Parallel diffusion through pores and solid material is equivalent to flow through two parallel resistors. Resistance to diffusion exists for both solid and pore, but the resistance offered by the pores is often smaller. More mass flows through the lower resistance.

porosity

(Figure 4.3.6). Since liquid or gas diffusion is much faster than solid diffusion, the effective area for diffusion is εA, where A is the total cross-sectional area of the porous solid. The value for ε can take on any value between zero and one, but a value of zero denotes a nonporous solid, with diffusion completely through solid material, and a value of one means that there is no solid material at all. Thus ε is practically limited to the region between 0.2 and 0.8.

tortuosity

The other parameter of importance is the tortuosity τ. This parameter is used to correct for the diffusion path longer than the width of the solid because the pores are often anything but straight. Tortuosity values can be any numbers greater than or equal to one, but are often between 1.5 and 5. Diffusion between two parallel planes has been presented as

$$\dot{m}_A = \frac{AD_{AB}(c_1 - c_2)}{(x_2 - x_1)} \tag{4.3.21}$$

where $\dot{m}_A$ is the mass flow rate of substance A (kg/sec); A, the cross-sectional area (m^2); c, the concentration (kg/m^3); x, the distance (m); and D_{AB}, the mass diffusivity of substance A in medium B filling the pores (m^2/sec).

Applying both area and path length corrections, as indicated previously,

$$\dot{m}_A = \frac{\varepsilon A D_{AB}(c_1 - c_2)}{\tau(x_2 - x_1)} \qquad (4.3.22)$$

leading to an equivalent form for mass diffusivity

$$D'_{AB} = D_{AB}(\varepsilon/\tau) \qquad (4.3.23)$$

Both ε and τ are usually experimentally measured, and the ratio (ε/τ) is often obtained by inference from experimental measures of D_{AB} and D'_{AB}.

It is important to note a conceptual point. Diffusion through the liquid- or gas-filled pores occurs in parallel with diffusion through the solid material. The previous analysis assumes negligible solid diffusion. If this is not the case, then the two diffusion resistances must be treated as a parallel combination (Figure 4.3.7). Thus total diffusion resistance is

parallel
resistance
paths

$$1/R_{\text{tot}} = 1/R_{\text{p}} + 1/R_{\text{s}}$$
$$= \frac{D_{AP}\varepsilon A}{\tau L} + \frac{D_{AS}A(1 - \varepsilon)}{L} \qquad (4.3.24)$$

where R_{tot} is the total diffusion resistance (sec/m^3); R_{p}, the resistance of pores (sec/m^3); R_{s}, the resistance of solid material (sec/m^3); τ, the tortuosity (unitless); L, the width of material (m); D_{AP}, the mass diffusivity of substance A in pore medium B (m^2/sec); ε, the void fraction (unitless); A, the total cross-sectional area (m^2); and D_{AS}, the mass diffusivity of substance A in solid material C (m^2/sec).

4.3.2.5 Knudsen Diffusion

If the pores are relatively small compared to the mean free path of the diffusing species, then another type of diffusion occurs wherein the rate of movement of each diffusing molecule is influenced strongly, perhaps even determined, by collisions with the walls of the pores instead of collisions with other molecules. Such a process is called Knudsen diffusion (Figure 4.3.8).

collisions with
walls

Diffusing liquids hardly ever undergo Knudsen diffusion; the mean free paths (the distances molecules travel between collisions) of liquid molecules are hardly ever more than a few angstroms (10^{-10} m). Gas molecules, however, especially at low pressures or high temperatures, can have mean free paths of several hundreds or thousands of angstroms.

important for
gases

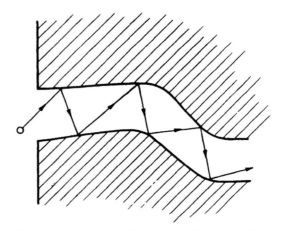

Figure 4.3.8. Molecules interacting with the walls of a pore. This is called Knudsen diffusion.

Gaseous mean free paths can be determined from the kinetic theory of gases as

mean free
path

$$L = \frac{3.2\mu}{p} \sqrt{\frac{RT}{2\pi M}}$$ (4.3.25)

where L is the mean free path (m); μ, the viscosity of the gas (N sec/m²); p, the total pressure (N/m²); R, the gas constant [8314.34 kg m²/sec² kg mol K)]; T, the absolute temperature (K); and M, the molecular mass (kg/kg mol).

Notice that there are two temperature dependencies in this equation for mean free path. The first is the obvious inclusion of T, the absolute temperature. The second is the strong temperature dependence of viscosity: gaseous viscosities increase greatly as temperature increases (see Section 2.4.1). Thus mean free path also increases greatly with temperature increases.

temperature
effects

The Knudsen number is

Knudsen
number

$$\text{Kn} = L/d$$ (4.3.26)

where Kn is the Knudsen number (dimensionless); L, the mean free path (m); and d, the pore diameter (m).

If the Knudsen number is very small (Kn $\leq$ 0.1), then normal Fickian diffusion occurs, perhaps with void fraction and tortuoisity corrrections. If the Knudsen number is very large (Kn $\geq$ 100), then Knudsen diffusion dominates. In this case, a special Knudsen mass diffusivity, D_{Kn}, is calculated and used with Fick's equation 4.3.1.

Knudsen
diffusivity

$$D_{\text{Kn}} = 48.5d\sqrt{T/M}$$ (4.3.27)

where D_{Kn} is the Knudsen diffusivity (m^2/sec); d, the pore diameter (m); T, the absolute temperature (K); and M, the molecular mass (kg/kg mol).

mixed
diffusion

When the Knudsen number is of the order of 1.0, neither Fickian nor Knudsen diffusion predominates. There are a number of means to calculate diffusion in this case. One possibility is to calculate a mass diffusivity that combines contributions from both the Knudsen diffusivity and Fick's mass diffusivity. The combined mass diffusivity is then used in the Fick equation 4.3.1. One possible combination scheme is to combine the two contributions vectorally

$$D = \{[D_{AB} \cos(\tan^{-1} \text{Kn})]^2 + [D_{Kn} \sin(\tan^{-1} \text{Kn})]^2\}^{1/2} \qquad (4.3.28)$$

where D is the combined mass diffusivity (m^2/sec).

Because D_{AB} is afffected by pressure, whereas D_{Kn} is not, D displays pressure dependence at low Knudsen numbers, but becomes independent of pressure at high Knudsen numbers (Figure 4.3.9).

EXAMPLE 4.3.2.5-1. Gas Chromatography

Gas chromatography (GC) is an important technique for identification of chemical compounds and concentrations (Figure 4.3.10). GC is used to detect food flavor compounds, to assay for steroids, enzymes, and metabolites in bodily fluids, and to ascertain pesticides present in water samples, for instance. One type of gas chromatograph (see Figure 4.3.11) uses a very long fused silica capillary tube ("column") that has a monolayer of polysiloxane polymer or other absorbent material fused to the inside of the silica. Upon injection of the sample, each unknown compound is absorbed and deabsorbed from the absorbent material; at the same time they are carried along by an inert carrier gas (usually hydrogen, helium, or nitrogen). At the other end of the column, a flame ionization detector (may also be other types of detectors) gives an electrical signal proportional to the concentration of chemical compound. Identification of the chemical depends on retention time, or the time between injection and detection. Chemical concentration is determined by the area under the peak (see Figure 4.3.12). Tube lengths and types, absorbent coatings (called the "stationary phase"), temperature, carriers, and detectors can be optimized for detections of specific compounds.

For a gas chromatograph used to detect ammonia in samples, with a capillary column 30 m × 250 µm (dia.), a carrier gas of dry nitrogen, and a column temperature of 50°C, check for Knudsen diffusion.

Solution

Using Eq. 4.3.15 and values from Table 4.3.2, we can estimate the diffusivity for a binary mixture of ammonia and nitrogen to be:

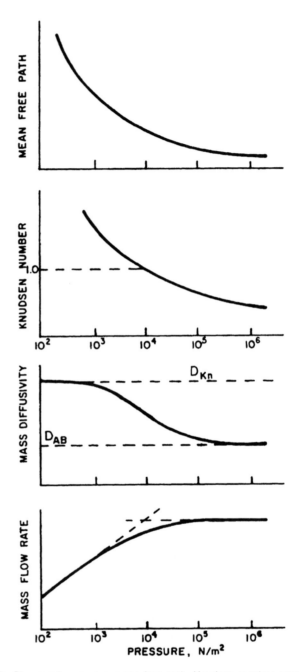

Figure 4.3.9. Changes in gaseous mean free path, Knudsen number, mass diffusivity, and mass flow rate with gas pressure. Mean free path and Knudsen number vary inversely with pressure. Mass diffusivity varies from Knudsen to Fick diffusivity as pressure increases. Mass flow rate is proportional to diffusivity times pressure, but Fick diffusivity varies inversely with pressure.

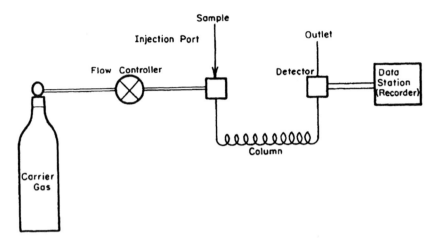

Figure 4.3.10. Gas chromatograph schematic.

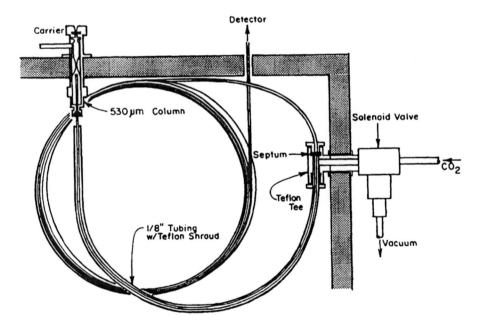

Figure 4.3.11. Capillary column used in one type of gas chromatograph. The 30-m column is coiled inside the machine.

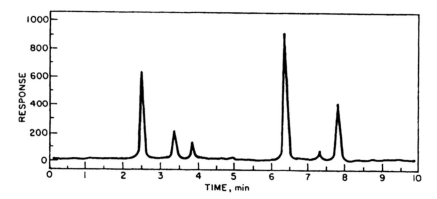

Figure 4.3.12. Response of a gas chromatograph. Identification of the compound comes from the chart position of the peaks; compound amounts are obtained from areas under each peak.

$$D_{NH_3-N_2} = \frac{0.0103(323 \text{ K})^{1.75}\left(\dfrac{1}{28} + \dfrac{1}{17}\dfrac{\text{kg mol}}{\text{kg}}\right)^{1/2}}{101300 \text{ N/m}^2[(17.9 \text{ m}^3)^{1/3} + (14.9 \text{ m}^3)^{1/3}]^2}$$

$$= 2.94 \times 10^{-5} \text{ m}^2/\text{sec}$$

From Eq. 4.3.25, and using viscosity values from Table 2.4.3, the mean free path for nitrogen is

$$L_{N_2} = \frac{3.2(1.88 \times 10^{-5} \text{ N sec/m}^2)}{101300 \text{ N/m}^2}\sqrt{\frac{(323 \text{ K})(8314.34 \text{ kg m}^2/(\text{sec}^2 \text{ kg mol K}))}{(2\pi)28 \text{ kg/kg mol}}}$$

$$= 7.34 \times 10^{-8} \text{ m}$$

and for ammonia

$$L_{NH_3} = \frac{3.2(1.09 \times 10^{-5} \text{ N sec/m}^2)}{101300 \text{ N/m}^2}$$

$$\times \sqrt{\frac{(323 \text{ K})(8314 \text{ kg m}^2/(\text{sec}^2 \text{ kg mol K}))}{(2\pi)17 \text{ kg/kg mol}}}$$

$$= 5.46 \times 10^{-8} \text{ m}$$

Calculated Knudsen numbers are

$$Kn_{N_2} = L_{N_2}/d = 7.34 \times 10^{-8} \text{ m}/250 \times 10^{-6} \text{ m} = 2.93 \times 10^{-4}$$

$$Kn_{NH_3} = L_{NH_3}/d = 5.46 \times 10^{-8} \text{ m}/250 \times 10^{-6} \text{ m} = 2.18 \times 10^{-4}$$

In both cases, Kn $\ll$ 1.0; so normal Fickian diffusion prevails.

Remarks

It is possible that one gas could undergo Knudsen diffusion and the other undergo Fickian diffusion if their two mean free paths were very much different. The Knudsen diffusivity for nitrogen can be obtained from Eq. 4.3.27

$$D_{Kn} = 48.5(250 \times 10^{-6} \text{ m})\sqrt{323 \text{ K}/28 \text{ kg/kg mol}}$$

$$= 4.12 \times 10^{-2} \text{ m}^2/\text{sec}$$

Diffusivity for nitrogen would be calculated from Eq. 4.3.28 as

$$D_{N_2} = \{[(2.94 \times 10^{-5} \text{ m}^2/\text{sec}) \cos(\tan^{-1} 2.93 \times 10^{-4})]^2$$

$$+ [(4.12 \times 10^{-2} \text{ m}^2/\text{sec}) \sin(\tan^{-1} 2.93 \times 10^{-4})]^2\}^{1/2}$$

$$= 2.94 \times 10^{-5} \text{ m}^2/\text{sec}$$

4.3.2.6 Food and Biological Materials

Food and biological materials are extremely complex structurally. They often cannot be considered to be homogeneous, and thus diffusion is complicated. Apples, for instance, release moisture through their skins to the atmosphere, but their skins are essentially impermeable to moisutre; water moves through pores scattered randomly around their surface. And how does the water reach the skin? Is it through water-filled capillaries or through gas-filled interstitial voids? Gas diffusivities are so much greater than liquid diffusivities that the latter are the preferred paths.

complex
structures

Food materials are made to be texturally interesting to eat, and so have many inclusions of different materials. Many of the pores in food are water filled, but there can be air-filled pores that can give rise to Knudsen diffusion.

To avoid having to deal with the intricacies of these diffusion processes, effective mass diffusivities have been measured. Thus, while avoiding many of the complexities of real diffusion paths, large variations in measured values would be expected to be encountered as physical or chemical compositions of the food and biological materials varied.

measured
diffusivities

In Table 4.3.7 are found some values for effective diffusivities in these types of materials. Values are all of the order of 10^{-11} m^2/sec, which makes

Table 4.3.7   **Effective Mass Diffusivities Through Foods**[a]

Diffusing Substance	Food	Temperature (°C)	Effective Mass Diffusivity (10^{-11} m^2/sec)
water	apple	30	26
		70	3.6–110
water	rice	40	0.56–6.6
		50	1.2–9.6
water	dried apple	25	0.24
water	raisin	25	4.2
water	roasted peanut	25	3.8
water	banana chip	25	2.9
water	sliced roated almond	25	1.6
water	carrot cubes	40	6.8
		80	18
water	wheat	20	6.9–33
water	pasta	40	0.8–1.5
		60	1.5–4.1
water	fish	30	1.3–3.9
water	potatoes	30	1.5
		60	3.9
water	flour	25	3.2
water	milk	40	140
		50	200
water	oatmeal cookie	25	4.0
water	shredded wheat	25	5.5
water	wheat bran	100	1.8
water	pepperoni	—	4.7–5.7
water	soybeans	50	8.7–10
water	pears	66	96
water	onion	62	4.2
water	alfalfa stems		
	radial	26	26
	axial	26	260
water	eggs	85	1.0
carbon monoxide	potato	10	3200
		20	4000
carbon dioxide	potato	10	14,000
		20	17,000
sulfur dioxide	potato	10	29,000
		20	77,000

[a] For example, the mass diffusivity of water through oatmeal cookies is 4.0×10^{-11} m^2/sec.
Sources: Okos et al. (1992); Sirivicha et al. (1990).

them less than most liquid diffusivity values in Tables 4.3.3 and 4.3.4, greater than the solid diffusivity values found in Table 4.3.6, and about the same as values for various solutes in aqueous solution found in Table 4.3.5.

Diffusivity values in food and biological materials are dependent on both moisture content and temperature. At moisture contents below about 10–20%

dry materials

dry basis (and thus high solute concentrations), effective diffusivity often decreases considerably. Values in Table 4.3.7 are generally for higher moisture contents where mass diffusivity values are relatively constant.

temperature dependence

Temperature dependencies of mass diffusivities of water in food materials is generally considered to follow the Arrhenius equation (Eq. 3.5.4), but expanded to

$$D_\theta = D_O e^{-E/RT} \qquad (4.3.29)$$

where D_θ is the mass diffusivity value corrected for temperature (m^2 sec); D_O, the mass diffusivity value at some reference temperature (m^2/sec); E, the activation energy (N m/kg mol); R, gas constant [8314.34 N m/(kg mol K)]; and T, the absolute temperature (K).

Activation energy values are moisture content dependent, and various values can be found in Okos et al. (1992). The value of Eq. 4.3.29 here, however, is in temperature correction of mass diffusivities at approximately constant food moisture contents.

4.3.3 Diffusion through Membranes and Films

Membranes and films are thin layers of materials that can be used more or less to control access of gases, liquids, or solids in solution to the volumes away from the solutions. Biological engineers with many different interests use membranes and films to produce different effects.

controlled access

There are important examples of diffusion in films and membranes. The kidney, the intestine, a red blood cell, and a plant root hair each have biological membranes that may become of interest. To know how to analyze the flow of substances across these membranes enables the biological engineer to control, replace, or understand membrane action. Thus membranes are used in medical devices to purify substances, in food apparatus to separate various components, and in environmental applications to remove impurities.

examples

We consider two general classes of membranes: porous and nonporous. The former act as molecular sieves, allowing particles smaller than the pore diameters to pass and retaining the remainder of the particles. Diffusion occurs through these membranes just as it would through any other porous solid. Some membranes act in a nonporous manner, where the solute dissolves in the membrane material on its way through. There are, in addition, other possible complexities. Many biological membranes, for example, are formed from lipoproteins (Figure 4.3.13) and have several layers. There may or may not be pores in these membranes, and some active transport may occur of ions and molecules too large to fit through the pores. Also, the solute may be ionic or nonionic, each acting differently from the other, especially when the membrane consists of at least one polar layer.

two kinds of membranes

Because porous and nonporous membranes are usually treated so differently, we consider each separately. We begin with nonporous membranes.

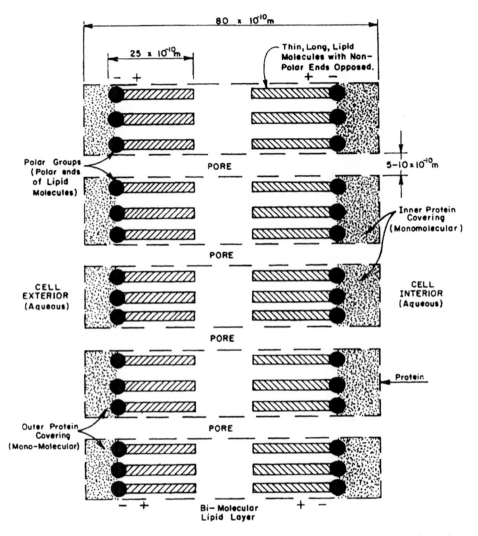

Figure 4.3.13. Biological cell membrane structure. Outer and inner monomolecular layers of protein are separated by a closely packed bimolecular lipid layer with polar ends to the outside. Pores allow for water and hydrated ions to penetrate the otherwise hydrophobic lipid structure.

4.3.3.1 Nonporous Membranes Diffusion through nonporous membranes requires two steps: First, the molecular species passing through the membrane must dissolve in the membrane material. At this point it usually forms a solid-within-solid solution, although it is possible that the membrane has liquid properties. Second, the solute must move by molecular diffusion through the membrane material. Because of the first step, it is often possible that the limitation to improvement of the diffusing species is not diffusion, but is instead due to limited solubility.

solute
dissolves

Partition Coefficient   When the solute dissolves in the membrane material,
concentration   there is usually a concentration discontinuity at the intereface between the
discontinuity   membrane and the surrounding medium (Figure 4.3.14). The equilibrium ratio
of the solute concentration in one medium to the solute concentration in an
immiscible surrounding medium is called the partition coefficient

$$K_{12} = c_1/c_2 \qquad (4.3.30)$$

where K_{12} is the partition coefficient for media 1 and 2 (dimensionless); c_1, the
equilibrium concentration in medium 1 (kg/m^3); and c_2, the equilibrium
concentration in medium 2 (kg/m^3).

There are partition coefficients for many liquid–liquid and liquid–solid
systems, including oil–water, ether–water, and others. The particular partition
coefficient of interest here applies for a membrane–water system, since most
biological membranes are surrounded by a solution with water as the solvent.

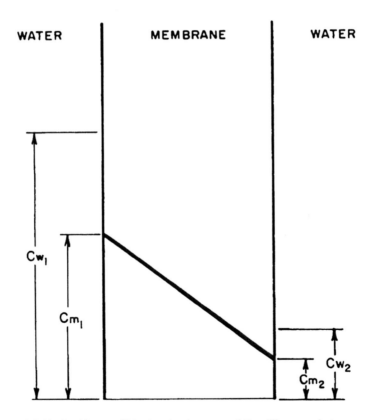

Figure 4.3.14. Partition coefficients arise because of the differences between concentrations immediately outside and inside membranes.

The partition coefficient in biological membranes, for nonporous membranes where the mobile species actually dissolves in the membrane while passing through, is nearly equal to the solubility of a saturated solution of the substance in oil divided by the solubility of a saturated solution of the substance in water (see Section 4.6.1)

$$K_{12} = \frac{\alpha_{oil}}{\alpha_{water}} \qquad (4.3.31)$$

where α_{oil} is the solubility coefficient of substance in oil (kg substance/kg oil); and α_{water}, the solubility coefficient of substance in water (kg substance/kg water).

Nonpolar substances usually have greater α_{oil} and lower α_{water} values, and membrane permeabilities to these substances (usually lipids) increase as K_{12} values increase.

Values for partition coefficient can vary greatly, depending on the solute moving through the water-based medium and the lipoid membrane. Hexane has an oil–water partition coefficient of about 500; propyl alcohol has a value of about 0.1; urea has a value of about 0.00015; and glycerol has a value of about 0.00007.

Partition coefficients can also be used to determine inhaled concentrations of volatile organic compounds (VOCs) in different tissues of the body. Several VOC partition coefficient values appear in Table 4.3.8 for the interfaces between air and blood, and fat and blood (Fisher et al., 1997). The latter should be nearly the same as the fat (or oil) to water partition coefficient. Also in Table 4.3.8 are octanol–water partition coefficients for some environmentally important organic compounds.

Referring to Figure 4.3.14

$$K_{mw} = \frac{c_{m1}}{c_{w1}} = \frac{c_{m2}}{c_{w2}} \qquad (4.3.32)$$

where K_{mw} is the partition coefficient for membrane and water (dimensionless); c_{mi}, the concentration in the membrane (kg/m^3); and c_{wi}, the concentration in the water phase (kg/m^3).

Diffusion As long as the membrane can be considered to be homogeneous (no pores, solute dissolved in membrane)

$$\dot{m} = D_{ms}A\frac{\Delta c_m}{L} = \frac{\Delta c_m}{R_m} \qquad (4.3.33)$$

where $\dot{m}$ is the mass flow rate (kg/sec); D_{ms}, the diffusion coefficient of the solute in the membrane (m^2/sec); Δc_m, the concentration difference within the

Table 4.3.8 Partition Coefficients for Volatile Organic Compounds in Human Tissues and the Environment

Chemical	Blood–Air	Fat–Blood	Octanol–Water
Anthracene			3.05×10^4
Benzene	7.79	52.11	
Benzidine			38.5
Benzo[a]pyrene			3.16×10^6
Bromochloromethane	50.25	6.42	
Carbofuran			39.8
Carbon tetrachloride	4.20	511.68	
Chlorobenzene	40.87	2.09	
2-Chloroethyl vinyl ether			19.1
Chloroform	10.68	26.21	85.1
DDT			25.1×10^5
Diethylether	11.17	52.65	
Dimethyl nitrosamine			1.15
2,4-Dimethyl phenol			288
1,4-Dioxane	2905.61	0.29	
Halothane	3.54	42.09	
n-Hexane	2.13	74.74	
Isoflurane	3.40	1.85	
Methylchloroform	4.23	66.62	
Methylene chloride	8.43	14.25	
Methyl ethyl ketone	195.44	0.52	
Parathion			6460
Pentachlorophenol			1.02×10^5
Perchloroethylene	16.67	87.10	
Styrene	69.74	20.64	
Trichloroethylene	10.62	66.63	
1,1,1,2-Tetrachloroethane	33.35	64.38	
Toluene	17.01	60.01	
o, p, m-Xylenes	44.78	41.28	

Sources: Fisher et al. (1997); Schnoor (1996).

membrane (kg/m^3); L, the membrane thickness (m); A, the membrane area (m^3); and R_m, the membrane resistance (sec/m^3).

The concentrations of the solute within the membrane are not easily measured. More easily measured are the concentrations within the water-based media on either side of the membrane. Thus

$$\dot{m} = D_{ms} A K_{mw} \frac{\Delta c_w}{L} = \frac{\Delta c_w}{R_{mw}} \qquad (4.3.34)$$

where Δc_w is the concentration difference in the water-based media on both sides of the membrane (kg/m^3); R_{mw}, the mass transfer resistance of the water–membrane–water system (sec/m^3; $= R_m/K_{mw}$); and K_{mw}, the partition coefficient (dimensionless).

Actually, the above equation is technically correct only insofar as the greatest resistance to mass transfer occurs within the membrane. There is a possibility, analogous to heat transfer, that there is a significant mass transfer resistance at the interface between membrane and the water-based medium that contacts it. We will return to this point in the section dealing with convective mass transfer (Section 4.4).

Permeability Many fresh fruits and vegetables retain quality for longer periods of time if stored in atmospheres containing high levels of carbon dioxide (up to 30% for some fruits and some temperatures) and low levels of oxygen (as low as 1%). A common modified atmosphere (MA) composition is 2–4% oxygen and 5–7% carbon dioxide. Such an atmosphere suppresses fungal growth, discourages insect activity, and slows the process of senescence, or aging. Removal of ethylene gas produced by fruit also slows ripening. Damage to the particular product can occur, however, if carbon dioxide levels become too high or too much water vapor is retained.

Respiration by living tissue consumes oxygen and produces carbon dioxide. Modified atmosphere storage of fresh fruits can thus be achieved by limiting access of ambient atmospheric gases to the interior of packages containing produce. The transparent plastic wrapping found on packages of produce on supermarket shelves serves not only to keep the produce clean and attractive, but also to retain interior high carbon dioxide levels and low oxygen levels.

There is also a common need to exclude oxygen from packaging. This is the case when it is essential to retain flavor, texture, color, nutrition, or shelf life of foods, drugs, or samples of biological materials that can deteriorate when exposed to oxygen.

Specialized films include biodegradable ones often produced from starch compounds, soluble films that dissolve when placed in water, and edible films for packaging food materials and that need not be removed before eating.

Diffusion through plastic films is customarily expressed by means of the permeability, or transmission rate. Permeability values published in the literature have units of cm^3 gas/(m^2 day) at 1 atmosphere pressure differential for a 0.0254 mm (1 mil) thick film at 22–25°C at relative humidities approaching zero. Some representative values have been transformed into the standard units used here and appear in Table 4.3.9.

To obtain the volume flow rate of gas through these films

modified atmosphere

retain carbon dioxide

exclude oxygen

specialized films

permeability values

volume flow rate

$$\dot{V} = \frac{PA}{L}(p_1 - p_2) \qquad (4.3.35)$$

where $\dot{V}$ is the volume flow rate of gas through the film (m^3/sec); P, the film permeability [m^4/(sec N)]; A, the film area (m^2); L, the film thickness (m);

Table 4.3.9 Permeabilities for Several Plastic Films[a]

Film Type	Permeability [10^{16} m⁴/(sec N)]		
	Oxygen	Carbon Dioxide	Water Vapor
Low-density polyethylene[b]	0.45–1.5	0.88–8.8	0.97–3.8
Linear Low-density polyethylene[b]	0.80–1.1	—	2.6–5.0
Medium-density polyethylene	0.30–0.95	0.88–4.4	1.3–2.4
High-density polyethylene	0.059–0.46	0.45–1.1	0.65–1.6
Polypropylene	0.15–0.73	0.88–2.4	0.65–1.7
Polyvinylchloride	0.071–0.26	0.49–0.93	—
Polyvinylchloride, plasticized[b]	0.0088–0.086	0.088–6.3	>1.3
Polystyrene	0.023–0.088	1.1–3.0	18–25
Ethylene vinyl acetate copolymer (12%)[b]	0.091–1.5	4.0–6.1	9.7
Ionomer	0.40–0.86	1.1–2.0	3.6–4.9
Ruber hydrochloride	0.015–0.15	0.059–0.59	>1.3
Polyvinylidine chloride[c]	0.00091–0.0030	0.0067	0.24–0.81

[a] For example, the permeability of oxygen through low-density polyethylene is 0.45–1.5×10^{-16} m⁴/(sec N). This permeability is to be used with a pressur-flow system.
[b] Commonly used to package living produce.
[c] Commonly used to seal out oxygen.

p_1, p_2, the partial pressures of the diffusing gas on both sides of the film (N/m²). This is similar to the Fick equation presented previously, but differs mainly in the calculation of diffusing volume flow rate rather than mass flow rate. Permeability is related to mass diffusivity through the solubility of the solute gas within the plastic film (see Table 4.3.6)

$$P_B = D_{AB}\alpha_{AB} \tag{4.3.36}$$

where P_B is the film permeability of material B [m⁴/(sec N)]; D_{AB}, the mass diffusivity of gas A through film material B (m²/sec); and α_{AB}, the solubility of gas A in film material B [(m³ gas A m²)/(m³ solid B N)].

It should be apparent that the gas dissolves in the film material on its way through the film, and does not pass through pores in film. Carbon dioxide permeabilities through most common plastic films are 3–4 times higher than those for oxygen because of the higher solubilities of carbon dioxide through hydrocarbon materials.

Permeability values are affected by temperature and relative humidity. Most moisture effects permeability values are higher at higher temperatures, but the difference is modest. Relative humidity, on the other hand, can greatly affect permeability. Cellophane is an excellent oxygen blocker when dry but becomes a very poor oxygen barrier when moist. Swelling of natural fibers at high relative humidities can reduce oxygen transmission by great amounts.

composite films

One material does not often possess all desired characteristics of strength, clarity, impermeability, stability, and others. An approach to impart all desired characteristics to one material is to laminate several materials together to form a composite. Since resistances presented to diffusion in the composite material are in series (Figure 4.3.15), the overall resistance is the sum of the individual resistances

volume resistance

$$R_{\text{tot}} = R_1 + R_2 + \cdots = \frac{L_1}{AP_1} + \frac{L_2}{AP_2} + \cdots \qquad (4.3.37)$$

where R_{tot} is the total volume transfer resistance (sec N/m^5); R_i, the individual film thickness (m); P_i, the individual film permeability [m^4/(sec N)]; and A, the film area (m^2).

The volume flow rate is obtained from

$$\dot{V} = \frac{(p_1 - p_2)}{R_{\text{tot}}} \qquad (4.3.38)$$

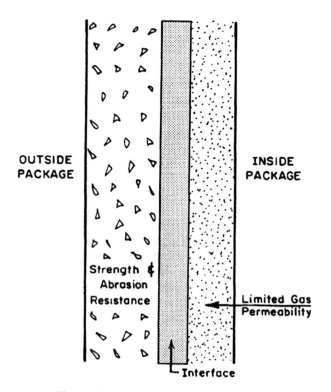

OUTSIDE PACKAGE

INSIDE PACKAGE

Strength & Abrasion Resistance

Limited Gas Permeability

Interface

Figure 4.3.15. Schematic of a composite film.

The resistance given in Eq. 4.3.37 is different from the resistance given in Eq. 4.3.33, which can be seen by studying the units on each. The resistance given here is used to calculate the volume flow rate of a gas rather than to calculate mass transfer rate. The two forms of resistance differ by the solubility of the gas within the membrane material, as indicated by Eq. 4.4.36.

The permeabilities listed in Table 4.3.9 are strictly valid for flat materials only. When formed into packages, flex cracks, seams, delaminations, and overlapping are likely to occur, and, therefore, the package permeability cannot exactly be predicted from that of the film material used.

When designing packaging systems using plastic films to limit gas transmission, the final objective is usually to maintain a modified atmosphere inside, an atmospheric composition known to extend shelf life. Starting from a mass balance on the gas constituent of interest (where, at steady state, gas transmission must equal respiration), a film material resistance can be calculated, and from that some combination of permeability and thicknesss can be chosen (Figure 4.3.16). Often there are several gases of interest (O_2, CO_2, and H_2O), and there may not be one best value of resistance to be obtained. Each gas permeability is independent of the others (except for the influence of

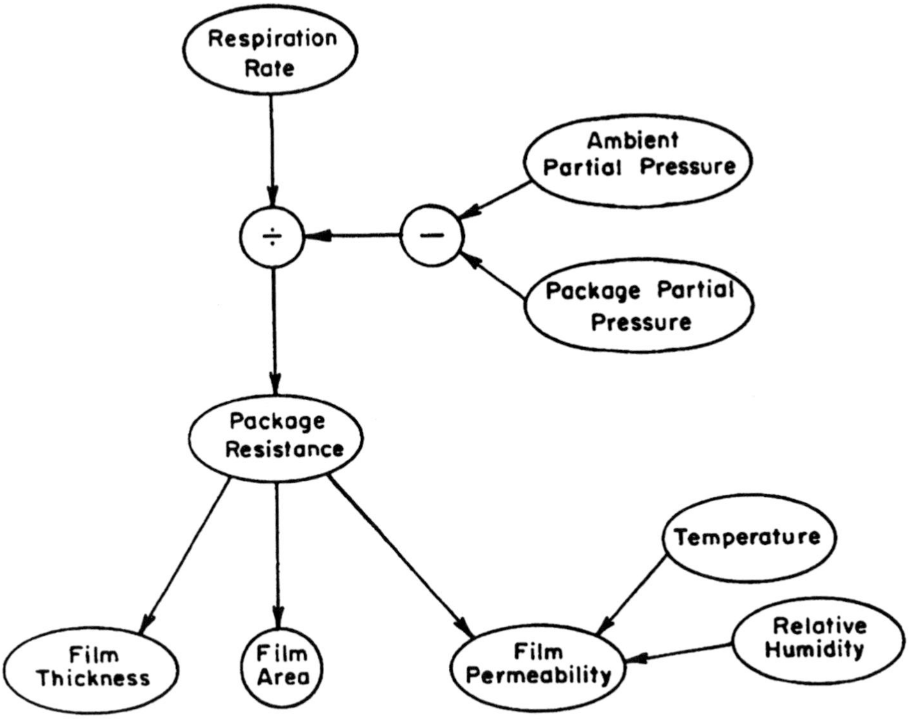

design
procedure

Figure 4.3.16. Schematic of film selection factors.

relative humidity); and so a more or less severe conflict may arise between specifications for different gases. In that case, the engineer must exercise her/his judgment to make a choice that is at least not harmful.

EXAMPLE 4.3.3.2-1. Packaging Peaches

Peaches held at 10°C in an atmosphere containing 1–2% oxygen and 3–5% carbon dioxide produce about $1.5 \times 10^{-9} \, m^3 \, CO_2/(kg$ fruit sec). Specify a plastic film package to hold four medium-sized peaches at a temperature of 10°C.

Solution

The mass of the enclosed peaches will first be estimated. Peaches are nearly spherical, and medium-sized peaches have diameters of about 8.3 cm. Using the volume of a sphere and the density of water gives an estimate of the mass of each peach

$$m = \frac{4}{3} \pi r^3 \rho_w = \frac{4\pi}{3} (4.15 \times 10^{-2} \, m)^3 (1000 \, kg/m^3)$$

$$= 0.3 \, kg$$

and four peaches have a total mass of 1.2 kg. The evolution of carbon dioxide is

$$\dot{V}_{CO_2} = (1.2 \, kg)[1.5 \times 10^{-9} \, m^3 \, CO_2/(kg \, sec)]$$

$$= 1.8 \times 10^{-9} \, m^3 \, CO_2/sec$$

Because the evolution of carbon dioxide is from respiration of hydrocarbons (or sugars) in the peaches, the same volume of oxygen (and the same number of moles) is used as the volume of carbon dioxide produced

$$\dot{V}_{O_2} = 1.8 \times 10^{-9} \, m^3 \, O_2/sec$$

The total resistance of the plastic film to carbon dioxide transport, assuming retention of 4% CO_2 inside the package, is

$$R_{CO_2} = \frac{p_i - p_o}{\dot{V}_{CO_2}} = \frac{(101,300 \, N/m^2)(0.04 - 0)}{(1.8 \times 10^{-9} \, m^3/sec)}$$

$$= 2.25 \times 10^{12} \, N \, sec/m^5$$

The total film resistance to oxygen transport, assuming 2% O_2 inside the package, is

$$R_{O_2} = \frac{p_i - p_o}{\dot{V}_{O^2}} = \frac{(101,300 \ N/m^2)(0.21 - 0.02)}{1.8 \times 10^{-9} \ m^3/sec}$$

$$= 1.07 \times 10^{13} \ N \ sec/m^5$$

The required ratio of R_{CO_2} to R_{O_2} is about 0.2. Because $R = L/AP$, the ratio of CO_2 permeability to O_2 permeability for the required plastic film should be about 5. From Table 4.3.9, we find that a low-density polyethylene film can perhaps meet this requirement. We will design for a film with

$$P_{O_2} = 1.0 \times 10^{-16} \ m^4/(sec \ N) \ and \ P_{CO_2} = 5.0 \times 10^{-16} \ m^4/(sec \ N).$$

Now to calculate the required thickness of the film, we assume the bottom of the package is made from an impermeable cardboard tray, and we assume that the top area is about $0.04 \ m^2$. Thus

$$L = RAP$$

$$L_{O_2} = R_{O_2}AP_{O_2} = (1.07 \times 10^{-13} \ N \ sec/m^5)(0.04 \ m^2)[1.0 \times 10^{-16} \ m^4/(sec \ N)]$$

$$= 0.00004 \ m = 0.04 \ mm$$

$$L_{CO_2} = R_{CO_2}AP_{CO_2} = (2.25 \times 10^{12} \ N \ sec/m^5)(0.04 \ m^2)[5.0 \times 10^{16} \ m^4(sec \ N)]$$

$$- 0.000045 \ m = 0.04 \ mm$$

The required plastic film is thus a low-density polyethylene with thickness 0.04 mm. If this is too thin to be strong enough, then a composite film may be used.

Remarks

1. The calculation of fruit mass was based on a number of more or less rough assumptions. However, the fruit size varies, its density varies, its shape varies, and its CO_2 production varies. Thus greater precision is not justified.

2. Choosing permeability values from Table 4.3.9 does not make them so. Effort must be spent to find a film manufacturer to supply a product with these permeabilities or the calculation process must be revised to account for products actually available. In the end, some compromise may have to be made in the modified atmosphere composition maintained inside the package.

3. Peaches are not ordinarily packaged in plastic films. The reason is that if the level of oxygen falls too low, or if the carbon dioxide level gets too high, then the fruit will be injured. The abilities of fruits and vegetables to

tolerate modified atmospheres, and the shelf-life advantages conferred by these atmospheres, vary considerably from one to another, and even drastically among varieties (cultivars) within a type. Thus there are few generalizations that can be made, and applying experimental data from one type of fruit or vegetable to another is extremely dangerous.

4. We did not consider moisture transport across the film. Some fruits and vegetables tolerate, or even require, high moisture contents inside the packages, whereas others cannot.

4.3.3.2 Porous Membranes

semi-
permeable
membranes

The previous section dealt with transport of material through a solid membrane, and a necessary step to that process is that the material moving through the membrane must dissolve (fit between atoms or molecules) in the membrane material. There are other membranes that are porous, but the pores are so small that they effectively block some molecules from moving through the membrane while allowing smaller molecules to pass relatively freely (Table 4.3.10). These are often called microporous membranes. When considering food and biological materials, these membranes are used to distinguish between water (as the solvent) and solute molecules or ions. Such membranes are called semipermeable.

pore size
definitions

Sometimes the terms microfiltration, ultrafiltration, and nanofiltration are used. These terms are used to designate membrane pore sizes, but exact definitions have not been agreed upon. In general, microfiltration is used to clarify slurries or remove suspended particulate matter; ultrafiltration is used to separate simple molecules, such as salts and sugars, from macromolecules; nanofiltration separates the smaller salt molecules from larger sugars and organic compounds; reverse osmosis separates water (or other solvent) from solutes (Figure 4.3.17). Two other terms that can be encountered are permeate and retenate, referring to materials passing or not passing through the membrane.

Because the material passing through the membrane during osmosis does not dissolve in the membrane, but passes through pores, partition coefficients previously introduced (Section 4.3.3.1) are not applicable. This is quantitative

Table 4.3.10 General Comparison of Membrane Types

	Filtration Membranes	Solution–Diffusion Membranes
Basis for separation	Molecular size	Solubility in membrane phase Diffusion of dissolved material through membrane

Characteristics	High transport rates
	Low selectivity
	Problems with concentration
	    polarization and fouling
		Low transport rates
		High selectivity
		No appreciable concentration
		    polarization and fouling

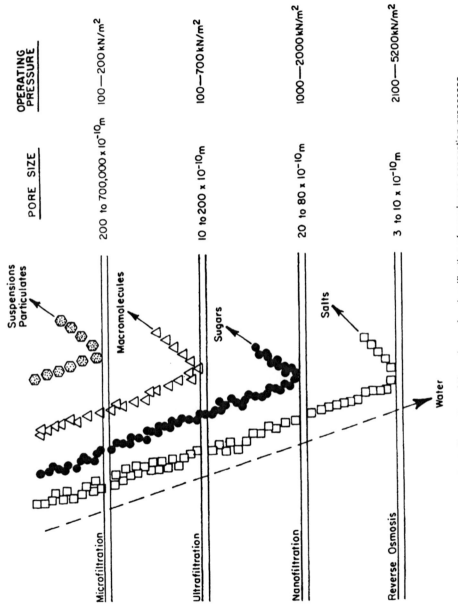

Figure 4.3.17. Illustration of the scheme for classification of membrane separation processes.

539

recognition of the different mechanisms involved with semipermeable membranes.

Ultrafiltration Membranes For a membrane that is partially solute permeable, solute particles flow through the membrane because of a concentration gradient across the membrane. Membrane pores are large enough, however, that solvent particles flow because of the pressure difference. They flow as a streamlined fluid through a porous medium and obey Darcy's law (Section 1.8.1)

Darcy flow

$$\dot{V} = \frac{-kA}{\mu} \frac{\partial p}{\partial L}$$ (1.8.2)

Thus the membrane resistance in a porous membrane is related to the resistance to flow in a porous medium: It is inversely proportional to cross-sectional area and directly proportional to membrane thickness.

tissue engineering

Ultrafiltration membranes are important, especially in biomedical applications involving implantation of foreign cells or tissues into the body to assume functions of failed natural cells or tissues. Examples are bioartificial pancreases and livers, where allographs (foreign tissue from the same species) or xenographs (foreign tisssue from different species) are common. One of the main difficulties with these devices is the immune reaction of the body to the tissues they contain. Ultrafiltration membranes effectively isolate antibodies in the blood from the tissues contained in the devices, yet allow passage of smaller molecules (such as insulin and glucose) so that the functions of these tissues can proceed unimpeded.

Dalton

The unit of measurement often used with this application of ultrafiltration is the Dalton, which is defined as the number of grams per gram mole (or kg/kg mol) of a molecule. This is equivalent to molecular mass. The minimum size of blood antibodies is usually taken to be 100 kilodaltons; so an ultrafiltration membrane that excludes compounds this large and larger protects the foreign tissues from attack and destruction by the new host.

maximizing area

One means to increase flow rate is to increase surface area. Membranes for all levels of filtration are manufactured with large surface areas included in very small volumes. Some membrane configurations are similar to those used for heat exchangers (Section 3.7.1), because in both instances the design goal is to maximize surface area and minimize volume. Thus there are tubular membranes, membrane plates, spiral-wound membranes, and pleated sheets. One popular configuration is the hollow fiber type (Figure 4.3.18), which is a modification of the shell-and-tube heat exchanger design (Amjad, 1993; Wu et al., 1995). Small diameter (40–70 μm inside diameter, 80–160 μm outside

hollow fibers

diameter) hollow fibers form a mass inside a protective shell. The high-pressure permeate is usually forced through the hollow fibers while the low-pressure solvent flows through the shell. This design is easy to troubleshoot,

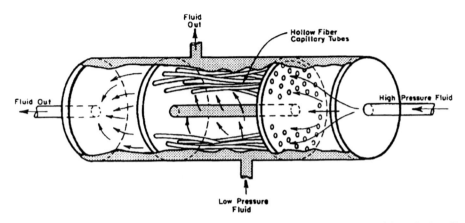

Figure 4.3.18. Hollow-fiber membrane device. Inside the shell is a bundle of many thin-walled capillary tubs with specific sizes of pores. The low-pressure feed liquid is pumped across these tubes, allowing small-diameter molecules to pass into the product liquid. This device can be used to purify brackish water (feed) into potable water (product) or to a separate whole blood (feed).

and the hollow fiber bundle is easy to change. Futher information on specific filtration membranes may be obtained from manufacturers.

membrane characteristics

Selected membrane characteristics appear in Table 4.3.11. Membrane resistances per unit area (m^2) are among the most important of these. These are given in the table as pressure difference across the membrane (N/m^2) divided by volume flow rate (m^3/sec). Permeate resistances are somewhat temperature dependent, decreasing about 3% per °C rise. Retentate resistances are of the order of 100 times permeate resistances for all tabled membrane types. In addition to membrane resistance, other factors are important:

practical considerations

1. Cleaning. Membranes easily foul with retentate particles and must be cleaned with solutions that do not damage the membranes.

2. Temperature. Membranes may decompose or lose their physical strength at high temperatures.

3. Chemical and biological resistance. Some membrane types are much more susceptible to chemical or biological activity than others.

4. Aging. There is an average increase of 4–6% in membrane permeate resistance for each year in service. Typical membranes have useful lives of 3 years.

5. Cost. Cellulose acetate membranes are particularly inexpensive.

6. Pressures. Two pressures are important: (a) the pressure drop along the flow path, which influences the size of the pump required and (b) the pressure drop across the membrane, determined by membrane strength, and which must be resisted by the membrane.

Table 4.3.11 Ultrafiltration Membrane Characteristics

Membrane Type	Typical Operating Pressure (kN/m^2)	pH Range	Permeate Resistance per m^2 Area $(N\,sec/m^3)$[a]	Chlorine Tolerance (ppm)	Temperature Range (°C)
Nanofiltration	—	—	4 to 14×10^{10}	—	—
Cellulose acetate	2760	3.0–6.5	17 to 50×10^{10}	0.2–0.5	0–35
Thin film composite	1720	2–11	13 to 65×10^{10}	0	0–45
Hollow fiber polyaramid	2760	4–11	150 to 300×10^{10}	0	0–35
High temperature	—	—	25×10^{10}	—	to 150°C

[a] Resistance decreases with surface area. Thus the resistance for 2 m^2 surface area is one-half the tabulated value in $N\,sec/m^5$. The pressure–flow resistance of a cellulose acetate membrane of 2 m^2 area is thus 8.5 to $25 \times 10^{10}\,N\,sec/m^5$.

4.3.3.3. Osmotic Pressure Osmosis is the passage of fluid from the side of the semipermeable membrane with low solute concentration to the side with high solute concentration. In effect, the fluid concentration is higher on the side with the more dilute solution, and so the fluid flows from the region of higher concentration to the region of lower concentration on the other side of the membrane. Since the solute is constrained from passing through the membrane, fluid movement is the means to attempt to equalize concentrations of both solvent and solute on both sides of the membrane. If the membrane is truly impermeable to the solute, and often the solutes are high-molecular-weight macromolecules that cannot move through the membrane, the movement of fluid alone cannot equalize concentrations on both sides of the membrane. There can be only two final results of this process: Either the space containing the solute ruptures, or else the boundaries of the space containing the solute must be strong enough to apply pressure to the fluid in the space to counteract the tendency of more fluid to fill the space.

osmosis
defined

Osmotic pressure for aqueous solutions can be calculated from the Gibbs equation

osmotic
pressure

$$\pi = \frac{RT}{V_w} \ln \mu_w \qquad (4.3.39)$$

where π is the osmotic pressure (N/m^2); R, the universal gas constant $[= 8314.34\, Nm/(kg\, mol\, K)]$; T, the absolute temperature (K); V_w, the molar volume of water $(m^3/kg\, mol)$; and μ_w, the mole fraction of water in the solution (moles water/total moles).

This equation was derived by considering differences in free energy for both solvent and solution. The ideal gas law was used to determine partial pressures of solvent above pure solvent and solution. Equation 4.3.39 is presented for aqueous solutions because of the importance of water as a solvent in food and biological materials.

Because the mole fraction of solute in a simple solution equals $(1 - \mu_w)$, Eq. 4.4.39 can be rewritten as

$$\pi = \frac{RT}{V_w} \ln(1 - \mu_{sol}) \qquad (4.3.40)$$

where μ_{sol} is the mole fraction of single solute (moles solute/total moles).

For the case where more than one solute is present in solution, the sum of mole fractions for all solutes must be substituted for μ_{sol} in Eq. 4.3.40.

For ionic solutes, the net osmotic pressure is the sum of contributions of individual ions. For a fully dissociated compound such as NaCl, the net osmotic pressure is an integer multiple of the osmotic pressure calculated assuming no dissociation (the multiple equals 2 for NaCl). For a partially dissociated compound, the net osmotic pressure depends on variables such as temperature and pressure.

dissociated
ions

Raoult's law can be used to relate partial pressure of water vapor to mole fraction for dilute solutions

Raoult's law

$$p_w = \mu_w p_{sat} \tag{4.3.41}$$

where p_w is the partial pressure of water vapor above a solution (N/m^2); p_{sat}, the saturated vapor pressure of water vapor (N/m^2); and μ_w, the mole fraction of water in solution [moles water/(moles water plus moles solute)].

Thus an alternative form for Eq. 4.3.39 is

$$\pi = \frac{RT}{V_w} \ln\left(\frac{p_w}{p_{sat}}\right) \tag{4.2.42}$$

Another equation sometimes encountered for calculating osmotic pressures is the Van't Hoff equation found in many texts. Experimental data show, however, that the Van't Hoff equation is not as accurate as the Gibbs equation (Eq. 3.3.39) and can only be used for very dilute solutions.

molar volume of water

Water is a nearly incompressible substance. The molar volume of water is nearly constant at $V_w = 0.028016 \, m^3/kg$ mol. Thus the term $RT/V_w = 138 \times 10^6 \, N/m^2$ at 298.15 K (25°C). This large magnitude (nearly 1360 atm) shows that osmotic pressures can be very large indeed.

salt solutions

Sea water is usually considered to have 3.45 weight percent of dissolved salts, and has a water vapor pressure 1.84% below pure water. Its osmotic pressure at 25°C is $-2.5 \times 10^6 \, N/m^2$. Blood plasma is considered to have an osmotic pressure equal to that of isotonic saline solution (0.154 N NaCl or 0.9 g NaCl/100 mL), which gives an osmotic pressure of $-8.0 \times 10^5 \, N/m^2$. Osmotic pressures of other materials are found in Table 4.3.12.

solvent attraction

Osmotic pressures are all calculated as negative values. That is, a solution with a high osmotic pressure value tends to draw solvent from the adjacent compartment rather than supply solvent to the compartment. In general, the higher the concentration of solute, the lower is the concentration of solvent, and the greater the solution will attract solvent from other sources.

concentration conversions

There are many ways commonly used to express concentrations, but mole fractions are required in Eqs. 4.3.39 and 4.3.40. Conversion equations between some of the most common concentration designations appear in Table 4.3.13. Concentrations, expressed as kg/m^3, are computationally equivalent to ψ, expressed as g/liter, in Table 4.3.13.

There are also a gram percent (g/100 mL) and milligram percent (mg/100 mL) that do not appear in Table 4.3.13. Grams percent is just $\psi/10$.

parts per million

There can be some confusion about the use of the concentration unit of parts per million (ppm). When applied to liquids and solids, ppm usually denotes (mass of solute/mass of solvent). This use of ppm has been designated as ppm in Table 4.3.13. Sometimes, especially for very dilute solutions, ppm may be incorrectly given as (mass of solute/mass of solution). When applied to gases, ppm usually denotes (volume of one gas/volume of

Table 4.3.12   Osmotic Pressures of Liquids at Room Temperature (with water as the solvent)

Material	Concentration	Osmotic pressure (kN/m^2)
Skim milk	9% solids, not fat (% wt/wt) (5% lactose, 0.7% salts, 3.2% protein)	-690
Whey	6% total solids (% wt/wt)	-690
Orange juice	11% total solids (% wt/wt)	$-1,600$
Apple juice	15% total solids (% wt/wt)	$-2,100$
Grape juice	16% total solids (% wt/wt)	$-2,100$
Coffee extract	28% total solids (% wt/wt)	$-3,500$
	1% wt/vol	-73
Lactose	5% wt/vol (ψ)	-380
Sodium chloride	1% wt/vol (ψ)	-860
Lactic acid	1% wt/vol (ψ)	-550
Albumin	1% wt/vol	-4
Sweet potato wastewater	22% total solids	$-6,000$
Mushroom blanch water		-261.15 (w% Tot. Solids)
	5% ppm total solids	-1300
Impure ground or surface source water to be purified with reverse osmosis	370 ppm dissolved solids	-29
Perilla anthocyanins	10.6% total solids	$-2,300$
Sea water	3.45 wt percent	$-2,500$
Blood plasma	(body temperature)	-800
Pure water		0
Lithium chloride	1% wt/wt	-1030
Sodium chloride	2.5% wt/wt	-2070
Ethanol	2.5% wt/wt	-1380
	5% wt/wt	-3030
Ethylene glycol	2% wt/wt	-690
	5% wt/wt	-2200
Magnesium sulfate	5% wt/wt	-1240
	10% wt/wt	-2620
Zinc sulfate	5% wt/wt	-900
	10% wt/wt	-1790
	15% wt/wt	-3060
Fructose	5% wt/wt	-760
	10% wt/wt	-1650
	15% wt/wt	-2620
Sucrose	5% wt//wt	-410
	10% wt/wt	-860
	25% wt/wt	-2930

Table 4.3.13 Conversions among Different Designations (Symbol Definitions at End of Table)[a]

	M	m	μ	N	ψ	ppm
M	$M = M$	$m = \dfrac{10^3 M}{\rho - MM_{sl}}$	$\mu = \dfrac{MM_{sv}}{\rho + M(M_{sv} - M_{sl})}$	$N = \kappa M$	$\psi = MM_{sl}$	$\text{ppm} = \dfrac{10^6 MM_{sl}}{\rho - MM_{sl}}$
m	$M = \dfrac{m\rho}{10^3 + mM_{sl}}$	$m = m$	$\mu = \dfrac{mM_{sv}}{10^3 + mM_{sv}}$	$N = \dfrac{m\rho\kappa}{10^3 + mM_{sl}}$	$\psi = \dfrac{\rho mM_{sl}}{10^3 + mM_{sl}}$	$\text{ppm} = 10^3 mM_{sl}$
μ	$M = \dfrac{\rho\mu}{M_{sl}\mu + (1-\mu)M_{sv}}$	$m = \dfrac{10^3 \mu}{M_{sv}(1-\mu)}$	$\mu = \mu$	$N = \dfrac{\rho\mu\kappa}{M_{sl}\mu + (1-\mu)M_{sv}}$	$\psi = \dfrac{\rho\mu M_{sl}}{\mu M_{sl} + (1-\mu)M_{sv}}$	$\text{ppm} = \dfrac{10^6 \mu M_{sl}}{(1-\mu)M_{sv}}$
N	$M = N/\kappa$	$m = \dfrac{10^3 N}{\rho\kappa - NM_{sl}}$	$\mu = \dfrac{NM_{sv}}{\rho\kappa + N(M_{sv} - M_{sl})}$	$N = N$	$\psi = \dfrac{NM_{sl}}{\kappa}$	$\text{ppm} = \dfrac{10^6 NM_{sl}}{\kappa\rho - NM_{sl}}$
ψ	$M = \psi/M_{sl}$	$m = \dfrac{10^3 \psi}{M_{sl}(\rho - \psi)}$	$\mu = \dfrac{M_{sv}}{\rho M_{sl} + \psi(M_{sv} - M_{sl})}$	$N = \dfrac{\psi\kappa}{M_{sl}}$	$\psi = \psi$	$\text{ppm} = \dfrac{10^6 \psi}{\rho - \psi}$
ppm	$M = \dfrac{\rho\,\text{ppm}}{(10^6 + \text{ppm})M_{sl}}$	$m = \dfrac{\text{ppm}}{10^3 M_{sl}}$	$\mu = \dfrac{\text{ppm}\,M_{sv}}{10^6 M_{sl} + \text{ppm}\,M_{sv}}$	$N = \dfrac{\kappa\rho\,\text{ppm}}{(10^6 + \text{ppm})M_{sl}}$	$\psi = \dfrac{\rho\,\text{ppm}}{10^6 + \text{ppm}}$	$\text{ppm} = \text{ppm}$
ppv	$M = \dfrac{\rho\,\text{ppv}}{10^6 M_{sv} + \text{ppv}\,M_{sl}}$	$m = \dfrac{\text{ppv}}{10^3 M_{sv}}$	$\mu = \dfrac{\text{ppv}}{10^6 + \text{ppv}}$	$N = \dfrac{\kappa\rho\,\text{ppv}}{10^6 M_{sv} + \text{ppv}\,M_{sl}}$	$\psi = \dfrac{\text{ppv}\,M_{sl}}{10^6 M_{sv} + \text{ppv}\,M_{sl}}$	$\text{ppm} = \dfrac{\text{ppv}\,M_{sl}}{M_{sv}}$
v%	$M = \dfrac{\rho_{sl}\,\text{v\%}}{10^2 M_{sl}}$	$m = \dfrac{10^3 \rho_{sl}\,\text{v\%}}{M_{sl}(10^2 \rho - \rho_{sl}\,\text{v\%})}$	$\mu = \dfrac{\text{v\%}\,\rho_{sl}M_{sv}}{10^2 \rho M_{sl} + \text{v\%}\,\rho_{sl}(M_{sv} - M_{sl})}$	$N = \dfrac{\kappa\rho_{sl}\,\text{v\%}}{10^2 M_{sl}}$	$\psi = \dfrac{\rho_{sl}\,\text{v\%}}{10^2}$	$\text{ppm} = \dfrac{10^6 \rho_{sl}\,\text{v\%}}{10^2 \rho - \rho_{sl}\,\text{v\%}}$
w%	$M = \dfrac{\rho\,\text{w\%}}{10^2 M_{sl}}$	$m = \dfrac{10^3\,\text{w\%}}{M_{sl}(10^2 - \text{w\%})}$	$\mu = \dfrac{\text{w\%}\,M_{sv}}{10^2 M_{sl} + \text{w\%}(M_{sv} - M_{sl})}$	$N = \dfrac{10\kappa\rho\text{w\%}}{10^2 M_{sl}}$	$\psi = \dfrac{\rho\,\text{w\%}}{10^2}$	$\text{ppm} = \dfrac{10^6\,\text{w\%}}{10^2 - \text{w\%}}$
X	$M = \dfrac{\rho X}{M_{sl}}$	$m = \dfrac{10^3 X}{M_{sl}(1-X)}$	$\mu = \dfrac{XM_{sv}}{M_{sl} + X(M_{sv} - M_{sl})}$	$N = \dfrac{\kappa\rho X}{M_{sl}}$	$\psi = \rho X$	$\text{ppm} = \dfrac{10^6 X}{1-X}$
α	$M = \dfrac{\rho\alpha}{(1+\alpha)M_{sl}}$	$m = \dfrac{10^3 \alpha}{M_{sl}}$	$\mu = \dfrac{\alpha M_{sv}}{M_{sl} + \alpha M_{sv}}$	$N = \dfrac{\kappa\rho\alpha}{(1+\alpha)M_{sl}}$	$\psi = \dfrac{\rho\alpha}{1+\alpha}$	$\text{ppm} = 10^6 \alpha$

	ppv	v %	w %	X	α
M	$ppv = \dfrac{10^6 M M_{sv}}{\rho - M M_{sl}}$	$v\% = \dfrac{10^2 M M_{sl}}{\rho_{sl}}$	$w\% = \dfrac{10^2 M M_{sl}}{\rho}$	$X = \dfrac{M M_{sl}}{\rho}$	$\alpha = \dfrac{M M_{sl}}{\rho - M M_{sl}}$
m	$ppv = 10^3 m M_{sv}$	$v\% = \dfrac{10^2 \rho m M_{sl}}{\rho_{sl}(10^3 + m M_{sl})}$	$w\% = \dfrac{10^2 m M_{sl}}{10^3 + m M_{sl}}$	$X = \dfrac{m M_{sl}}{10^3 + m M_{sl}}$	$\alpha = \dfrac{m M_{sl}}{10^3}$
μ	$ppv = \dfrac{10^6 \mu}{1 - \mu}$	$v\% = \dfrac{10^2 \rho \mu M_{sl}}{\rho_{sl}[\mu M_{sl} + (1 - \mu) M_{sv}]}$	$w\% = \dfrac{10^2 \mu M_{sl}}{\mu M_{sl} + (1 - \mu) M_{sv}}$	$X = \dfrac{\mu M_{sl}}{\mu M_{sl} + (1 - \mu) M_{sv}}$	$\alpha = \dfrac{\mu M_{sl}}{(1 - \mu) M_{sv}}$
N	$ppv = \dfrac{10^6 N M_{sv}}{\kappa \rho - N M_{sl}}$	$v\% = \dfrac{10^2 N M_{sl}}{\kappa \rho_{sl}}$	$w\% = \dfrac{10^2 N M_{sl}}{\kappa \rho}$	$X = \dfrac{N M_{sl}}{\kappa \rho}$	$\alpha = \dfrac{N M_{sl}}{\kappa \rho - N M_{sl}}$
ψ	$ppv = \dfrac{10^6 \psi M_{sv}}{(\rho - \psi) M_{sl}}$	$v\% = \dfrac{10^2 \psi}{\rho_{sl}}$	$w\% = \dfrac{10^2 \psi}{\rho}$	$X = \dfrac{\psi}{\rho}$	$\alpha = \dfrac{\psi}{\rho - \psi}$
ppm	$ppv = \dfrac{ppm\, M_{sv}}{M_{sl}}$	$v\% = \dfrac{10^2 \rho\, ppm}{(10^6 + ppm)\rho_{sl}}$	$w\% = \dfrac{10^2\, ppm}{10^6 + ppm}$	$X = \dfrac{ppm}{10^6 + ppm}$	$\alpha = \dfrac{ppm}{10^6}$
ppv	$ppv = ppv$	$v\% = \dfrac{10^2 \rho\, ppv\, M_{sl}}{(10^6 M_{sv} + ppv\, M_{sl})\rho_{sl}}$	$w\% = \dfrac{10^2\, ppv\, M_{sl}}{10^6 M_{sv} + ppv\, M_{sl}}$	$X = \dfrac{ppv\, M_{sl}}{10^6 M_{sv} + ppv\, M_{sl}}$	$\alpha = \dfrac{ppv\, M_{sl}}{10^6 M_{sv}}$
v %	$ppv = \dfrac{10^6 \rho_{sl} v\%\, M_{sv}}{(10^2 \rho - \rho_{sl} v\%) M_{sl}}$	$v\% = v\%$	$w\% = \dfrac{\rho_{sl} v\%}{\rho}$	$X = \dfrac{\rho_{sl} v\%}{10^2 \rho}$	$\alpha = \dfrac{\rho_{sl} v\%}{10^2 \rho - \rho_{sl} v\%}$
w %	$ppv = \dfrac{10^6 w\%\, M_{sv}}{(10^2 - w\%) M_{sl}}$	$v\% = \dfrac{\rho w\%}{\rho_{sl}}$	$w\% = w\%$	$X = \dfrac{w\%}{10^2}$	$\alpha = \dfrac{w\%}{10^2 - w\%}$
X	$ppv = \dfrac{10^6 X M_{sv}}{(1 - X) M_{sl}}$	$v\% = \dfrac{10^2 \rho X}{\rho_{sl}}$	$w\% = 10^2 X$	$X = X$	$\alpha = \dfrac{X}{1 - X}$
α	$ppv = \dfrac{10^6 \alpha M_{sv}}{M_{sl}}$	$v\% = \dfrac{10^2 \rho \alpha}{(1 + \alpha)\rho_{sl}}$	$w\% = \dfrac{10^2 \alpha}{1 + \alpha}$	$X = \dfrac{\alpha}{1 + \alpha}$	$\alpha = \alpha$

M, molarity, moles solute/liter solution; m, molality, moles solute/kg solvent; μ, mole fraction, moles solute/moles solution; N, normality, gram equivalents/liter solution; ψ, grams/liter, grams solute/liter solution or kg solute/m^3 solution; ppm, parts per million, grams solute/10^6 grams solvent; ppv, liters solute/10^6 liters solvent; v%, volume percent, liters solute/100 liters solution; w%, weight percent, grams solute/100 grams solution; X, mass fraction, grams solution; ρ, solution density, kg solution/m^3 solution; ρ_{sl}, solute density, kg solute/m^3 solution; M_{sv}, solvent molecular weight, grams solvent/mole solvent; M_{sl}, solute molecular weight, grams solute/mole solute; κ, ionic charge, gram equivalents/mole solute; and α, solubility (saturated solution), kg solute/kg solvent.

other gas). This has been designated as ppv in Table 4.3.13. There is often no completely sure way to determine exactly whether the use of parts per million is ppm or ppv. The two are related by the ratio of molecular weights of the two substances.

Reverse Osmosis Mass Transfer The result of placing a semipermeable membrane between a liquid consisting of solvent only on one side of the mobile solvent membrane and a liquid consisting of solvent and solute on the other side of the membrane is that solvent moves through the membrane from the solution side to the pure solvent side. If the two compartments are open to the atmosphere and can accommodate increased height of liquid, then equilibrium in this system (no further net solvent flow) occurs when the difference in height between the two compartments equals the difference in osmotic pressures of the two original liquids (Figures 4.3.19 and 4.3.20). If both compartments are not open to the atmosphere, then a physical pressure difference develops between the two compartments.

Transport of material across a semipermeable membrane is usually determined on the basis of relative pressures on both sides. This is because osmotic pressure can either be enhanced or counteracted by phsyical pressure applied to the fluid in contact with the membrane.

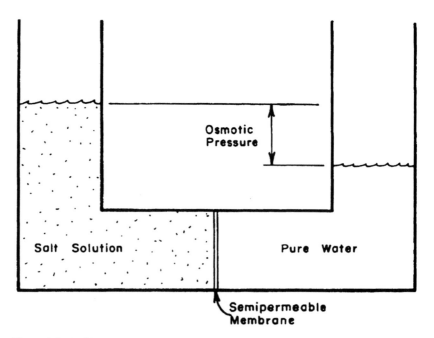

Figure 4.3.19. Placing a semipermeable membrane between an aqueous solution and pure water results in a movement of water from the pure water side to the salt solution side that is halted when the difference in heights of the two liquids counterbalances the osmotic pressure.

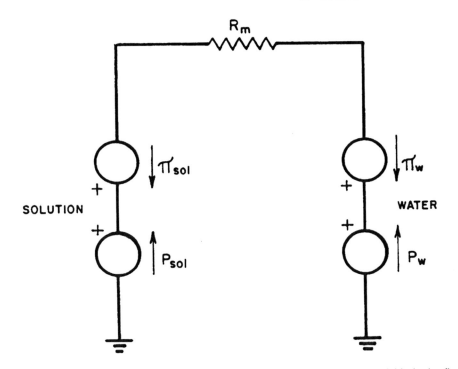

Figure 4.3.20. Systems diagram for the apparatus in Figure 4.3.19. The flow variable is the flow of water, and the effort variable is pressure. Two pressure sources appear in series on both sides of the membrane. One pressure source of each pair is osmotic pressure (π) and the other is mechanically applied pressure (p). In all cases $\pi_w > \pi_{sol}$, and, therefore, to stop the flow requires that $p_{sol} > p_w$. The membrane resistance R_m limits the flow rate when pressures on the two sides are unbalanced.

external
pressure

Reverse osmosis takes this process a step further. Applying additional pressure to the fluid in the space with the concentrated solute can send additional fluid back through the membrane and increase the concentration of the solute. If the intent of the process is to concentrate the solution, it is called reverse osmosis. If the attention of the process is on the purification of the fluid, then we call it filtration.

Filtration is used to remove salt from sea water to produce potable water for human consumption. Reverse osmosis is used to concentrate various fruit juices and liquid food products.

When physical pressure is applied to the liquid on one side of the semipermeable membrane system, solvent is driven through the membrane away from the pressure. The rate at which mass transport occurs begins with

mass
transport

$$\dot{m} = \frac{(\Delta p - \tau \, \Delta \pi)}{R_m} \qquad (4.3.43)$$

where $\dot{m}$ is the rate of solvent mass transfer (kg/sec); Δp, the difference in physical pressures imposed on liquids on both sides of the membrane (N/m²); $\Delta\pi$, the difference in osmotic pressures for liquids on both sides of the membrane (N/m²); τ, the transmission coefficient (dimensionless); and R_m, the membrane resistance [N sec/(kg m²) or (sec m)$^{-1}$].

Or, in terms of volume flow rate,

volume
transport

$$\dot{V} = \frac{(\Delta p - \tau\,\Delta\pi)}{R_m}$$

where $\dot{V}$ is the volume flow rate (m³/sec); and R_m, the membrane resistance (N sec/m⁵).

Note that the units of resistance change depending on whether mass or volume flow rate is of interest.

Some membranes are not entirely impermeable to all solute particles. If the membrane is permeable to the solvent but completely impermeable to the solute, the transmission coefficient value is 1.0. At the other extreme, if the

transmission
coefficient

membrane is as permeable to the solute as it is to the solvent, then the transmission coefficient value is zero. In the latter case an osmotic gradient is not maintained across the membrane. Practical reverse osmosis procedures can be maintained even if the membrane is not completely impermeable to the solute just as long as the solvent is sufficiently more easily transmitted across the membrane than is the solute. This is because reverse osmosis is often used to concentrate the solute, but not completely eliminate the solvent. Flowing systems are often designed where new feedstock (dilute solution) is brought to one side of the membrane while concentrated solution is simultaneously removed from the same side. Pure solvent is pumped across the face of the other side of the membrane to eliminate the accumulation of solute with concomitant lowering of osmotic pressure difference across the membrane (Figure 4.3.21). In a system such as this, concentrations on both sides of the membrane can be maintained despite the partial membrane permeability to solute. When in doubt about the value of the transmission coefficient. It can be assumed to be 1.0 because of the flow conditions on both sides of the membrane.

The rate of mass transfer was stated to begin with that in Eq. 4.3.43. However, resistance to mass transfer in a reverse osmosis apparatus also

concentration
polarization

appears outside the membrane (Figure 4.3.22). Escape of solvent ions from the high concentration side of the membrane raises the local solute concentration near the membrane face, raising the local osmotic pressure as well. This tends to reduce solvent flow through the membrane. This action has been termed "concentration polarization," and is reponsible for slowing mass transfer from one side of the membrane to the other (Figure 4.3.23). Concentration polarization is corrected largely by the solution flowing across the membrane face, as mentioned earlier. Since the rate of solvent flow through the membrane is often determined more by the convective mass transfer of

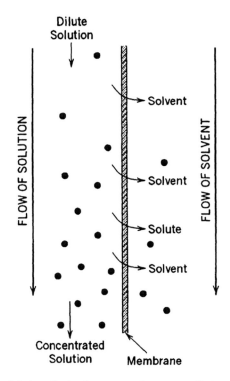

Figure 4.3.21. To minimize effects of some membrane porosity to solute molecules, pure solvent is used to wash solute molecules from the solvent side of the membrane.

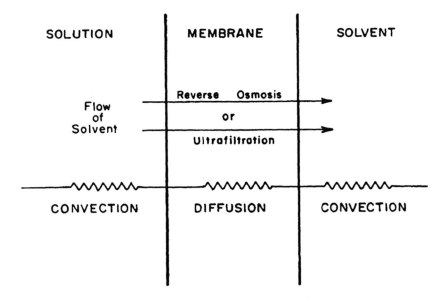

Figure 4.3.22. Resistances to solvent flow occur on both sides as well as inside the membrane.

concentrated solution away from the membrane than by the membrane itself, turbulent flow is often used to promote mixing.

The rate of solvent flow through the membrane is theoretically zero until a pressure is applied that is greater than the osmotic pressure of the solvent. Flow rates then increase proportionately to pressure applied to one side of the membrane. When concentration polarization occurs, solvent flow becomes nearly independent of applied pressure (Figure 4.3.23).

Reverse osmosis membranes are very thin (0.1 to 0.25 μm) and require a porous support layer to resist the large imposed physical pressure differences used. Original membranes were made from a single polymer, but composite materials are being developed to refine membrane characteristics (similar to Figure 4.3.15).

Because strength of a reverse osmosis membrane is about $4 \times 10^6 \, \text{N/m}^2$, there is a limit to increasing flow rate by increasing pressure. As a matter of fact, there is even a limit to the concentration of solution that can be treated by reverse osmosis because, if the osmotic pressure of a solution exceeds the strength of the membrane, sufficient applied pressure to cause solvent flow would break the membrane.

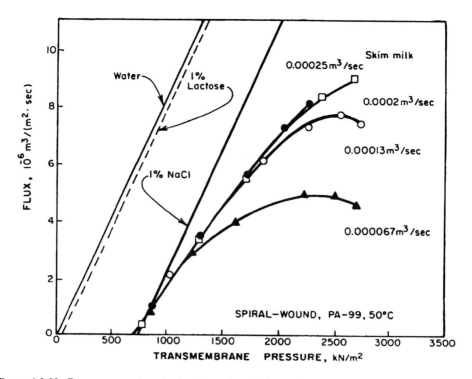

Figure 4.3.23. Reverse osmosis water flow through a spiral wound membrane. Flow depends on applied pressure until concentration polarization occurs. Here the flow of water from skim milk decreases from linearity at high pressures because of concentration polarization and membrane fouling.

EXAMPLE 4.3.3.3-1. Molecular Weight from Osmotic Pressure

Determine the molecular wright of fructose from its osmotic pressure.

Solution

From Table 4.3.12 we find that the osmotic pressure for a 10% by weight solution of fructose is $-1650\,kN/m^2$. Using Eq. 4.3.39,

$$\pi = -1650000 = -138 \times 10^6 \ \ln \mu_w$$

$$\mu_w = 0.988$$

Thus

$$\mu_{sol} = 1 - \mu_w = 1 - 0.988 = 0.012$$

From Table 4.3.13 we find the relationship between weight percent (w%) and mole fraction (μ) to be

$$\mu = \frac{w\% \ M_{sv}}{10^2 M_{sl} + w\%(M_{sv} - M_{sl})}$$

From the definitions at the bottom of the table we find that this is the mole fraction of solute. Solving for the molecular weight of solute, we obtain

$$M_{sl} = \frac{w\%(1 - \mu_{sol})M_{sv}}{(10^2 - w\%)\mu_{sol}} = \frac{w\% \ \mu_w M_{sv}}{(10^2 - w\%)\mu_{sol}}$$

$$= \frac{(10)(0.988)(18 \ kg/kg \ mol)}{(90)(0.012)}$$

$$= 166$$

Remarks

The actual molecular weight of fructose ($C_6H_{12}O_6$) is 180.2 kg/kg mol.

If we were to perform the same calculation for a 2.5% sodium chloride solution, we would obtain a molecular weight value of 30.5. The actual molecular weight is 58.5, which is about twice the calculated value. The difference is the dissociation of NaCl into sodium and chloride ions, which is not quite 100%.

EXAMPLE 4.3.3.3-2. Correction of Osmotic Pressure at Different Concentration

Determine from the value at 5% by weight the expected osmotic pressure at 2.5% by weight of ethanol.

Solution

From Table 4.3.12 we find the osmotic pressure for a 5% by weight solution of ethanol to be -3030 kN/m^2. From Eq. 4.3.39 we obtain

$$\pi = -3030 \times 10^3 \frac{N}{m^2} = 138 \times 10^6 \frac{N}{m^2} \ln \mu_w$$

$$\mu_w = 0.978$$

and

$$\mu_{sol} = 1 - \mu_w = 0.0217$$

From the relationship between weight percent and mole fraction found in Table 4.3.13, we can see that halving the weight percent will not exactly halve the mole fraction. Thus we must first find the molecular weight of ethanol in the same way as in Example 4.3.3.3-1.

$$M_{sl} = \frac{w\% \, \mu_w M_{sv}}{(10^2 - w\%)\mu_{sol}} = \frac{(5)(0.978)(18 \text{ kg/kg mol})}{(10^2 - 5)(0.0217)}$$

$$= 42.68 \text{ kg/kg mol}$$

For a 2.5% by weight solution of ethanol, we calculate

$$\mu_{sol} = \frac{w\% \, M_{sv}}{10^2 M_{sl} + w\%(M_{sv} - M_{sl})}$$

$$= \frac{(2.5)(18 \text{ kg/kg mol})}{100(42.68 \text{ kg/kg mol}) + 2.5(18 - 42.68 \text{ kg/kg mol})}$$

$$= 0.0107$$

Thus

$$\pi = 138 \times 10^6 \frac{N}{m^2} \ln(1 - 0.0107)$$

$$= -1480 \text{ kN/m}^2$$

Remark

The value of osmotic pressure found in Table 4.3.12 for this concentration of ethanol is $-1380 \, kN/m^2$. The difference between calculated and observed values is probably due to an accumulation of errors (roundoff, experimental observation, and lack of total agreement between theory (Eq. 4.3.39) and experiment). The tabled value for molecular weight of ethanol (C_2H_5OH) is $46.1 \, kg/kg \, mol$.

EXAMPLE 4.3.3.3-3. Implantable Drug Delivery System

The DUROS system produced by Alza Corp. (Palo Alto, CA) is a small (4 cm long by 4 mm dia.) titanium tube that can be implanted under the forarm skin within 5 min during outpatient surgery (Figure 4.3.24), and that can deliver medications at standard rates of release. The device consists of two chambers separated by a moveable piston. One chamber contains the drug and the other contains a saturated salt solution. Osmotic pressure draws water from the interstitial space of the arm through a cellulose acetate membrane to increase the salt solution volume. This causes the piston to move and forces drug through the 3-mm-long, 50-μm-diameter orifice. As long as the salt solution remains saturated, the drug is delivered at a constant rate.

The DUROS system can be used to deliver a medication called leuprolide used for treatment of advanced prostate cancer. The drug must be delivered at a constant rate of 125 $\mu g/day$. The density of the drug as given (40% wt/wt) is 0.5 g/mL. Determine the size of the cellulose acetate membrane to produce the required drug delivery rate.

Solution

The system concept is this: Osmotic pressure of the saturated salt solution inside the device acts to draw water into the device. Opposing the movement

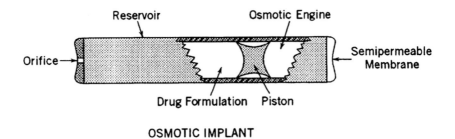

OSMOTIC IMPLANT

Figure 4.3.24. Drawing of the Alza DUROS implantable device for controlled drug release into the body.

of water are: (1) osmotic pressure of body interstitial fluid, (2) pressure drop across the resistance of the membrane, (3) pressure drop due to hydraulic flow of the drug moving through the orifice, (4) back pressure exerted by tissues caused to swell as the drug is injected, and (5) pressure drop as a result of drug moving through and into the tissues (Figure 4.3.25). Resistance of the orifice can be estimated from the Hagen–Poiseuille equation 2.5.40. Use of this equation is justified because the length-to-diameter ratio of the orifice is 60:1, and laminar flow must prevail at such a small-diameter tube.

$$R = \frac{128\mu L}{\pi D^4} = \frac{128[6.984 \times 10^{-4} \text{ kg/(m sec)}](3 \times 10^{-3} \text{ m})}{\pi(50 \times 10^{-6} \text{ m})^4}$$

$$= 1.37 \times 10^{13} \text{ N sec/m}^5$$

The flow rate of leuprolide is

$$\dot{V} = \left(125 \times 10^{-6} \frac{\text{g}}{\text{d}}\right)\left(\frac{\text{mL}}{0.5 \text{ g}}\right)\left(\frac{10^{-6} \text{ m}^3}{\text{mL}}\right)\left(\frac{\text{d}}{24 \text{ h}}\right)\left(\frac{\text{h}}{3600 \text{ sec}}\right)$$

$$= 2.89 \times 10^{-15} \text{ m}^3/\text{sec}$$

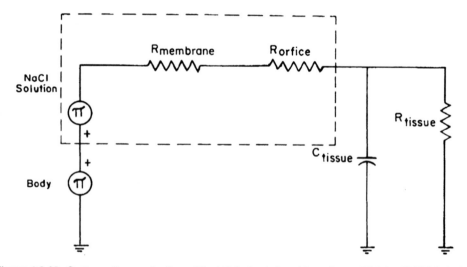

Figure 4.3.25. Systems diagram for flow of liquid into (water) and from (leuprolide) the DUROS device. The reason both fluids may be treated as equals is because both flow rates are equal and both are driven by the same osmotic pressure difference. At a flow rate of 2.9×10^{-15} m³sec, pressure drops through the orifice and in the tissue become negligible.

The pressure drop across the orifice is

$$\Delta p = R\dot{V} = (1.37 \times 10^{13} \text{ N sec/m}^5)(2.89 \times 10^{-15} \text{ m}^3/\text{sec})$$

$$= 3.95 \times 10^{-2} \text{ N/m}^2$$

which is negligible compared to osmotic pressures. Likewise, tissue resistance and compliance will cause negligible pressure drops.

The osmotic pressure of a saturated solution of sodium chloride must now be calculated. The solubility of NaCl in water at 37°C is 0.366 kg salt/kg water (Table 4.6.1). From Table 4.3.13, we find the relationship between solubility in a saturated solution (α) and mole fraction (μ_{sol}) to be

$$\mu_{sol} = \frac{\alpha M_{sv}}{M_{sl} + \alpha M_{sv}}$$

$$= \frac{(0.366 \text{ kg salt/kg H}_2\text{O})(18 \text{ kg H}_2\text{O/kg mol H}_2\text{O})}{(58.5 \text{ kg/kg mol salt}) + (0.36 \text{ kg salt/kg H}_2\text{O})(18 \text{ kg/kg mol H}_2\text{O})}$$

$$= 0.101 \text{ kg mol salt/kg mol solution}$$

Because NaCl dissociates into Na^+ and Cl^-,

$$\mu_w = 1 - 2(0.101) = 0.798$$

Using Eq. 4.3.38,

$$\pi_{salt} = \frac{[8314.34 \text{ N m/(kg mol K)}](310 \text{ K})}{0.018016 \text{ m}^3/\text{kg mol}} \ln 0.798$$

$$= -32,400 \text{ kN/m}^2$$

We find from Table 4.3.12 that the osmotic pressure of interstitial fluid, almost identical to blood plasma, is -800 kN/m^2.

Then the net osmotic pressure is

$$\pi_{body} - \pi_{salt} = -800 \text{ kN/m}^2 + 32,400 \text{ kN/m}^2$$

$$= 33.2 \times 10^6 \text{ N/m}^2$$

The required membrane resistance to produce the desired flow rate is

$$R = \frac{\Delta p}{\dot{V}} = \frac{33.2 \times 10^6 \text{ N/m}^2}{2.89 \times 10^{-15} \text{ m}^3/\text{sec}} = 1.15 \times 10^{22} \text{ N sec/m}^5$$

From Table 4.3.11 we pick a typical value of cellulose acetate resistance to be

$$R_{CA} = 40 \times 10^{10} \text{ N sec/m}^3$$

where A is the area (m^2). Thus the required area of the membrane is

$$A = \frac{40 \times 10^{10} \text{ N sec/m}^3}{1.15 \times 10^{22} \text{ N sec/m}^5} = 34.9 \times 10^{-11} \text{ m}^2$$

The diameter of the circular membrane is

$$d = \sqrt{\frac{4A}{\pi}} = \sqrt{\frac{(4)(3.44 \times 10^{-11} \text{ m}^2)}{\pi}}$$

$$= 6.7 \times 10^{-6} \text{ m}$$

Remarks

This is an extremely small membrane!! A more practical approach would be to cover the entire end of the device with a membrane of sufficient resistance to limit the flow to the desired rate. With a diameter of 4 mm, the area of the end of the tube is 0.125 cm^2. A membrane could be sought that has a resistance of $(1.15 \times 10^{22} \text{ N sec/m}^5)$ $(0.125 \times 10^{-4} \text{ m}^2) = 1.45 \times 10^{17}$ N sec/m^3, or 1.45×10^{17} N sec/m^5 per 1 m^2 area. Such a high-resistance membrane is likely to be nonporous, with water movement thorough the membrane by slow diffusion.

With a total enclosed volume of 0.5 mL, the DUROS device has a capacity of up to 2010 days of drug at the intended delivery rate.

4.3.3.4 Ionic Equilibria

Living cells contain semipermeable membranes that separate ionic components. It is common that large protein macromolecules within cells carry negative charges (act as anions). These macromolecules are trapped inside the cells as long as the cell membranes maintain their integrity (Figure 4.3.26).

macro-molecules

Other ionic components are present in fluids inside and outside the cells (extracellular, or interstitial). In animals, potassium and sodium are the major cations, and chloride is the major mobile anion. In plants, calcium is of major importance.

several equilibria

Several equilibria (or balances) must be maintained in the presence of these ions. First, if all ions are freely mobile through the membrane, concentrations of each ion species must be the same inside and outside the cell. Second, there must be no net charge difference across the membrane. Third, osmotic pressures must be equal on both sides of the membrane to maintain cell integrity.

Since the cell membrane is semipermeable, allowing ions such as chloride and potassium to pass freely, sodium to pass with some difficulty, and proteins to not pass at all, all three of the above equilibria cannot be maintained independently of one another.

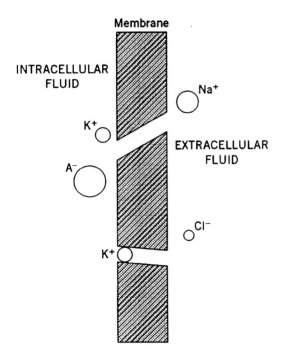

Figure 4.3.26. Diagram of the nerve cell membrane at rest. Pore diameters are presumably smaller than proteins (A^-) and hydrated sodium cations (Na^+), but larger than hydrated potassium (K^+) and chloride cations (Cl^-).

Donnan
equilibria

    The Donnan equilibria relates to the distribution of diffusible ions in the presence of a nondiffusible ion: at equilibrium the product of the molar concentrations of the diffusible ions on one side of the membrane equals the product of the molar concentrations of the diffusible ions on the other:

$$mc_{K^+} mc_{Cl^-} mc_{Na^+}|_{inside} = mc_{K^+} mc_{Cl^-} mc_{Na^+}|_{outside} \qquad (4.3.45)$$

immobile
anions

where mc_i is the molar concentration of i (moles/m^3). In order that electrochemical neutrality be satisfied, the presence of immobile protein anions inside the cell causes

$$mc_{K^+}|_{inside} > mc_{K^+}|_{outside}$$

$$mc_{Na^+}|_{inside} > mc_{Na^+}|_{outside}$$

$$mc_{cl^-}|_{inside} < mc_{cl^-}|_{outside}$$

But electrochemical neutrality may not be satisfied. The Nernst equation allows calculation of the electrical gradient that exists across the cell membrane (relative to the outside of the cell) in the presence of a concentration difference from one side to the other:

Nernst
equation

$$E_j = \frac{-RT}{F\kappa} \ln \frac{c_{ji}}{c_{jo}}$$ (4.3.46)

where E_j is the equilibrium electrical potential for ion (N m/coulomb, or volts); R, the universal gas constant $[= 8314.34 \text{ N m/(kg mol K)}]$; T, the absolute temperature (K); F, the Faraday $(= 96.487 \times 10^6$ coulombs/kg mole); κ, the ionic charge (dimensionless); c_{ji}, the concentration of ion j inside the cell (kg/m^3); and c_{jo}, the concentration of ion j outside the cell (kg/m^3). Since $mc_j = c_j/M_j$ (where M_j is the molecular weight kg/kg mole), then $c_{ji}/c_{jo} = mc_{ji}/mc_{jo}$.

trans-
membrane
concentration
differences

Measured concentrations of four different ions appear in Table 4.3.14 along with equilibrium potentials calculated from the Nernst equation at 37°C. Actual typical equilibrium potential measured across a neural membrane is about -70×10^{-3} N m/coul, suggesting that the ratio of concentrations for chloride ion is nearly at equilibrium, but that the ratios of concentrations of sodium, potassium, and proteins are not at equilibrium. Concentrations of these ions is maintained in a nonequilibrium state by various mechanisms. The concentration of protein inside the cell is maintained as discussed previously, by the impermeability of the membrane to proteins. Sodium and potassium ion concentrations are maintained at the expense of metabolic energy expended in the membrane for a so-called "ion pump." Without this pumping action, the voltage across the membrane would change significantly.

EXAMPLE 4.3.3.4-1. Equilibrium Potential for Chloride Ion

Calculate the equilibrium potential for chloride ion at 37°C using the Nernst equation.

Table 4.3.14 Typical Ionic Concentrations and Equilibrium Potentials for Vertebrate Nerve Cells

	Relative Concentration		Calculated Equilibrium Potential (N m/coulomb)
Ion	Inside Cell	Outside Cell	
Na$^+$	15	150	$+60 \times 10^{-3}$
K$^+$	150	5.5	-90×10^{-3}
Cl$^-$	9	125	-70×10^{-3}
(protein)$^-$	65	0	∞

Solution

From Eq. 4.3.46 and the ratio of chloride ion concentrations found in Table 4.3.13

$$E = \frac{-(8314.34 \text{ N m/(kg mol K)})(310 \text{ K})}{(96.487 \times 10^6 \text{ coul/kg mol})(-1)} \ln\left(\frac{9}{125}\right)$$

$$= -70.3 \times 10^{-3} \text{ N m/coul}$$

4.3.3.5 *Skin Permeability*   Skin is the tissue meant to separate an organism from its environment. Skin performs this function by being highly impermeable to many environmental fluids. However, skin is not totally impermeable to all substances, and this can be useful as well as harmful.

skin nearly impermeable

Skin is not completely impermeable to either air or water. Some animals can obtain a reasonably large fraction of their oxygen requirements through the skin. Humans can obtain about 2% of their resting oxygen consumption through the skin (Johnson, 1991). Passive diffusion of water vapor through the skin is about 6% of maximum sweat capacity (Johnson, 1991). Water can also be absorbed through the skin. The mass diffusivity value of water through the skin is about 2.5 to 25×10^{-12} m^2/sec, with a resistance value of 20–200 sec/m^3 for 1 m^2 of skin.

air and water permeability

The skin has two major layers: the nonvascular epidermis and the highly vascularized dermis. The outer layer of the epidermis is the stratum corneum, composed of compacted dead cells in a layer about 10 μm thick, which is the greatest barrier to permeability of penetrating substances. The stratum corneum provides the greatest barrier against hydrophilic compounds, while the viable epidermis and dermis are most resistant to lipophilic compounds.

stratum corneum

Substances that contact the skin may evaporate or drop off. Some may penetrate the skin and react with local enzymes to produce nontoxic compounds. Still others may pass through the dermis into the blood stream (USEPA, 1992).

In some cases, permeability of skin causes problems. Common solvents, for instance, can pass through the skin into the blood stream and cause intoxication. Mechanics, after working for long hours with gasoline and other petroleum solvents, are more likely to have impaired judgement and accidents than they are at the beginning of the day. My wife, Cathy, always claims to feel the effect of the alcohol after hand mixing a batch of brandy ball cookies at Christmas time. Other chemical pesticides can also enter the body through the skin; the hands have been found to be two to three times less permeable to pesticides than the face (USEPA, 1992). Impermeable clothing and skin protectants can be used to enhance skin protection.

solvent permeability

In other cases, attempts have been made to increase the permeability of the skin. Various drugs are much more safe and convenient to apply to human skin than to inject or give in pill form. Some anthelmintic drugs (deworming) to remove internal parasites can be poured on the skin of animals, and sufficient material diffuses through the skin to be effective. Delivery of many herbicides or systemic fungicides and insecticides requires that the chemicals be able to penetrate the relatively impermeable hydrophobic waxy outer layers of plant epidermis (called the cuticle). For this reason, detergent materials are sometimes mixed with pesticides to enhance penetration.

enhancing penetration

In Table 4.3.15 are shown permeabilities of female breast epidermis (300 to 600×10^{-6} m thick) to various commercial solvents (Ursin et al., 1995). The units of permeability are kg/(m^2 sec), which are different from those used previously in Section 4.3.3.2. Other values of permeability are found in USEPA (1992) and have units of cm^3/(cm^2 h), which differ from the units of those in Table 4.3.15 by the densities of the various compounds.

4.3.3.6 Drug Delivery   Delivery of drugs to correct medical deficiencies is an area of active and ongoing development. Recombinant DNA biotechnology is now used to produce on a commercial scale a variety of therapeutic products including interleukins, interferons, antibodies, insulin, human and animal hormones, anticoagulants, antiinflammation drugs, and contraceptives. These agents are often peptides and proteins that are often highly unstable both chemically and physically. Changes in chemical or physical structure of these compounds can render them biologically unusable (Burgess, 1993). The challenge then, is to deliver these drugs as close as possible to their intended sites of action, without significant drug degradation,

unstable drugs

Table 4.3.15 Skin Permeability to Various Commercial Solvents

Solvent	Permeability $[10^{-7}\,kg/(m^2\,sec)]$
Dimethyl sulfoxide (DMSO)	489
N-methyl-2-pyrrolidone	475
Dimethyl acetamide	297
Dimethyl formamide	272
Methyl ethyl ketone	147
Methylene chloride	66.7
Water	41.1
Ethanol	31.4
Butyl acetate	4.4
Gammabutyrolactone	3.1
Toluene	2.2
Propylene carbonate	1.9
Sulfolane	0.6

For example, the permeability of dimethyl sulfoxide is 4.89×10^{-5} kg/(m^2 sec).
Source: Ursin et al. (1995).

and to release them at a controlled rate that may last for time periods of hours to years.

The traditional method of drug delivery, taking the drug by mouth, perhaps in pill form, requires that the drug be made available for absorption by the blood supply to the gastrointestinal tract. There are at least several mass transfer processes that occur, including diffusion of the pill material to the stomach or intestinal contents, solution of the drug in the surrounding liquid, solubility of the drug in the epithelial mucosa (partition coefficient), and diffusion across the epithelial layers into blood capillaries (Figure 4.3.27). If these processes do not occur fast enough, the drug may be hydrolyzed or otherwise altered before significant absorption occurs. Large-molecular-weight compounds are very poorly absorbed and are largely lost from the body.

Hyperdermic injection has been used for years to adminster such drugs as insulin and antibiotics. This method can avoid the problems of poor absorption of large molecules, but has the disadvantage that drugs so delivered are not available at anything like a constant rate. Movements of the drug through tissue, into the blood, and from the blood to the sites of action do attenuate the otherwise wide fluctuations of drug concentrations that could accompany injections directly into the intended site of action. However, any drugs delivered through the blood stream are subject to metabolism in the liver. Thus injection is not an ideal method of drug delivery.

There are some techniques for improved drug delivery that are both ingenious and highly technical. Drug encapsulation, microsphere transport, retroviral gene delivery, and penetration enhancement are examples of these (Burgess, 1993).

pill form

hypodermic injection

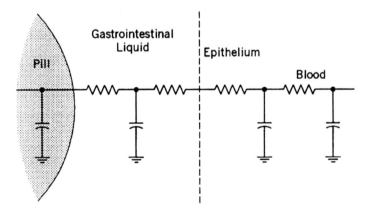

Figure 4.3.27. A simplified systems diagram for the absorption of a drug given by a swallowed pill. Diffusion processes are given as resistances and solution (or suspension) is given as capacity elements. There are several additional rate limitations, including drug solubilities in the liquid, in the epithelium, and in the blood.

transdermal
delivery

One drug delivery technique that is receiving a great deal of attention is transdermal delivery: the use of drug patches. This method is advantageous because it is noninvasive, easy to adminster, and relatively inexpensive. Agents delivered this way include methyl nicotinate, caffeine, aspirin, corticosteroids, and insulin (Ghosh and Banga, 1993a).

patch delivery

There are two diffusion processes with the use of drug patches, both of which require biological engineering attention. The first is the capture and release of the drug in the patch material matrix. There are several methods used, three of which are diagrammed in Figure 4.3.28. A drug core may be surrounded by a polymer that allows the drug to diffuse through it; the drug may be dissolved or dispersed in the polymer material, and then is released by diffusion; or the drug may be dissolved or dispersed in the polymer material, which then disperses over time (Langer, 1990). Of these, the first method, polymer surrounding a drug core, is most likely to result in uniform drug delivery over time.

diffusion
enhancement

The second patch diffusion process is skin diffusion. We have already learned that the stratum corneum, the hard outside layer of skin, is largely impermeable to hydrophilic compounds. The normal diffusion constant of the stratum corneum to certain compounds can increase as much as tenfold when the layer is hydrated (Ghosh and Banga, 1993a). Hydrating agents and surfactants may be added to patches to enhance drug movement through the stratum corneum (Ghosh and Banga, 1993a,b).

Dimethyl sulfoxide (DMSO) is a very popular diffusion-enhancing solvent that is miscible with both water and organic solvents. Other sulfoxides, azone, amines and amides, alcohols, and other compounds may also be used to enhance diffusion through the skin.

EXAMPLE 4.3.3.6-1. Dissolving a Placebo Pill

At what rate does sucrose dissolve from a 5-mm-diameter spherical pill into water?

Solution

We will assume several things: (1) What is required is the instantaneous rate of dissolution (as the pill dissolves it gets smaller and the dissolution rate decreases): (2) that mass transfer depends on diffusion and not on convection; (3) that sucrose dissolves from the surface into a saturated solution, which decreases to a concentration of zero at an infinite radius.

The rate of mass transfer is

$$\dot{m} = \frac{\Delta c}{R}$$

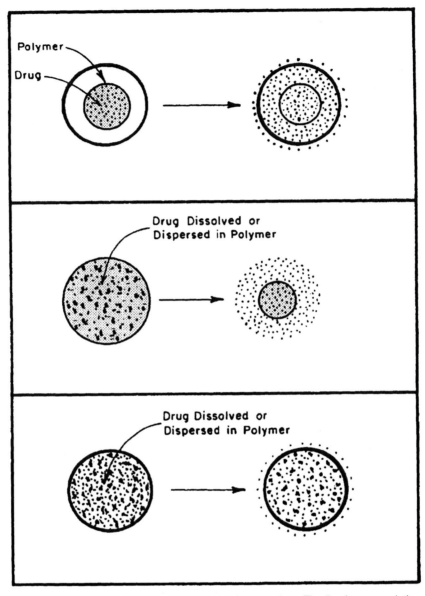

Figure 4.3.28. Three methods to devlier drugs from patches. The first is encapsulation, the second is polymeric dispersal, and from the third is polymer ablation.

where $R = (1/r_1 - 1/r_2)/4\pi D_{AC}$ from Eq. 1.8.33 for a hollow sphere.

The diffusion constant for sucrose in water can be found in Table 4.3.5.

$$R = \frac{(1/0.0025 - 1/\infty)\text{m}^{-1}}{4\pi(0.697 \times 10^{-9} \text{ m}^2/\text{sec})} = 45.7 \times 10^9 \; \frac{\text{sec}}{\text{m}^3}$$

Saturated sucrose concentration (c_o) is 1522 kg/m^3 (Table 4.6.1). The sucrose concentration at $r_2 = \infty$ is zero. Thus

$$\dot{m} = \frac{(1522 - 0)\ \text{kg/m}^3}{45.7 \times 10^9\ \text{sec/m}^3} = 3.33 \times 10^{-9}\ \text{kg/sec}$$

This is a conservative estimate, since there is likely to be significant convection occurring, especially if the pill is in the digestive system.

EXAMPLE 4.3.3.6-2. Time to Dissolve Completely

How much time is required for the pill in the previous problem to dissolve completely.

Solution

Using the previous result of mass flow rate from the pill gives one estimate of the time to dissolve. Total mass of sugar in the pill, using the density of sucrose as 1590 kg/m^2 from Table 2.5.4, is

$$m = \left(\frac{4\pi r^3}{3}\right)\rho = \frac{4\pi(0.0025m)^3}{3}\,1590\,\frac{\text{kg}}{\text{m}^3} = 1.04 \times 10^{-4}\ \text{kg}$$

$$\text{time} = m/\dot{m} = 1.04 \times 10^{-4}\ \text{kg}/3.33 \times 10^{-8}\ \text{kg/sec} = 3125\ \text{sec} = 52\ \text{min}$$

A better time estimate can be obtained, however, by realizing that, as the pill dissolves, its surface area decreases, the mass flow resistance increases, and the mass flow rate decreases. To illustrate these effects, we estimate each value for an arbitrary time interval, calculate the new pill radius at the end of the time interval, and repeat the procedure. A time interval of 750 sec was chosen based upon the rough estimate of dissolving time calculated above. Smaller time intervals give more accurate answers.

At $t = 0$, pill radius (r_0) is 0.0025 m, resistance (R) is 45.7 × 10^9 sec/m^3 as calculated in the previous example, and mass flow rate is 3.33 × 10^{-8} kg/sec, also as calculated in the previous example. The new radius at the end of the 750 sec interval is

$$\dot{m}\,\Delta t = \frac{4\pi}{3}(r_0^3 - r_{750}^3)$$

$$r_{750} = \sqrt[3]{r_0^3 - 3\dot{m}\,\Delta t/4\rho\pi}$$

$$= \sqrt[3]{(0.0025\text{ m})^3 - \frac{3(3.33\times 10^{-8}\text{ kg sec})(750\text{ sec})}{4(1590\text{ kg/m}^3)\pi}}$$

$$= 0.00228\text{ m}$$

This procedure was then repeated using the new radius as the radius at the beginning of the interval. Putting the results in tabular form

t (sec)	r_0 (m)	R (sec/m^3)	$\dot{m}$ (10^{-8} kg/sec)	r_{750} (m)
0	0.00250	45.7×10^9	3.33	0.00228
750	0.00228	50.1×10^9	3.30	0.00201
1500	0.00201	56.8×10^9	2.68	0.00172
2250	0.00172	66.4×10^9	2.29	0.00136
3000	0.00136	84.0×10^9	1.81	0.00078
3750	0.00078	146.5×10^9	1.04	≈ 0

Or the time to dissolve the pill is about 62 min rather than the 52 min calculated assuming a constant mass flow rate.

Remarks

Because the mass flow rate becomes ever more slow when the pill becomes very small, an uneven time step should be considered. Larger time steps at the beginning and small time steps at the end would combine the efficiency of larger steps with the accuracy of smaller steps to yield an acceptably accurate solution with as few steps as possible.

How small should the time steps be? One way to judge this is to set an acceptable error bound, say, 1%. Halve the time step and compare the calculated dissolving time with the dissolving time from the original time step. If the results from the two time steps agree to within 1%, then the time step chosen is probably sufficiently small. If the two time steps do not agree to within 1%, then halve the time interval again and compare results.

This problem was solved assuming no convection. If convection is appreciable, then the problem should be solved in a manner similar to Example 4.4.1-1.

4.3.4 Diffusion Resistance

nonporous
membranes

The systems schematic for mass diffusing through a nonporous membrane is given in Figure 4.3.29. Mass flows by diffusion due to a concentration difference $(c_1 - c_2)$. It is impeded in its flow by resistances R_1 and R_2 in the surrounding fluids and by membrane resistance R_m. The expression for R_m has been given by Eq. 4.3.33 in terms of membrane concentrations, or by Eq. 4.3.34 in terms of water concentrations, or by

$$R_m = \frac{L\alpha_{AB}}{AP_B} = \frac{L}{AD_{BA}} \qquad (4.3.47)$$

where R_m is the membrane resistance (sec/m^3); L, the membrane thickness (m); α_{AB}, the solubility of substance A in membrane material B $[(\text{m}^3 A\,\text{m}^2)/(\text{m}^3 B\,\text{N})]$; A, the membrane surface area (m^2); P_B, the permeability of membrane material B $[\text{m}^4/(\text{sec N})]$; and D_{BA}, the mass diffusivity of substance A in membrane material B (m^2/sec).

Partition coefficients have been diagrammed as concentration sources arranged to oppose each other. These sources are not constant, but maintain a constant ratio from one side to another. In this respect they differ from the effort sources described in Chapter 1. In electrical terms they are closer to amplifiers with gains equal to the partition coefficient, but we do not wish to detail them too closely here.

The thin membrane has a very small amount of capacity (C_m). Placement of
membrane
capacity
the capacity element within the membrane is open to some question because

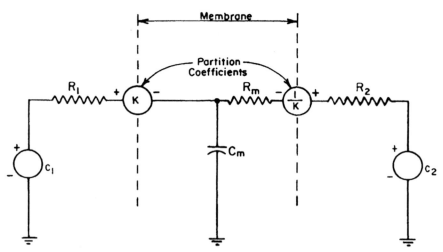

Figure 4.3.29. Systems diagram for diffusion through a membrane. Mass flows from a concentration source c_1 to a lesser concentration source c_2. R_1 and R_2 represent resistances in the fluid surrounding the membrane. Partition coefficients are given by the two concentration sources at the membrane surface. Inside the membrane there is a resistance R_m and a small mass capacity C_m.

the resistance and capacity in the membrane are really distributed together. Perhaps C_m should have been placed between the two resistors each of value $R_m/2$. The combination of R_m and C_m gives a time constant value of $R_m C_m$ that characterizes an exponential response within the membrane.

This diagram at least conveys some of the essential concepts for the systems approach to diffusion mass transfer in nonporous membranes.

porous membranes

Resistances for porous membranes are calculated somewhat differently, as given in Section 4.3.3.2. Resistances for these membranes are usually calculated from Darcy's law, Eq. 1.8.2

$$\dot{V} = \frac{-kA}{\mu} \frac{dp}{dL} \qquad (1.8.2)$$

with volume flow rate calculated due to a pressure difference.

Thus Figure 4.3.29 should be modified for use with porous membranes: There should be no partition coefficients, and the internal membrane capacity, C_m, is usually neglected.

The various forms of resistance are summarized in Table 4.3.16. Since there are two possible effort variables (concentration and pressure), and two possible flow variables (mass flow rate and volume flow rate), there are four possible resistance forms. Of these, the resistances $R = c/\dot{m}$ and $p/\dot{V}$ are used most often. The rightmost form, $R = c/\dot{V}$, is rarely, if ever, used.

If the resistance form $R = c/\dot{m}$ is considered the basic diffusion resistance, then it can be transformed into the $R = p/\dot{V}$ form by multiplying by pressure (p). The basic resistance from can be transformed into the $R = p/\dot{m}$ form by multiplying by the product of the gas constant, absolute temperature, and inverse of the molecular mass (RT/M), for a gas only. The basic resistance can be reformed into the $R = c/\dot{V}$ form by multiplying by gas density.

4.4 CONVECTION

Resistance to mass transfer does not reside solely in the penetrated medium. In a similar fashion to heat transfer, mass transfer can occur by diffusion (similar to conduction heat transfer) or by convection (similar to convective heat transfer). Convective mass transfer occurs when mass is moved in bulk, either swept along as a component of some moving medium, or moving by itself from a region of high concentration or pressure to a region of lower concentration or pressure [in a microgravity environment, surface tension differences can even cause Marangoni convection (Pearson, 1958)]. An example of the former is moisture moving with air, and an example of the latter is the concentration polarization mentioned in the discussion of membrane diffusion (Section 4.3.3.3). In the case of water vapor moving with surrounding air, convection mass transfer assists moisture transfer because the alternative is moisture movement by diffusion, a mechanism that

bulk mass movement

Table 4.3.16 Various Forms of Resistance and Capacity

Effort variable	concentration (c)	pressure (p)	pressure (p)	concentration (c)
Flow variable	mass flow rate $(\dot{m})$	volume flow rate $(\dot{V})$	mass flow rate $(\dot{m})$	volume flow rate $(\dot{V})$
Resistance	$c/\dot{m}$	$p/\dot{V}$	$p/\dot{m}$	$c/\dot{V}$
Resistance units	sec/m^3	$N\,sec/m^5$	$(m\,sec)^{-1}$	$kg\,sec/m^6$
Basic resistance modifier (gas)	1	p	RT/M^a	ρ^b
Capacity	$m/c = V$	V/p	m/p	V/c
Capacitance units	m^3	m^5/N	$m\,sec^2$	m^6/kg
Basic capacitance modifier (gas)	1	$1/p$	M/RT	$1/\rho$
Uses	molecular movement, impermeable membrane	permeability, porous membrane	convection	rarely used

$^a RT/M = $ (gas constant) (absolute temperature)/(molecular mass).
$^b \rho$, gas density.

is usually much slower than convection (consider wet clothes drying outside with and without a breeze blowing). In the case of concentration polarization, inclusion of convective mass transfer may indicate the rate-limiting step in membrane mass transfer, because the alternative is an ideal concentration of the diffusing species at the membrane face.

4.4.1 Analogies with Heat Transfer

complicated
process

Convective mass transfer is no easier to analyze than is convective heat transfer. It is not surprising, then, that the microscopic processes contributing to convection are replaced by an equation similar to convective heat transfer

$$\dot{m} = k_G A(c_1 - c_2) \tag{4.4.1}$$

where $\dot{m}$ is the mass transfer (kg/sec); k_G, the mass transfer coefficient (m/sec); A, the cross-sectional area (m^2); and c, the concentration (kg/m^3). Once the mass transfer coefficient is known, resistance to convective mass transfer can be calculated from

mass transfer
resistance

$$R_m = 1/k_G A \tag{4.4.2}$$

where R_m is the mass transfer resistance (sec/m^3). The trick is to determine the value of the mass transfer coefficient.

Before this determination is shown, it needs to be noted that there are other forms of Eq. 4.4.1. There are times when gas pressures or partial pressures are used instead of concentrations. In this case, Eq. 4.4.1 becomes

partial
pressure
formulation

$$\dot{m} = k'_G A(p_1 - p_2) \tag{4.4.3}$$

where $\dot{m}$ is the mass transfer (kg/sec); k'_G, the modified mass transfer coefficient (sec/m); A, the area (m^2); and p, the pressure [N/m^2 or kg/(m sec^2)].

From the ideal gas law

$$c = \frac{m}{V} = \frac{Mn}{V} = \frac{Mp}{RT} \tag{4.4.4}$$

where m is the mass (kg); M, the molecular weight (kg/kg mol); n, the number of moles (kg mol); R, the universal gas constant [$= 8314.34$ N m/(kg mol K); T the absolute temperature (K).

Thus

$$k'_G = k_G M/RT \tag{4.4.5}$$

for gases. For liquids, there is no direct relation between concentration and pressure. The connection between these two depends on Bernoulli's equation, and there is no simple relation between k_G and k'_G for liquids. From Darcy's equation (Eq. 1.8.2), however

liquid in
porous
medium

$$\dot{m} = \rho \dot{V} = \frac{k\rho}{\mu L} A(p_1 - p_2) \tag{4.4.6}$$

$$k'_G = \frac{k\rho}{\mu L} \tag{4.4.7}$$

where k is the Darcy permeability (m^2); ρ, the density (kg/m^3 or N sec^2/m^4); μ, the fluid viscosity (N sec/m^2); and L, the material thickness (m).

Sometimes, especially with air–water vapor mixtures, mass fraction differences are used in the convection mass transfer equation. Humidity ratios are used when considering air–water vapor mixtures

mass fraction
formulation

$$\dot{m} = k''_G A(X_1 - X_2) = k''_G A(\omega_1 - \omega_2) \tag{4.4.8}$$

where k''_G is the modified mass transfer coefficient [kg/(sec m^2)]; X, the mass fraction (kg/kg); and ω, the humidity ratio (kg H$_2$O/kg dry air). From thermodynamics and the ideal gas law

absolute
humidity

$$\omega = \frac{m_{H_2O}}{m_{air}} = \frac{M_{H_2O} p_{H_2O} V/RT}{M_{air} p_{air} V/RT} \tag{4.4.9}$$

where M is the molecular weight (kg/kg mole); p, the partial pressure (N/m^2); V, the volume (m^3); R, the universal gas constant [8314.34 N m/(kg mol K)]; and T, the absolute temperature (K).

Since temperature and volume are the same for the mixture

$$\omega = \frac{M_{H_2O} p_{H_2O}}{M_{air} p_{air}} = 0.622 \frac{p_{H_2O}}{p_{air}} \approx 0.622 \frac{p_{H_2O}}{p_{tot}} \tag{4.4.10}$$

$$k''_G \approx \frac{0.622}{p_{tot} RT} k_G \tag{4.4.11}$$

This applies only to air–water vapor mixtures.

There is a strong parallelism between convective heat and convective mass transfers. To develop this analogy, further, three additional dimensionless numbers will be defined. The first is the Schmidt number

Schmitt
number

$$Sc = \frac{\mu}{\rho D} \tag{4.4.12}$$

where Sc is the Schmidt number (dimensionless); μ, the fluid viscosity (N sec/m^2); ρ, the fluid density (kg/m^3 or N sec^2/m^4); and D, the mass diffusivity of dilute solute through solvent (m^2/sec).

The Schmidt number is the ratio of liquid (momentum) diffusivity to mass diffusivity, and is analogous to the Prandtl number in heat transfer.

The second dimensionless number is the Sherwood number

Sherwood number

$$\text{Sh} = \frac{k_G L}{D} \tag{4.4.13}$$

where Sh is the Sherwood number (dimensionless); k_G, the mass transfer coefficient (m/sec); L, the significant length (m); and D, the mass diffusivity (m^2/sec).

The Sherwood number is analogous to the Nusselt number in heat transfer. The significant length may be a distance along a straight path of mass transfer, or it may be a diameter or radius for a circular or spherical geometry.

Convective heat and mass transfer occur with a great deal of similarity. Both are limited by molecular transport across surface boundary layers. The relationship between mass transfer and heat transfer boundary layer thicknesses is given by the nondimensional Lewis number, defined as:

Lewis number

$$\text{Le} = \frac{\text{Pr}}{\text{Sc}} = \frac{c \rho D}{k} \tag{4.4.14}$$

where Le is the Lewis number (dimensionless); Pr, the Prandtl number (dimensionless); Sc, the Schmidt number (dimensionless); c_p, specific heat of fluid [N m/(kg °C)]; ρ, fluid density (kg/m^3); D, the mass diffusivity (m^2/sec); and k, the thermal conductivity of fluid [N m/(sec m °C)].

For most heat and mass transfer conditions, the Lewis number is assumed to be unity (Johnson and Kirk, 1981). This assumption becomes better as turbulence is approached.

As long as the surface of the material through which convection mass transfer occurs is wetted by the moving material (that is, a mass boundary layer exists at the surface), and the Lewis number is unity, then convection mass transfer equations can be obtained by replacing the Nusselt number with the Sherwood number and the Prandtl number with the Schmidt number. Thus

$$\text{Nu} = A\,\text{Re}^p\,\text{Pr}^q \tag{4.4.15a}$$

convective heat and mass transfer

in heat transfer becomes, in mass transfer

$$\text{Sh} = A\,\text{Re}^p\,\text{Sc}^q \tag{4.4.15b}$$

where the coefficient A and exponents p and q maintain the same values from one equation to the next. Hence the many equations for forced convection heat transfer, for both laminar and turbulent conditions, may be used to obtain numerical values for convection mass transfer.

Additionally, Eqs. 4.4.15a and 4.4.15b can be manipulated to give

$$\frac{Sh}{Nu} = \frac{A \, Re^P \, S_c^q}{A \, Re^P \, Pr^q} = \frac{k_G L/D}{h_c L/k} \tag{4.4.16}$$

$$\frac{k_G k}{h_c D} = \left(\frac{S_c}{Pr}\right)^q = Le^q = 1^q \tag{4.4.17}$$

mass transfer coefficient from heat transfer coefficient

Thus

$$k_G = h_c D/k \tag{4.4.18}$$

where k_G is the mass transfer coefficient (m/sec); h_c, the convection heat transfer coefficient [N m/(m^2 sec °C)]; D, the mass diffusivity (m^2/sec); and k, the thermal conductivity [N m/(m sec °C)].

If a value for h_c has been determined, then a value for k_G can be obtained from it.

There is no simple analogy between heat transfer by natural convection and natural convective mass transfer. Although natural convection does occur in mass transfer, and this process is due to density differences between regions with higher and lower concentrations of solute within a solvent, there seems to be little attention paid to this mode of convection.

EXAMPLE 4.4.1-1. Drying of Cranberries

Cranberries are often harvested in flooded bogs by beating the plants with a reel. The buoyant fruits pop to the surface, where they are herded to a dike and elevated into trucks. Surface moisture is removed from the wet berries as they roll over a fine screen through which warm air passes. Ambient air at the bog is assumed to be at 4°C and saturated with water vapor. Heated air moving at 2 m/sec is assumed to be at 30°C. If the average diameter of the berries is 1.5 cm and the water layer is 0.2 mm thick, how long will it take to dry the berries?

Solution

Saturated water vapor pressure in 4°C air is 813 N/m^2 (see Section 4.8.1.2). When the air is heated, no moisture is added; so the water vapor pressure in the heated air before it contacts the berries remains at 813 N/m^2. If the air leaving the berries is saturated with water vapor at 30°C, its vapor pressure is 4240 N/m^2. The driving pressure difference for evaporation is thus $4240 - 813 = 3430$ N/m^2.

We need values for the Reynolds number and Schmidt number. To calculate the Reynolds number requires the viscosity of air $[1.87 \times 10^{-5}\,\text{kg}/(\text{m sec})$ from Table 2.4.1] and density of air ($1.168\,\text{kg/m}^3$ from Table 2.5.4) at 30°C. Normally, we would need physical properties of the air–water vapor mixture, but, even for air saturated with water vapor, the amount of vapor in the air is small and the physical properties of the mixture are not considerably different from those for dry air. The Reynolds number is therefore

$$\text{Re} = \frac{dv\rho}{\mu} = \frac{(0.015\ \text{m})(2\ \text{m/sec})(1.168\ \text{kg/m}^3)}{[1.87 \times 10^{-5}\ \text{kg}/(\text{m sec})]} = 1874$$

From Table 4.3.1, the mass diffusivity for water vapor in air at 25°C can be found to be $0.260 \times 10^{-4}\,\text{m}^2/\text{sec}$. This should be corrected using Eq. 4.3.16 to the value at 30°C

$$D_{\text{H}_2\text{O}-\text{Air}} = 0.260 + (0.028)(303^{1.75} - 298^{1.75})/(315^{1.75} - 298^{1.75})$$

$$= 0.268 \times 10^{-4}\ \text{m}^2/\text{sec}$$

The Schmidt number is, from Eq. 4.4.12,

$$\text{Sc} = \frac{\mu}{\rho D_{\text{H}_2\text{O}-\text{Air}}} = \frac{1.87 \times 10^{-5}\ \text{kg}/(\text{m sec})}{(1.168\ \text{kg/m}^3)(0.268 \times 10^{-4}\ \text{m}^2/\text{sec})}$$

$$= 0.597$$

We make use of the analogy between forced convection heat and mass transfer. Equation 3.3.32d gives

$$\text{Nu} = 0.41\ \text{Re}^{0.6}\ \text{Pr}^{0.3}$$

which leads to

$$\text{Sh} = 0.41\ \text{Re}^{0.6}\ \text{Sc}^{0.3}$$

$$= 0.41(1874)^{0.6}(0.597)^{0.3} = 32.3$$

Using Eq. 4.4.13 gives

$$k_{\text{G}} = \frac{\text{Sh}\ D}{d} = \frac{(32.3)(0.268 \times 10^{-4}\ \text{m}^2/\text{sec})}{(0.015\ \text{m})} = 5.77 \times 10^{-2}\ \text{m/sec}$$

From Eq. 4.4.5

$$k'_G = \frac{k_G M_{H_2O}}{RT} = \frac{(5.77 \times 10^{-2} \text{ m/sec})(18 \text{ kg/kg mol})}{[8314.34 \text{ N m/(kg mol K)}](303.15 \text{ K})}$$

$$= 4.12 \times 10^{-7} \text{ kg/(N sec) or sec/m}$$

The area of the sphere from which water evaporates is

$$A = \pi d^2 = \pi (0.015 \text{ m})^2 = 7.07 \times 10^{-4} \text{ m}^2$$

The rate of moisture removal is, according to Eq. 4.4.3

$$\dot{m} = k'_G A(p_1 - p_2) = [4.12 \times 10^{-7} \text{ kg/(N sec)}](7.07 \times 10^{-4} \text{ m}^2)(3430 \text{ N/m}^2)$$

$$= 9.99 \times 10^{-7} \text{ kg/sec}$$

The volume of water surrounding each berry to be evaporated is

$$\text{volume} = \frac{\pi}{6}(d_o^3 - d_i^3) = \frac{\pi}{6}[(0.0152 \text{ m})^3 - (0.0150 \text{ m})^3]$$

$$= 7.16 \times 10^{-8} \text{ m}^3$$

Water at 30°C has a density of 996 kg/m^3. Thus the time to dry is

$$\text{time} = (7.16 \times 10^{-8} \text{ m}^3)(996 \text{ kg/m}^3)/(9.99 \times 10^{-7} \text{ kg/sec})$$

$$= 71.4 \text{ sec}$$

This is a realistic time, especially if the berries are rolling freely across the blast of air.

4.4.2 Packed Beds

many small particles

Packed beds are containers filled with many small particles to give a very large surface area in a small volume. They are used where surface area phenomena are exploited, for example, the adsorption of vapors and gases on activated charcoal or the oxidation of ammonia to nitrite and/or nitrate by bacteria growing on the surface of the particles. The growth of microorganisms on

biofilms

surfaces is called a "biofilm" (Figure 4.4.1).

Take, for example, the conversion of nitrogen forms by bacterial biofilms. In the presence of oxygen, ammonia is used by the *Nitrosomonas* group of bacteria to supply its energy needs

nitrification

$$2NH_3O_2 \xrightarrow{\text{Nitrosomonas}} 2NO_2^- + 2H^+ + 2H_2O + 3.3 \times 10^5 \text{ N m} \quad (4.4.19)$$

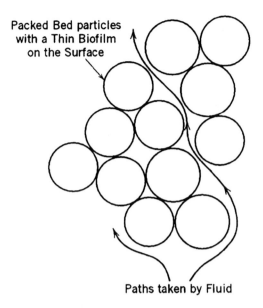

Packed Bed particles
with a Thin Biofilm
on the Surface

Paths taken by Fluid

Figure 4.4.1. Diagram of a packed bed.

Nitrobacter is a second group of bacteria that oxidizes nitrite to nitrate

$$2NO_2^- + O_2 \xrightarrow{\text{Nitrobacter}} 2NO_3^- + 1.8 \times 10^5 \text{ N m} \tag{4.4.20}$$

Because these reactions require oxygen to proceed, poorly aerated media cannot nitrify ammonia; so there is an advantage to growing these bacteria as thin films on surfaces where oxygen diffusion paths are short. A disadvantage of these biofilms is that water diffusion paths are also short, and drying is a constant problem. In the absence of oxygen, other bacteria can reverse the above chemical reactions and form ammonia from nitrate.

Packed beds are useful only when there is a convective flow through them. This fluid flow supplies raw materials and carries products away. In the case of contaminant removal by charcoal, the raw materials are the contaminant gases and vapors along with an air carrier; the product is the purified air. In the case of nitrification, the raw material is ammonia dissolved in water; the product is nitrate ions also dissolved in water.

When the flow through the bed becomes very great, drag forces on the particles begin to suspend the particles in the fluid, and thus becomes a
fluidized bed   "fluidized bed." Fluidized beds are used in many processes in the chemical industry, but they are analyzed somewhat differently from packed beds.

The void fraction (ε) for a packed bed is

void fraction
$$\varepsilon = \frac{\text{volume of void space}}{\text{total volume}} \tag{4.4.21}$$

Void fraction values range from 0.3 to 0.5. In order that the packed bed be effective, flow channeling, that is, flow through preferentially low resistance paths that circumvents much contact with particle surfaces, is to be avoided.

Assuming the packed bed contains spherical particles, the volume of the spheres is $(1 - \varepsilon)V_{tot}$. An equivalent surface area of the particles in the packed bed thus becomes

surface area
$$\frac{A_{tot}}{(1 - \varepsilon)V_{tot}} = \frac{A_{sphere}}{V_{sphere}} \qquad (4.4.22)$$

or

$$A_{tot} = \frac{A_{sphere}}{V_{sphere}}(1 - \varepsilon)V_{tot} = \frac{\pi d_s^2}{\pi d_s^3/6}(1 - \varepsilon)V_{tot}$$
$$= \frac{6(1 - \varepsilon)V_{tot}}{d_s} \qquad (4.4.23)$$

where d_s is the equivalent sphere diameter for packed bed particles (m). The rate of mass transfer in the packed bed can be obtained from the convection mass transfer equation

mass transfer
rate
$$\dot{m} = k_G A_{tot}\, \Delta c \qquad (4.4.1)$$

The mass transfer coefficient, k_G, is determined from experimental correlations. For the flow of gases within a Reynolds number range of 10 to 10^4 (Geankoplis, 1993)

Sherwood
number, gas
$$Sh = \frac{k_G d_s}{D_{AB}} = \frac{0.4548 v d_s}{\varepsilon\, Re^{0.4069}\, Sc^{2/3}\, D_{AB}} \qquad (4.4.24)$$

where Sh is the Sherwood number (dimensionless); v, the superficial average velocity (m/sec); d_s, the diameter of spherical particles (m); k_G, the mass transfer coefficient (m/sec); ε, the void fraction (dimensionless); D_{AB}, the mass diffusivity of substance A (that interacts with the packed bed particles) within the gas B (m^2/sec); Re, the Reynolds number (dimensionless); and Sc, the Schmidt number (dimensionless).

For liquids flowing within the packed bed (Geankoplis, 1993)

Sherwood
number, liquid
$$Sh = \frac{1.09 v d_s}{\varepsilon\, Re^{2/3} Sc^{2/3}\, D_{AB}}, \qquad \text{for } 0.0016 < Re < 55 \text{ and } 165 < Sc < 70600$$
$$(4.4.25)$$

$$Sh = \frac{0.250 v d_s}{\varepsilon\, Re^{0.31}\, Sc^{2/3}\, D_{AB}}, \qquad \text{for } 55 < Re < 1500 \text{ and } 165 < Sc < 10690$$
$$(4.4.26)$$

$$Sh = \frac{0.4548 v d_s}{\varepsilon\, Re^{0.4069}\, Sc^{2/3}\, D_{AB}}, \qquad \text{for } 10 < Re < 1500 \qquad (4.4.27)$$

where D_{AB} is the mass diffusivity of solute A in solvent B (m^2/sec).

The superficial velocity, v, is obtained as the volume flow rate of the fluid divided by the total cross-sectional area of the packed bed, as if it were not filled with particles. The Reynolds number is also calculated using the superficial velocity

$$\text{Re} = \frac{d_s v \rho}{\mu} \qquad (4.4.28)$$

where ρ is the density of fluid (kg/m^3); μ, the viscosity of fluid [kg/(m sec)]; d_s, the particle diameter (m); and v, the superficial velocity (m/sec).

If the particles in the packed bed are nearly spherical, then the value for diameter, d_s, is obtained straightforwardly. If the particles are irregularly shaped, then an equivalent diameter can be obtained by using an inverted form of Eq. 4.4.23

$$d_s = \frac{6(1 - \varepsilon) v_{\text{tot}}}{A_{\text{tot}}} \qquad (4.4.29)$$

as long as there is some estimate of the total surface area inside the bed.

mass diffusivity cancels

Although the mass diffusivity, D_{AB} appears in Eqs. 4.4.24–4.4.27, it may not matter what value is used because the parameter of interest is usually not the Sherwood number, but instead is the mass transfer coefficient, k_G. Note that the mass diffusivity in Eqs. 4.4.24–4.4.27 cancels with the mass diffusivity in Eq. 4.4.13 when solving for k_G. Once k_G is found, it can be inserted into Eq. 4.4.1 to obtain mass flow rate. The correct diffusivity value must, however, be used to calculate the Schmidt number.

changing concentration difference

The concentration difference inside the bed is constantly changing. The concentration within the fluid is greatest at the entrance and least at the exit. An analogous situation existed for a heat exchanger in cross flow or parallel flow (Section 3.7.1). Thus, in a similar manner, the logarithmic mean concentration difference is used

logarithmic concentration difference

$$\Delta c = \frac{(c_{AS} - c_{A1}) - (c_{AS} - c_{A2})}{\ln\left(\dfrac{c_{AS} - c_{A1}}{c_{AS} - c_{A2}}\right)} \qquad (4.4.30)$$

where c_{AS} is the concentration of substance A at the surface of the packed bed particles (kg/m^3); c_{A1}, the concentration of substance A in the fluid as it enters the packed bed (kg/m^3); and c_{A2}, the concentration of substance A in the fluid as it leaves the packed bed (kg/m^3). Because the condition around the particles is often (when emitting the solute) that of a saturated solution of substance A in the fluid, c_{AS} can be obtained as the solubility of substance A in fluid B (see Table 4.6.1). See Budavari (1989), Lide (1995), and Stephen and Stephen (1963) for solubility values.

A mass balance on the packed bed also gives

$$\dot{m} = \dot{V}(c_{A1} - c_{A2}) \tag{4.4.31}$$

where $\dot{V}$ is the volume flow rate (m³/sec). Thus unknown concentrations or flow rates can be determined and packed beds can be designed for particular operations of substance removal, substance addition, or substance transformation (as in the case of nitrification).

limitations to logarithmic concentration difference

When concentrations inside the packed bed no longer change along the length of the bed, the meaning of c_{A2} in Eq. 4.4.30 must also change. Such cases arise when substance is being removed from the fluid flowing through the packed bed and concentration effectively reaches zero before the end of the bed, or when substance is being added to the fluid as it flows through the bed and concentration saturates. The first case is analogous to a heat transfer

analogies to heat exchangers

system where the contained fluid reaches absolute zero temperature and the second case is analogous to the contained fluid reaching the temperature of the walls of the pipe. For both of these cases, the point where concentration no longer changes must be identified and the effective mass transfer area of the bed must be reduced, sometimes drastically.

EXAMPLE 4.4.2-1. Removal of Airborne Toluene with Charcoal Filtration

Toluene (C_7H_8) is a common solvent found in industrial settings, often at a concentration of about 300 ppm. The highest allowable level of toluene in breathable air (called the threshold limit value) is 50 ppm (ACGIH, 1991). Thus some kind of respiratory protection must be worn.

An air-purifying respirator mask uses activated charcoal to filter airborne contaminants for inspired air. One type of filter contains 250 cm³ of charcoal in a canister 10 cm in diameter. The charcoal is packed under pressure to avoid the formation of channels through which inhaled air may bypass the charcoal. Charcoal particle diameters are in the range of 0.83 to 1.65 mm, with an average of about 1.2 mm. The void fraction is about 0.2. At a temperature of 25°C and an airflow rate of 65 L/min, it takes about 40 min for toluene to appear in the air leaving the charcoal. Calculate the mass flow rate of toluene attaching to the charcoal particles. The molecular weight of toluene is 92.13.

Solution

There are several ways to approach this problem. We will first use the method outlined in Eqs. 4.4.1, 4.4.24, 4.4.28, and 4.4.30. The canister cross-sectional area is

$$A = \pi(0.1 \text{ m})^2/4 = 0.00785 \text{ m}^2$$

Then

$$v = \frac{\dot{V}}{A} = \frac{(65 \text{ L/min})(10^{-3} \text{ m}^3/\text{L})(\text{min}/60 \text{ sec})}{0.00785 \text{ m}^2} = 0.138 \text{ m/sec}$$

Because there is so little toluene in the air, density and viscosity of the air will be used; $\rho_{air} = 1.19 \text{ kg/m}^3$ and $\mu_{air} = 0.1079 \times 10^{-3} \text{ kg/(m sec)}$. The Reynolds number (Eq. 4.4.28) is thus

$$\text{Re} = \frac{(0.0012 \text{ m})(0.138 \text{ m/sec})(1.19 \text{ kg/m}^3)}{0.0179 \times 10^{-3} \text{ kg/(m sec)}} = 11.0$$

The mass diffusivity of toluene in air can be estimated from Eq. 4.3.15 with help from Table 4.3.2. The atomic diffusion volume for toluene, containing a benzene ring, is

$$V = 7V_c + 8V_H - V_{ring} = 7(16.5) + 8(1.98) - 20.2$$

$$= 111 \text{ m}^3$$

$$D_{\text{Tol-Air}} = \frac{0.01013(298 \text{ K})^{1.75}\left(\dfrac{1}{29} + \dfrac{1}{92.13}\right)^{1/2}}{(101300 \text{ N/m}^2)[(20.1 \text{ m}^3)^{1/3} + (111 \text{ m}^3)^{1/3}]^2}$$

$$= 8.04 \times 10^{-6} \text{ m}^2/\text{sec}$$

The Schmidt number is, from Eq. 4.4.12,

$$\text{Sc} = \frac{0.0179 \times 10^{-3} \text{ kg/(m sec)}}{(1.19 \text{ kg/m}^3)(8.04 \times 10^{-6} \text{ m}^2/\text{sec})} = 1.87$$

We can use Eq. 4.4.24 to calculate the Sherwood number

$$\text{Sh} = \frac{(0.4548)(0.138 \text{ m/sec})(0.0012 \text{ m})}{(0.2)(11.0)^{0.4069}(1.87)^{2/3}(9.04 \times 10^{-6} \text{ m}^2/\text{sec})} = 11.6$$

From this we obtain

$$k_G = \frac{\text{Sh } D}{d_s} = \frac{(11.6)(8.04 \times 10^{-6} \text{ m}^2/\text{sec})}{0.0012 \text{ m}} = 0.0779 \text{ m/sec}$$

The toluene concentration in the entering air is

$$c = \frac{m_{Tol}}{V_{air}} = \frac{pM_{Tol}(ppv)}{RT} = \frac{(101300 \text{ N/m}^2)(92.13 \text{ kg/kg mol})}{(8314.34 \text{ N m/(kg mol K))}(298 \text{ K})}\left(\frac{300}{10^6}\right)$$

$$= 1.13 \times 10^{-3} \text{ kg/m}^3$$

We might be tempted to use Eq. 4.4.30 to calculate a logarithmic mean concentration difference. However, to use this equation here would be incorrect, because the concentration of toluene near the charcoal is not saturation, but zero. The charcoal efficiently adsorbs toluene from the air and holds it tight. Its vapor pressure is nearly zero. If the charcoal were giving off toluene, then Eq. 4.4.30 could be used. We can probably use one-half the value of toluene concentration just calculated as the mean concentration difference. We use Eq. 4.4.23 to obtain A_{tot}

$$A_{tot} = \frac{6(1 - \varepsilon)V_{tot}}{d_s} = \frac{6(1 - 0.2)(250 \times 10^{-6} \text{ m}^3)}{(0.0012 \text{ m})}$$

$$= 1 \text{ m}^2$$

Thus

$$\dot{m} = k_G A_{tot} \, \Delta c = (0.0779 \text{ m/sec})(1 \text{ m}^2)(0.565 \times 10^{-3} \text{ kg/m}^3)$$

$$= 4.40 \times 10^{-5} \text{ kg/sec}$$

This is the answer to the problem using one approach.

The second approach is much simpler. We can use Eq. 4.4.31 to calculate

$$\dot{m} = \dot{V}(c_{in} - c_{out}) = \left(\frac{65 \times 10^{-3} \text{ m}^3}{60} \frac{}{\text{sec}}\right)(1.13 \times 10^{-3} \text{ kg/m}^3)$$

$$= 1.22 \times 10^{-6} \text{ kg/sec}$$

Remark

There is a forty-fold difference between the two answers. The explanation probably lies in the calculation for A_{tot}. There is probably a much smaller area of the charcoal that is contributing to the absorption than the entire 1 m². When charcoal absorbs a chemical, it is usually very efficient about it; the charcoal on the upstream side of the front where absorption is occurring is saturated with toluene; charcoal downstream from the front is nearly toluene-free.

There is, in addition, a huge increase in absorption area that is developed in charcoal when it is prepared to become filtration material. From geometric considerations alone, the area is calculated to be $1 \, m^2$. Processing the charcoal makes each particle very porous, and this $250 \, cm^3$ of charcoal would be expected, by experiment, to have about $169,000 \, m^2$ of area available to absorb the toluene. Of course, the actual area at the absorption front would still have to be very small, but this shows that the absorption front must be extremely abrupt.

4.5 MASS GENERATION

materials production

Strictly speaking, mass cannot be generated or destroyed unless we consider atomic fission or synthesis, which, of course, is not within the bounds of this book. From an engineering perspective, however, the production of mass species is as important as the production of mass itself, since some mass species can either be extremely desirable or undesirable. In either case, the species appears and must be dealt with. From a biological perspective, life itself depends on the production or removal of specific materials, and so the appearance of these materials is extremely critical.

Materials generation can come about as a result of physical processes, as with evaporation of moisture into the air or absorption of contaminants by charcoal filters; materials generation can also come about through simple chemical reactions, as with the oxidation of glucose to form water and carbon dioxide.

4.5.1 Enzymatic Reactions

biological catalysts

The biological engineer would probably also be interested in the more complex physico-chemical process of enzyme-catalyzed reactions important in biotechnology as well as in living organisms. For the specialized class of biological engineers called "metabolic engineers," enzymatic reactions within living cells are the objedct of attempts at manipulation to change cellular products.

A general reversible chemical reaction can be given by

$$nA + mB \rightleftharpoons A_n B_m \tag{4.5.1}$$

where n, m are coefficients; and A, B, are chemical constituents.

equilibrium constant

The equilibrium constant for this reaction, which is the ratio of the rate of the forward reaction to the rate of the reverse reaction, is (Murray et al., 1990)

$$K_{eq} = \frac{K_{fwd}}{K_{rev}} = \frac{c_{A_n B_m}}{c_A^n c_B^m} \tag{4.5.2}$$

where K_{eq} is the equilibrium constant; K_{fwd}, the rate of forward reaction; K_{rev}, the rate of reverse reaction; and c_i, the concentration of chemical constituent i (units are usually molar or moles/L = kg mol/m³).

If $K_{eq} > 1$, then the reaction proceeds spontaneously from nA and mB toward A_nB_m. If $K_{eq} < 1$, the reaction proceeds spontaneously in the reverse direction. The standard change in free energy is given by

free energy
$$\Delta G^\circ = -RT \ln K_{eq} \tag{4.5.3}$$

where ΔG^0 is the standard change in free energy (N m/kg mol); R, the gas constant [8314.34 N m/(kg mol K)]; and T, the absolute temperature (K).

If ΔG^0 is negative, the reaction proceeds spontaneously toward the formation of A_nB_m.

The preceding tells the direction of the reaction, but does not tell the rate.

reaction rates Reaction rates depend on energy barrier magnitudes, not on the magnitude of ΔG^0. For a serial string of chemical reactions that begin with certain substrates and end with other products, the reaction rate depends on the rate-limiting step.

Most biochemical processes proceed at required rates because they involve enzymes that speed reaction rates by increasing local substrate concentrations. Thus

$$\text{enzyme} + \text{substrate} \rightleftharpoons \text{enzyme} + \text{product} \tag{4.5.4}$$

and

$$K_{eq} = \frac{c_{enzyme} c_{prod}}{c_{enzyme} c_{substr}} = \frac{c_{prod}}{c_{substr}} \tag{4.5.5}$$

Realistic concentration values for living systems are 10^{-9} kg mol/m³ for enzymes and 10^{-3} kg mol/m³ for substrates.

One of the simplest enzymatic reactions involves only a single enzyme and a single substrate forming a single product. Much more complicated multienzyme and multisubstrate reactions are possible, for example, in the production of ethanol from biomass or in the citric acid cycle present in cell mitochondria. In a single enzyme–single substrate reaction, there is often a limit on the number of active sites on an enzyme to which substrate molecules can bind. Since an enzyme operates by holding reactants close to each other to form products more rapidly and easily than would otherwise be the case, a binding sites fixed limit on the number of binding sites results in a limitation on the rate of product formation. This limit is called saturation.

4.5.1.1 Enzyme–Substrate Kinetics Kinetics of simple enzyme–substrate reactions are often described by the Michaelis–Menten

equation (Rubinow, 1975)

Michaelis–
Menten
equation

$$\dot{m}_{gen} = \frac{\dot{m}_{max} c_s}{K_m + c_s}$$ (4.5.6)

where $\dot{m}_{gen}$ is the rate of product formation or rate of substrate disappearance (kg/sec); $\dot{m}_{max}$, the maximum rate of the reaction (kg/sec); c_s, the concentration of substrate (kg mol/m^3); and K_m, the Michaelis–Menten constant (kg mol/m^3).

maximum
reaction rate

The maximum reaction rate, $\dot{m}_{max}$, is mathematically equal to the product of molecular weight, reaction rate of product formation from the enzyme–substrate complex, and initial molar concentration of enzyme. However, the value for $\dot{m}_{max}$ is almost exclusively determined by experiment. Changing the initial concentration of substrate in the reactor has no effect on $\dot{m}_{max}$; changing the initial concentration of enzyme changes $\dot{m}_{max}$.

Michaelis–
Menten
constant

The Michaelis–Menten constant represents the value of substrate concentration (c_s) that results in a reaction rate one-half of the maximum, as shown in Figure 4.5.1. Its value can also be related to the dissociation constant of the enzyme–substrate complex, but is also usually determined experimen-

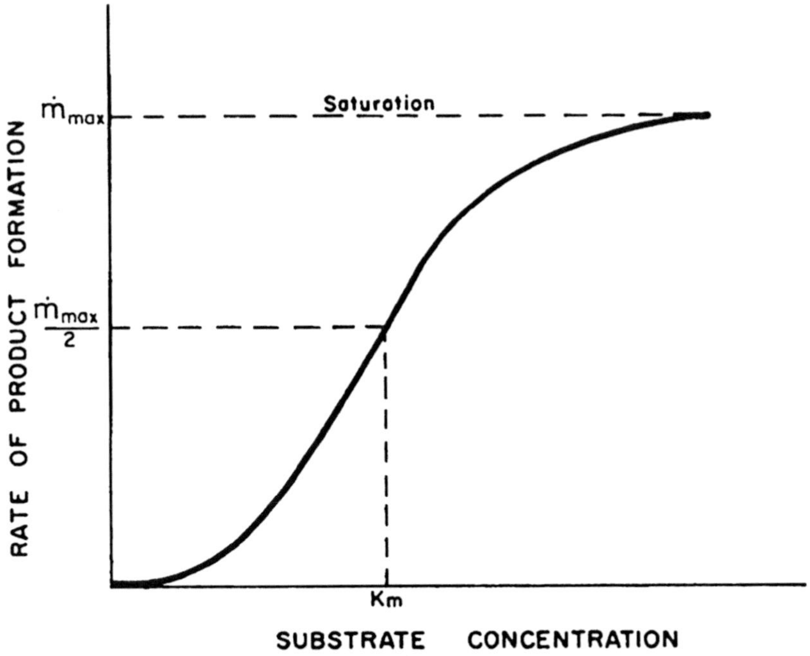

Figure 4.5.1. Graph of Michaelis–Menten kinetics. The rate of product formation depends on the concentration of the substrate from which it is made. Enzyme concentration is assumed to be adequate so that it does not limit the reaction rate.

tally. Lower values for K_m suggest higher affinity of the enzyme for the substrate. K_m values of 10–100 nanomolar concentration of substrate represent high-affinity systems; normal values of K_m are in the range of 10–100 millimolar concentration; a typical enzyme–substrate system has a K_m value of about 5 millimolar at room temperature (Tao, 1996). These values may be converted into units of kg/m^3 by multiplying with the molecular weight of the substrate (see Table 4.5.1). Temperature and pH of the surroundings can greatly influence K_m values.

The use of the Michaelis–Menten equation to determine the rate of substance formation would require some special circumstances. First, there must be a simple enzyme and substrate reaction occurring. Biologically important examples are cholinesterase–acetylcholine, invertase–sucrose, urease–urea, amylase–starch, and cellulase–cellulose. Second, the rate of activity of the enzyme may well depend on temperature (usually through an Arrhenius equation) and acidity level. Third, if the enzyme is immobilized (see Figure 4.5.3) so that it is not lost to the reactor as the product is harvested, diffusion of substrate into the enzyme carrier may limit the reaction rate. Fourth, there may be an initial transient period during which the reaction does not proceed as rapidly as predicted by the Michaelis–Menten equation. Fifth, the class of protein molecules we know as enzymes can contain many different variants. Enzymes are known by names related to the substrates they act upon (for example, lactase and maltase), but there may be many different molecules that perform the same function and are therefore known by the same name. Each different molecule type is likely to have different values of $\dot{m}_{max}$ and K_m. Thus it is difficult to characterize the kinetics of one type of enzyme completely unless the exact form of the enzyme is identified.

4.5.1.2 Immunoassays

These tests have become extremely important in the detection of biochemicals, exposure to diseases, and determination of various environmental compounds causing active allergic responses. The basis for these tests is the specific antigen–antibody reactions that occur within higher-order living organisms to ward off intrusion by foreign bodies. All antibodies belong to a special category of proteins called immunoglobulins; these are usually abbreviated Ig. There are several types of these, belonging to classes IgA, IgD, IgE, IgG, and IgM. Antibodies that are able to bind to a number of antigens are called polyclonal; those that are specific to one particular antigen are monoclonal antibodies.

antigen–antibody reactions

Immunoassays are used to test for either antigens or antibodies. Included in the former are bee venom, various helminthic parasites, and viruses (Wardley and Crowther, 1982). Included in the latter are antibodies to particular disease microorganisms as indicators of exposure to, or carriers of, diseases of interest. When immunoassays are used, a precipitate of some kind is usually formed, and the visual presence of the precipitate is usually sufficient evidence of detection.

Table 4.5.1 Some Michaelis–Menten Constant Values

Enzyme	Importance	Substrate	K_m (kg mol/m³)
Chymotrypsin	Digestion of proteins	Acetyl-L-tryptophanamide	5.3×10^{-3}
		Phenylalanine methyl ester	9.6×10^{-3}
		Clycyl-L-tyrosinamide	1.22×10^{-1}
		Acetyl-L-tyrosinamide	3.1×10^{-2}
Trypsin	Digestion of proteins		6.85×10^{-3}
Lipsozyme	Antibacterial in mucus	Hexa-N-acetylglucosamine	6×10^{-6}
β-Galactosidase	Hydrolyzes lactose into galactose and glucose	Lactose	4×10^{-3}
Threonine deaminase	Amino acid synthesis	Threonine	5×10^{-3}
Carbonic anhydrase	Converts CO_2 to bicarbonate	Carbon dioxide	8×10^{-3}
Penicillinase	Destroys penicillin, produced by bacteria	Benzylpenicillin	5×10^{-5}
Pyruvate carboxylase	Gluconeogenesis	Pyruvate	4×10^{-4}
		Bicarbonate ion (HCO_3^{-})	1×10^{-3}
		Adenosine triphosphate (ATP)	6×10^{-5}
Arginine–tRNA synthetase	Synthesizes tRNA	Argenine	3×10^{-6}
		Transfer ribonucleic acid (tRNA)	4×10^{-7}
		Adenosine triphosphate (ATP)	3×10^{-4}
Maltase	Produces glucose from maltose	Maltose	2.1×10^{-1}
Lactate dehydrogenase	Reduces pyruvate to lactate	Pyruvate	3.5×10^{-5}
Peroxidase	Transfers oxygen from hydrogen peroxide to an acceptor substance (found mostly in plants)	Hydrogen peroxide	4×10^{-7}

See Williams and Lansford (1967), Stryer (1988), White et al. (1959), and Schomburg and Salzman (1990).

ELISA

One quick and popular immunoassay is the enzyme-linked immunosorbent assay, or ELISA (Kemeny, 1991). In ELISA, a surface is coated with antigen or antibody and dried. An antibody is then added to the test, usually suspended in water. Chemically linked to the antibody is an enzyme, usually one of three: horseradish peroxidase, alkaline phosphatase, or β-D-galactosidase. When the sample to be tested is added, an antibody–antigen complex is formed and becomes fixed to the surrace. Excess enzyme-labeled antibody is then washed from the surface. A chromogen substrate is added that combines with the enzyme linked to the antibody. When it does, it forms a colored compound that can be quantified by optical density. The optical density is thus dependent on the amount of enzyme remaining in the assay, and that amount, in turn, depends on the concentration of antibody in the sample tested.

faster test

This assay usually takes several hours to complete, but there is interest in speeding it up. The time taken by the assay is composed of diffusion time, antigen–antibody binding (very quick), and enzyme–substrate kinetics of the Michaelis–Menten type (except that the substrate concentration does not limit the rate). Thus a faster assay will come only as a result of minimizing each component of time.

4.5.1.3 Bionsensors

Biosensors is a term that usually applies to sensors of biochemicals. Biosensors are used for environmental monitoring, food technology, and biomedicine. There are biosensors based on the same principles as immunoassays, based on the production of light from luciferin in the presence of ATP as catalyzed by luciferase (the same bioluminescence reaction that appears in the American firefly), based on sensing of the heats of formation of biochemical reactions, based on microbial activity, and based on specific chemical and physical reactions with electronic or electrode elements (Blum and Coulet, 1991). The most popular class of bionsensors is based upon enzymatic reactions that take place in the presence of one or more substrates to produce products that can be sensed electrochemically.

Clark electrode

The basic sensor used with many of these biosensors is the Clark electrode (Wise, 1989) constructed from a central platinum electrode surrounded by a tubular coaxial silver/silver chloride electrode. If the Pt electrode is used as the cathode and the Ag/AgCl electrode is used as the anode, then this arrangement can be used as an oxygen sensor by the following reaction

$$\tfrac{1}{2}O_2 + 2H^+ + 2e^- \xrightarrow{\text{Pt}} H_2O \tag{4.5.7}$$

An external voltage source must be used to keep the Pt electrode about 0.7 V lower than the Ag/AgCl electrode, and current is monitored as the signal.

If the external voltage source is reversed, making the Pt electrode the anode, then the sensor is used to detect hydrogen peroxide

$$H_2O_2 \xrightarrow{\text{Pt}} O_2 + 2H^+ + 2e^- \tag{4.5.8}$$

Either of these arrangements may be used.

glucose
sensor

One bionsensor that continues to be important is the Clark glucose sensor based on the Pt–Ag/AgCl electrode pair. In the presence of the enzyme glucose oxidase

$$\text{glucose} + O_2 + H_2O \xrightarrow{\text{glucose oxidase}} \text{gluconic acid} + H_2O_2 \qquad (4.5.9)$$

and either the oxygen consumption can be monitored with the Pt cathode or hydrogen peroxide can be monitored with the Pt anode (Figure 4.5.2), depending on the polarity of the external voltage source.

The Clark glucose sensor is sufficiently precise to make useful glucose measurements and has been considered for use in diabetes mellitus. It has a

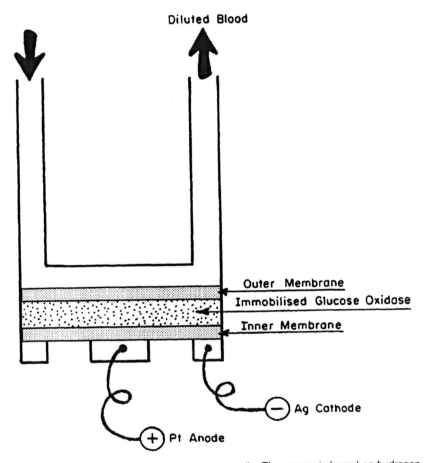

Figure 4.5.2. Glucose sensor used extracorporeally. The sensor is based on hydrogen peroxide detection and is used to detect glucose in blood.

response that is 90% complete in 20 sec, a time that reflects the miniature size of the sensor and short paths for convective and diffusive mass transfer (Cass, 1990).

other sugars

The glucose sensor can also be used for other sugars in food materials. Sucrose can be measured by incorporating the enzyme invertase that converts sucrose into glucose, and lactose can be coverted to glucose by the enzyme β-galactosidase (Wagner and Guilbault, 1994).

immobilized enzyme

Glucose oxidase or other enzyme does not just stay on the electrode. Some means must be used to immoblize the enzyme (Figure 4.5.3), especially because many enzymes are water soluble. One immobilization technique that may be used is covalent attachment to the surface of a water-insoluble material such as porous glass, ceramic, synthetic polymers, cellulose, nylon, or alumina. Different foundation materials are used for different enzymes.

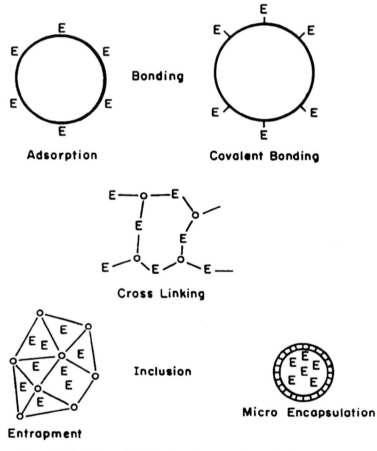

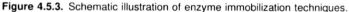

Figure 4.5.3. Schematic illustration of enzyme immobilization techniques.

Another method is absorption of the enzyme on solid surfaces. Carbon is a common adsorbent used for this, but alumina, cellulose, clay, glass, hydroxylapatite, and various silicaceous materials have also been used. Desorption of the enzyme may take place in the presence of competing substances found in the substrate or product.

Entrapment of enzymes in gel matrixes can be used with little to no effect on enzyme activity. Difficulties with this method are continuous loss of enzyme through gel pores and the limitation on enzyme activity because substrates and products must diffuse into the gel.

Enzymes may be cross-linked with other agents to form insoluble particles. Enzymes themselves are sometimes incorporated covalently into other copolymers to form a three-dimensional network of enzyme molecules.

A last immobilization technique is encapsulation in membranes permeable to low-molecular-weight substrates and products but impermeable to macromolecules. Collodian, cellulose derivatives, polystyrene, and nylon are membranous materials used for this purpose.

EXAMPLE 4.5-1. Hexokinase Michaelis Constants

There are four hexokinases that occur in tissues. Type I hexokinase occurs in the brain and has a Michaelis constant value of about 5×10^{-5} kg moles/m^3 (or molar). Type IV hexokinase predominates in the liver parenchymal cells and has a Michaelis constant value of about 2×10^{-2} kg moles/m^3 (McGilvery, 1983). The concentraiton of glucose in the blood supply to the brain usually ranges around 5×10^{-3} kg moles/m^3, but may increase to 9×10^{-3} kg moles/m^3 after consumption of a large amount of carbohydrate, or may fall to 3×10^{-3} kg mole/m^3 during fasting or heavy muscular work. In the metabolism of glucose in brain tissue

$$\text{ATP} + \text{glucose} \rightarrow \text{ADP} + \text{glucose 6-phosphate}$$

determine the range of rates of formation of glucose 6-phosphate assuming (1) Type I hexokinase is present, or (2) Type IV hexokinase is present.

Solution

From Eq. 4.5.6, we obtain

$$\frac{\dot{m}_{gen}}{\dot{m}_{max}} = \frac{1}{\dfrac{K_m}{c_s} + 1}$$

For the two types of hexokinase,

	Type I		Type IV	
Glucose concentration in blood	$\dfrac{K_m}{c_s}$	$\dfrac{\dot{m}_{gen}}{\dot{m}_{max}}$	$\dfrac{K_m}{c_s}$	$\dfrac{\dot{m}_{gen}}{\dot{m}_{max}}$
$3 \times 10^{-3}\,\dfrac{\text{kg mole}}{\text{m}^3}$	$\dfrac{5 \times 10^{-5}}{3 \times 10^{-3}}$	0.984	$\dfrac{2 \times 10^{-2}}{3 \times 10^{-3}}$	0.130
5×10^{-3}	$\dfrac{5 \times 10^{-5}}{5 \times 10^{-3}}$	0.990	$\dfrac{2 \times 10^{-2}}{5 \times 10^{-3}}$	0.200
9×10^{-3}	$\dfrac{5 \times 10^{-5}}{9 \times 10^{-3}}$	0.994	$\dfrac{2 \times 10^{-2}}{9 \times 10^{-3}}$	0.312

Or, for Type I hexokinase, changing the concentration of glucose in the blood by three times changes the rate of glucose metabolism by less than 1%. If Type IV hexokinase were present, then the threefold change in blood glucose level would result in a 91% ($+56\%$, -35%) change in glucose metabolic rate. This is why Type IV hexokinase is not found in the brain: The brain requires the more constant environment supplied by Type I hexokinase.

Remarks

The liver is served by the same systemic arterial blood coming directly from the intestines. This latter blood often has very high glucose concentrations. When the portal blood glucose concentration is high, the liver metabolizes glucose; when the blood concentration is low, the liver supplies glucose to the blood. The relatively high K_m value for Type IV hexokinase makes this system very sensitive to blood glucose levels.

Both Type I and Type IV hexokinases have K_m values for ATP of about $10^{-4}\,\text{kg mole/m}^3$. Since the concentrations of ATP in liver and brain are held at about $10^{-3}\,\text{kg mole/m}^3$, ATP concentration does not limit glucose metabolism by either hexokinase.

4.5.2   Plant Root Nutrient Uptake

nitrate into roots

Wheeler and colleagues (1994a) have presented an application of the Michaelis–Menten relationship with a different twist. They were concerned with plant root uptake of nitrate and performed a set of controlled experiments on hydroponically grown lettuce. They found evidence for superposition of active and passive nitrate ion transport across the cell membrane controlling uptake, the plasmalemma. The active system appears to transport nitrate across the membrane similar to the sodium ion pump described in Section 4.3.3.4. Since there is a limit to the number and rate of nitrate transporting sites, there is a similarity between this system and the enzyme system described earlier. Wheeler at al. (1994a,b) found evidence that the active transport of nitrate across root hairs can be described by the Michaelis–Menten equation.

The ability to apply the Michaelis–Menten relationship to active transport systems moves this relationship beyond mere product generation to an active counterpart to passive diffusion. The Michaelis–Menten relationship can then be used as a mathematical description of the action of an ion pump that depends on energy expenditure and has saturating characteristics. Pumps such as these are very important on a cellular level.

4.5.3 Bacterial Growth Rate

It has been found empirically that the rate of microbial growth in biofilms and bioreactors follows a kinetic equation similar to the Michaelis–Menten equation 4.5.6. This equation is known as the Monod equation, and appears to be identical in form to Eq. 4.5.6. In the Monod equation, however, the rate of product formation is replaced by the specific growth rate of microorganisms

$$\dot{\mu} = \frac{\mu_{max} c_s}{K_s + c_s} \qquad (4.5.10)$$

where μ is the specific growth rate of microorganisms [number of microbes/(sec number of microbes)]; μ_{max}, the maximum specific growth rate of microorganisms [number of microbes/(sec number of microbes)]; c_s, the substrate concentration (kg/m^3); and K_s, the half-saturation constant (kg/m^3).

This equation relates the microbial growth rate to concentrations of various substrates necessary for growth. Examples of these substrates are oxygen, glucose, various forms of nitrogen, and trace elements. The Monod equation modifies the microbial growth rates discussed in Section 3.5.3.1, but does not change the fact that microbial growth rates depend on the number of microbes present. The specific growth rate, μ, has been discussed already in Eq. 3.5.5.

Although the Monod equation appears to be a carbon copy of the Michaelis–Menten equation, the Monod equation is considered to be more empirical and less firmly based on theory compared to the Michaelis–Menten equation.

4.6 MASS STORAGE

There are two cases of mass storage to be considered here. The first is the simple case where mass moves by diffusion into a vacuum. The amount of mass storage in that case is just

diffusion into
vacuum

$$m = \int \dot{m} \, dt \qquad (4.6.1)$$

where m is the stored mass (kg); $\dot{m}$, the mass flow rate (kg/sec); and t, the time (sec).

This is a very simple storage system where mass diffuses from a higher concentration to a lower concentration and accumulates somewhere.

4.6.1 Mass Storage in Solution

The second mass storage case is a bit more complex, and involves a solute that dissolves in a solvent. The solvent is usually a liquid in biological systems. That liquid is usually water.

An example that typifies mass storage of this type is oxygen that dissolves in water. We learned in Section 3.8.2 that oxygen, a typical gas, is more soluble in water at lower temperatures. Some solids are more soluble at higher temperatures.

oxygen
solubility

The solubility of oxygen in blood at human body temperature of $37°C$ is called the Bunsen coefficient, and is given the value of $0.023 \text{ m}^3 \, O_2/\text{m}^3$ blood at an oxygen partial pressure of 1 atm (101.3 kN/m^2). The Bunsen coefficient defines the maximum amount of oxygen found in a given volume of blood. Thus

maximum
oxygen
storage

$$V_{O_2} = \frac{\alpha_{O_2} p_{O_2}}{p_{atm}} V_{blood} \qquad (4.6.2)$$

where V_{O_2} is the volume of oxygen dissolved in the blood (m^3); V_{blood}, the blood volume (m^3); α_{O_2}, the Bunsen coefficient [$0.023 \text{ m}^3 \, O_2/(\text{m}^3$ blood)]; p_{O_2}, the partial pressure of oxygen in the blood (N/m^2); and p_{atm}, the atmospheric pressure (101.3 kN/m^2).

For the more general case of any solute in any solvent, α becomes the solubility of the solute, and it is usually temperature dependent.

If the partial pressure of oxygen in the blood equals the partial pressure of oxygen in the atmosphere with which it is in equilibrium, then Eq. 4.6.2 expresses the maximum volume of oxygen dissolved in the blood. If for some reason dissolved oxygen in the blood is not equilibrated with the atmosphere, as when the partial pressure of atmospheric oxygen is greater than partial pressure of dissolved oxygen, then the volume of dissolved oxygen in the blood no longer equals the maximum. Such a case can arise for extremely severe exercise in healthy individuals, where the transit time of blood through pulmonary capillaries is so short that equilibrium is not always obtained (Johnson, 1991).

We desire to express the mass storage as a function of concentration. For this we twice use the ideal gas law. Oxygen concentration can be obtained from its partial pressure by

oxygen concentration

$$c_{O_2} = \frac{m_{O_2}}{V_{O_2}} = \frac{p_{O_2} M_{O_2}}{RT} \tag{4.6.3}$$

where c_{O_2} is the oxygen concentration (kg/m^3); M_{O_2}, the oxygen molecular weight ($32\,kg/kg\,mole$); R, the universal gas constant [$8314.34\,N\,m/(kg\,mole\ K)$]; and T, the absolute temperature (K).

The oxygen volume that dissolves is just

dissolved volume

$$V_{O_2} = \frac{m_{O_2} RT}{p_{atm} M_{O_2}} \tag{4.6.4}$$

Thus, combining Eqs. 4.6.2, 4.6.3, and 4.6.4

$$m_{O_2} = \alpha_{O_2} c_{O_2} m_{blood} / \rho_{blood} \tag{4.6.5}$$

where m_{O_2} is the mass of oxygen to dissolve (or to be stored) in the blood (kg); m_{blood}, the mass of blood (kg); and ρ_{blood}, the density of blood ($1050\,kg/m^3$).

Oxygen, carbon monoxide, and other gases can react with constituents in the blood, such as hemoglobin. Blood can carry more than 100 times the amount of oxygen dissolved in the blood because of hemoglobin. Thus stored mass may involve more complex phenomena than just solution.

In Table 4.6.1 are found solubilities for selected substances in terms of kg

solubilities

solute per kg of water for various temperatures. Solubilities in units of (kg solute/kg solvent) are related to solubilities in units of (m^3 solute/m^3 solvent) by the ratio of (solvent density/solute density). Solubilities in blood can be found by dividing by the specific gravity of blood, 1.05. Because it is important sometimes to know the volume of dissolved gas, volume solubilities are given also in Table 4.6.1. Such a case arises for divers who suffer from

decompression sickness

decompression sickness (a condition called "the bends") when they rise to lower surrounding pressures: Gases dissolved in the blood at higher pressures come out of solution and form gas bubbles in the blood and tissues at lower pressures (Billings, 1973; Diaz, 1996). This can be extremely painful and even deadly. It can be seen from the table that much lower volumes of helium gas dissolve in the blood compared to other gases. Thus helium–oxygen breathing mixtures are standard for diving deeply.

It can be seen that gas solubilities decrease as temperature increases. Solubilities for salts and amino acids generally increase with temperature.

Table 4.6.1 Table of Solubilities in Water (kg solute/kg water)[a]

	Temperature (°C)														
Substance	0	5	10	15	20	25	30	35	37	38	40	45	50	75	100

Gases (solubilities in m³ gas/m³ water given in parentheses)

Substance	0	5	10	15	20	25	30	35	37	38	40	45	50	75	100
Air	4.14×10^{-5} (0.032)				2.42×10^{-5} (0.020)										1.13×10^{-5} (0.012)
Ammonia	9.48×10^{-1} (1250)				4.95×10^{-1} (700)										
Argon										3.90×10^{-5} (0.0262)					
Carbon dioxide	3.36×10^{-3} (1.713)	2.74×10^{-3} (1.424)	2.26×10^{-3} (1.194)	1.90×10^{-3} (1.019)	1.61×10^{-3} (0.878)	1.36×10^{-3} (0.759)	1.18×10^{-3} (0.665)	1.03×10^{-3} (0.592)	9.80×10^{-4} (0.567)	9.47×10^{-4} (0.550)	9.07×10^{-4} (0.530)	8.07×10^{-4} (0.479)	7.23×10^{-4} (0.436)		3.74×10^{-4} (0.26)
Carbon monoxide	4.87×10^{-5} (0.039)				2.91×10^{-5} (0.025)										
Chlorine	1.59×10^{-2} (5.0)				7.37×10^{-3} (2.5)										
Helium										1.36×10^{-6} (0.0087)					
Hydrogen	1.87×10^{-6} (0.021)	1.75×10^{-6} (0.020)	1.72×10^{-6} (0.020)	1.61×10^{-6} (0.019)	1.49×10^{-6} (0.018)	1.47×10^{-6} (0.018)	1.36×10^{-6} (0.017)	1.34×10^{-6} (0.017)	1.33×10^{-6} (0.017)	1.33×10^{-6} (0.017)	1.25×10^{-6} (0.016)	1.23×10^{-6} (0.016)	1.21×10^{-6} (0.016)		1.04×10^{-6} (0.016)
Hydrogen chloride	9.10×10^{-1} (560)				7.28×10^{-1} (480)										
Hydrogen sulfide	7.60×10^{-3} (5.0)				3.97×10^{-3} (2.8)										9.68×10^{-4} (0.87)
Nitrogen	2.99×10^{-5} (0.024)	2.57×10^{-5} (0.021)	2.29×10^{-5} (0.019)	2.02×10^{-5} (0.017)	1.74×10^{-5} (0.015)	1.61×10^{-5} (0.014)	1.46×10^{-5} (0.013)	1.44×10^{-5} (0.013)	1.32×10^{-5} (0.012)	1.31×10^{-5} (0.012)	1.30×10^{-5} (0.012)	1.18×10^{-5} (0.011)	1.17×10^{-5} (0.011)		9.60×10^{-6} (0.0105)
Oxygen	6.99×10^{-5} (0.049)	6.03×10^{-5} (0.043)	5.23×10^{-5} (0.038)	4.60×10^{-5} (0.034)	4.13×10^{-5} (0.031)	3.66×10^{-5} (0.028)	3.34×10^{-5} (0.026)	3.03×10^{-5} (0.024)	3.01×10^{-5} (0.024)	2.88×10^{-5} (0.023)	2.87×10^{-5} (0.023)	2.70×10^{-5} (0.022)	2.53×10^{-5} (0.021)		1.93×10^{-5} (0.0185)
Sulfur dioxide	2.28×10^{-1} (80)		1.62×10^{-1} (59)		1.13×10^{-1} (42)		7.80×10^{-2} (30)				5.40×10^{-2} (22)		4.50×10^{-2} (18.6)		
Sulfur trioxide	3.11×10^{-1} (87)				1.43×10^{-1} (43)										

Salts

Substance	0	5	10	15	20	25	30	35	37	38	40	45	50	75	100
Aluminum sulfate	3.13×10^{-1}												5.21×10^{-1}		8.91×10^{-1}
Aluminum potassium	3.00×10^{-2}												1.70×10^{-1}		1.540

Ammonium bicarbonate	1.19×10^{-1}					5.04×10^{-1}	7.60×10^{-1}
Ammonium chloride	2.97×10^{-1}					3.44	8.710
Ammonium nitrate	1.183					8.47×10^{-1}	1.033
Ammonium sulfate	7.06×10^{-1}					4.36×10^{-1}	5.87×10^{-1}
Barium chloride	3.17×10^{-1}					1.72×10^{-1}	3.45×10^{-1}
Barium nitrate	5.00×10^{-2}		1.8×10^{-5}				8.80×10^{-4}
Calcium carbonate	5.94×10^{-1}						1.576
Calcium chloride	1.77×10^{-3}					3.561	6.70×10^{-4}
Calcium hydroxide	9.31×10^{-1}					2.06×10^{-3}	3.626
Calcium nitrate	1.76×10^{-3}					3.34×10^{-1}	1.69×10^{-3}
Calcium sulfate	1.40×10^{-1}					3.160	7.53×10^{-1}
Copper sulfate	7.30×10^{-1}					8.20×10^{-1}	5.369
Ferric chloride						4.82×10^{-1}	1.060
Ferrous chloride			6.44×10^{-1}			1.67×10^{-2}	3.33×10^{-2}
Ferrous hyroxide	6.70×10^{-6}						1.255
Ferrous sulfate	1.56×10^{-1}					9.03×10^{-1}	7.83×10^{-1}
Lead chloride	6.73×10^{-3}					5.00×10^{-1}	7.10×10^{-1}
Lead nitrate	4.03×10^{-1}					1.216	1.562
Lead sulfate			4.20×10^{-5}			4.35×10^{-1}	5.66×10^{-1}
Magnesium carbonate		1.30×10^{-4}				1.414	1.773
Magnesium chloride	5.24×10^{-1}					8.51×10^{-1}	2.477
Magnesium hydroxide		9.00×10^{-6}				1.65×10^{-1}	2.41×10^{-1}
Magnesium nitrate	6.65×10^{-1}						
Magnesium sulfate	2.69×10^{-1}						
Potassium carbonate	8.93×10^{-1}	3.10×10^{-1}					
Potassium chloride	2.84×10^{-1}						
Potassium hydroxide	9.71×10^{-1}		3.40×10^{-1}	3.70×10^{-1}	4.00×10^{-1}		
Potassium nitrate	1.31×10^{-1}						
Potassium sulfate	7.40×10^{-2}						
Silver nitrate	1.22	1.70	2.22	3.00	3.76	4.55	
Sodium bicarbonate	6.90×10^{-2}	3.58×10^{-1}					
Sodium carbonate	2.04×10^{-1}					1.45×10^{-1}	
Sodium chloride	3.57×10^{-1}		3.60×10^{-1}	3.63×10^{-1}	3.66×10^{-1}	4.75×10^{-1}	4.52×10^{-1}
Sodium hydroxide	4.20×10^{-1}					3.66×10^{-1}	3.92×10^{-1}
Sodium nitrate	7.33×10^{-1}					1.448	3.388
Sodium sulfate	4.90×10^{-2}					1.148	1.755
Zinc chloride	2.044					4.66×10^{-1}	4.22×10^{-1}
Zinc nitrate	9.47×10^{-1}					4.702	6.147

(continued)

Table 4.6.1 (Continued)

Substance	0	5	10	15	20	25	30	35	37	38	40	45	50	75	100
Zinc sulfate	4.19×10^{-1}												7.68×10^{-1}		8.07×10^{-1}

Amino acids

Substance	0	5	10	15	20	25	30	35	37	38	40	45	50	75	100
L-Alanine	1.27×10^{-1}					1.66×10^{-1}							2.31×10^{-1}	2.85×10^{-1}	3.73×10^{-1}
L-Aspartic acid	2.09×10^{-3}					5.00×10^{-3}							2.00×10^{-2}	2.88×10^{-2}	6.89×10^{-2}
L-Glutamic acid	3.41×10^{-3}					8.64×10^{-3}							2.19×10^{-2}	5.53×10^{-2}	1.40×10^{-1}
Glycine	1.42×10^{-1}					2.50×10^{-1}							3.91×10^{-1}	5.44×10^{-1}	6.72×10^{-1}
L-Histidine						4.19×10^{-2}									
L-Isoleucine	1.83×10^{-2}					2.23×10^{-2}							3.03×10^{-2}	4.61×10^{-2}	7.80×10^{-2}
L-Leucine	7.97×10^{-3}					9.91×10^{-3}							1.41×10^{-2}	2.28×10^{-2}	4.21×10^{-2}
L-Phenylalanine	1.98×10^{-2}					2.96×10^{-2}							4.43×10^{-2}	6.62×10^{-2}	9.90×10^{-2}
L-Proline	1.274					1.623							2.067		
L-Tryptophan	8.23×10^{-3}					1.14×10^{-2}							1.71×10^{-2}	2.80×10^{-2}	4.99×10^{-2}
L-Tyrosine	1.96×10^{-4}					4.53×10^{-4}							1.05×10^{-3}	2.44×10^{-3}	5.65×10^{-3}
L-Valine	8.34×10^{-2}					8.85×10^{-2}							9.62×10^{-2}		

Other

Substance	0	5	10	15	20	25	30	35	37	38	40	45	50	75	100
Sucrose							1.522								

[a] Gas solubilities are given for 1 atm (101.3 kN/m²) total pressure where dissolving gases constitute 100% of the atmosphere. For gases that constitute less than 100% of the atmosphere, tabulated gas solubilities should be multiplied by the mass fractions of those gases in the atmosphere.

There is an interaction that can occur between solutes, causing solubilities for individual substances to be lower in the presence of other solubilities. Thus the volume solubility of carbon dioxide, which is $0.550\,m^3\,CO_2/m^3$ water, becomes $0.537\,m^3\,CO_2/m^3$ solution in Ringer's solution (a standard solution of NaCl, KCl, and $CaCl_2$ used topically on burns and wounds) and $0.510\,m^3\,CO_2/m^3$ serum in blood serum (Dawson et al., 1986). For a wider range of substance solubilities, see Stephen and Stephen (1963).

solubility interactions

EXAMPLE 4.6.1-1. Maintaining a Saturated Salt Solution

The Alza DUROS system (Figure 4.3.24) was described in Example 4.3.3.3-3. In order to maintain a constant rate of drug delivery throughout its entire lifetime, the salt solution must remain saturated. As water from the body passes through the membrane into the osmotic engine, it dilutes the salt solution, which would eventually be watered enough to reduce significantly its osmotic pressure and the rate of leuprolide delivery. To overcome dilution, extra salt must be placed in the osmotic engine chamber. Determine how much salt is required.

Solution

The leuprolide flow rate is $125\,\mu g/day$, or $250 \times 10^{-6}\,mL/day$. This daily liquid volume is replaced by water at a rate of

$$\dot{m}_{water} = \left(250 \times 10^{-6}\,\frac{mL}{d}\right)(1000\ kg/m^3)\left(10^{-6}\,\frac{m^3}{mL}\right)(1000\ g/kg)$$

$$= 250 \times 10^{-6}\,\frac{g}{d}$$

From Table 4.6.1 we find that the solubility of sodium chloride is $0.366\,kg\,NaCl/kg$ water. Thus the rate of salt to be dissolved for the solution to remain saturated is

$$\dot{m}_{salt} = (250 \times 10^{-6}\ g\ H_2O/d)(0.366\ g\ NaCl/g\ H_2O)$$

$$= 91.5\ \mu g/d$$

If the DUROS is to remain operable for a period of 360 d, the salt solution reservoir must be at least $0.09\,mL$ and contain $0.033\,g$ NaCL.

EXAMPLE 4.6.1-2. Cell Cultures in Roller Bottles

Cultures of human immunodeficiency viruses (HIV) are to be grown for virus research and testing. HIV are grown in a liquid medium containing lymphocyte cells serving as the hosts for the viruses. Tightly sealed plastic roller bottles (11 cm dia. × 22 cm high) with 2000 mL capacity are filled with 500 mL of split culture and placed on their sides. They are then rolled from side to side at the rate of 1 revolution per minute for a total of 3–4 d incubation time to maintain a thoroughly mixed medium and to keep the cells from settling. Write a mass balance for oxygen for this system to see whether the oxygen contained in the bottle can be supplied to the cells at an adequate rate.

Solution

There are many uncertainties with this problem, as there are with many problems involving biological systems. During the incubation time the number of cells and the resulting oxygen usage will be increasing. Also, it is known empirically that the oxygen contained in the bottle is adequate for the cells to thrive throughout the incubation. We make two assumptions here: (1) that the oxygen consumption is constant throughout the incubation, and (2) that for the cells to thrive the fraction of oxygen in the air in the bottle does not fall below 0.18. The estimate of the rate of oxygen usage is then

$$\dot{V}_{O_2} = (1500 \text{ mL air})(10^{-6} \text{ m}^3/\text{mL})/(0.21-0.18 \text{ m}^3 \text{ O}_2/\text{m}^3 \text{ air})/4 \text{ d}$$

$$= 1.125 \text{ m}^3 \text{ O}_2/\text{day} = 1.30 \times 10^{-10} \text{ m}^3 \text{ O}_2/\text{sec}$$

$$\dot{m}_{O_2} = \frac{p\dot{V}M}{RT} = \frac{(101300 \text{ N/m}^2)(1.30 \times 10^{-10} \text{ m}^3/\text{sec})(32 \text{ kg/kg mole})}{[8314.34 \text{ N m/(kg mol K)}](293 \text{ K})}$$

$$= 1.73 \times 10^{-10} \text{ kg O}_2/\text{sec}$$

This value represents a negative oxygen generation term.

Oxygen is only available to the cells if the oxygen is in solution. Thus the mass (or materials) balance on oxygen will be written for the solution.

Oxygen enters the solution by dissolving from the atmosphere in the bottle. Once in solution in a thin film on the surface of the liquid, the oxygen is thoroughly mixed into the remainder of the liquid by the rolling action of the bottles. Here we make other assumptions: (1) We assume that the film thickness for a saturated solution of oxygen in the liquid is about 3 mm on the surface of the liquid in the bottom of the bottle; (2) we assume that the film thickness on the top of the bottle is about 0.1 mm thick; and (3) the top film is saturated with oxygen (Figure 4.6.1).

To find the volumes represented by each of these films, the value for the angle θ must be determined. The area of a segment of a circle (the shaded area

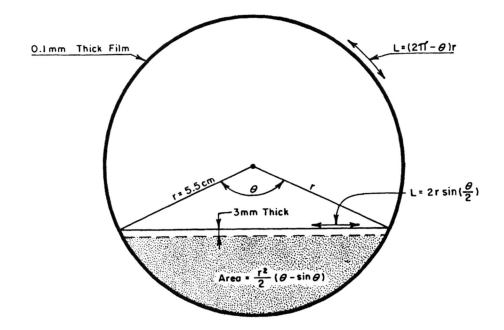

Figure 4.6.1. Diagram of the roller bottle on its side. Thicknesses of films for saturated oxygen solutions are given for the bottom liquid surface and for the liquid that adheres to the walls of the bottle.

in Figure 4.6.1) is

$$A = \frac{r^2}{2}(\theta - \sin \theta) = 500 \text{ mL}$$

or

$$\frac{500 \text{ mL}}{2000 \text{ mL}} = \frac{r^2(\theta - \sin \theta)}{2\pi r^2} = \frac{1}{4}$$

The value of θ can be found by iterations of the equation

$$\theta = \sin \theta + \pi/2$$

to be $\theta = 0.7352\pi$. Length of the chord across the bottle is

$$L = 2r \sin(\theta/2) = (0.11 \text{ m}) \sin(0.3676\pi) = 0.101 \text{ m}$$

Thus the volume of the assumed oxygen-saturated layer at the top of the medium is

$$V = (0.101 \text{ m})(0.22 \text{ m})(0.003 \text{ m}) = 6.64 \times 10^{-5} \text{ m}^3$$

The length of the arc around the upper side of the bottle is

$$L = (2\pi - \theta)r = (2\pi - 0.7352\pi)(0.055 \text{ m}) = 0.219 \text{ m}$$

The volume of the assumed film inside the upper part of the bottle is

$$V = (0.219 \text{ m})(0.22 \text{ m})(0.0001 \text{ m}) = 4.81 \times 10^{-6} \text{ m}^3$$

Total assumed oxygen-saturated volume is

$$V_{tot} = 6.64 \times 10^{-5} \text{ m}^3 + 4.81 \times 10^{-6} \text{ m}^3 = 8.12 \times 10^{-5} \text{ m}^3$$

From Table 4.6.1, we see that the solubility of oxygen is $4.13 \times 10^{-5} \text{ kg O}_2/\text{kg}$ water for an atmosphere of 100% oxygen, or, for 21% oxygen

$$\alpha_{O_2} = (0.21)(4.13 \times 10^{-5} \times 10^{-5} \text{ kg O}_2/\text{kg water})$$

$$= 8.67 \times 10^{-6} \text{ kg O}_2/\text{kg water}$$

The amount of oxygen added to the culture medium as the bottle rolls around is

$$m_{O_2} = \alpha_{O_2} V_{tot} \rho_{H_2O}$$

$$= (8.67 \times 10^{-6} \text{ kg O}_2/\text{kg water})(7.12 \times 10^{-5} \text{ m}^3)\left(1000 \ \frac{\text{kg water}}{\text{m}^3}\right)$$

$$= 6.18 \times 10^{-7} \text{ kg}$$

Because the bottle makes one revolution per minute, this calculated mass of oxygen is mixed with the fluid each minute. Thus

$$\dot{m}_{O_2} = 6.18 \times 10^{-7} \text{ kg}/60 \text{ sec} = 1.03 \times 10^{-8} \text{ kg/sec}$$

The oxygen mass balance is thus

	rate of mass in	=	1.03×10^{-8} kg O_2/sec
$-$	rate of mass out	=	0
$+$	rate of mass generation	=	-1.73×10^{-10} kg O_2/sec
$=$	rate of mass accumulation	=	0

Remarks

The fact that the mass balance was not balanced was not surprising in view of the number of assumptions made. Actually, the rate of oxygen leaving the culture medium could justifiably be made equal to the difference between rate of O_2 in and rate of O_2 used. The rate of mass accumulation was set equal to zero, because, even if some O_2 had accumulated in the fluid, this amount would have been limited by the partial pressure of oxygen in the air inside the bottle (the value of the effort variable inside the fluid cannot exceed the value of the effort variable in the air as long as there are no additional sources).

What this balance tells us is that the rate of oxygen added to the culture medium is more than enough to sustain the oxygen needs of the cells.

4.6.2 Mass Capacitance

Mass capacitance has been defined as

$$C = \frac{1}{c} \int_0^t \dot{m} \, dt \tag{4.6.6}$$

where C is the mass capacitance (m^3); c, the concentration (kg/m^3); $\dot{m}$, the mass flow rate (kg/sec); and t, the time (sec).

Thus mass capacitance is the relationship between concentration and accumulated mass. For a simple diffusion system, mass capacitance must be the volume that contains the mass. Depending on whether the shape of the volume is planar, cylindrical, or spherical, mass capacitance assumes the mathematical form of AL, $\pi L(r_2^2 - r_1^2)$, or $4\pi(r_2^3 - r_1^3)$, where A is the cross-sectional area.

volume containing mass

For the solution system given in the previous section, mass capacitance is again the relationship between concentration and the maximum amount of mass that can be contained

$$C = \frac{\alpha m_{sv}}{\rho_{sv}} = \alpha V_{sv} \tag{4.6.7}$$

where C is the mass capacitance (m^3); α, the solute solubility (m^3 solute/m^3 solvent); m_{sv}, the mass of solvent (kg); ρ_{sv}, the solvent density (kg/m^3); and V_{sv}, the solvent volume (m^3).

For reactive solutes, additional capacity may need to be accounted for.

Mass capacitance for a nonporous membrane system, where the effort variable is concentration and the flow variable is mass flow rate, is given by the volume of mass as given above. Mass capacitance for a porous membrane must usually be treated differently, however. Since the effort is variable is pressure and the flow variable is volume flow rate, and the resistance for a porous membrane is calculated in a manner similar to that of a fluid flow system with a porous medium, mass capacitance for a porous membrane is given by the pressure–volume (or pressure–mass) relationship of the container accepting or losing mass (see Table 4.3.16). This description was given in detail in Section 2.5.3.1. In a system where membrane permeate is removed as fast as it is formed, effective volume to accept the mass is close to infinite, and mass capacitance is likewise unbounded.

4.7   MIXED-MODE MASS TRANSFER

internal
diffusion,
external
convection

As mentioned earlier, mass may commonly be transferred by diffusion within an object and convection from the surfaces, similar to that diagrammed in Figure 4.3.21. There are several topics to be explored that deal with mixed-mode mass transfer.

4.7.1   Extended Surfaces

fins

Convective mass transfer from a surface can be enhanced by increasing the surface area analogously to enhancement of convective heat transfer. Fins are used to increase surface area exposed to a flowing stream causing mass convection. Diffusion is assumed to deliver the mass to the surface from the interior of the fin (Figure 4.7.1).

In a manner similar to the use of fins for convective heat transfer (Section 3.7.3) from a surface, fins used for mass transfer could decrease rather than increase mass transfer if they significantly interfere with the stream of flowing fluid and reduce convection. Thus increased mass transfer from a fin can only be expected if

$$\frac{k_G A}{PD} \le 1 \tag{4.7.1}$$

where k_G is the convective mass transfer coefficient (m/sec); A, the surface area (m^2); P, the fin cross-sectional perimeter (m); and D, the diffusivity of the mass flowing inside the material constituting the fin (m^2/sec).

small intestine

A biologically important application of fins occurs in the small intestine, where absorption of nutrients from food takes place. The surface of the small intestine wall is increased first by infoldings of the mucosal tissue that increase the surface area available for absorption by three times over the area of a simple cylinder. Next are slender protrusions called villi that project from the surface and increase the epithelial area exposed to intestinal contents by another ten times. In addition, there are subcellular microvilli that increase the available surface area another twenty times. The result is a total small intestinal surface area of 200 m^2, more than 100 times the surface area of the body (1.8 m^2).

This great number of infoldings and protrusions would be vastly less effective if it were not for the constant motion of the entire intestine and the villi inside the intestine. This movement ensures that intestinal contents remain mixed and do not become stagnant.

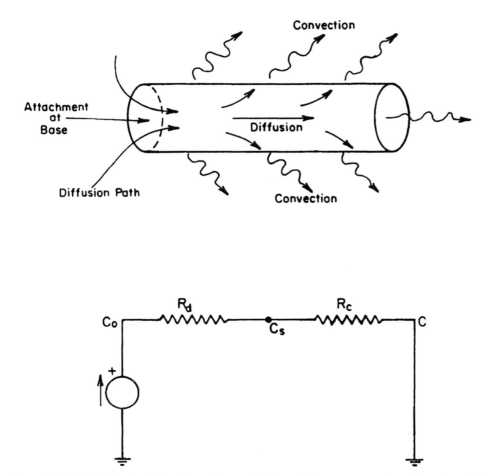

Figure 4.7.1. A pin fin attached to the surface at the base delivers mass to the surface through the interior of the fin by diffusion. Mass is removed from the surface by convection. Below the drawing is shown a simplified systems schematic for the fin that neglects diffusion in several directions simultaneously. Symbols are: c_0, mass concentration at base; c_s, surface concentration; c_∞, ambient concentration; R_d, diffusion resistance; and R_c, convection resistance.

Additionally, a high rate of blood flow through capillaries just below the intestinal surface rapidly carries nutrients away, thus helping to maintain a high concentration difference between the intestinal lumen and the intestinal blood flow. This, of course, also maintains high rates of mass transfer (Figure 4.7.2).

Mass transfer through intestinal walls is similar to the cases of packed beds (Section 4.4.2) and heat transfer from long pipes (Section 3.7.12) considered previously. Nutrient absorption along the length of the intestine depletes nutrients inside the intestine. Nutrient concentration thus decreases along the length of the intestine except where absorption may be locally enhanced due to

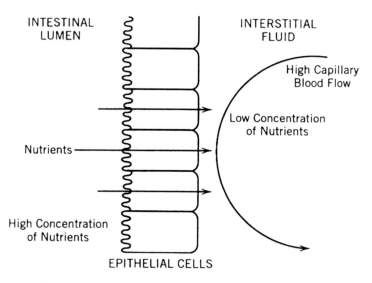

Figure 4.7.2. The large surface area of the intestine and the high concentration difference between the lumen and intestine wall makes a very efficient mass transfer system.

active transport or enzyme mediation. Also, when the concentration reaches zero, no further absorption can, of course, take place. The equivalent concentration difference to use for intestinal diffusion is thus the logarithmic mean concentration difference defined in Eq. 4.4.30. Nutrient concentration on the blood side of the intestine can often be considered to be constant, and the point inside the intestine where concentration effectively reaches zero must be identified. Beyond this point no further absorption takes place.

4.7.2 Simultaneous Diffusion and Convection

The analysis in Sectoin 4.3 was concerned with diffusion through a stagnant medium. In Section 4.4 we considered mass convection by itself. In the next section (Section 4.7.3) will be developed a combination of diffusion and convection when there is a concentration gradient in a direction different from the direction of flow. In this section we consider diffusion in response to a concentration gradient in the direction of flow.

conditions

Conditions that must exist for this condition are that: (1) substance M is moving within a space by convection (or bulk) flow, (2) the velocity of this flow is unidirectional and is everywhere the same, (3) there is a region within M through which substance N is well mixed, and (4) the transition between pure M and mixed N and M is abrupt and moving at a velocity v_{dN} relative to the moving bulk fluid.

Diffusion of N occurs by Fick's first law

diffusion flow
$$\dot{m}_{dN} = v_{dN}c_N A = -D_{NM}A\,\frac{dc_N}{dz} \qquad (4.7.2)$$

where $\dot{m}_{dN}$ is the mass movement of substance N by diffusion (kg/sec); v_{dN}, the velocity of diffusion relative to the moving fluid (m/sec); c_N, the concentration of N mixed in fluid M (kg/m^3); and A, the cross-sectional area (m^2).

Simultaneous convection carries N (and M also)

convection flow
$$\dot{m}_{cN} = v_c c_N A \qquad (4.7.3)$$

where $\dot{m}_{cN}$ is the mass movement of substance N by convection (kg/sec); and v_c, the velocity of convective movement relative to a stationary reference (m/sec).

Total mass movement of substance N is

total flow
$$\dot{m}_N = \dot{m}_{cN} + \dot{m}_{dN} = v_c c_N A - D_{MN}A\,\frac{dc_N}{dz} \qquad (4.7.4)$$

To discover the value of v_c, we note

$$\dot{m}_{tot} = \dot{m}_N + \dot{m}_M = v_c c_{tot} A \qquad (4.7.5)$$

where c_{tot} is the total concentration of substances M and N (kg/m^3).

From which we obtain

$$v_c = \frac{\dot{m}_N + \dot{m}_M}{c_{tot}A} \qquad (4.7.6)$$

Thus

$$\dot{m}_N = (\dot{m}_N + \dot{m}_M)\frac{c_N}{c_{tot}} - D_{MN}A\,\frac{dc_N}{dz} \qquad (4.7.7)$$

For equal mass counter diffusion, $\dot{m}_N = -\dot{m}_M$, and the convective term becomes zero, leaving only diffusion. Actually, counter diffusion (within a gaseous medium) is more likely to be equimolar rather than equal mass, resulting in a net flow of mass.

one substance stagnant
 The special case where one substance is stagnant arises where the other substance evolves at a surface, either by vaporization or by reaction. There is then a net outflow of the evolved substance. Of course, condensation or

absorption can also be considered this way. We assume substance M is stagnant

$$\dot{m}_N = (\dot{m}_N + 0)\frac{c_N}{c_{tot}} - D_{MN}A\frac{dc_N}{dz} \tag{4.7.8}$$

$$\dot{m}_N \, dz = \frac{-D_{MN}A \, dc_N}{(1 - c_N/c_{tot})} \tag{4.7.9}$$

$$\dot{m}_N \int_{z_1}^{z_2} dz = -D_{MN}A \int_{c_{N1}}^{c_{n2}} \frac{dc_N}{(1 - c_N/c_{tot})} \tag{4.7.10}$$

$$\dot{m}_N = \frac{-d_{MN}Ac_{tot}}{(z_2 - z_1)} \ln\left(\frac{c_{tot} - c_{N2}}{c_{tot} - c_{N1}}\right) \tag{4.7.11}$$

resistance

Resistance to mass flow is

$$R_m = \frac{(c_{N1} - c_{N2})}{\dot{m}_N} = \frac{(c_{N1} - c_{N2})(z_2 - z_1)}{-D_{MN}Ac_{tot}} \ln\left(\frac{c_{tot} - c_{N1}}{c_{tot} - c_{N2}}\right) \tag{4.7.12}$$

where R_m is the mass flow resistance ($\sec/m^3$).

special conditions

Note that the treatment of mass convection in this section is much different from the treatment of mass convection in Section 4.4. The difference arises from the special nature of the conditions treated here. Methods in Section 4.4 are more general and are to be used whenever details of the convection process cannot be specified as they were here.

EXAMPLE 4.7.2-1. The Benefits of Mulch

Mulching is a common practice in gardening and farming that is used to: (1) conserve moisture, (2) maintain steady soil temperature, (3) discourage weed growth, and (4) add to the organic matter of the soil through decomposition. Compare the resistance to moisture loss through mulch relative to that of bare earth. Assume the soil surface in both cases is saturated with water. Assume that bare earth has a surface water vapor boundary layer of 10 mm and that the mulch is 100 mm thick. The mulch should be assumed to have a porosity of 0.1%. Temperature of the bare soil surface will be assumed to be 20°C and of the soil beneath the mulch to be 15°C.

Solution

The solution begins by conceptualizing it as in Figure 4.7.3. Liquid water lies at the bottom of an open-topped tube that represents a channel through the mulch from the soil to the open air. Stagnant air is present throughout the tube. Water evaporates at the liquid surrace with a vapor pressure (and concentration) determined by the temperature of the water surface. Water

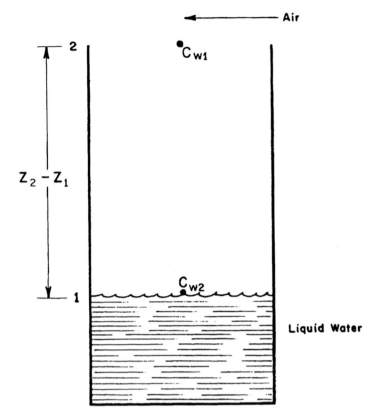

Figure 4.7.3. Simplified schematic of a channel for moisture movement through mulch material. The concentration of water at the soil surface (c_{W2}) is saturated at the temperature of the soil; the concentration of water at the opening of the channel (c_{W1}) is equal to ambient concentration. Because water flows one-way from the soil to the atmosphere, there is a net mass movement that we call convection.

vapor moves because of a concentration difference to the top of the tube, where the concentration is assumed to be negligible.

Saturated water vapor pressure at 15°C is 1705 N/m^2 and at 20°C is 2341 N/m^2 (Table 4.8.2). Total atmospheric pressure is $1.013 \times 10^5 \text{ N/m}^2$. Concentrations can be obtained from the ideal gas law

$$c = \frac{m}{v} = \frac{pM}{RT}$$

Thus for bare earth

$$c_{\text{tot}} = \frac{(1.013 \times 10^5 \text{ N/m}^2)(29 \text{ kg/kg mole})}{[8314.34 \text{ N m/(kg mol K)}](293 \text{ K})}$$

$$= 1.21 \text{ kg/m}^3$$

Using an average temperature $(20°C + 15°C)/2 = 17.5°C$ for the mulch gives $c_{tot} = 1.22\ kg/m^3$.

For the bare earth, the water vapor concentration is

$$c_{W2} = \frac{(2341\ N/m^2)(18\ kg/kg\ mole)}{[8314.34\ N\ m/(kg\ mol\ K)](293\ K)}$$

$$= 0.01730\ kg/m^3$$

For the mulch, there is a temperature of 288 K and $c_{W2} = 0.01282\ kg/m^3$.

Mass diffusivity for water vapor in air is obtained from Table 4.3.1 and corrected for temperature using Eq. 4.3.16. This results in diffusivity values of $0.250 \times 10^{-4}\ m^2/sec$ at 20°C and $0.246 \times 10^{-4}\ m^2/sec$ at 17.5°C. Resistance for a 1 m² surface area of bare earth gives (using Eq. 4.7.12)

$$R_m = \frac{(0.01730 - 0\ kg/m^3)(0.01\ m)}{-(0.250 \times 10^{-4}\ m^2/sec)(1\ m^2)(1.21\ kg/m^3)}$$

$$\times \ln\left(\frac{1.21 - 0.01730\ kg/m^3}{1.21 - 0\ kg/m^3}\right)$$

$$= 0.0824\ sec/m^3$$

For the mulch, the area is $(0.1\%)\ (1\ m^2) = 0.001\ m^2$, and

$$R_m = \frac{(0.01282 - 0\ kg/m^3)(0.1\ m)}{-(0.246 \times 10^{-4}\ m^2/sec)(0.001\ m^2)(1.22\ kg/m^3)}$$

$$\times \ln\left(\frac{1.22 - 0.01282\ kg/m^3}{1.22 - 0\ kg/m^3}\right)$$

$$= 451\ sec/m^3$$

These results are summarized in the following table

	Mulch	No Mulch
A	0.001 m²	1 m²
θ	15°C	20°C
D_{WA}	$0.246 \times 10^{-4}\ m^2/sec$	$0.250 \times 10^{-4}\ m^2/sec$
$(z_2 - z_1)$	100 mm	10 mm
c_{W1}	0 kg/m³	0 kg/m³
c_{W2}	0.01282 kg/m³	0.01730 kg/m³
c_{tot}	1.22 kg/m³	1.21 kg/m³
R_m	451 sec/m³	0.0824 sec/m³

Adding the mulch increases the resistance to moisture vapor loss by nearly 5500 times. This is a very significant increase and demonstrates why mulch is so effective in reducing soil moisture loss.

4.7.3 Dispersion

radial diffusion, axial convection

Dispersion comes about as an interaction between radial diffusion and axial convective flow. One simple way to understand dispersion is to imagine a steady flow of fluid through a tube (Figure 4.7.4). If the flow regime is fully developed laminar, then the velocity is the greatest at the center of the tube; center velocity is twice the average velocity. If a dye were injected into the center of the tube, it would be carried downstream very rapidly, but mostly in the center of the tube. Those parts of the fluid next to the center would not receive high dye concentrations from upstream as rapidly as the fluid exactly in the tube center. Thus there would be a very high radial concentration gradient that would tend to cause diffusion to occur very rapidly (Eq. 4.3.1).

large concentration gradient

If the fluid were stagnant, the region of high dye concentration would still spread out, but the radial concentration gradient in the tube center, or anywhere in the tube for that matter, would not be as high as when the fluid flows. Thus diffusion would occur more slowly in the stagnant fluid. This interaction between axially flowing fluid and radial diffusion causes dispersion to be a very rapid and a very important phenomenon.

biological importance

Dispersion is important to biological engineers on many different levels. Dispersion causes global dissemination of volcanic ash (ecological level); dispersion causes the rapid movement of smoke or toxic particulates

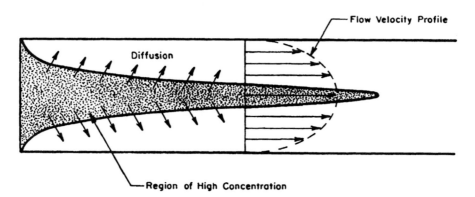

Figure 4.7.4. Because the velocity in the center of the tube is greater than anywhere else, the region of high concentration is carried down the center of the tube most rapidly. Fluid adjacent to the center will have low concentration. Thus there is always a very high concentration gradient in the radial direction that causes maximum diffusion. If the fluid in the pipe were stagnant, the radial concentration gradient at the center would become much smaller, with lower amounts of diffusion as a result.

(environmental level); dispersion causes the rapid mixing of gases in the respiratory airways (organismic level). Dispersion is also a useful tool for analyzing the spread of odor from a compost pile, the dissemination of noxious fumes from a chemical spill, or the scattering of petroleum from a tanker accident.

4.7.3.1 Static Dispersion

spreading
tendency

Beginning with the static case of a nonflowing pulse of material particles, the particles will diffuse outward from the original pulse (Figure 4.7.5). Because the pulse represents stored material, which neither increases nor decreases in magnitude, the original mass must be conserved while it spreads out. Thus the peak concentration must decrease as the material spreads.

Comparing the shapes of these curves of concentration with distance (Figure 4.7.5) to the curves describing biological variation (Figure 1.4.5) shows that both phenomena can be considered to be Gaussian. Concentration can be expressed as (Cussler, 1994)

concentration

$$c = \frac{m/A}{\sqrt{4\pi D t}} e^{-x^2/4Dt}$$ (4.7.13)

where c is the material concentration (kg/m^3); m, the original amount of mass (kg); A, cross-sectional area at the source through which diffusion occurs (m^2);

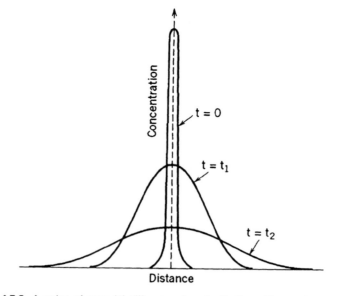

Figure 4.7.5. A pulse of material diffuses outward with time. The peak concentration decreases but the horizontal spread is much wider. The shapes of the curves are Gaussian and the areas underneath all curves must equal the original area.

D, the mass diffusivity (m^2/sec); t, the time after release of material (sec); and x, the distance from original pulse (m).

Comparison of Eq. 4.7.13 with 1.4.18 shows that the mean of the concentration curves is at $x = 0$, and the standard deviation $\sqrt{2Dt}$ m. Materials with larger mass diffusivities or that diffuse for longer times will have more spread.

4.7.3.2 *Dispersion in Flowing Fluid*

To this static case we now impose a flowing fluid. As illustrated in Figure 4.7.4, the material particles in the center of the tube are swept along with the fluid, and act similarly to the static material pulse. Thus there are two processes simultaneously occurring: Particles are being swept longitudinally by the fluid convection and are spreading radially by molecular diffusion.

The method for dealing with dispersion is deceptively simple: The mass diffusivity is replaced by a dispersion coefficient in the Fick equation. Because time is involved, we use Fick's second law, Eq. 4.3.12

$$\frac{\partial^2 c}{\partial x^2} = \frac{1}{\mathscr{D}} \frac{\partial c}{\partial t} \tag{4.7.14}$$

where c is the mass concentration (kg/m^3); $\mathscr{D}$, the dispersion coefficient (m^2/sec); t, the time (sec).

dispersion coefficient

Péclet number

Values for the dispersion coefficient are very difficult to predict from first principles, and, for most circumstances, they must be measured. Important in the determination of the dispersion coefficient is the Péclet number giving the relative importance of axial convection and radial diffusion

$$\text{Pe} = \frac{vr}{D} \tag{4.7.15}$$

where Pe is the Péclet number (dimensionless); v, the fluid flow velocity (m/sec); r, the tube radius (m); and D, the mass diffusivity (m^2/sec). This formulation for the Péclet number is meant to be applied to flow in tubes. For other situations v may be replaced by wind velocity or external fluid velocity, and r may be replaced by a significant length. The Péclet number in Figure 4.7.9 has been calculated using pipe diameter rather than radius.

laminar flow

4.7.3.3 *Taylor Dispersion*

The process for laminar flow in a tube is called Taylor dispersion. This is an extremely simple case, unusual in occurrence, with a simple result. The dispersion coefficient for Taylor dispersion has been explicitly shown to be (Cussler, 1984)

$$\mathscr{D} = \frac{(rv)^2}{48D} \tag{4.7.16}$$

where $\mathscr{D}$ is the dispersion coefficient (m^2/sec); r, the tube radius (m); v, the flow velocity (m/sec); and D, the mass diffusivity (m^2/sec).

The resemblance of this equation to Eq. 4.7.15 defining the Péclet number is apparent.

greater
diffusion
means lesser
dispersion

This equation gives a surprising result: There is an *inverse* relationship between the dispersion coefficient and the mass diffusivity. Those substances that diffuse faster disperse slower. Looking again at Figure 4.7.4 might help to understand why this should be. Diffusion occurs radially in laminar flow; axial movement is caused by convection. The faster that material diffuses radially, the smaller is the concentration in the center of the tube and the less likely it is that material will be moved downstream by convection. Thus the material that spreads radially more rapidly will also spread more slowly axially.

If a pulse of material is injected into a round tube with laminar flow, that pulse will spread in two directions at once. It will spread radially by diffusion, and it will spread axially by convection. The previous consideration of the spreading of a static pulse is thus made more complicated in a tube with laminar flow. If the concentration of material in the tube is averaged in the radial direction, that is, across the cross section of the tube, then the axial concentration distribution more closely resembles the one-dimensional spread of the pulse as given previously (Cussler, 1984).

average tube
concentration

$$c_{avg} = \frac{m/\pi r^2}{\sqrt{4\pi \mathscr{D} t}} e^{-(x-vt)/4\mathscr{D} t} \qquad (4.7.17)$$

where c_{avg} is the average concentration in the tube cross section (kg/m^3); m, the total mass in the pulse (kg); r, the radius of the tube (m); $\mathscr{D}$, the dispersion coefficient (m^2/sec); t, the time after injection of the pulse (sec); x, the axial distance along the tube (m); and v, the fluid velocity (m/sec).

Comparing this equation to Eq. 4.7.2 shows that they are the same, with the exception that the mass diffusivity has been replaced by the dispersion coefficient and that the stationary point of reference $(x_0 = 0)$ has been replaced by a moving point of reference $(x_0 = vt)$.

This form for Taylor dispersion has been applied to the transport of gases and particles in the respiratory system (Ultman, 1981), and can also be used to analyze drug flow in the vasculature.

The twisting, turning, and branching of fluid conduits in respiratory and cardiovascular systems, however, mean that fully developed laminar flow is hardly ever achieved. The entrance lengths for most respiratory airway segments is longer than the segments themselves (Johnson, 1991). Thus
real limitations respiratory and cardiovascular dispersion occurs more by turbulent dispersion than by Taylor dispersion. Some modification of the Taylor dispersion coefficient can be made to account for the turbulent velocity profile (Ben Jebria, 1984), but the result is of only limited usefulness, because inhaled

particulate matter (as opposed to inhaled gases) is relatively heavy; particles impact the airways walls more because of inertia than because of dispersion.

EXAMPLE 4.7.3.3-1. Dispersion in Gas Chromatography

The gas chromatograph (GC) was introduced in Example 4.3.2.5-1. GC peaks should be as narrow as possible to be able to resolve differences between similar compounds. For a GC with capillary column 30 m × 250 μm (dia.), a carrier gas of dry nitrogen moving at 50 cm/sec, ammonia as the sample gas, and a column temperature of 50°C, what is the minimum peak width to be expected?

Solution

Since the solute, or injected compound, moves between stationary and mobile phases (that is, absorbed by the walls and carried along by the nitrogen stream), there is an effect called band broadening that occurs with the transitions between these two phases as the solute is carried down the column. The minimum peak width would be obtained if an inert compound, one that is not absorbed into the stationary phase, is injected into the carrier. In that case, the only peak-widening process occurs because of dispersion. We will thus check for the width of a pulse of gas (width $= 0$) as it disperses toward the detector end of the capillary column.

The diffusivity of ammonia in nitrogen was already calculated as 2.94×10^{-5} m^2/sec in Example 4.3.2.5-1. Using Eq. 4.7.16, the dispersion coefficient is

$$\mathscr{D} = \frac{(rv)^2}{48\, D_{NH_3-N_2}} = \frac{[(125 \times 10^{-6}\ \text{m})(50 \times 10^{-2}\ \text{m/sec})]^2}{(48/\text{m}^2)(2.94 \times 10^{-5}\ \text{m}^2/\text{sec})}$$

$$= 2.77 \times 10^{-6}\ \text{m}^2/\text{sec}$$

The time for the carrier gas to tranverse the column is

$$t = \frac{30\ \text{m}}{50 \times 10^{-2}\ \text{m/sec}} = 60\ \text{sec}$$

The standard deviation for the pulse is, from Eq. 4.7.13,

$$\sigma = \sqrt{2\mathscr{D}t} = \sqrt{2(2.77 \times 10^{-6}\ \text{m}^2/\text{sec})(60\ \text{sec})}$$

$$= 0.0182\ \text{m}$$

It was mentioned in Section 1.4.6 that 95.4% of the readings are within $\pm 2\sigma$, making the 95% pulse width 4σ. Thus the total minimum pulse width is

0.0729 m, or 7.3 cm. Since the gas is flowing at 50×10^{-2} m/sec, this represents 1.5 sec.

Remarks

Good peak separation for a GC is about 15 sec; so this width is not significant. The conclusion is that dispersion does not limit the sensitivity of the device.

4.7.3.4 Turbulent Dispersion in Pipes
Dispersion coefficients for turbulent flow in pipes have been measured and are normally correlated with Reynolds number (Figure 4.7.6). Axial dispersion at Reynolds numbers above 10,000 is about 300 times greater then radial dispersion.

Dispersion is also important in packed beds (see Section 4.4.2), where it interferes with separation between materials. Various measurements have been made, and the reader is referred to Cussler (1984) for more information.

4.7.3.5 Atmospheric Dispersion
Dispersion of atmospheric discharge of a pollutant is often mathematically described by assuming Gaussian dispersion of gaseous or particulate matter. The problem is complicated by the

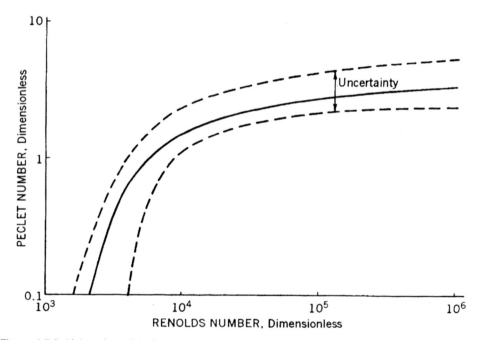

Figure 4.7.6. Values for axial dispersion during turbulent flow in pipes. At high Reynolds numbers. $(dv\rho/\mu)$ the Péclet number $(dv/\mathscr{D})$ approaches a constant value; so the dispersion coefficient becomes proportional to flow velocity.

three
dimensional

fact that pollutant movement occurs simultaneously in three dimensions: down wind, vertically, and horizontally (cross wind). The problem is usually to try to predict how fast or how far the pollutant will travel before dispersing below some threshold value.

A plume originates at some effective emission height (z_0) that can be the actual emission height or could be higher if propelled vertically from the source by an air blast or the chimney effect of rising hot air (Figure 4.7.7). Assuming a Gaussian distribution of pollutant concentration in the vertical direction, we have, from Eq. 1.4.18,

vertical
dispersion

$$f(z) = \frac{1}{\sigma_z \sqrt{2\pi}} e^{-(z-z_0)^2/2\sigma_z^2} \qquad (4.7.18)$$

where $f(z)$ is the concentration distribution in the vertical direction (m^{-1}); σ_z, the standard deviation of concentration in the vertical direction (m); z, the vertical distance from the ground (m); and z_0, the vertical distance to plume source height (m).

Likewise, in the horizontal direction,

horizontal
dispersion

$$f(y) = \frac{1}{\sigma_y \sqrt{2\pi}} e^{-(y-y_0)^2/2\sigma_y^2} \qquad (4.7.19)$$

The overall pollutant distribution at any distance down wind from the source is just the product of the above two equations. The pollutant concentration is that product times the source strength divided by the wind speed

two-
dimensional
concentration

$$c = \dot{m}f(y)f(z)/v \qquad (4.7.20)$$

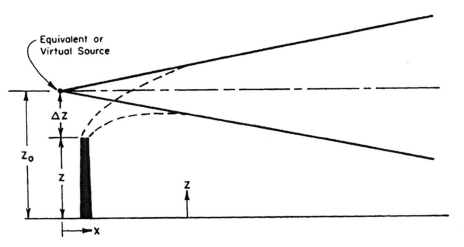

Figure 4.7.7. Dispersion from a virtual source at an effective stack height z_0.

where c is the pollutant concentration (kg/m^3); $\dot{m}$, the mass flow rate at source (kg/sec); $f(z)$, the vertical concentration distribution (m^{-1}); $f(y)$, the horizontal (cross wind) concentration distribution (m^{-1}); and v, the wind speed (m/sec).

It is often the case that the y_0 term in Eq. 4.7.19 is taken to be zero, and all horizontal measurements referenced to the horizontal location of the plume source. In the vertical direction, however, distances are usually measured from ground level. Thus z_0 cannot usually be taken to be zero.

ground
reflection

The presence of the ground changes the boundary conditions of the vertical dispersion problem. Since the ground usually represents a surface that neither transmits nor absorbs pollutant particles, it is often assumed that pollutant particles contacting the ground from above reflect back in the upward direction (Wark and Warner, 1981). These particles add to those dispersing from above to increase the concentration of pollutant near the ground.

Reflection is mathematically equivalent to adding the vertical contribution of a source located at a distance beneath the ground of $-z_0$ (Figure 4.7.8). The expression for concentration at any point down wind from the source thus becomes

$$c = \frac{\dot{m}}{2\pi v \sigma_y \sigma_z} e^{-y^2/2\sigma_y^2} \left(e^{-(z-z_0)^2/2\sigma_z^2} + e^{(z-z_0)^2/2\sigma_z^2} \right) \qquad (4.7.21)$$

where c is the pollutant concentration (kg/m^3); $\dot{m}$, the mass flow rate at source (kg/sec); v, the wind speed (m/sec); σ_y, the standard deviation in the cross wind direction (m); σ_z, the standard deviation in the vertical direction (m); y,

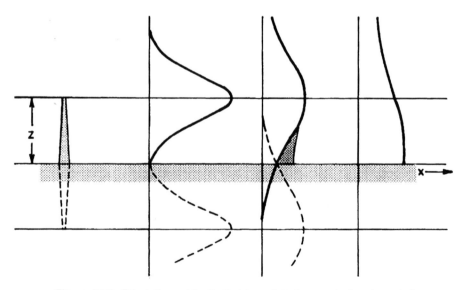

Figure 4.7.8. Effect of ground reflection on pollutant concentration downwind.

the cross wind distance referenced to source (m); z, the vertical distance referenced to ground (m); and z_0, the height of the source above ground (m).

Measurements of dispersion coefficients have been made that relate dispersion to the standard deviation of the presumed Gaussian distribution of the particles in the fluid. Taking into account the fact that distance, velocity, and time are all related in a flowing fluid

standard
deviation

$$\sigma = \sqrt{2\mathscr{D}x/v} = \sqrt{2\mathscr{D}t} \qquad (4.7.22)$$

where σ is the standard deviation (m); $\mathscr{D}$, the dispersioin coefficient (m^2/sec); x, the distance (m); v, the velocity (m/sec); and t, the travel time (sec).

This would indicate that the amount of spread of the plume of material is directly related to distance. In Figures 4.7.9 and 4.7.10 are shown widely accepted (read this as "no one has any better information, and it is difficult to

standard
deviation
curves

generate better data") charts for standard deviation values in the vertical and

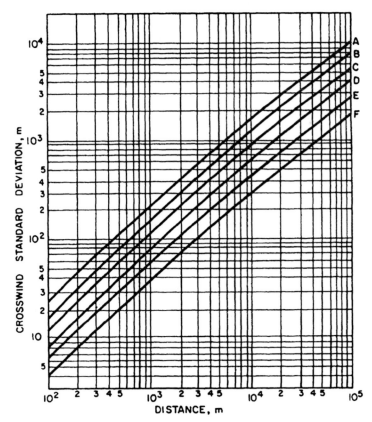

Figure 4.7.9. Standard deviation, σ_y, in the crosswind direction as a function of distance downwind.

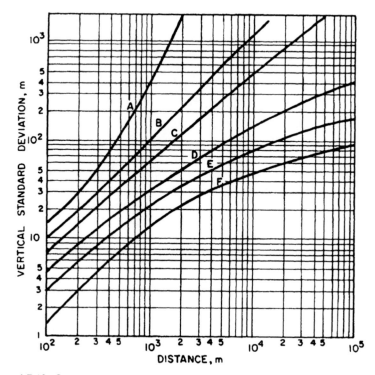

Figure 4.7.10. Standard deviation, σ_z, in the vertical direction as a function of distance downwind.

Table 4.7.1   Key to Stability Categories

Wind speed at 10 m height (m/s)	Day			Night	
	Incoming Solar Radiation			Cloud cover	
	Strong	Moderate	Slight	Mostly Overcast	Mostly Clear
Class[a]	(1)	(2)	(3)	(4)	(5)
< 2	A	A–B	B	E	F
2–3	A–B	B	C	E	F
3–5	B	B–C	C	D	E
5–6	C	C–D	D	D	D
> 6	C	D	D	D	D

[a] The neutral class, D, should be assumed for overcast conditions during day or night. Class A is the most unstable, and class F is the most stable, with class B moderately unstable and class E slightly stable.

horizontal directions when dealing with dispersion downwind from a point source. The choice of the curve to be used in each graph depends on atmospheric conditions, and these appear in Table 4.7.1. Once values for σ_y and σ_z are determined from the graphs, use an average wind velocity across the plume (if such is available) to determine the dispersion coefficient from Eq. 4.7.22. For further details, see Wark and Warner (1981), Turner (1969), and Cussler (1984).

insensitive to materials

Dispersion coefficient values for turbulent flow are not very sensitive to the type of material being dispersed. They depend more on particle sizes (Table 4.7.2) and atmospheric conditions than on either the material being dispersed or the fluid through which the material travels.

Gaussian assumption

One difficulty with the application of dispersion in the open atmosphere is the assumption of a Gaussian dispersion form. This assumption is probably more correct for the horizontal direction than for the vertical direction, because of the contact that the plume makes with the ground. Better choices of probability distribution functions to describe the vertical plume shape are probably log-normal or Poisson, described in most probability or statistics texts. There are other atmospheric (for example, inversions) and environmental conditions (for example, forests) that can influence the shape of the plume.

Odor Substances dispersing in the air may or may not be toxic. The odor of the substance may often be a more restrictive criterion of acceptability than any other, and thus some knowledge about odor can be helpful.

odor qualities

Exactly what causes odor is not well known, and odorous substances often, but not always, are those with high vapor pressures (are volatile). As an organoleptic quantity, odor has attributes of intensity, pervasiveness, persistance, quality, and acceptability (Wark and Warner, 1981). Of these, intensity and pervasiveness are influenced by substance concentration.

Odor intensity appears to follow the Weber–Fechner relationship

Weber–
Fechner law

$$I = k \log c/c_0 \tag{4.7.23}$$

Table 4.7.2 Sizes of Common Particles

Particle Type	Size (μm)
Simple molecules	0.00015–0.006
Viruses	0.002–0.06
Atmospheric dusts	0.001–20
Lung-damaging dust	0.6–6
Insecticide dust	0.6–10
Bacteria	3–30
Red blood cell	7.5–0.3
Pollen	10–100
Beach sand	90–2000

Source: Fasman (1989).

Table 4.7.3 Odor Thresholds in Air at 20°C

Chemical	Odor Threshold (kg/m^3)
Acetic acid	1.2×10^{-6}
Acetone	1.2×10^{-4}
Amine monomethyl	2.5×10^{-8}
Amine trimethyl	2.5×10^{-9}
Ammonia	5.6×10^{-5}
Carbon disulfide	2.5×10^{-7}
Chlorine	3.8×10^{-7}
Diphenyl sulfide	5.7×10^{-9}
Dimethyl sulfide	1.7×10^{-5}
Ethanol	1.6×10^{-6}
Ethanethiol	1.3×10^{-8}
Formaldehyde	1.2×10^{-6}
Hydrogen sulfide	5.7×10^{-10}
Methanethiol	9.5×10^{-7}
Methanol	1.2×10^{-4}
Methylene chloride	2.6×10^{-4}
Phenol	5.7×10^{-8}

where I is the perceived intensity (units of measurement); k, the constant with magnitude from 0.3 to 0.6 (units of I); c, the concentration (kg/m^3); and c_o, the reference concentration (kg/m^3).

odor reduction difficult

The fact that odor intensity is perceived based upon logarithmic concentration, but concentration diminishes roughly with the square of the distance from the source, means that the biological engineer must expend a great deal of effort to reduce odor perception by a small amount.

threshold

Odor threshold is the lowest substance concentration that can be detected. Threshold would be expected to vary among individuals and is usually determined by panels of experts. In Table 4.7.3 are found some odor thresholds. Other values can be found in Leonardos et al. (1969) and Hellman and Small (1974).

EXAMPLE 4.7.3.5-1. Atmospheric Dispersion

Among other gases, dimethyl sulfide is being released from a compost pile at the rate of $600 \, \mu g/(m^2 \, sec)$. The surface area of the pile is $150 \, m^2$. Assume that the emission acts as if it were from a point source. Will the concentration of $(CH_3)_2S$ be above or below the odor threshold at a distance of 1000 m directly downwind from the source? Wind speed is 2.5 m/sec, and the sun is shining brightly.

Solution

The strength of the source is

$$\left(\frac{600\ \mu g}{m^2\ sec}\right)(150\ m^2)\left(\frac{1\ kg}{1000\ g}\right) = 9.00 \times 10^{-5}\ kg/sec$$

From Table 4.7.1, we see that for a wind speed of 2–3 m/sec and strong solar radiation, classes A or B should be chosen. We choose class B because this will result in smaller standard deviations and higher concentrations. From Figure 4.7.9 we see that $\sigma_y = 150$ m, and from Figure 4.7.10, we find $\sigma_z = 100$ m.

If the compost pile is located on the ground, and the point of measurement of odor is also on the ground, then Eq. 4.7.21 becomes

$$c = \frac{\dot{m}}{2\pi v \sigma_y \sigma_z} e^{-y^2/2\sigma_y^2}\left(e^{-(z-z_0)^2/2\sigma_z^2} + e^{-(z+z_0)^2/2\sigma_z^2}\right)$$

$$= \frac{(9.00 \times 10^{-5}\ kg/sec)}{2\pi(2.5\ m/sec(150\ m)(100\ m)} e^0(e^0 + e^0)$$

$$= 7.64 \times 10^{-10}\ kg/m^3$$

From Table 4.7.3 we see that this concentration is much below the odor threshold.

4.7.3.6 *Dispersion within Soil*

soil is complex

The movement of water and contaminants through soil is an important, but complex, process. Soil is, after all, a medium that is anything but simple, or even homogeneous. There are physical, chemical, and biological interactions that can occur in the soil that make simple analyses inadequate. Yet we have already considered several of these processes; so we begin by a review of pertinent past discussions.

The movement of water and solutes in soil is similar to the case of simultaneous diffusion and convection considered in section 4.7.2. Soil hydrologists, however, usually use the term *advection* for the process called *convection* here.

advection

Darcy's law (Section 1.8.1) can be used to calculate fluid velocity (advection) through the soil, but actual velocity in pores must be calculated by dividing the velocity obtainable from Darcy's law by the porosity of the soil (Best et al., 1993).

Movement of water and solutes in soil includes elements of the discussion on diffusion through porous solids, given in Sectiopn 4.3.2.4. There can also be actual dispersion of solutes, due to differences in the velocities of movement across soil pores, as discussed in Section 4.7.3.1. Although this dispersion can occur in two or three different directions, unidimensional

porous soil

dispersion (from the surface downward) is often assumed as a reasonable approximation over a limited area.

In addition, the soil can physically adsorb ions; clays have a special affinity for potassium ions. Soils can serve as the base medium for the growth of biofilms; this is important for denitrification and other bioremediation. Some ions can react chemically with soil particles, especially H^+ with $CaCO_3$ (limestone) to form water and carbon dioxide. ($2H^+ + CaCO_3 \rightarrow Ca^{2+} + H_2CO_3 \rightarrow Ca^{2+} + H_2O + CO_2$). Fixed cationic charges in soil organic matter can produce electrical double layers locally in groundwater that influence the mobility of anions and cations (Russell, 1961). In all these respects, soil acts similarly to the packed beds discussed in Section 4.4.2.

biological interactions

Let us not forget the fact that water movement affects the presence of earthworms, grubs, burrowing insects and other animals, and plant roots that directly affect soil porosity.

Soil permeability is an important parameter that is often manipulated to produce desired results. Soils may be plowed or disked to increase porosity (and thus permeability) for improved plant growth, or soils may be purposely compacted to contain contaminants existing in garbage landfills or in manure storage facilities. Such a case was considered in Example 1.6.2-2.

soil manipulation

Finally, consider that the analytical solutions to various hydrologic models have a great deal in common with the solutions to transient heat transfer problems, as given in Section 3.7.2.2. Although the analytical basis for the Heisler charts was not given in that section, the graphical solution is applicable, and solutions in mathematical form can be obtained from Carslaw and Jaeger (1959).

analogy to heat transfer

4.7.3.7 Dispersion Impedances
The nature of dispersion is such that the most appropriate systems simulation model would use distributed impedance elements. This is because the plume spreads in all directions. Nonetheless, we have seen that it is useful to use concentration averaged over the tube cross section for Taylor dispersion. If we use average concentrations, then it is possible to propose lumped elements for our systems model.

The similarity between diffusion and dispersion mathematical formulations allows us to write, for axial flow resistance

axial resistance

$$R_{axial} = \frac{L}{\mathscr{D}A} \qquad (4.7.24)$$

where R_{axial} is the axial dispersion resistance (sec/m^3); L, the significant length (m); $\mathscr{D}$, the dispersion coefficient (m^2/sec); and A, the appropriate cross-section area (m^2).

The significant length may be the total length of a pipe, or it may be distance over which the fluid would flow in some length of time (vt) of interest. The cross-sectional area would be the cross-sectional area of a pipe or the area corresponding to the size of the plume, depending on the situation.

Despite the ambiguity associated with the exact meanings of the symbols, the form for axial dispersion resistance is that for resistance in a region bounded by two parallel flat surfaces, as given in Section 1.8.3. This resistance applies to dispersion flow as a result of concentration averaged over the cross section of a pipe or some arbitrary cylindrical volume.

radial
resistance

For the case of Taylor dispersion within a pipe, radial dispersion resistance is of little consequence. This is because the radius of the pipe is some practically small value that limits the spreading of the plume in the radial direction. We can conceptualize the source of the plume as a flow source that fills the capacity of the pipe inexorably. So radial resistance has no real effect.

Dispersion capacity is calculated as a volume in the pipe, because capacity is the relationship between mass and concentration. Thus

capacity

$$C = \pi r^2 L \qquad (4.7.25)$$

where C, is the mass capacity (m^3); r, the pipe radius (m); and L, the length of pipe or of a portion of the pipe (m).

inertia

As discussed previously (Section 1.4.4), there is a small amount of inertia associated with mass transport. Since dispersion involves fluid flow, there is a significant amount of inertia associated with the flowing fluid, but, for relatively low concentrations of substance particles being dispersed, the term $d\dot{m}/dt$ cannot usually become large enough to cause large inertial mass transfer effects.

The problem with this simple-minded concept of dispersion is that the point of reference for resistance and capacity keeps changing. As soon as part of the pipe fills with the dispersed mass, another part is already in the process of filling. Thus it is difficult to imagine what value for length (L) should be used to calculate R_{axial} and C.

Taylor dispersion can be envisioned as in Figure 4.7.11. Here a flow source fills a series of capacity elements C through a series of resistors R_{axial}. The

distributed
elements

repetitive pattern of C and R_{axial} elements continues to the end of the pipe. The

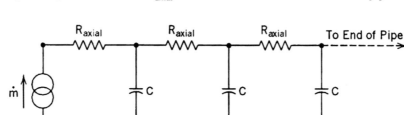

Figure 4.7.11. Taylor dispersion actually involves a repetitive pattern of resitances and capacity elements arranged along the length of the tube. As soon as one capacity is full, the next begins filling. The mass flow source supplying the plume assures that all capacity elements are filled eventually. Resistance and capacity elements actually overlap and are really distributed throughout the pipe.

value for L to be used in calculation of each type of element can be made as small or large as one wishes; the smaller the value of L, the closer the model becomes to a distributed parameter model, and the more realistic it is. This type of model has been extensively analyzed for the case of electrical transmission lines.

We have seen that Taylor dispersion, occurring as it does within a tube with fluid in laminar flow, is actually a simple case of dispersion. Much more complicated is the case of dispersion in the open atmosphere. Here, radial dispersion adds to the complexity of the transmission line analogy just discussed for Taylor dispersion.

We would like to be able to calculate a radial dispersion resistance using the expression from Eq. 1.8.32

$$R_{radial} = \frac{\ln(r_2/r_1)}{2\pi L \mathscr{D}} \qquad (4.7.26)$$

where R_{radial} is the radial dispersion resistance (sec/m^3); r_1, some inner radius (m); and r_2, some outer radius (m).

mistake!

Do not do it! This expression was developed for a hollow cylinder, and when $r_1 = 0$, as it must because there is mass flow from the center outward, the resistance becomes infinite. Another approach is necessary.

We could try to work through this difficulty, but we would soon realize that this approach is close to fruitless. We would have a transmission line analogy in at least two different directions, and there would not be any simplification by reducing the ambient dispersion to resistance and capacity elements. Dispersion is one phenomenon that tests the boundaries of the simple systems approach to transport processes and finds it lacking. A mathematical approach should be followed.

4.7.4 Non-Unsteady-State Mass Transfer

There are many instances of non-steady-state mass transfer that may be of interest to the biological engineer. By the analogies presented in Chapter 1,

diffusion and conduction

these can be treated in the same ways as other non-steady-state transport processes. Diffusion mass transfer is directly analogous to conduction heat transfer, and many conduction problems are solved in closed mathematical form in Carslaw and Jaeger (1959).

surface convection

Mixed-mode mass transfer specifically involves diffusion to a surface and convection at the surface. This type of mass transfer problem is a very important one that occurs when drying foods, absorbing bloodborne substances within artificial kidneys, or releasing drugs from implants, to name a few examples.

We have earlier seen that non-steady-state heat transfer involving conduction and convection used several dimensionless parameters. Among them were the temperature difference ratio $[(\theta - \theta_\infty)/(\theta_o - \theta_\infty)]$ Fourier number $(\alpha t/L^2)$, and Biot number $(h_c L/k)$. Non-unsteady-state mixed-mode mass transfer uses three analogous dimensionless numbers

analogies

$$\left(\frac{c - c_{\text{equil}}}{c_0 - c_{\text{equil}}}\right) \sim \left(\frac{\theta - \theta_\infty}{\theta_0 - \theta_\infty}\right) \tag{4.7.27}$$

$$\left(\frac{Dt}{L^2}\right) \sim \left(\frac{\alpha t}{L^2}\right) \tag{4.7.28}$$

$$\left(\frac{k_G L}{D}\right) \sim \left(\frac{h_c L}{k}\right) \tag{4.7.29}$$

where θ is the temperature (°C); c, the concentration (kg/m^3); c_{equil}, the equilibrium concentration (kg/m^3); c_0, the initial concentration (kg/m^3); D, the mass diffusivity of inside material (m^2/sec); t, the time (sec); L, the significant length (m); α, the thermal diffusivity (m^2/sec); k_G, the convection mass transfer coefficient (m/sec); k, the thermal conductivity of inside material [N m/(sec m °C)]; and h_c, the convection coefficient [N m/(sec m^2 °C)].

In the case of a process to dry materials, moisture contents are substituted for concentrations, and the dry matter density is used in place of substrate density. The moisture content terms will be discussed in more detail later in Section 4.8.2.

As long as the species of matter that is moving in the process is being depleted from the interior of the object (for example, falling-rate drying, discussed in Section 4.8.2.3), and as long as the geometry of the mass transfer

Heisler charts process conforms to the required shapes of solid spheres, cylinders, or plates, then the Heisler charts (Figures 3.7.17, 3.7.18, and 3.7.19) previously used for mixed-mode non-steady-state heat transfer can be used for mixed-mode non-steady-state mass transfer as well. Substitution of the analogous mass transfer dimensionless terms for their heat transfer counterparts is required before the Heisler charts apply.

Heisler charts do not apply when the mass transfer is quasi-steady-state, as when analyzing sweating. Here moisture is continually being delivered to the skin and the source of moisture is not being effectively depleted from the underlying body.

EXAMPLE 4.7.4-1. Drying Fruit Cake

Cathy's fruit cake is especially yummy when she soaks it in rum or brandy. However, it is not always eaten right away, and when left out, it dries. When it reaches about 20% moisture, it is not fit for man, although there may be some very hungry beasts who will touch it. How dry will the fruit cake be if it is left uncovered for a month. The fruitcake begins as a rectangular parallelpiped shape $6.4 \times 12.7 \times 23$ cm.

Solution

First, the mass transfer coefficient value must be determined. The simplest means to do this is to use Eq. 4.4.18 with estimates of the values for convection coefficient, mass diffusivity of moisture through air, and thermal conductivity of air. Still air will be assumed in the room, with an air movement of 0.15 m/sec. From Table 3.3.3, a value of 3.1 N m/(sec m² °C) is obtained for convection coefficient. From Table 4.3.1 we obtain a value of 0.260×10^{-4} m²/sec for diffusion of water through air. From Table 3.2.2 we obtain a value of thermal conductivity of air as 0.0255 N m/(sec m °C). Thus

$$k_G = \frac{(3.1 \text{ N m/(sec m}^2 \text{ °C)})(0.260 \times 10^{-4} \text{ m}^2/\text{sec})}{0.0255 \text{ N m/(sec m °C)}}$$

$$= 3.16 \times 10^{-3} \text{ m/sec}$$

From the value of water diffusivity through oatmeal cookies at 25°C, we estimate the water diffusivity through moist fruit cake to be about 4.0×10^{-11} m²/sec. Inserting the half-thicknesses of slabs with the three dimensions as given above

$$\left(\frac{k_G L}{D}\right)_1 = \frac{(3.16 \times 10^{-3} \text{ m/sec})(3.2 \times 10^{-2} \text{ m})}{4.0 \times 10^{-11} \text{ m}^2/\text{sec}} = 2{,}528{,}000$$

$$\left(\frac{k_G L}{D}\right)_2 = \frac{(3.16 \times 10^{-3} \text{ m/sec})(6.35 \times 10^{-2} \text{ m})}{4.0 \times 10^{-11} \text{ m}^2/\text{sec}} = 5{,}016{,}500$$

$$\left(\frac{k_G L}{D}\right)_3 = \frac{(3.16 \times 10^{-3} \text{ m})(11.5 \times 10^{-2} \text{ m})}{4.0 \times 10^{-11} \text{ m}^2/\text{sec}} = 9{,}085{,}000$$

Inverting each of these to use with the Heisler chart (Figure 3.7.19) yields all results indistinguishable from zero.

The analogous Fourier numbers must be determined. A time value of (30 d) (24 h/d) (3600 sec/h) = 2,592,000 sec will be used, as specified.

$$\text{Fo}'_1 = \frac{(4.0 \times 10^{-11} \text{ m}^2/\text{sec})(2{,}592{,}000)}{(3.2 \times 10^{-2} \text{ m})^2} = 0.101$$

$$\text{Fo}'_2 = \frac{(4.0 \times 10^{-11} \text{ m}^2/\text{sec})(2{,}592{,}000 \text{ sec})}{(6.35 \times 10^{-2} \text{ m})^2} = 0.0257$$

$$\text{Fo}'_3 = \frac{(4.0 \times 10^{-11} \text{ m}^2/\text{sec})(2{,}592{,}000 \text{ sec})}{(11.5 \times 10^{-2} \text{ m})^2} = 0.00784$$

Looking on the Heisler chart for average moisture difference ratios

$$\frac{\Delta c}{\Delta c}\bigg|_1 = 0.66$$

$$\frac{\Delta c}{\Delta c}\bigg|_2 = 0.95$$

$$\frac{\Delta c}{\Delta c}\bigg|_3 = 0.98$$

The overall moisture difference ratio is the product of these

$$\frac{\Delta c}{\Delta c} = \left(\frac{c - c_{\text{equil}}}{c_0 - c_{\text{equil}}}\right) = (0.66)(0.95)(0.98) = 0.61$$

Initial moisture content of the fruit cake is about 45% (Watt and Merrill, 1963). We assume the air indoors in winter to be very dry, and that the fruit cake equilibrium moisture content, if dried for a long enough time, is about the same as for bread sticks, or 5%. Therefore,

$$c = 0.61(45\% - 5\%) + 5\% = 29\%$$

This is the average moisture content in the fruit cake after 30 d.

Remark

The fruit cake will remain tantalizingly on the counter where all can see and should be eaten long before it dries. Beware, however, that each time a slice is taken the problem changes, and the last slice is likely to be so dry that the hungry beast will feast.

4.7.5 Mass Exchangers

We learned in Section 3.7.1 about heat exchangers, devices that allow heat to pass from one fluid to another without the fluids mixing. Mass exchangers allow a solute to pass from one fluid to another without mixing the fluids. You probably will not see the name "mass exchangers" anywhere, but you will see the names "lung," 'kidney," and "placenta" (Middleman, 1972). Flavor extraction in food processing is accomplished with a similar arrangement of membranes and flowing fluids. These devices can be analyzed similarly to heat exchangers.

mass
exchange

Heat stored in a flowing fluid was given in Eq. 3.6.14 as

$$\dot{q} = \dot{m}c_p \, \Delta\theta \tag{3.6.14}$$

Using Eq. 4.6.7, a similar expression for the storage of a solute in a flowing field would be

solute storage

$$\dot{m}_{sl} = \dot{m}_{sv}\alpha \, \Delta c/\rho_{sv} \tag{4.7.30}$$

where $\dot{m}_{sl}$ is the mass flow rate of the solute (kg/sec); $\dot{m}_{sv}$, the mass flow rate of the solvent (kg/sec); α, the solubility of solute (m^3 solute/m^3 solvent); Δc, the solute concentration difference (kg/m^3); and ρ_{sv}, the solvent density (kg/m^3).

limiting fluid

We introduced in Section 3.7.1 the concept of the limiting fluid, or that fluid that limits the exchange of heat from one side of the heat exchanger to the other. Mass exchangers, where two fluids are separated by a membrane permeable to the moveable solute, can also have a limiting flow rate, but instead of calculating the limiting flow rate as $\dot{m}c_p$, it is calculated as $\dot{V}_{lim} = \dot{m}_{sv}\alpha/\rho_{sv}$. For the case with the same solvent at the same temperature on both sides of the membrane, the limiting flow rate is the one with the lower mass flow rate.

Effectiveness curves for parallel flow (Figure 3.7.8), counter flow (Figure 3.7.7), and cross flow (Figure 3.7.9) heat exchangers can also be used for mass exchangers. Mass exchanger effectiveness is given as

effectiveness

$$E = \frac{(c_{i_{lim}} - c_{o_{lim}})}{(c_{i_{lim}} - c_{i_{nonlim}})} \tag{4.7.31}$$

The abscissa variable for the mass exchanger curves analogous to the heat exchanger curves is either $\Delta p/R_m \dot{V}_{lim}$ or $\Delta p \, AP_m/L\dot{V}_{lim}$; the first is for membrane resistance (R_m) obtained from Table 4.3.11; the second is to be used with membrane permeabilities (P_m) obtained from Table 4.3.9. In both cases, Δp is the pressure difference across the membrane. L is the membrane thickness. The curves parameter is $\dot{V}_{nonlim}/\dot{V}_{lim}$. Once effectiveness is known, solute mass exchange can be calculated from

analogous
parameters

$$\dot{m}_{sl} = E\dot{V}_{lim}(c_{i_{lim}} - c_{i_{nonlim}}) \tag{4.7.32}$$

The volume flow rate of the solute moving through the membrane of the mass exchanger can be easily obtained as

$$\dot{V}_{sl} = \frac{\Delta p}{R_m} = \frac{\dot{m}_{sl}}{\rho_{sl}} \tag{4.7.33}$$

where $\dot{V}_{sl}$ is the solute volume flow rate (m^3/sec); Δp, the pressure difference of the solute across the membrane (N/m^2); R_m, the membrane resistance (N sec/m^5); and $\dot{m}_{sl}$, the solute mass flow rate (kg/sec); and ρ_{sl}, the solute density at the membrane (kg/m^3).

The pressure difference in the above equation is often expressed as a difference of gas partial pressures (when appropriate) on both sides of the membrane. These can be obtained from concentrations by using the ideal gas law

$$c = \frac{m}{V} = \frac{pM}{RT} \tag{4.7.34}$$

where c is the concentration of solute (kg/m^3); p, the solute partial pressure (N/m^2); R, the universal gas constant [8314.34 N m/(kg mol K)]; T, the absolute temperature (K); and M, the molecular mass of solute (kg/kg mol).

For a liquid, the pressure difference is the difference in hydrostatic pressures for each side.

To know the UA of heat exchangers is sufficient to select from among various manufacturers' offerings and is a suitable end point for heat exchanger design. Like heat exchangers, membrane mass exchangers are often available in standard commercial sizes (Section 4.3.3.3). The reader is referred to texts such as Middleman (1972) for further details of the design process. In general, a material permeability (P) replaces the overall heat transfer coefficient (U) in mass exchanger design.

EXAMPLE 4.7.5.1. Urea Dialysis

Urea accumulates in the blood as a waste product of protein metabolism (Galetti et al., 1995). Urea is normally removed from the blood into the urine by ultrafiltration in the kidney, but when the kidneys fail, urea must be removed externally using dialysis.

A dialyzer (or artificial kidney) is an ultrafiltration membrane that has circulating blood on one side and a dialyzing fluid on the other. This fluid contains small molecules (for example, sodium and potassium salts and glucose) in exactly the concentrations that are to be maintained in the blood. Large molecules that will not pass through the membrane are added to the dialyzer fluid to increase osmotic pressure to make it isotonic with the blood. Thus urea is the main constituent moving from the blood to the dialyzer fluid (Longmore, 1968).

Countercurrent mass exchange is usually used with the hollow fiber device illustrated in Figure 4.3.18. Urea concentration in the blood as it enters the dialysis unit is about 2×10^{-4} g/100 mL. The desired urea concentration of blood urea as it leaves the unit is 5×10^{-5} g/100 mL. Blood flow is normally 200–350 mL/min, and dialysate flow is 500 mL/min. The dialysate is kept at about 13 to $67 \, \text{kN/m}^2$ lower than blood pressure to: (1) help move urea and some water through the membrane, and (2) assure that any breaks in the membrane will result in blood leaking into the dialysate rather than vice versa. Determine the required membrane resistance for the dialysis unit.

Solution

This problem fits neatly into the boundaries of mass exchangers. We know that the new dialyzer fluid contains no urea, but we do not know the dialysate urea concentration when it leaves the unit. First, we must decide which is the limiting fluid. If the blood flow rate is $350 \, \text{mL/min} = 5.83 \times 10^{-6} \, \text{m}^3/\text{sec}$, and the dialysate flow rate is $500 \, \text{mL/min}$, then the blood is clearly the limiting flow rate. Thus we can calculate mass exchanger effectiveness from Eq. 4.7.31

$$E = \frac{c_{i_{\text{lim}}} - c_{o_{\text{lim}}}}{c_{i_{\text{lim}}} - c_{i_{\text{nonlim}}}} = \frac{(2 \times 10^{-4}) - (5 \times 10^{-5})}{(2 \times 10^{-4}) - 0} = 0.75$$

Also

$$\frac{\dot{V}_{\text{nonlim}}}{\dot{V}_{\text{lim}}} = \frac{500 \, \text{mL/min}}{350 \, \text{mL/min}} = 1.43$$

Consulting Figure 3.7.8 for counter flow heat exchangers gives an abscissa value of 2.2. Thus

$$2.2 = \frac{\Delta p}{R_m \dot{V}_{\text{lim}}}$$

or assuming a pressure difference across the membrane of $33,330 \, \text{N/m}^2$

$$R_m = \frac{\Delta p}{2.2 \dot{V}_{\text{lim}}} = \frac{33,330 \, \text{N/m}^2}{2.2(5.83 \times 10^{-6} \, \text{m}^3/\text{sec})}$$

$$= 2.60 \times 10^9 \, \text{N sec/m}^5$$

Remarks

For a total area of $1 \, \text{m}^2$, a membrane resistance of $2.60 \times 10^9 \, \text{N sec/m}^3$ would be required. This is an order of magnitude or two lower than those appearing in Table 4.3.11, but the operating pressure for the dialysis unit is much lower than for reverse osmosis; so the membrane can be much thinner.

This resistance includes significant convection resistances ($k_G A$) appearing on each side of the membrane. Manufacturers would likely have tested dialysis units to determine exact resistances of their manufactured units. Some devices have convolutions in their membranes to induce turbulence and reduce convection resistances.

4.8 SIMULTANEOUS HEAT AND MASS TRANSFER

There are some very important processes that involve simultaneous heat and mass movement. Most of these involve a change of state of the substance. Examples of this come in the processes of evaporative cooling, drying, evaporative mass removal, condensation, and burning. In other cases, there is not a change of state, but heat is removed as the mass is transferred, as in simultaneous convective movement of a high- (or low-) temperature fluid, ablation, or manual movement of hot objects into an oven for cooking.

4.8.1 Psychrometrics

air–water
vapor
mixtures

Some of the most important biological processes involving simultaneous heat and mass transfer are those involving air–water vapor mixtures. This leads naturally to the study of psychrometrics, or the measurement of properties of such mixtures.

standard
atmosphere
often not
maintained

4.8.1.1 Ideal Atmosphere As long as the composition of the air does not change during a process, then psychrometric relationships can be considered to be those of a binary mixture of dry air and water vapor. The composition of standard dry air is given in Table 4.8.1. The difficulty with the assumption of a constant dry air composition is that biological activity often causes changes in other atmospheric constituents besides water vapor. Respiration, for example, produces carbon dioxide as well as water vapor,

Table 4.8.1 Standard Composition of Dry Air

Constituent	Molecular Mass (kg/kg mol)	Volume Fraction in Air (m^3/m^3)
Air	29.0	1.0000
Ammonia	17.0	0.0000
Argon	39.9	0.0093
Carbon dioxide	44.0	0.0003
Carbon monoxide	28.0	0.0000
Helium	4.0	0.0000
Hydrogen	2.0	0.0000
Nitrogen	28.0	0.7808
Oxygen	32.0	0.2095
Water vapor	18.0	—

and depletes oxygen; microbial activity during composting may produce sulfurous gases and nitogenous gases; photosynthesis produces oxygen and depletes carbon dioxide, and, sometimes, water vapor. Thus analysis and design of systems and processes based upon standard atmospheric dry air composition may require modifications in order to be accurate. On the other hand, many engineering designs still require adjustments following the calculation stage; so these modifications may not need to be so severe.

Atmospheric gases come reasonably close (usually well within 2%) to obeying the ideal gas law

ideal gas law

$$pV = nRT \qquad (4.8.1a)$$

where p is the pressure (N/m^2); V, the gas volume (m^3); n, the number of moles (kg mol); R, the universal gas constant [8314.34 N m/(kg mol K)]; and T, the absolute temperature (K).

Such is the case as long as the ideal gas law is applied well away from the gas condensation point, where there are large interactions between molecules. The ideal gas law may even be applied to vapors, including water vapor, although errors up to 5% may be incurred with saturated vapors.

It is often useful to use the ideal gas law in terms of mass rather than kg moles

$$pV = mRT/M \qquad (4.8.1b)$$

where m is the total mass of gas (kg); and M, the molecular weight of gas (kg/kg mole).

partial pressure

The partial pressure of a gas is the pressure that it would exert in a given volume if it were the only constituent present. Partial pressure is a useful concept when talking about respiratory gases, blood gases, and atmospheres in plant or animal production systems. Dalton's law states that the total pressure is the sum of the partial pressures of a mixture

$$p = \sum_{i=1}^{N} p_i \qquad (4.8.2)$$

where p_i is the partial pressure of the ith constituent (N/m^2); N, the total number of constituents; and p, the total pressure (N/m^2).

The value for total atmospheric pressure is normally $101.325\,kN/m^2$. With psychrometric calculations, the two constituents are usually considered to be dry air and water vapor.

water–vapor lightens air

When air is very humid and unpleasant, we often characterize it as "heavy" or "thick." However, if total atmospheric pressure is $101.325\,kN/m^2$, then the partial pressure of dry air decreases as the partial pressure of water vapor increases. The molecular mass of dry air is (from Table 4.8.1) 29.0 kg/kg mol, whereas the molecular mass of water vapor is 18.0 kg/kg mol. Thus, when water vapor displaces dry air, the density of air actually decreases (because of this there is less lift on an airplane or an insect's wings when the air is more humid). This counterintuitive trend is included here to illustrate the caution that should be exercised when trying to make judgments about physical things from psychophysical sensations.

Partial pressures can be determined from volume fractions (as given in Table 4.8.1) because the temperature of a mixture is everywhere the same, and Avagadro's principle states that different gases at the same temperature and pressure contain equal numbers of molecules. Thus

partial
pressures
from volume
fractions

$$\frac{p_i}{p} = \frac{V_i}{V} \qquad (4.8.3)$$

Collection of water vapor and other gases at different environmental conditions can cause confusion when samples are to be compared. For this reason, there are two standard conditions usually specified for comparisons.

STPD

The STPD (standard temperature and pressure, dry) condition may be used for microbial, plant, or animal gas collections. STPD conditions are specified as temperature of $0°C$, pressure of $101.325 \, kN/m^2$, and water vapor removed. To correct from arbitrary environmental conditions to STPD conditions, gas volumes are given as

STPD
correction

$$V_{STPD} = V_{amb} \left(\frac{273}{273 + \theta}\right) \left(\frac{p - p_{H_2O}}{101.325}\right) \qquad (4.8.4)$$

where V_{STPD} is the gas volume at STPD (m^3); V_{amb}, the gas volume at ambient conditions (m^3); θ, the actual ambient temperature $(°C)$; p, the actual barometric pressure (kN/m^3); and p_{H_2O}, the water vapor pressrue (kN/m^2).

BTPS

Gas volumes from human beings may be converted to BTPS (body temperature and pressure, saturated) conditions. BTPS conditions are defined as temperature of $37°C$ (nominal human body temperature), pressure of $101.325 \, kN/m^2$, and water vapor pressure of $6.28 \, kN/m^2$ (saturated at body temperature). The conversion from arbitrary environmental conditions to BTPS is accomplished by

BTPS
correction

$$V_{BTPS} = V_{amb} \left(\frac{310}{273 + \theta}\right) \left(\frac{p - p_{H_2O}}{p - 6.28}\right) \qquad (4.8.5)$$

where V_{BTPS} is the gas volume at BTPS (m^3). The conversion between STPD and BTPS conditions is just

$$V_{STPD} = V_{BTPS} \left(\frac{273}{310}\right) \left(\frac{101.325 - 6.28}{101.325}\right) = 0.826 V_{BTPS} \qquad (4.8.6)$$

4.8.1.2 *Saturated Water Vapor Pressure* Saturation is defined where the partial pressure of the water vapor in the air equals the partial pressure of

equal
pressure in
liquid and gas

the water vapor in the liquid at the same temperature as the air. There is thus
no net tendency toward further evaporation or condensation. The vapor
pressure of water at saturation is for all purposes independent of total pressure
(there is only about a 0.07 percent increase in vapor pressure with total
pressure at 1 atmosphere pressure). Saturation vapor pressure is thus
considered to be a function of temperature only, and values are given in
Table 4.8.2.

Calculation of saturated vapor pressure of water can be made with the use
of two equations (ASAE, 1979)

saturated
vapor
pressure

$$p_{sat} = \exp\left(31.9602 - \frac{6270.3605}{T} - 0.46057 \ln T\right),$$

$$\text{for } 255.38 \leq T \leq 273.16 \text{ K} \qquad (4.8.7a)$$

Table 4.8.2   Values of Water Vapor Saturation Pressure
at Various Temperatures

Temperature (°C)	Vapor Pressure (N/m²)
− 100	0.001333
− 90	0.009332
− 80	0.05333
− 70	0.2666
− 60	1.067
− 50	3.866
− 40	12.93
− 30	38.13
− 20	103.5
− 10	260.0
0	610.6
10	1,228
15	1,705
20	2,337
25	3,166
30	4,242
35	5,622
37	6,266
40	7,374
45	9,580
50	12,332
55	15,737
60	19,915
70	31,156
80	47,353
90	70,102
100	101,325

where p_{sat} is the saturation vapor pressure of water (N/m^2); and T, the absolute temperature (K, $= \theta + 273.16°C$).

high temperatures

$$p_{sat} = c_o \exp\left(\frac{a_0 + a_1 T + a_2 T^2 + a_3 T^3 + a_4 T^4}{b_1 T + b_2 T^2}\right)$$

for $273.16 \le T \le 533.16$ K   (4.8.7b)

where $c_o = 22,105,649.25 \, N/m^2$, $a_0 = -27,405.526$, $a_1 = 97.5413 \, K^{-1}$, $a_2 = -0.146244 \, K^{-2}$, $a_3 = 1.2558 \times 10^{-4} \, K^{-3}$, $a_4 = -4.8502 \times 10^{-8} \, K^{-4}$, $b_1 = 4.34903 \, K^{-1}$, and $b_2 = -3.9381 \times 10^{-3} \, K^{-2}$.

This latter equation can be extremely useful for calculating saturated water vapor pressure by computer. However, because it is an empirical equation that uses small differences between terms of nearly equal magnitude, errors can be magnified greatly. To minimize this problem, Eq. 4.8.7b should be used in the following form

computational formula

$$p_{sat} = c_o \exp\left(\frac{T\{T[T(a_4 T + a_3) + a_2] + a_1\} + a_0}{T(b_1 + b_2 T)}\right)$$ (4.8.7c)

When water evaporates (or sublimates below freezing temperatures), heat energy is required for the change of state. ASAE (1979) gives the latent heat of sublimation as

sublimation heat

$$H = 2,839,683.144 - 212.56384 \, (T - 255.38)$$

for $255.38 \le T \le 273.16$ K   (4.8.8a)

where H is the latent heat of sublimation ($N \, m/kg$), and the latent heat of evaporation as

$$H = 2,502,535.259 - 2,385.76424 \, (T - 273.16),$$

evaporation heat

for $273.16 \le T \le 338.72$ K   (4.8.8b)

$$H = (7,329,155,978,000 - 15,995,964.08T^2)^{1/2}$$

for $338.72 \le T \le 533.16$ K   (4.8.8c)

These equations apply to free water in air. These equations will underestimate (by 20% or more) the heat of vaporization if the water is bound by chemical forces to the underlying substrate, such as water existing as moisture in a product to be dried. As hygroscopic materials are dried, the remaining water is more tightly bound, and more energy is required to remove it.

bound water

4.8.1.3 *Measurement of the Amount of Water in the Air*   There are two such measures important in psychometrics: the humidity ratio and relative

humidity. The humidity ratio (sometimes confused with absolute humidity or specific humidity) is the ratio of the mass of water vapor to the mass of dry air in any given volume

absolute humidity

$$\omega = \frac{m_{H_2O}}{m_{da}}$$ (4.8.9)

where ω is the humidity ratio (kg H_2O/kg dry air); m_{H_2O}, the mass of water vapor (kg); and m_{da}, the mass of dry air (kg).

The humidity ratio is the mass fraction of water vapor to dry air.

The humidity ratio can also be expressed in terms of partial pressure of water vapor in the air (see also Eq. 4.4.10)

$$\omega = \frac{0.6219 \, p_{H_2O}}{p - p_{H_2O}}$$ (4.8.10)

where p_{H_2O} is the partial pressure of water vapor in air (N/m^2); and p, the total atmospheric pressure (N/m^2).

The other two terms often confused with humidity ratio are the absolute humidity, defined as kg moisture per unit volume of dry air, and the specific humidity, defined as kg moisture per kg of moist air.

Relative humidity is the ratio of the partial pressure of water vapor in the air to the saturation vapor pressure at the same temperature

relative humidity

$$\phi = \frac{p_{H_2O}}{p_{sat}}$$ (4.8.11)

where ϕ is the relative humidity (dimensionless water); and p_{sat}, the water vapor saturation pressure (N/m^2).

Unlike the humidity ratio, relative humidity is affected by temperature and can decrease without a change in the amount of water vapor in the air just by a temperature increase. Since the tendency of water vapor to move from one place to another depends on a concentration gradient, and, therefore, a pressure

effort variable

difference, relative humidity expresses a tendency for water to evaporate into the air. Relative humidity can thus be classified as an effort variable. When $\phi < 1$, water will evaporate; when $\phi = 1$, no additional water will evaporate, and the air is said to be saturated; when $\phi > 1$, the mixture is supercooled and condensation is imminent: Dew happens.

dry bulb temperature

4.8.1.4 Temperature Measures Of these, the most common is dry bulb temperature (θ_{db}), taken with a dry temperature sensor in thermal equilibrium with the surrounding air. The sensor should be shielded from external radiation.

wet bulb temperature

The wet bulb temperature (θ_{wb}) is an approximation to the adiabatic saturation temperature. This temperature is measured by a sensor wetted with pure water, shielded from external radiation, and with air blowing across it at 5 m/sec or more.

adiabatic saturation temperature

The adiabatic saturation temperature is determined as in Figure 4.8.1. Air evaporates water vapor from the insulated chamber and leaves saturated. All energy required to supply the latent heat of evaporation comes from the incoming air. The warmer the incoming air, the warmer will be the exiting air, and the drier the entering air, the cooler will be the exiting air.

If the liquid in the chamber and on the surface of the wet bulb temperature sensor is not pure water, a higher amount of energy will be required to evaporate the same amount of liquid and the measurement will be in error. Thus tap water is not pure enough to be used to obtain wet bulb temperature.

dew point temperature

The dew point temperature (θ_{dp}) is the temperature at which condensation of water vapor in the air begins. At the dew point temperature, the relative humidity becomes 1.0 and the wet bulb temperature, dry bulb temperature, and dew point temperature all assume the same value.

There are some additional temperature measurements related to psychometrics that are sometimes measured. These are important when environmental exposure of humans is an issue. The natural wet bulb temperature (θ_{nwb}) is similar to the thermodynamic wet bulb temperature mentioned earlier, except that there is no attempt to blow air across the temperature sensor or to shield it from radiation. The natural wet bulb temperature value is usually higher than the thermodynamic wet bulb temperature, and is considered to integrate environmental aspects impinging on a sweating human.

natural wet bulb temperature

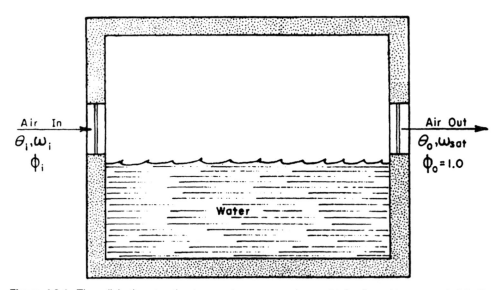

Figure 4.8.1. The adiabatic saturation temperature occurs when water is allowed to evaporate into the air without any additional heat exchange. The air leaves saturated and cooler.

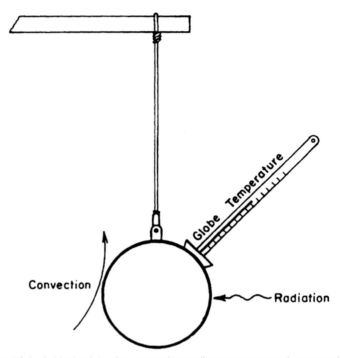

Figure 4.8.2. A black globe for measuring radiant temperature important for human comfort.

<table>
<tbody>
<tr><td>black globe temperature</td><td>The black globe temperature (θ_{bg}) is measured with a temperature sensor inside a 15 cm copper globe (Figure 4.8.2) hung about 120 cm above the floor. The surface of the globe is painted flat black to maximize radiation exchange. It has usually been considered that the radiation coefficient (h_r) of the black globe is nearly the same as that of a clothed human (Section 3.4.6).</td></tr>
<tr><td>wet bulb globe temperature</td><td>The above measurements can be combined into a wet-bulb globe temperature (WBGT) measurement as an index of human heat stress in a hot environment. The WBGT is calculated as</td></tr>
</tbody>
</table>

$$\text{WBGT} = 0.1\theta_{db} + 0.2\theta_{bg} + 0.7\theta_{nwb} \qquad (4.8.12)$$

where WBGT is the wet bulb globe temperature (°C); θ_{db}, the dry bulb temprature (°C); θ_{bg}, the black globe temperature (°C); and θ_{nwb}, the natural wet bulb temperature (°C).

There are additional heat and cold stress indices that are also useful for human application (Hertig, 1973).

4.8.1.5 Enthalpy   Enthalpy is composed of internal energy plus the product of pressure and volume. Specific enthalpy for a mixture of dry air and water vapor is usually expressed on the basis of a mass of dry air

specific
enthalpy
$$h_{en} = u + pv \tag{4.8.13}$$

where h_{en} is the specific enthalpy (N m/kg dry air); u, the specific internal energy (N m/kg dry air); p, the pressure (N/m^2); and v, the specific volume (m^3/kg dry air).

Specific enthalpy is a state function only; that is, it does not depend on the particular pathway or history taken between two states. Specific enthalpy can
state process   be evaluated at each state of interest, and differences in specific enthalpy depend only on the differences that appear in values at two states. An example of a nonstate function is work that is calculated as the integral over time of pressure times flow rate ($W = \int p\dot{V}\, dt$). The amount of work calculated depends on the relationship between pressure and flow rate during the transition from one point to another. However, if pressure is held constant, then work becomes independent of the pathway taken.

For a quasistatic isobaric process, one that would be encountered if the process proceeds slowly enough and at atmospheric pressure, specific enthalpy undergoes a change of

$$(h_{en_2} - h_{en_1})_p = (u_2 - u_1)_p + p(v_2 - v_1) \tag{4.8.14}$$

where 1 and 2 denote the beginning and end points of the process. But by the first law of thermodynamics, heat can be equated to physical work and changes in internal energy

heat energy
$$q = c_p(\theta_2 - \theta_1) = (u_2 - u_1) + w$$

$$= (u_2 - u_1) + p(v_2 - v_1) \tag{4.8.15}$$

where q is the heat transferred to the system per unit mass (N m/kg mixture); c_p, the specific heat at constant pressure [N m/(°C kg mixture)]; θ, temperature (°C); and w, the work per unit mass (N m/kg mixture).

Since the mass of water vapor in the mixture is a very small fraction of the total mass, heat added to a psychometric mixture comprises the difference of enthalpy of the mixture

$$q = (h_{en_2} - h_{en_1}) \tag{4.8.16}$$

air–water
mixture
The specific enthalpy of the mixture of dry air and water vapor is the sum of (Figure 4.8.3):

1. Enthalpy of the dry air;
2. Enthalpy of water (or ice) at the dew-point temperature;
3. Enthalpy of evaporation (or sublimation) at the dew-point temperature; and
4. Enthalpy added to the water vapor after vaporization.

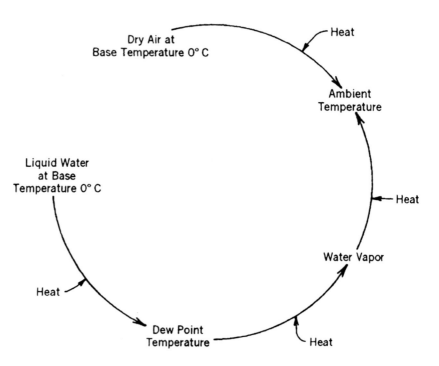

Figure 4.8.3. Contributions to total enthalpy of an air–water vapor mixture.

Each term can be taken separately. The enthalpy of dry air is, by Eq. 4.8.15, the heat added above the base, or reference temperature. The enthalpy of water at the dew-point temperature is similarly determined, except that the mass fraction of water vapor in the air must be used to determine the amount of water of interest. The enthalpy of evaporation uses the latent heat of water vapor, and the enthalpy added to the water vapor is calculated as heat added from the dew-point temperature. Thus

$$h_{en} = c_{p_{da}}\theta + c_{p_w}\omega\theta_{dp} + H\omega + c_{p_{wv}}\omega(\theta - \theta_{dp}) \qquad (4.8.17)$$

where h_{en} is the specific enthalpy of the mixture (N m/kg dry air); $c_{p_{da}}$, the specific heat at constant pressure of dry air [N m/(°C kg dry air)]; c_{p_w}, the specific dry heat of liquid water [N m/(°C kg water)]; $c_{p_{wv}}$, the specific heat of water vapor [N m/(°C kg water vapor)]; θ, the temperature (°C); θ_{dp}, the dew-point temperature (°C); H, the latent heat of evaporation (N m/kg water); ω, the humidity ratio (kg H$_2$O/kg dry air).

specific heat values   The value for specific heat of dry air is constant at 1006 N m/(°C kg air) over the range of 0 to 40°C. The specific heat of liquid water varies from 4229 to 4183 over the same range. ASAE (1967) used a value of 4186.8 N m/(°C kg water). The specific heat of water vapor also varies slightly with temperature.

ASAE (1967) used a value of $1875.6864 \, \text{N m}/(^\circ \text{C kg}$ water vapor). In addition, the latent heat of evaporation [about $2400 \, \text{kN m}/(^\circ \text{C kg}$ water)] also varies slightly with temperature (see Eqs. 4.8.8b and 4.8.8c).

Differences of enthalpy of air–water vapor mixtures above the dew-point temperature where no water is added or removed are more simply given as

specific
enthalpy
difference

$$\Delta h_{\text{en}} = (c_{p_{\text{da}}} + c_{p_{\text{wv}}} \omega) \, \Delta \theta \qquad (4.8.18)$$

where $\Delta \theta$ is the change in temperature ($^\circ$C). For sublimating ice, ASAE (1967) gives

$$h_{\text{en}} = 1006.925400 \, \theta - \omega(333432.1 - 2030.5980 \theta_{\text{dp}})$$

$$+ H\omega + 1875.6864\omega(\theta - \theta_{\text{dp}}), \qquad (4.8.19)$$

$$\text{for } 255.38 \le T = \theta + 273.16 \le 273.16 \text{ K}$$

The latent heat of sublimation (H) is given by Eq. 4.8.8a.

4.8.1.6 Psychrometric Charts

air–water
vapor
mixtures

Three properties of the atmosphere must be specified before all additional properties are known. These are usually: (1) composition, (2) temperature, and (3) pressure. The psychrometric chart (Figures 4.8.4 and 4.8.5) has been drawn for standard dry air composition and standard atmospheric pressure. Temperature is the variable along the horizontal axis, and the humidity ratio is the variable appearing on the right side vertical axis. Lines of constant property values are drawn to specify the state of the air–water vapor mixture.

splayed lines

Splayed from the left, opening to the right, are lines of constant relative humidity (Figure 4.8.6). The curved top of the chart is the 100% relative humidity line, and temperatures along this line are dew-point temperatures.

inclined lines

There are three types of inclined lines on the chart. The most vertical of the three types displays constant values of specific volume. Less vertical lines display constant wet-bulb temperatures. Constant values of specific enthalpy are on lines with nearly the same inclination as those for constant wet bulb temperature; these are not extended into the charts in Figures 4.8.4 and 4.8.5.

Psychrometric charts can be useful for visualizing as well as determining quantitative parameters associated with processes important to biological engineering. Some of these follow.

Sensible Heating or Cooling

no water
added

This is heating or cooling of the air without the addition or removal of water (Figure 4.8.7). The process follows a horizontal line along a constant humidity ratio value. Two parameters must be specified to locate the initial point in the process. Only one additional condition (except humidity ratio) must be specified to determine the final point in the process because the humidity ratio at the final point is the same as it is at the initial point.

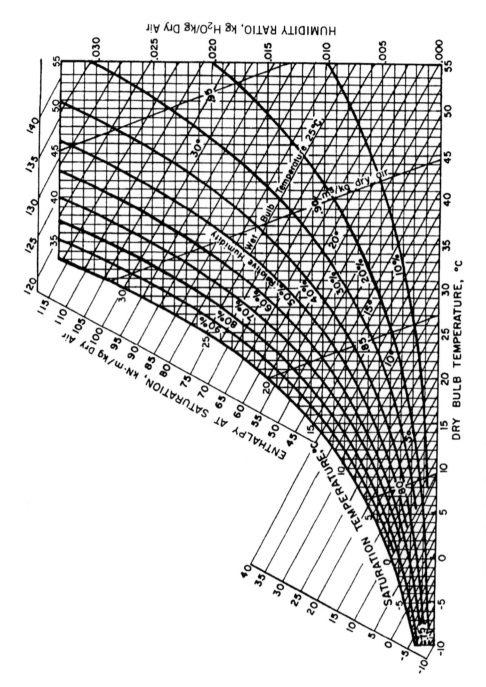

Figure 4.8.4. Psychrometric chart for lower temperatures. Barometric pressure is 101.325 kN/m².

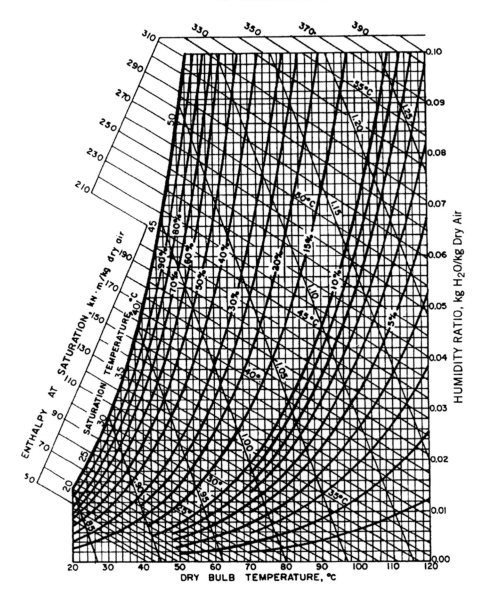

Figure 4.8.5. Psychrometric chart for higher temperatures.

Since no moisture is added or removed, the heat required for the process is just Δh_{en}.

riddle   *Adiabatic Saturation*   There is an old riddle that asks, "what gets wetter as it dries?" The answer is "a towel." When a product is dried, the air that takes part in the drying process is both cooled and wetted: The air is like the towel

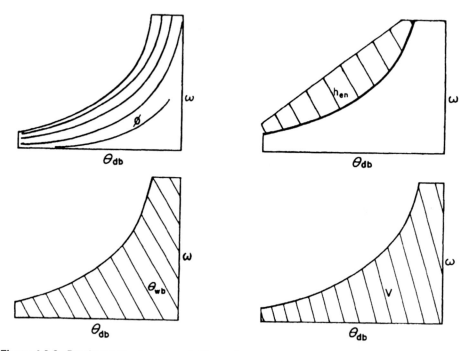

Figure 4.8.6. Psychrometric chart layout. Lines of constant relative humidity (ϕ) are curved. Lines of constant specific enthalpy (h_{en}) are inclined but extend off the chart. Lines of constant wet-bulb temperature (θ_{wb}) are nearly coincident with lines of constant specific enthalpy. Lines of constant specific volume (v) are nearly vertical.

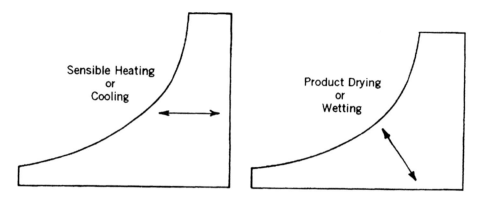

Figure 4.8.7. Sensible heating or cooling without addition of water vapor follows a constant humidity ratio line. Product drying wets and cools the air along a constant wet-bulb temperature line.

that becomes wetter as it dries. One can see on the psychrometric chart that following a constant wet-bulb temperature during the addition of moisture also cools the air (θ_{db} decreases).

drying and wetting

Comparing this process to the adiabatic saturation process described earlier reveals many similarities. Indeed, a process designed to dry a product is in many ways the same as adiabatic saturation. Thus this process follows a constant wet-bulb temperature line. Again, two properties are needed to specify the initial point, but only one additional property (except wet-bulb temperature) is needed to specify the final point. The amount of moisture removed from the product is $\dot{m}(\omega_2 - \omega_1)$, where $\dot{m}$ is the mass flow rate of air and $(\omega_2 - \omega_1)$ is the difference of humidity ratios from air entering to air leaving the drying zone.

evaporative cooling

Sometimes air is cooled by moisture evaporation. Evaporative cooling is a relatively inexpensive means of air cooling, as long as sufficient moisture is available and the air enters with a low relative humidity. This process appears on the chart the same as a drying process, and also follows a constant wet bulb temperature line. The object of attention here is the resulting air rather than the resulting dried product.

Mixing   When mixing two volumes of air, each at different temperatures and pressures, the result is air that must have temperature and moisture properties that lie somewhere between the temperatures and humidities of the original two unmixed samples. Thus the resultant mixture is one that appears on the psychrometric chart on the straight line that connects the properties of the two original masses of air (Figure 4.8.8).

The question of where on the line are located the properties of the mixture is answered by how much of each original sample is mixed with the other. The

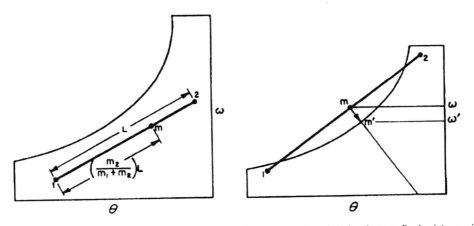

Figure 4.8.8. Mixing two masses of air at two original states (1 and 2) leads to a final mixture with properties at state *m*, which is closer to the original properties of the larger air mass. If warm, humid air is mixed with cold, condensation can easily occur.

mixed air mass properties will lie closer to the properties of the original sample that contributes relatively more mass to the mixture.

This is expressed mathematically as

$$\frac{m_1}{m_2} = \frac{\omega_m - \omega_2}{\omega_1 - \omega_m} = \frac{h_{en_m} - h_{en_2}}{h_{en_1} - h_{en_m}} = \frac{\theta_m - \theta_2}{\theta_1 - \theta_m} \tag{4.8.20}$$

where m is the mass (kg); ω, the humidity ratio (kg H_2O/kg dry air); h_{en}, the specific enthalpy (N m/kg dry air); θ, the temperature (°C); and 1, 2, and m refer to original (unmixed) air with properties at states 1 and 2, and to air properties of the mixed mass.

For a continuous-flow mixing process, the mass flow rates may be substituted for masses in the above equation.

In graphical terms, if the length of the line connecting the two states 1 and 2 is given as L, then the state of the mixed air is found a length of $[m_2/(m_1 + m_2)] L$ from state 1.

mixing
outcomes
There are three possible outcomes of the mixing process. In the first, the line connecting original properties 1 and 2 lies completely in the interior of the chart, and the result is air that is not moisture saturated. The second possibility happens when the line just touches the curved saturation line at the top of the psychrometric chart. Since the connecting line touches the 100% relative line, it is possible that mixed air could leave the process at saturation and with eondensation imminent.

The third possibility locates part of the connecting line above the saturation line. This happens when very warm air with high moisture content is being mixed with cold air, also with relatively high moisture content. If the mixed air properties are predicted to lie above the saturation line, the mixture is condensation thermodynamically unstable. Condensation will occur, liberating latent heat. This drying and heating of the air is the opposite of the drying process explained earlier, and follows a constant wet-bulb temperature line from the originally predicted point on the connecting line to the line of saturation (100% RH). The air will leave the process saturated and produce an amount of water equal to $m_m(\omega'_m - \omega_m)$, where m_m is the mass of the mixture, ω_m is the originally predicted humidity ratio, and ω'_m is the humidity ratio where the wet-bulb temperature intersects the line of saturation. Some means will usually have to be provided to dispose of this water that may form as fog and rain. Other properties, such as temperature, specific volume, enthalpy, etc., can be found from the chart.

Dehumidification   Air is often dehumidified by cooling below the dew-point temperature and then reheating. The first stage of the process appears on the cooling and reheating psychrometric chart as sensible cooling, as described earlier (Figure 4.8.9). Once the constant humidity ratio line intersects with the saturation line, further cooling results in condensation along the saturation line. This water is then

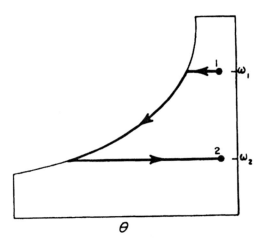

Figure 4.8.9. Dehumidification by cooling below the dew-point temperature and reheating to the original temperature. The amount of moisture removed is $m(\omega_1 - \omega_2)$.

removed and the air reheated along a humidity ratio line lower than the original. In a continuous process, the heating and cooling can occur simultaneously in a countercurrent heat exchanger (see Section 3.7.1). There is always more heat to be removed than added, however.

no
generalization

Ventilation Ventilation is a general term that can involve almost any kind of air property manipulation. The psychrometric chart can be useful to design ventilation systems because the amount of heating, cooling, moisture addition, or moisture removal can be graphically presented. Heating or cooling is determined from enthalpy differences and moisture addition or removal can be determined from humidity ratio differences.

Ventilation air is often specified in terms of volume flow rate, which can be determined by obtaining the specific volume from the psychrometric chart and multiplying by the required mass flow rate of dry air

$$\dot{V} = v\dot{m} \qquad (4.8.21)$$

where $\dot{V}$ is the volume flow rate of air (m^3/sec); v, the specific volume of air (m^3/kg dry air); and $\dot{m}$, the mass flow rate (kg dry air/sec).

human
comfort

Ventilation processes must be designed for many different types of systems that often include living organisms. Ventilation of spaces occupied by human beings is a very important class of problems. Humans usually require that the environment be comfortable for living or working (Figure 4.8.10), and this comfort zone changes in summer or winter, or depending on the individual and the thermal history of her or his recent experience. Humans usually require about four to seven days to acclimate fully to environmental temperature challenges, and acclimation can be lost at about half the rate required to

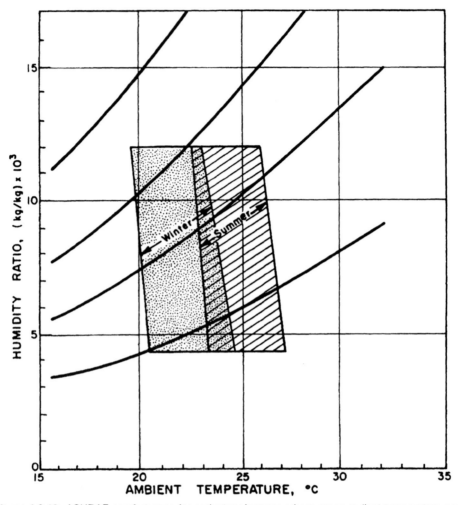

Figure 4.8.10. ASHRAE comfort zone for sedentary humans where mean radiant temperature equals dry-bulb temperature and clothing thermal conductance is $10.8\,N\,m/(m^2\,°C\,sec)$. The comfort zone is drawn on a standard metric psychrometric chart. (Adapted from ASHRAE, 1981.)

establish it (Johnson, 1991). Various physiological mechanisms are involved in acclimation, such as changes in thermogenic hormones, sweat production, blood volume, and vasodilation or vasoconstriction.

Humans are usually very particular about the environments in which they live and work, and thus ventilation systems must be designed to satisfy their perceptions of comfort. Not supplying these needs can reduce workplace productivity or even cause health impairments. Temperature and humidity are the most important variables to be controlled, but sometimes other conditions can become important. With the trend toward closed buildings with

health
considera-
tions

recirculated air, fungal and microbial growth in ventilation systems can cause health problems to occur ranging from allergic reactions to Legionnaire's Disease. Components outgassed from carpets, furniture, building materials, and office machines must also be able to be removed before they can cause health problems for the occupants (sick-building syndrome).

animal growth
rates

Ventilation systems provided for plants and animals are not related so much to comfort as to productivity. Weight gain of animals is directly related to environmental temperature (Figure 4.8.11). There is usually an increasing average daily weight gain as temperature increases until some optimal temperature, and then a decreasing weight gain with temperature increases until a lethal temperature is reached. Different animals have different optimal temperatures.

biologic gases

Of special importance in animal production is the ratio of body weight gain to feed weight consumed. The optimal growth temperature is not always the optimal gain/feed temperature. There may also be a detrimental effect of gases resulting from excreta, such as ammonia (important for birds), hydrogen sulfide (important for swine), and methane (swine). These must be removed by the ventilation system.

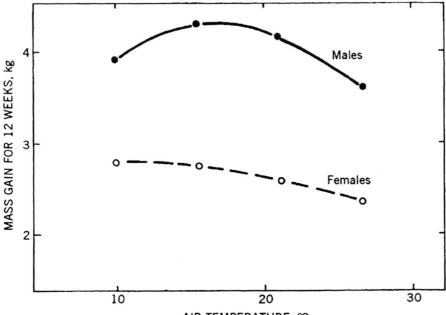

Figure 4.8.11. Effect of constant air temperature on weight gain and feed efficiency conversion of broad-breasted white and broad-breasted bronze turkeys between 12 and 24 weeks of age. Relative humidity is about 50 percent with 16 h daylength. (Adapted from Hellickson et al., 1966.)

plants

Plant production requires careful light and gas control as well as temperature and moisture. Plants also have optimal environmental requirements, and some gases present in greenhouses (such as ethylene that promotes ripening and senescence) can have severe effects. Sometimes the most severe constraint on a designed ventilation system must be the removal of unwanted gases. And the ventilation system itself may be the source of plant pathogen introduction; so careful biosecurity must be observed.

storages

Storage systems are usually designed to maintain product quality. Since enzymatic processes usually increase exponentially with temperature (see Section 3.5.5.2), and these processes usually degrade product quality, product storages are often maintained at low temperatures approaching freezing (below freezing temperatures must be avoided because they cause disruption of cellular integrity through ice crystal formation; see Sections 3.8.1 and 3.8.3). Some tropical fruits, such as bananas, mangoes, and tomatoes, cannot be stored at such low temperatures. Modified atmospheres consisting of low oxygen and high carbon dioxide levels often are used to extend product life further (see Section 3.5.3.4 and 4.3.3.2). Some products must be stored in very high humidities, whereas other products can discolor or decompose at such high humidities.

modified atmospheres

heat and moisture

It is not obvious to realize that these inert-looking stored products are still living and respiring, giving off heat, moisture (Table 4.8.3), and carbon dioxide. Damaged areas in the product may heal through normal biological mechanisms, but, in the process of healing, produce higher than normal levels of heat and carbon dioxide. Ventilation systems must be designed to accommodate these qualities of living tissue.

metabolic requirements

As a parenthetical note, when I spent a year at the Baltimore Shock-Trauma Center, I regularly observed technicians measuring oxygen consumption and carbon dioxide production of patients recovering from severe trauma. The healing process requires excess energy, and there is no procedure to predict how much extra food energy to supply a patient. Thus nutrient intakes for these patients were determined from direct measurements of their metabolic requirements.

composting

Environments for microbial systems can also be maintained through ventilation. Composting is one process that illustrates this requirement. Sufficient oxygen must be supplied by the ventilation system to maintain aerobic conditions. Excess heat must also be removed. Not too much air can be supplied, however, or temperatures will fall below those required to kill pathogens in the compost, or the compost may dry too much for optimal microbial growth. Other microbial systems, such as in bioreactors, usually require nutrient supply and metabolite removal by means of liquids rather than air. Nonetheless, the principles are similar to those for air ventilation.

control volume

Ventilation systems are usually designed on the basis of steady-state balances as described in Section 1.7. A closed control volume is either constructed or drawn around the system to be ventilated and various balances written (Figure 4.8.12). One of the main parameters to be maintained in the

Table 4.8.3 Product Moisture Loss in Storage

Commodity	Temperature (°C)	Humidity (%)	Rate of Moisture Loss $[10^{-9}\,kg\,H_2O/(kg$ product sec)]
Apples	0	90	2.2
	3	95	2.5
	3	75	8.1
	15	95	5.8
	15	75	19.0
Beef: 1 kg steak		90	104
		80	208
Beef: 18 kg roast		90	46
		80	92
Beef: 73 kg quarter		90	23
		80	58
Beef	Normal storage	Normal storage	11.6
Beets	0	98	92.6
Brussels sprouts	0	98	38.6
	0	85–90	34.7
Cabbage: curled	−1	85	7.4
Cabbage: red	−1	85	5.8
Cabbage: white	−1	85	5.3
Carrots	0	98	5.8
Cauliflower	0	98	42.4
Celery	−1	90	4.7
Cheese	16	75	6.4
Cucumbers	0	98	23.1
Currants	−1	85	11.6
Eggs	0	99	2.6
	0	88	9.0
Gooseberries	−1	85	11.6
Lima beans	0	98	15.4
Onions	−1	85	3.9
Peaches	−1	90	7.7
Pears	0	90	3.5
Peppers	0	98	11.6
Plums	−1	85	9.6
Spinach	−1	90	2.7
Squash	0	98	3.6
Strawberries	1	90	15.4
For comparison			
Cat, 2.9 kg			11900
Chicken, 1.8 kg	8		580
	12		780
	18		810
	28		1060

(continued)

Table 4.8.3   (Continued)

Commodity	Temperature (°C)	Humidity (%)	Rate of Moisture Loss $[10^{-9}\,kg\,H_2O/(kg\ product\ sec)]$
Dog, 18.6 kg			11400
Guinea pig, 0.45 kg			9200
Human, 70 kg (minimum)			430
(maximum)			7140
Monkey, 409 kg			11900
Mouse, 0.02 kg			44700
Rat, 0.130 kg			29200
Steer, 584 kg			10800
Turkey, 0.1 kg	36		43.3
0.2 kg	32		27.2
0.4 kg	29		20.0
0.6 kg	27		11.4
1.0 kg	24		6.7
15 kg	25		4.7

[a] For example, the maximum sweat rate of a typical 70 kg human male is $7140 \times 10^{-9}\,kg\,H_2O/(kg$ body sec).

Sources: Spectro (1956), Jordan and Priester (1956), Johnson (1991), and ASAE (1989).

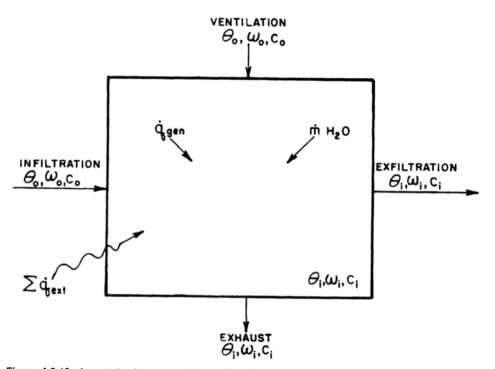

Figure 4.8.12. A control volume around a ventilated space should include all parameters pertinent to a steady-state mass or energy balance.

control volume is temperature. In order for temperature to be maintained by ventilation, a steady-state heat balance must be written

rate of heat in − rate of heat out + rate of heat generated = 0

$$\dot{m}_i c_p \theta_o + \sum \dot{q}_{ext} - \dot{m}_o c_p \theta_i + \sum \dot{q}_{gen} = 0 \qquad (4.8.22)$$

where $\dot{m}_i$ is the mass flow rate of incoming air (kg/sec); c_p, the specific heat of air–water vapor mixture [N m/(°C kg), $= c_{p_{da}} + c_{p_{wv}} \omega$]; θ_o, the temperature of outside air (°C); $\sum \dot{q}_{ext}$, the sum of heat coming from external sources (N m/sec); $\dot{m}_o$, the mass flow rate of outgoing air (kg/sec); θ_i, the temperature of inside air (°C); and $\sum \dot{q}_{gen}$, the sum of heats generated inside the ventilated space (N m/sec).

Incoming air must include infiltration (uncontrolled air leakage into the space), and outgoing air must include exfiltration (negative infiltration). Most infiltration and exfiltration in older structures comes from leaks and cracks; most infiltration and exfiltration from newer and tighter structures comes from opening doors. Without better information as a guide, infiltration volume rate can be estimated from

air leakage

$$\dot{V}_{in} = \frac{HWLG}{3600} \qquad (4.8.23)$$

where $\dot{V}_{in}$ is the infiltration volume flow rate (m^3/sec); H, the room height (m); W, the room width (m); L, the room length (m); and G, the wall factor ($= 1$, one wall facing a different environment; $= 1.5$, two walls facing a different environment; and $= 2$, three or more walls facing different environment).

The sum of heat coming from external sources includes heat transferred through the structure by conduction and through windows by radiation (including sunshine).

heat generation

Heat generated inside a room comes from the occupants, electric lights, chemical reactions, and machinery, as well as others. Remember that the rate of heat production by humans depends on what they are doing (see Section 3.5.3.2), and the rate of heat generated by plants, animals, and microbes depends on the temperature (Figure 4.8.13).

If infiltration is ignored, then the rate of ventilation air to maintain a certain temperature is just

temperature criterion

$$\dot{m}_i = \frac{\sum \dot{q}_{ext} + \sum \dot{q}_{gen}}{c_p(\theta_i - \theta_o)} = \frac{\sum \dot{q}_{ext} + \sum \dot{q}_{gen}}{(c_{p_{da}} + c_{p_{wv}} \omega)(\theta_i - \theta_o)} \qquad (4.8.24)$$

where $\dot{m}_i$ is the mass flow rate of ventilation air (kg/sec). Usually, the outside temperature is assumed based on weather data and the inside temperature is determined based on design criteria.

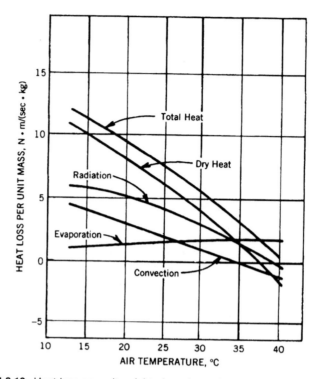

Figure 4.8.13. Heat loss per unit weight of newborn pigs vs. air temperature. (Adapted from Butchbaker and Shanklin, 1964.)

Excess moisture is often removed by ventilation. A steady-state moisture balance on the control volume gives:

$$\text{rate of moisture in} - \text{rate of moisture removed}$$
$$+ \text{rate of moisture generated} = 0 \tag{4.8.25}$$

From which

$$\dot{m}_i \omega_o - \dot{m}_o \omega_i + \dot{m}_{H_2O} = 0 \tag{4.8.26}$$

Resulting in

moisture
criterion

$$\dot{m}_i = \frac{\dot{m}_{H_2O}}{(\omega_i - \omega_o)} \tag{4.8.27}$$

where $\dot{m}_i$, is the mass flow rate of incoming air (kg/sec); $\dot{m}_0$, the mass flow rate of outgoing air (kg/sec); $\dot{m}_{H_2O}$, the rate of moisture generation (kg H_2O/sec; ω_o, the outside humidity (kg H_2O/kg dry air); and ω_i, the inside humidity ratio (kg H_2O/kg dry air). The amount of moisture released into the air, translated into heat by the latent heat of vaporization, is called the *latent heat*. Heat added directly to air to cause an immediate temperature rise is called *sensible heat*.

Latent heat only becomes a heat burden for the ventilation system if condensation occurs.

If ventilation must be designed to remove a particular gaseous component, such as methane, then a steady-state materials balance is formulated for that component

$$\text{rate of gas in} - \text{rate of gas removed} \atop + \text{rate of gas generated} = 0 \tag{4.8.28}$$

or

$$\dot{m}_i c_o - \dot{m}_o c_i + \dot{m}_{gas} = 0 \tag{4.8.29}$$

from which

gas criterion

$$\dot{m}_i = \frac{\dot{m}_{gas}}{(c_i - c_o)} \tag{4.8.30}$$

where $\dot{m}_{gas}$ is the rate of gas generated inside the control volume (kg gas/sec); c_i, the gas concentration inside (kg gas/kg dry air); and c_o, the gas concentration outside (kg gas/kg dry air). The rate of incoming air has again been considered to equal the rate of outgoing air.

It is unlikely that ventilation mass flow rates, independently determined to meet different balance criteria, will all be the same. It is not unusual, however, to require the ventilation system to meet several independent design goals. When the various mass flow rates are not the same, then some compromises must be made. Under the control of the design engineer is not only amount of ventilation air, but also barriers to heat (insulation), water vapor, and gaseous transmission, and also generation or removal of extra heat, moisture, or gas within the control volume. Each of these affects a different term of the steady-state balance.

EXAMPLE 4.8.1.6-1. Ethylene Removal

Ethylene (C_2H_4) is a physiologically active gas produced by all higher-order plants. As little as 0.1 ppm (by volume) of ethylene in the atmosphere has an effect on the plant, and 10–100 ppm causes ripening and senescence of fruits. If there are 2500 plants in a greenhouse, each producing $8.33 \times 10^{-13} \text{ m}^3$ ethylene/sec (LaRue and Johnson, 1989), what ventilation rate would be required to maintain ethylene concentration within the greenhouse to 1 ppm or less?

Solution

The total rate of ethylene production at 25° is

$$\dot{m} = \frac{pM\dot{V}}{RT}$$

$$= \frac{(101{,}300 \text{ N/m}^2)(28 \text{ kg/kg mol})[8.33 \times 10^{-13} \text{ m}^3/(\text{plant sec})](2500 \text{ plants})}{(8314.34 \text{ N m/(kg mol K)})(298 \text{ K})}$$

$$= 2.38 \times 10^{-9} \text{ kg/sec}$$

A concentration of 1 ppm (by volume), is, from Table 4.3.12

$$\frac{\text{kg eth}}{\text{kg air}} = \alpha = \frac{\text{ppv } M_{eth}}{10^6 \, M_{air}} = \frac{1(28 \text{ kg/kg mol})}{10^6(29 \text{ kg/kg mol})} \approx 1 \times 10^{-6} \frac{\text{kg eth}}{\text{kg air}}$$

From Eq. 4.8.30, the mass ventilation rate is

$$\dot{m}_i = \frac{2.38 \times 10^{-9} \text{ kg eth/sec}}{(1 \times 10^{-6} - 0) \text{ kg eth/kg air}}$$

$$= 2.4 \times 10^{-3} \text{ kg air/sec}$$

From the psychrometric chart, Figure 4.8.4, at 25°C and 80% relative humidity, we find the specific volume of air to be about $0.87 \text{ m}^3/\text{kg da}$. Thus

$$\dot{V} = \dot{m}_i v = (2.4 \times 10^{-3} \text{ kg air/sec})(0.87 \text{ m}^3/\text{kg da})$$

$$= 2.1 \times 10^{-3} \text{ m}^3/\text{sec}$$

Remark

We might also check the carbon dioxide level in the greenhouse. Actively photosynthesizing plants use CO_2 at a rate of about $1.75 \times 10^{-8} \text{ kg mol}$ $CO_2/(\text{m}^2 \text{ sec})$, which, for a plant with 650 cm^2 leaf area, is about $5 \times 10^{-8} \text{ kg } CO_2/(\text{plant sec})$. Outside CO_2 concentration is nearly constant at 350 ppm (by volume). Thus

$$c_o = \frac{350 \text{ ppv (44 kg } CO_2/\text{kg mol)}}{10^6(29 \text{ kg air/kg mol})} = 513 \times 10^{-4} \frac{\text{kg } CO_2}{\text{kg air}}$$

Using Eq. 4.8.30

$$c_i = 5.3 \times 10^{-4} \frac{\text{kg } CO_2}{\text{kg air}} + \frac{[-5 \times 10^{-8} \text{ kg } CO_2/(\text{plant sec})](2500 \text{ plants})}{2.4 \times 10^{-3} \text{ kg air/sec}}$$

$$= -5.2 \times 10^{-2} \text{ kg } CO_2/\text{kg air}$$

The plants will easily use all the CO_2 introduced into the greenhouse by the ventilation system. The ventilation rate must be increased substantially to satisfy the plants' requirements. Fortunately, increasing the ventilation rate will decrease inside ethylene concentration. Unfortunately, the air will likely have to be heated and a great deal of heat energy will be used.

4.8.2 Drying

reduces
spoilage

Drying is a process used for preservation and/or weight reduction. Drying of food materials is commonly performed before they are stored for long periods of time, and this storage can occur without need for refrigeration or aseptic packaging. Transportation of dried foods is more efficient than for wet foods because, in addition to the absence of refrigeration, energy need not be spent to transport water, and dried materials usually occupy less space than their wetted counterparts. This apparent energy advantage could be overwhelmed by the energy required to dry the materials in the first place; so one of the major challenges in drying is to perform it as economically as possible.

spoilage
micro-
organisms

There are a number of classes of microorganisms that are important in spoilage of food and other agricultural materials. Some of these appear in Table 4.8.4, along with figures related to moisture content that show the minimum relative dryness at which they will grown. Bacteria, molds, yeasts, and fungi all appear. Not all of these classes of microorganisms grow in all materials, but many of them are common in food spoilage.

The transformation from a susceptible to a totally spoilage-resistance condition occurs over a very narrow moisture range. Even a 1% increase in

Table 4.8.4 Microorganism Groups Important in Spoilage and Their Minimum Moisture Conditions

Group	Minimal Equilibrium Water Vapor Pressure
Normal bacteria	0.91
Normal yeasts	0.88
Normal molds	0.80
Halophilic bacteria	0.75
Xerophilic fungi	0.65
Osmophilic yeasts	0.60

moisture content in the range of 13–15% can greatly increase mold growth. For this reason, moisture content is only an inexact indication of the susceptibility of a product to microbial spoilage. One of the main reasons for the narrow transition zone from susceptible to resistant is that oxidative reactions of microorganisms generate water from carbohydrates precisely at the location where it is needed most for their growth. Thus it is very difficult to arrest spoilage once it has started, and limited drying is not effective once a microbial invasion has become established.

abrupt transition

environmental factors

Various environmental factors can assist with the struggle against spoilage. Very high or very low temperatures retard or arrest spoilage; very alkaline or very acid conditions do likewise; very high osmotic pressures, especially where essential nutrients for microbial growth are absent, can also reduce spoilage. In these instances, drying need not be as complete as in other cases.

products

Drying is done on a number of different products. Food materials have been mentioned, and they are very important. Grains, not necessarily all for food, because some are used for seed and others for industrial products, are also dried (Table 4.8.5). Those grains used for seed have germinability as well as spoilage as criteria for determining details of the drying process; those used for flour and industrial products have various grinding and baking criteria; both are subject to mechanical damage that can become worse for very dry or very wet kernels. Composted waste is often dried before it is stored, shipped, or marketed. Drying of wastes may eliminate or reduce unpleasant odors associated with the material. Municipal wastes may be dried before undergoing separation to remove recyclable materials and/or to burn the remainder. The manufacture of paper requires removal of excess water from cellulosic fibers. Many other manufactured goods (such as paper) that use water as a facilitation medium must also be dried before the finished product is formed.

food drying

The discussion to follow focuses mainly on food and grain drying. Most drying technology was historically developed for these materials. The general principles are, however, applicable to a wider range of materials. The biological engineer should not only know the physical principles of drying, but

Table 4.8.5 Approximate Maximum Moisture Content for a One-Year Safe Storage Period

Grain	Moisture Content (wet basis)
Barley	0.135
Corn, shelled	0.135
Oats	0.140
Rice	0.140
Sorghum	0.120
Soybeans	0.110
Wheat	0.140

also should be familiar with the biology of the microorganisms that cause spoilage (for example, *Aspergillus restrictus*, *A. repens*, *A. amstelodami*, *A. ruber*, *A. candidus*, *A. ochraceus*, *A. flavus*, *Penicillium*, and *Pullularia*) and the biological nature of the product (for example, its nutritional, germinability, respiratory, and survivability attributes, as appropriate). In this way, design goals for the drying process can be set to be appropriate for the final use of the product, and constraints on the manner in which drying is accomplished will be clearly known. For example, heating is often used for drying, but heating can destroy nutritional content or germinability. If retaining nutrition or germinability is a design goal, then this constrains the use of excessive heat in the drying process.

4.8.2.1 Moisture Content

Product moisture contents may be expressed on either a wet basis or a dry basis. Wet basis expresses moisture content as a fraction of total weight:

wet basis
$$M_{WB} = \frac{m_{H_2O}}{m_{tot}} = \frac{m_{H_2O}}{m_{H_2O} + m_{DM}} \qquad (4.8.31)$$

where M_{WB} is the moisture content (wet basis, dimensionless); m, the mass (kg); and H_2O, DM, tot stand for water, dry matter, and total, respectively.

Wet-basis moisture content is particularly easy to use in practice because total weight is often known and water weight can be easy to obtain. See Table 4.8.6 for selected values.

Moisture content can also be expressed on a dry basis

dry basis
$$M_{DB} = \frac{m_{H_2O}}{m_{DM}} \qquad (4.8.32)$$

where M_{DB} is the moisture content (dry basis). Dry-basis water content is useful when considering transport of water through the product, and is often used to convey information related to water movement. Conversion between the two bases is made from

conversion
$$M_{DB} = \frac{M_{WB}}{1 - M_{WB}} \quad \text{or} \quad M_{WB} = \frac{M_{DB}}{1 + M_{DB}} \qquad (4.8.33)$$

As long as dry matter is not converted into water (for example, through oxidation of carbohydrates), dry matter and water can be considered to be separate and identifiable substances.

4.8.2.2 Equilibrium Moisture Content

There are two important questions in the design of a drying process: The first is, "How do I know what the final moisture content will be?" and the second is, "How fast will the product dry?" The answer to the first is simply expressed by means of the equilibrium moisture content.

Table 4.8.6 Percentage Moisture Content (Wet Basis) for Selected Biological Materials

Foods	
Beef	
Flank	45%
Loin	57
Rib	59
Chuck	65
Shank	70
Lamb	
Loin	53
Leg	64
Pork	
Ham (fresh)	54
Loin	52–60
Shoulder	51
Veal	74
Vegetables	
Asparagus	94
Cabbage	91
Carrots	83
Lettuce	94
Peas	75
Potatoes, white	73
Potatoes, sweet	69
Radishes	92
Tomatoes	94
Fruits	
Apples	83
Cherries	82
Grapefruit	88
Lemons	89
Oranges	87
Peaches	87
Pears	83
Strawberries	90
Human body	
Total	60
Adipose tissue	10
Blood	83
Bone	22
Skeletal muscle	76
Skin	72
Soils (maximum)	
Clay	25
Loam	20
Sand	
Coarse	20
Fine	22

hygroscopic
materials

Many food and grain products are hygroscopic materials. They have a definite and characterizable relationship between moisture content and moisture in the surrounding atmosphere. If the material enters wetter than the equilibrium moisture content, it will lose moisture. If it enters drier than the equilibrium moisture content, then it will absorb moisture. Many materials exhibit this property.

Equilibrium moisture contents have been measured for many materials. Graphs of moisture contents as functions of air relative humidity often appear as rotated sigmoids (Figure 4.8.14), showing that moisture contents of the materials are directly related to the relative humidity of the surrounding air. There is a small temperature dependence of each of these curves, but the answer to the first question, "How do I know what the final moisture content

relation to
relative
humidity

of the product will be?," is simply that if drying air is supplied at the correct relative humidity, then it will result in a product with the desired moisture content.

The equilibrium moisture content curve actually expresses the locus of points where the vapor pressure of the water inside the product is equal to the water vapor pressure in the air. This implies several things: (1) Since water vapor pressure in the air at any given relative humidity depends on the saturated vapor pressure (Eq. 4.8.11), which, in turn, depends on absolute temperature, equilibrium moisture content curves also exhibit some temperature dependence, and (2) vapor pressure in the product is not linearly related to saturated vapor pressure of free water.

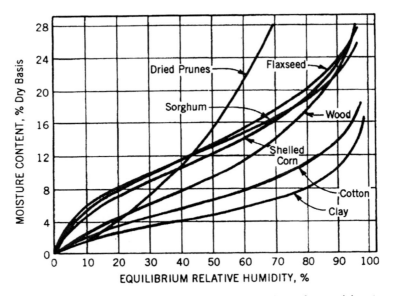

Figure 4.8.14. Equilibrium moisture curves of a number of materials at room temperature, approximately 25°C.

hysteresis

It has also been reported (Brooker et al., 1974; Heldman, 1975) that the desorption (or drying) curves are higher than the adsorption (or wetting) curves, indicating the presence of hysteresis. Several explanations for this hysteresis have been advanced, but none has been entirely satisfactory.

Attempts have been made to form mathematical equations to predict equilibrium moisture contents (Brooker et al., 1974). Many of these are either based upon restrictive theory or are less than satisfactory in their predictions; so none will be presented here. Equilibrium moisture content data are so important, however, that experimental data for many products have been obtained. These are presented in tabular (Tables 4.8.7 and 4.8.8) or graphical (Figure 4.8.14) forms, and are sufficiently accurate to form the basis for dryer design.

Table 4.8.7 Grain Equilibrium Moisture Content (percent, wet basis)

Material	Temp. (°C)	Relative humidity (percent)									
		10	20	30	40	50	60	70	80	90	100
Shelled corn,	−7				10.4	11.8	13.3	15.0	16.6		
yellow dent	0				10.1	11.3	12.6	14.0	16.8		
	10				9.2	10.7	12.1	13.6	15.5		
	21			7.1	8.3	9.8	11.4	13.2			
	25	5.1	7.0	8.4	9.8	11.2	12.9	14.0	15.6	19.6	23.8
	71	3.9	6.2	7.6	9.1	10.4	11.9	13.9	15.2	17.9	
	4	6.3	8.6	9.2	11.0	12.4	13.8	15.7	17.6	21.5	
	27	5.5	7.8	9.0	10.3	11.3	12.4	13.9	16.3	19.8	
	30	4.4	7.4	9.2	9.0	10.2	11.4	12.9	14.8	17.4	
	38	4.0	6.0	7.3	8.7	9.0	11.0	12.5	14.2	16.7	
	50	3.6	5.5	8.7	8.0	9.2	10.4	12.0	13.6	16.1	
	60	3.0	5.0	8.0	7.0	7.9	8.8	10.3	12.1	14.6	
Shelled popcorn	25	5.6	7.4	8.5	9.8	11.0	12.2	13.1	14.2	18.4	23.0
Sorghum	25	4.4	7.3	8.6	9.8	11.0	12.0	13.8	15.8	16.2	21.9
Soybeans	25		5.5	6.5	7.1	6.0	8.3	11.5	14.8	18.8	
	25				7.0	8.0	10.1	12.2	16.0	20.7	
Sugar beet	4			10.0	11.5	12.7	13.9	15.3	17.6	22.6	
seeds	16			9.0	10.0	11.5	13.5	14.1	16.2	19.9	
	27			8.0	9.1	10.4	11.6	12.9	14.7	18.0	
	38			7.0	8.3	9.2	10.4	11.5	13.2	15.8	
Wheat, soft	−7				11.3	12.8	14.1	15.6	17.0		
red winter	0				11.0	12.2	13.5	14.7	18.2		
	10				10.2	11.7	13.1	14.4	16.0		
	21				9.7	11.0	12.4	14.0			
	25	4.3	7.2	8.6	9.7	10.9	11.9	13.6	15.7	19.7	25.6
Wheat, white	25	5.2	7.5	8.6	9.4	10.5	11.8	13.7	16.0	19.7	26.3
Wheat, durum	25	5.1	7.4	8.5	9.4	10.5	11.5	13.1	15.4	19.3	26.7

Table 4.8.8 Agricultural Materials Equilibrium Moisture Contents (percent, wet basis)

Material	Temp. (°C)	10	20	30	40	50	60	70	80	90	100
Bran	21–27							14.0	18.0	22.7	33.0
Bread	25	0.9	1.8	3.1	4.4	6.1	7.7	10.3	12.3	16.0	
Bone meal	21–27							14.1	10.8	12.4	22.0
Cabbage	0	3.1		4.1		7.3		15.5	22.1		
air-dried	25	1.3		4.7		9.1		16.0	23.3		
scalded	37	1.7		4.7		8.6		16.5	22.7		
savoy	37	1.7		4.7		8.6		16.5		22.7	
Carrot	10	3.2	3.7	4.5	6.3	8.8	12.5	17.4	24.7		
scalded	25	2.1		4.4		9.5		18.2	26.6		
air-dried	60	1.2	2.3	4.4	8.4	9.4	13.0	19.1	29.2		
Cotton cloth	25	2.4	3.6	4.7	4.8	6.3	7.8	9.1	10.4	12.0	
Crackers	25	1.5	2.4	3.3	4.0	4.9	7.0	7.8	10.0	11.9	
Eggs	10	2.8	3.8	5.1	6.1	7.3	8.7	10.7			
spray-dried	37	2.3	3.4	4.3	5.0	6.5	8.3	10.0			
	80	1.2	2.6	2.7	3.4	4.5	5.8	6.7			
Fiber	21–27				4.5	5.5	6.5	9.0	12.2	26.5	
Flax	25	2.5	3.9	4.8	5.9	7.0	7.8	9.1	11.1	13.1	
Flour	25	2.0	3.6	5.2	5.7	7.5	9.6	11.2	13.7	16.0	
Linen	25	1.7	2.9	3.9	4.8	5.7	6.6	6.9	8.1	9.5	
Linseed coke	21–27							13.5	17.5	23.5	40.5
Locust beans	21–27							14.5	21.9	35.0	51.0
Lumber	25	2.8	4.2	5.7	7.0	8.4	10.1	12.2	14.8	18.0	23.1
Lumber white spruce	25	3.5	9.2	6.5	8.3	9.8	12.6	12.0	15.6	20.0	25.4
Macaroni	25	4.8	6.6	8.1	9.5	10.8	12.0	13.8	15.6	18.1	
Manila hemp	25	2.7	4.3	5.9	7.0	7.8	9.2	10.6	12.0	13.9	
Milk, powdered	10	2.7	3.0	3.4	4.8	7.0	6.5	7.6			
full-cream	37	2.6	3.3	4.1	4.0	4.5	6.5	7.9			
spray-dried	80	1.2	1.8	1.6	2.3	2.4	4.3	7.7			
Oats	21–27;							13.1	15.4	18.5	31.4
Peaches, dried	24	0.7	1.5	2.6	4.6	7.3	11.0	16.8	23.1		
Peas, whole		6.6		9.0		11.2		14.1	17.1		
Potatoes	10	4.9		7.3		10.7		14.3	19.8		
scalded	37	4.2		6.9		9.6		13.4	16.7		
air-dried	80	3.1		5.5		7.2		10.6	13.8		
Prunes, dried	24	2.1	3.9	6.6	9.5	12.7	16.1	20.7	27.7		
Raisins	24	2.9	5.4	8.2	11.3	14.5	18.6	23.1			
Reclaimed rubber	25		0.8	0.9	1.0	1.1	1.2	2.0	2.9	3.7	
Scotch beans	21–27							13.8	17.0	22.0	33.9
Sheepskin	25		7.8	9.8	11.1	12.2	13.3	16.7	22.4		

(continued)

Table 4.8.8 *(Continued)*

Material	Temp. (°C)	Relative humidity (percent)									
		10	20	30	40	50	60	70	80	90	100
Sisal hemp	25	3.2	4.7	5.4	6.6	7.4	8.7	10.2	11.5	13.4	
Starch	21–27				10.0	11.0	12.2	14.1	15.5	17.7	26.5
Straw	21–27				9.0	10.5	12.0	14.0	17.0	24.5	36.0
Tanned leather	25	7.9	10.4	11.7	13.1	13.9	14.7	16.1	18.5	23.1	
Tobacco, cigarette	25	6.4	8.3	10.0	11.6	13.8	16.1	19.3			

4.8.2.3 Drying Rate

The second question posed earlier concerning the rate at which drying occurs leads to a complicated answer. Depending on the initial condition of the product and its physical arrangement, quite different **different types** drying rates may result. For example, product drying can occur as either thin layer or thick layer, with thin-layer drying occurring when product moisture content is considered constant across the layer, but thick-layer drying occurring when there is a moisture gradient. These layers, by the way, may be horizontal, vertical, or anything between, depending on the flow direction of the drying air.

Constant-Rate Drying Removal of moisture occurs in two phases: (1) constant-rate period, and (2) falling-rate period (Figure 4.8.15). Constant-rate drying removes moisture that is largely present on exterior surfaces. There is very little impediment to removal of this water, and it is lost readily. Several **easily dried** different forms of a moisture balance on the material demonstrate the connection of the constant-rate drying period to information presented in previous sections. First, a control volume is drawn around the material being dried and a steady-state water balance is formulated

$$\text{moisture in} + \text{moisture generated} = \text{moisture out} \qquad (4.8.34)$$

In this case the moisture generated is water that is added to the drying air that comes from the product

$$\dot{m}_a \omega_i + \dot{m}_{H_2O} = \dot{m}_a \omega_o \qquad (4.8.35)$$

where $\dot{m}_a$ is the mass flow rate of ratio drying air (kg air/sec); ω, the humidity ratio (kg H_2O/kg air); $\dot{m}_{H_2O}$, the drying rate (kg H_2O/sec); and (i, o denote input and output conditions.

$$\dot{m}_{H_2O} = \dot{m}_a(\omega_o - \omega_i) \qquad (4.8.36)$$

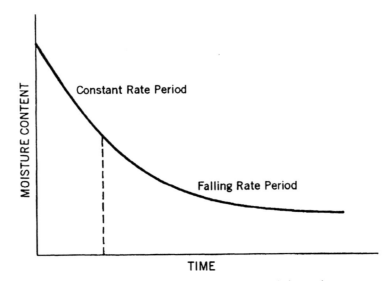

Figure 4.8.15. Product drying rates illustrating constant-rate drying and falling-rate drying.

During the constant rate period, $(\omega_o - \omega_i)$ is constant as long as the input conditions for the drying air are constant.

Another materials balance on each particle being dried assumes that the particle surface is coated with water, and that the absolute humidity at the surface is that corresponding to moisture saturation

$$\dot{m}_{H_2O} = k_G'' A(\omega_{sat} - \omega_{amb}) = \frac{h_c A(\theta_{amb} - \theta_{wb})}{H} \qquad (4.8.37)$$

where k_G'' is the mass transfer coefficient for water vapor [kg/(sec m^2)]; A, the surface area (m^2); ω_{sat}, the humidity ratio of water saturated air (kg H$_2$O/kg air); ω_{amb}, the humidity ratio of ambient air (kg H$_2$O/kg air); h_c, the convection coefficient [N m/(sec m^2 °C)]; θ_{amb}, the ambient dry-bulb temperature (°C); θ_{wb}, the wet-bulb temperature (°C); and H, the latent heat of vaporization of water (N m/kg). The two terms in this mass balance demonstrate the correspondence between heat transfer and mass transfer in a drying process. As liquid water evaporates and is removed from the product, air wets and cools — energy is consumed, usually in the form of heat. Thus drying air becomes wetter and cooler as it passes by the product.

Sweating The constant-rate drying period is usually relatively short because the surface moisture can be dried quickly. For the case of human sweating, animal panting, or plant evapotranspiration, however, surface moisture is continually replenished from inside the organism. The purpose of sweating and panting is surface cooling rather than drying. There may be a cooling

surface
moisture
replenishment

function associated with plant evapotranspiration, but it is more likely that moisture removal is an inadvertent side effect to carbon dioxide acquisition, in a way similar to human or animal respiration.

heat lost

Of major interest in sweating is not the rate of moisture lost, but rather the rate of heat lost, since this is the intended function of sweating. The rate of heat loss is

$$\dot{q}_{evap} = k''_G AH(\omega_{sat} - \omega_{amb}) = h_{va}A(p_{sat} - p_{H_2O}) \tag{4.8.38}$$

where $\dot{q}_{evap}$ is the rate of heat loss due to sweating (N m/sec); h_{va}, the heat transfer coefficient for evaporation (m/sec); A, the sweating surface area (m^2); p_{sat}, the saturated water vapor pressure (N/m^2); p_{H_2O}, the ambient water vapor pressure (N/m^2); H, latent heat of vaporization of sweat (2.6×10^6 N m/kg); k''_G, the mass transfer coefficient for water vapor in air [kg/(sec m^2)]; and ω, the humidity ratio of ambient air (kg H$_2$O/kg air).

Lewis
relationship

Due to the Lewis relationship between convective heat and mass transfer discussed earlier (Section 4.4.1), once the value of the convection coefficient is known, a value for h_{va} may be calculated. Johnson (1991) gives an alternate form of the Lewis number (Eq. 4.4.14) as

$$Le = \frac{h_c}{k''_G c_p} \tag{4.8.39}$$

where c_p is the specific heat of air at constant pressure [N m/(kg °C)]. From Eqs. (4.8.38) and (4.8.39)

$$h_{va} = \frac{h_c H(\omega_{sat} - \omega_{amb})}{Le\, c_p(p_{sat} - p_{H_2O})} \tag{4.8.40}$$

Making use of Eq. 4.4.10 gives

$$h_{va} = \frac{0.622\, h_c H}{Le\, c_p p_{amb}} \tag{4.8.41}$$

where p_{amb} is the ambient atmospheric pressure (N/m^2). Johnson and Kirk (1981) give

$$H = 2.502 \times 10^6 - 2.376 \times 10^3 \theta, \qquad 0 < \theta < 50°C \tag{4.8.42}$$

$$c_p = 1000 + 1880\left(\frac{\omega_{sat} + \omega_{amb}}{2}\right) \tag{4.8.43}$$

Inserting values from these equations gives

$$h_{va} \approx 0.150 h_c \tag{4.8.44}$$

ASHRAE (1977) gives

$$h_{va} = 0.165 h_c \qquad (4.8.45)$$

Either of these can be used with previously given tabulated values, of convection coefficient in order to determine evaporative heat loss due to sweating or similar evaporative processes.

Falling-Rate Drying   Falling-rate drying occurs when migration of the water through the solid is slower than evporation from the surface, and moisture is held more tightly as drying progresses. This type of drying is characterized by moisture removal that becomes slower as drying continues. Moisture may move from the interior to the surface as: (1) liquid through capillary ducts (porous solid), (2) liquid diffusing because of concentration differences (nonporous solid), (3) liquid moving by surface diffusion through pores, (4) water vapor diffusion in air-filled pores, or (5) water vapor formed within the material and diffusing outward. Drying in those cases in which water vapor is formed within the material may be limited by the rate of heat transfer into the material. Otherwise, liquid water movement can be very slow.

The rate of drying is usually limited by the falling rate period. Thin-layer drying will be covered under the next section. Material undergoing thick-layer drying may not begin to dry immediately if it is far from the air entrance, and, because these materials are hygroscopic, some thick-layer material downstream from the region of active drying may actually become wetter until the upstream material dries sufficiently.

Thick-Layer Drying   The drying rate in a thick layer appears as in Figure 4.8.16. On the ordinate of this diagram is the moisture ratio, defined as

$$MR = \frac{M_{DB} - M_{equil}}{M_o - M_{equil}} \qquad (4.8.46)$$

where MR is the moisture ratio (dimensionless); M_{DB}, the moisture content (dry basis, kg H_2O/kg dry matter); M_{equil}, the equilibrium moisture content at air relative humidity (dry basis, kg H_2O/kg dry matter); and M_o, the initial moisture content of the product being dried (dry basis, kg H_2O/kg dry matter).

The equilibrium moisture content represents the final moisture content during drying and is the same as M_{equil} found from equilibrium moisture content data (Figure 4.8.14 and Tables 4.8.5 and 4.8.6). This equation is similar to the concentration difference ratio appearing in Eq. 4.7.16, with moisture content (M_{DB}) replacing concentration (c).

On the abscissa of the diagram are time units. Converting time units to actual elapsed time depends on geometry, the type of product being dried and its physical properties, the latent heat of water (which increases as the product dries), air properties, and air flow rate.

Margin notes:
- tightly bound water
- limits drying time
- moisture ratio
- time units

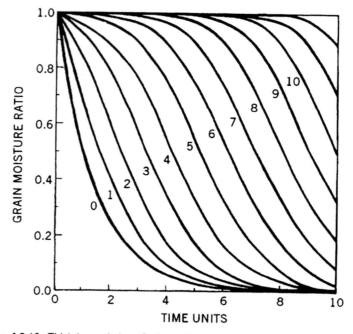

Figure 4.8.16. Thick-layer drying. Grain moisture reduces as either time increases or depth decreases.

depth factor   The parameter distinguishing the different curves is a depth factor, which can be converted to real dimensions depending on many of the same factors as the time units.

Looking at the diagram at a constant number of time units (a vertical line) gives a snapshot of the moisture content with depth at any one time. Moisture content of the material near the entrance for the drying air ($D = 0$) is the lowest of all, perhaps nearly completely dried, depending on the time chosen. Material a large number of depth units away from the entrance has not yet begun to dry.

Any product that cannot remain wet for the time required for the drying front to reach the farthest depth from the air entrance is bound to spoil. Once spoilage begins in a deep-bed dryer, it is extremely difficult to stop.

Looking at the moisture content along a line of constant depth units gives the moisture history at one particular place. Moisture content begins at M_o (MR = 1), changes slightly for a while, falls rapidly, and then again falls slowly toward M_{equil} (MR = 0).

The curves in Figure 4.8.16 are plotted from

$$MR = \frac{2^D}{2^D + 2^T - 1}$$

(4.8.47)

where D is a depth unit (dimensionless); T, a time unit (dimensionless); and

$$D = \frac{x \rho_s A H (M_o - M_{\text{equil}}) k}{c_p \dot{m} \ln 2(\theta_o - \theta_g)} \tag{4.8.48}$$

where x is the distance from the point of entry of air into the product (m); ρ_s, the density of product dry matter (kg dry matter/m³); A, the cross-sectional area of the dryer (m²); H, the latent heat of vaporization of moisture in the product (N m/kg water); c_p, the specific heat of dry air [N m/(kg air °C)]; $\dot{m}$, the mass flow rate of the air (kg air/sec); k, the thin-layer drying constant of the product (sec⁻¹); θ_o the air temperatuere as it enters the product (°C); θ_g, the dry-bulb temperature corresponding to the relative humidity of air in equilibrium with the product at its initial moisture content and at the wet-bulb temperature of the entering air (°C); M_o, the initial moisture content, dry basis (kg H₂O/kg dry matter); and M_{equil}, the moisture content of the product in equilibrium with the entering air, dry basis (kg H₂O/kg dry matter).

Deep-bed drying is used almost exclusively on grains dried on farm. The deep-bed driers can often serve the dual purpose of storage. Commercial dryers are often fabricated as thin-layer dryers because of the improved control that they offer.

Thin-Layer Drying Thin-layer dryers are distinguished by the nearly uniform moisture content of the product across the entire layer. There is no delay in the onset of drying because the drying air is nearly unchanged from its original condition even at the farthest point in the thin layer from the air entrance.

uniform
moisture
content

Heldman (1975) gives the following empirical equation for thin-layer drying of most biological solids

$$\dot{M} = \frac{\pi^2 D_{H_2O}}{4L^2} (M - M_{\text{equil}}) \tag{4.8.49}$$

where $\dot{M}$ is the rate of change of moisture content [kg H₂O/(sec kg dry matter)]; D_{H_2O}, the mass diffusivity of water through the solid being dried (m²/sec); L, the thickness of material (m); M, the moisture content (kg H₂O/kg dry matter); and M_{equil}, the equilibrium moisture content (kg H₂O/kg dry matter).

Notice that the rate of moisture loss depends on the difference between present and final moisture contents; as this difference decreases, so does the rate of moisture loss, leading to a falling-rate drying.

diffusivity may
not be
constant

Mass diffusivity is considered to be constant in the equation above. While this may be so, it is often also the case that mass diffusivity decreases as the product dries (see Section 4.3.2.6).

Heldman (1975) also presents an equation for falling-rate drying in thin-layered granular material

$$\dot{M} = \frac{h_c(\theta - \theta_{wb})}{2\rho LH} \left(\frac{M - M_{equil}}{M_o - M_{equil}} \right) \tag{4.8.50}$$

where h_c is the convection coefficient [$N\,m/(°C\,sec\,m^2)$]; θ, the ambient temperature (°C); θ_{wb}, the ambient wet-bulb temperature (°C); ρ, the density of granular material (kg dry matter/m^3); L, the thickness of material (m); H, the latent heat of water ($N\,m/kg$ water); and M, the moisture content (kg H_2O/kg dry matter). This equation should be applied primarily when $[(M - M_{equil})/(M_o - M_{equil}] > 0.6$.

Thin-layer drying often occurs with convection outside the material and diffusion inside the solid. It thus occurs in exactly the means described earlier in Section 4.7.4 on transient mass transfer. The technique described there, establishing an analogue between heat and mass transfer, allows the Heisler charts to be used as a reliable means to determine the rate of drying in a thin layer during the falling-rate period.

Heisler chart analogy

4.8.2.4 Shrinkage It has been suggested (Charm, 1971) that volume shrinkage during the early stages of drying, at least with vegetables, very nearly equals the volume of water lost, but in later stages the volume shrinkage is disproportionately smaller. No substantial further decrease in volume occurs as the pieces dry below about 15–20% moisture. Volume shrinkage for dried shelled corn appears in Table 4.8.9.

loss of moisture

EXAMPLE 4.8.2-1. Deep-Bed Corn Drying

Shelled corn harvested in early September in Maryland can have a moisture content of 30% (wet basis). Design a drying system in order to be able to store the corn for one year. The deep-bed dryer (Figure 4.8.17) is 6.1 m in diameter and 3.66 m high. Normal average dry-bulb temperature for September is 21.4°C, and average wet-bulb temperature is 18.3°C.

Table 4.8.9 Volume Shrinkage of Shelled Corn Dried to 12% Moisture Content

Original Moisture Content (wet basis)	Volume Shrinkage (percent)
30	29.1
25	22.5
20	14.5
17	9.1

Source: Brooker et al. (1974)

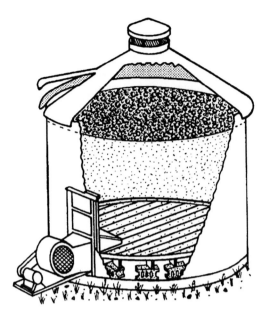

Figure 4.8.17. Diagram of deep-bed corn dryer.

Solution

From Table 4.8.5, we find that the corn must eventually reach 13.5% moisture (wb, or wet basis) to be safely stored for one year. That is the target for our drying system.

Notice in the Hukill diagram (Figure 4.8.16) that significant drying at any specific depth factor does not occur until the moisture ratio reaches about 0.95. At lower moisture contents the rate of drying is great enough to discourage further microbial growth. At higher moisture contents the grain dries so slowly that microbial growth can continue nearly unabated.

Choose any convenient depth factor, say, depth factor 8. Where this depth factor reaches a moisture ratio of 0.95 cannot be any longer than the safe storage period of the grain. Thus, for a depth factor of 8, the safe storage period cannot be any longer than 5 time units. This determines the time represented by each time unit. The dryer must now be designed to assume that the grain farthest from the drying air entrance is no greater than 8 depth factors and 5 time units.

Similarly, drying occurs very slowly at a moisture ratio lower than 0.05. At 8 depth factors this corresponds to 12 time units. It is therefore good design practice to dry with air at a lower relative humidity than would otherwise be necessary so that drying can be completed before the equilibrium moisture content is reached. Notice that when the grain at depth factor 8 reaches 0.05, the grain at lower depth factors is drier. If the grain is to be mixed after drying, then drying can be stopped when the grain at depth factor 8 is 0.2, and the average moisture content is 0.05 [average of moisture contents over all 8 depth factors $= (0.2 + 0.11 + 0.05 + 0.03 + 0.02 + 0.01 + 0 + 0)/8$]. This occurs at time unit 10. Otherwise, drying must continue until moisure ratio of grain at depth factor 8 is 0.05. This occurs at time unit 12.

The safe storage period for corn at 30% moisture wb and 24°C is found to be 2 days (Table 4.8.10). With 8 depth units chosen, a moisture ratio of 0.95 is reached in 5 time units. Each time unit is therefore equal to $2/5 = 0.4$ day. The total drying time, to a moisture ratio of 0.05, is 12 time units, or 4.8 days.

From the equilibrium relative humidity data for corn (Table 4.8.7) we find that a relative humidity of 71% is required for the air if 13.5% is the final moisture content. Our target, however, is a lower equilibrium moisture content.

If drying is to stop at a moisture ratio of 0.05, then a new equilibrium moisture content must be determined. For this calculation, all moisture contents must be on a dry basis. The conversion is made from using Eq. 4.8.33. Thus the initial moisture content is

$$M_{DB} = \frac{0.30}{1 - 0.30} = 0.43$$

and the final moisture content is 0.156% db. From Eq. 4.8.46, if

$$\frac{M - M_{equil}}{M_o - M_{equil}} = 0.05$$

then

$$M_{equil} = \frac{M - 0.05 M_o}{0.95} = \frac{15.6 - (0.05)(43)}{0.95} = 14.2\% \text{ db}$$

$$= 12.4\% \text{ wb}$$

Table 4.8.10 Safe Storage Period for Corn in Days

Storage Air Temperature (°C)	Moisture Content (%, wb)			
	15	20	25	30
24	115	12	4	2
22	142	15	5	3
20	175	18	6	3
18	211	21	7	4
16	252	26	9	5
14	304	32	11	6
12	374	39	13	8
10	466	48	17	10
8	652	67	24	14
6	803	83	30	18
4	948	98	35	21
2	1115	115	41	24

Source: Brooker et al. (1974).

Returning to the equilibrium moisture content data, a drying air relative humidity of 63% is required.

The psychrometric chart (Figure 4.8.4) indicates that at the specified ambient conditions, a dry-bulb temperature of 21.4°C and a wet-bulb temperature of 18.3°C, the ambient relative humidity is about 75%. Also from the chart, we find that incoming air must be heated to 24.0°C from 21.4°C to lower its relative humidity from 75 to 63%.

Heating the air from 21.4–24.0°C is a very minor amount of heating, and will cause no difficulty. If the air was required to be heated to 48°C or higher, the critical kernel temperature may have been exceeded, and the corn might not have been able to be used for seed, its baking quality might have been damaged, and the corn might have been subject to overdrying. This can cause surface checking and mechanical stress that may harm the kernels. Besides, there is an energy and economic penalty for heating air to too high a temperature.

To assure that the grain farthest from the entrance of the air is no more than 8 depth units, one of the parameters under our control must have a value chosen to meet the depth criterion. Depth factors can be determined from Eq. 4.8.48. Of the parameters appearing in the equation, grain depth, cross-sectional area, and mass flow rate of the air are the only ones directly controllable. Since the dimensions of the dryer were already given, mass flow rate of the air is the only parameter we can manipulate here.

The density of the corn solids is $700 \, \mathrm{kg/m^3}$. The latent heat of vaporization depends on the moisture content (see Table 4.8.11). As the grain dries, more heat is required to evaporate an equal mass of water. Since the grain in the dryer has a range of moisture contents, there are many values for latent heat (H). The highest value is used because the maximum number of depth factors depends directly on H, and using the highest H value ($3023 \, \mathrm{kN \, m/kg}$ at 14.2% moisture db) gives the most conservative calculation.

From information below Eq. 4.8.17, the specific heat of the air at the entrance is

$$c_p = 1006 + 1876\omega$$

where ω is the humidity ratio. Ambient air at 21.4°C and 75% relative humidity has an absolute humidity of $0.0120 \, \mathrm{H_2O/kg}$ air (Figure 4.8.4).

Table 4.8.11   Heat of Vaporization (kN m/kg) of Corn at Different Moisture Contents (% dry basis) and Temperatures

	Temperature (°C)			
Moisture Content	0	10	20	30
5	3428	3395	3366	3333
10	3219	3188	3164	3131
15	3034	3005	2978	2950
20	2795	2771	2747	2720

Source: Brooker et al. (1974).

Heating the air does not change the amount of moisture it contains

$$c_p = 1006 + 1876(0.0120) = 1029 \text{ N m}/(\text{kg }°C)$$

The thin-layer drying constant really is not constant (k), but depends on original moisture content and temperature. From Table 4.8.12, a k value of 0.132/h was chosen. The means to obtain θ_g is diagrammed in Figure 4.8.18. θ_o is the incoming air temperature, which was decided to be heated to 24°C. From the equilibrium moisture content curve for the grain, a relative humidity value ($\phi_o = 98\%$) is obtained that corresponds to the initial moisture content of the corn (30%). The initial point for this process is found on the psychrometric chart at $\theta_o = 24.0°C$ and $\omega = 0.0120 \text{ kg H}_2\text{O}/\text{kg air}$. The value for $\theta_g(19.4°C$ is obtained by following a constant wet-bulb temperature line on the psychrometric chart from the initial point to a relative humidity of $\phi_o(= 98\%)$.

Therefore, the required mass flow rate of air is

$$\dot{m} = \frac{x\rho_s AH(M_0 - M_{\text{equil}})k}{c_p D \ln 2 \,(\theta_o - \theta_g)}$$

$$= \frac{(3.66 \text{ m})(698 \text{ kg/m}^3)(29 \text{ m}^2)(3023 \text{ kN m/kg})(0.43 - 0.142)(0.132/\text{h})}{[1029 \text{ N m}/(\text{kg }°C)](8)(0.693)(24.0 - 19.4°C)}$$

$$= 324,000 \text{ kg/h} = 5407 \text{ kg/min}$$

Table 4.8.12   Thin-Layer Drying Constant (h^{-1}) for Shelled Corn

Original Moisture Content (%, wet basis)	Temperature (°C)				
	20	30	40	50	60
35	0.124	0.133	0.161	0.217	0.289
30	0.124	0.147	0.187	0.231	0.267
25	0.117	0.141	0.169	0.217	0.239
20	0.107	0.112	0.161	0.158	0.173

Source: Brooker et al. (1974).

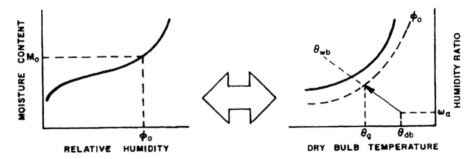

Figure 4.8.18. The parameter θ_g in Eq. 4.8.48 is related to the ambient conditions found in the psychrometric chart on the right and to the equilibrium relative humidity curve found on the left.

With an air density value of $1.270 \, kg/m^3$ (Figure 4.8.4), the volume flow rate is

$$\dot{V} = (5407 \text{ kg/min})/(1.270 \text{ kg/m}^3) = 4260 \text{ m}^3/\text{min}$$

Shelled corn has an airflow resistance of about $14.7 \, kN/m^2$ per meter of depth at this flow rate (airflow resistance of piled grains is nearly linearly related to air speed; thus the pressure drop of air flowing through grain is nearly proportional to the square of the air speed). The total pressure drop for the entire 3.66 m depth is

$$p = (14.7 \text{ kN/m}^3)(3.66 \text{ m}) = 53.8 \text{ kN/m}^2$$

and the power specification for the fan is

$$P = (53.8 \text{ kN/m}^2)(4260 \text{ m}^3/\text{min})(1 \text{ min}/60 \text{ sec})$$

$$= 3820 \text{ kNm/sec}$$

Since fan efficiencies are about 50% (about 60% peak), and large motors have efficiencies of about 75%, the motor running the fan must be rated at

$$P = \frac{3820 \text{ kN m/sec}}{(0.5)(0.75)} = 10,200 \text{ kN m/sec}$$

The rate of air heating required is

$$\dot{q} = \dot{m}c_p \, \Delta\theta$$

$$= (5407 \text{ kg/min})[1029 \text{ N m/(kg °C)}](24.0 - 21.4°\text{C})(1 \text{ min}/60 \text{ sec})$$

$$= 241 \text{ kN m/sec}$$

Remarks

In considering this solution, the engineer should be aware of several things:

1. The designed airflow rate is extremely high.
2. Consequently, the motor power specification is extremely high.
3. The heat generated by the motor due to its electrical inefficiency is more than enough to heat the air [(10,200 kN m/sec) (0.25) = 2550 kN m/sec > 241 kN m/sec].
4. At this high airflow rate the packed bed of corn may begin to float.

Better design solutions can be created by using a larger-diameter dryer, or heating the air more and stopping the drying process sooner.

4.9 DESIGN OF MASS TRANSFER SYSTEMS

bio-
compatibility

Whether designing an artificial kidney to purify the blood, a reverse osmosis machine to concentrate food flavor components, or processes to ameliorate the effects of environmental pollutants, the basic mass balance, as given by Eq. 4.2.1, must be satisfied. Each of these applications is subject to practical constraints that the biological engineer should appreciate. Especially when dealing with living organisms, care must be taken that materials used are biocompatible and that methods of movement are gentle enough to leave the organism undamaged. One of the biggest problems with materials that the blood contacts is thrombus formation, and dealing with this problem may be more important than satisfying the desired mass balance. Or a particular vegetable might be very sensitive to water vapor together with carbon dioxide. Satisfying particular requirements for this vegetable may be the hardest part of the design. Cellular damage can easily occur during necessary pumping operations, and this can be undesirable if the cells must thrive or be desirable if an extraction process is to follow the pumping operation.

mass balance

In Figure 4.9.1 is diagrammed a simple mass balance. The rates of mass input and output often depend on flow rates and concentrations. Even a simple diffusion process across a membrane usually involves substance delivery by a flow stream.

Mass is not really generated, for all practical purposes, but the particular substance of interest may be generated by chemical or biological activity. Thus there may often be environmental effects to be accounted for. Along with the increased metabolic rate with temperature, for instance, often comes an increase in carbon dioxide production (aerobic) or lactic acid formation (anaerobic).

self-inhibition

If the above three terms are not in balance, then mass accumulates. It is often true in biological systems that the accumulation of a certain substance acts as an environmental poison, which, in turn, slows the generation of that substance. This is true, for example, in the production of alcohol by yeast. There is thus an effect of accumulation on production, as indicated by the dotted line on the figure.

substance
interactions

Accumulation can also influence the rate of mass transfer into or out of the biological system. Cell membranes can become more permeable in the presence of environmental toxins. The presence of additional carbon dioxide in the air can lead to smaller leaf stomatal openings. In each of these cases there is a direct effect of the presence of one substance on the rates of transfer of that and other substances into and out of the biological system. This makes biological systems very interesting, but turns a simple algebraic mass balance equation into an interactive procedure.

Little more can be said in general for the design of mass transfer systems. As true for fluid and heat systems, manufacturers' specifications must be consulted for information about real constraints.

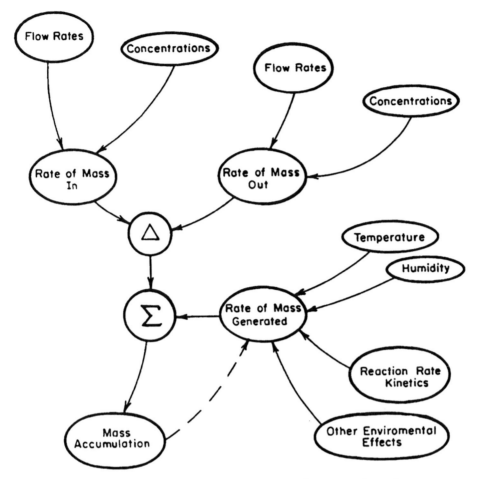

Figure 4.9.1. Schematic diagram for design of mass transfer systems.

Problems

4.1-1. **Mass generation.** Give three examples of generation of a particular species of mass in biological systems.

4.1-2. **Enzyme action.** Enzymes are thought to work by bringing precursor materials together physically and vastly increasing their local concentrations. Why should this be an important biological mode of action?

4.1-3. **Systems concepts.** Define effort and flow variables, resistance, and capacity in mass transfer terms.

4.2-1. **Beef trimmings.** The trimmings from beef briskets contain 50% fat. They are to be added to ground bull meat (about 6–8% fat) to produce hot dogs with 20% fat. What is the ratio of brisket trimmings to bull meat?

4.2-2. **Cardiac output by dilution.** Dilution methods for measuring cardiac output proceed by injecting a known mass of some tracer material (oxygen, dye, radioactive tracer, cold saline, or inert gas) into the bloodstream just before the heart. The diluted tracer concentration is measured downstream from the heart. Write a materials balance for the tracer, and determine how to calculate cardiac output from the indicated measurements.

4.2-3. **Constructed wetland.** Give a water balance for a concentrated wetland. Be specific about terms.

4.2-4. **Abalone population.** Abalone growing off the coast of California are harvested by fishermen, but sea otters also prefer to eat these creatures. As a result, the population of wild abalone has declined dramatically. Write a balance equation for the abalone population, and draw a systems diagram for this balance. What are some ways that the numbers of abalone can be increased? Give your answer in the context of the systems diagram.

4.2-5. **Resource recovery in space.** Despite the best attempts at resource recovery and recyling, some nutrient components are lost gradually over time. Carbon, for instance, appears as human waste that can be used by growing plants to produce food for human space travellers. The unavailability of a certain percentage of the component leads to an exponential decline over time of the amount of that component available for use by the system. Assume that 90% of the carbon in a spaceship is recoverable and reusable during a three-month cycle time (needed to grow plants, process them into food, and be consumed). How much extra carbon (in the form of carbohydrates) must be stocked in the spaceship for a two-year voyage to Mars?

4.3-1. **Dense gas.** Oxygen (20%) and sulfur hexafluoride (80%) mixtures are sometimes used for respiratory mechnics experiments in humans and animals when the objective is to see the effect of breathing dense gas. Calculate the diffusivity of this mixture at 37°C and 1 atm pressure. Compare the diffusivity of O_2-SF_6 to that of air under the same conditions.

4.3-2. **Diffusivity of air–carbon dioxide.** From values in Table 4.3.1, estimate the mass diffusivity of air–CO_2 at 37°C.

4.3-3. **Diffusion volume of toluene.** Estimate the atomic diffusion volume of toluene (molecular weight = 92.13 kg/kg mol).

4.3-4. **Exhaled air.** Exhaled air is approximately 17% O_2, 3.5% CO_2, 5.5% H_2O, and 74% N_2. If these are mole fractions, calculate constituent diffusivities.

4.3-5. **Temperature correction of diffusivity in gas.** The diffusion coefficient for water through air is $0.256 \, \text{cm}^2/\text{sec}$, measured at $15°C$ and $101.3 \, \text{kN}/\text{m}^2$ pressure. What is the value for the diffusion coefficient at $50°C$ and $200 \, \text{kN}/\text{m}^2$ pressure?

4.3-6. **Tobacco mosaic virus.** Electron microscopy shows the tobacco mosaic virus (TMV) to be cylindrically shaped with a diameter of $150 \times 10^{-10} \, \text{m}$ and a length of $3000 \times 10^{-10} \, \text{m}$. Its molecular weight is 40×10^6. The experimentally determined value of the TMV diffusion coefficient is $3 \times 10^{-12} \, \text{m}^2/\text{sec}$. Estimate the effective solute volume of TMV. Why is this an improper use of the Stokes–Einstein equation?

4.3-7. **Lactic acid.** Estimate the diffusion coefficient for lactic acid (molecular weight 90.08) in: (1) air at room temperature and pressure, (2) milk in the refrigerator, and (3) the plastic wall of a milk container.

4.3-8. **Diffusion pressure.** If total pressure on a diffusing gaseous medium is doubled, how does the diffusion rate compare to the original? What if the diffusing medium were liquid?

4.3-9. **Size of urea molecule.** One way to determine the approximate size of a molecule in solution is to measure the diffusivity of the solute and solve for solute radius. What is the approximate radius of urea in ethanol at $10°C$?

4.3-10. **Temperature correction of diffusivity in solution.** The diffusion coefficient for sucrose in water is $0.697 \times 10^{-9} \, \text{m}^2/\text{sec}$, measured at $20°C$ and $101.3 \, \text{kN}/\text{m}^2$. What is the value for the diffusion coefficient at $50°C$ and $200 \, \text{kN}/\text{m}^2$ pressure?

4.3-11. **Brandy alcohol.** Brandy contains 40% alcohol by volume. Determine the ethanol diffusivity in brandy at $20°C$.

4.3-12. **Gas chromatography.** Check for Knudsen diffusion in the gas chromatograph of Example 4.3.2.5-1, except with helium as the carrier gas.

4.3-13. **Knudsen diffusivity of water vapor.** Calculate the Knudsen diffusivity for water vapor at $20°C$ and 1 atmosphere pressure if the Knudsen number is 1.0.

4.3-14. **Drying of pasta.** Lasagna is long, thin, and flat pasta ($25 \, \text{cm}$ long $\times 5 \, \text{cm}$ wide $\times 1 \, \text{mm}$ thick) that must be dried to 10% moisture from approximately 85% moisture when first made. Assume all the excess moisture lies on the center plane of the lasagna piece. How long will it take the required amount of moisture to migrate from the center to the surface, assuming that the ambient air is perfectly dry and at $22°C$.

4.3-15. Sugar in Jello[R]. There are 19 grams of sugar in a 22-gram package of Jello gelatin dessert. Enough water is normally added to make 500 mL. However, especially firm Jello can be made if only enough water is added to make 375 mL. When set, the concentration of sugar in the less dilute Jello will be higher than in the more dilute Jello. If the two kinds of Jello are placed side by side, and if a 1-cm buffer region of Jello is assumed to exist at the interface, how long would it take for the excess sugar to diffuse into the less concentrated portion? Assume an interfacial area of 25 cm^2.

4.3-16. Diffusion of caffeine. Concentration of caffeine in a cup of coffee (0.24 L) is about 200 mg/L. This concentration is probably diluted by stomach contents to 150 mg/L. If intestinal diffusion of caffeine is limited by a liquid film 60 μm thick on the inside of the intestine, what would be initial rate of flux of caffeine across the membrane? Assume the diameter of the small intestine is 4 cm, and estimate the distance from the entrance of the intestine to the point where caffeine concentration inside the membrane becomes effectively zero.

4.3-17. Packaging hot dogs. Hot dogs with buns are to be packaged in plastic film in order to keep fresh for 21 days or more in refrigeration. Gas in the film is replaced with 100% nitrogen. What permeability and thickness film is needed? No more than 0.005% of oxygen can be present to ensure freshness.

4.3-18. Avleolar membrane resistance. Deoxygenated blood from the right side of the heart passes through the lungs such that only 0.4–2.0 μm of membrane separates the air-carrying alveoli from the pulmonary capillaries. Estimate the magnitude of resistance to diffusion of oxygen offered by this membrane. Alveolar air has 14% O_2; returning blood has about 5.3% O_2; O_2 consumption at rest is about 0.3 L/min.

4.3-19. Packaging cereal. Individual 19-gram servings of hot and cold cereal are being packaged in plastic container bowls with laminated foil lids. To ensure freshness for the 6-month shelf life of the cereal, the contents of the packages are flushed with dry nitrogen before sealing. The plastic material must allow no more than 0.5% of moisture by weight of cereal and 0.005% of oxygen to permeate into the cereal. Give your recommendation for the type of plastic to be used and its thickness.

4.3-20. Packaging grapes. Grapes use oxygen at the rate of 2.81 $\times$ 10^{-11} moles/sec per kg of fruit. What thickness of plasticized polyvinylchloride plastic film should be used to cover the top of a cardboard package 15 $\times$ 20 cm containing 0.5 kg of grapes to maintain an internal oxygen concentration of 5%?

4.3-21. Polyethylene film. The permeability of polyethylene to oxygen is 4.17×10^{-8} mL O_2/(sec cm^2 atm/cm). If the produce inside the package uses oxygen at the rate of 57×10^{-10} m^3/sec, what is the partial pressure of the oxygen inside the package? The surface area of the package is 100 cm^2 and the thickness of the film is 0.5 mm.

4.3-22. Osmotic pressure of isotonic saline. Calculate the osmotic pressure for isotonic saline solution.

4.3-23. Osmotic flow. One side of a semipermeable membrane with a resistance of 1.25×10^{-8} N sec/(kg m^2) is in contact with an 8% (by weight) sugar solution. The other side has a 0.2M urea solution under a pressure of 250 N/m^2. How much mass is flowing, and in which direction?

4.3-24. Molecular weight of lactose. Estimate the molecular weight of lactose from its osmotic pressure. Look up its actual molecular weight (*Handbook of Chemistry and Physics*) and compare to the calculated value.

4.3-25. Membrane cleaning. Speculate on methods used to clean membranes that do not damage the membranes. Give reasons for your speculations.

4.3-26. Dilute grape juice. Estimate the osmotic pressure of 10% total solids grape juice.

4.3-27. Cooled blood plasma. Estimate the osmotic pressure of human blood plasma cooled to 10°C.

4.3-28. Isotonic saline. Devise an apparatus involving a semipermeable membrane to be able to determine the proper salt concentration to make isotonic saline solution.

4.3-29. Drug dose delivery decreasing with duration. What modification in the DUROS system would allow it to deliver a decreasing dosage of drug with time? Describe the shape of the dosage–time curve.

4.3-30. Dissociation in solution. The osmotic concentration of a 0.25 mole fraction solution of an unknown substance is $-59{,}400$ kN/m^2. Is the substance completely dissociated, partially dissociated, or not dissociated in solution? Show how you arrived at the answer.

4.3-31. Equilibrium potential for sodium ion. Calculate the equilibrium potential for sodium ions at 37°C using the Nernst equation.

4.3-32. Dissolving salt tablet. Salt tablets are sometimes prescribed for chronic heat exposure. How long would it take for a spherical salt tablet 4 mm in diameter to completely dissolve, assuming no appreciable convection?

4.3-33. Separation of monoclonal antibodies. The objective of this process is to separate monoclonal antibodies from a suspension of murine hybridoma cells. Hollow-fiber cartridges are used and cleaned first with 0.1% peracetic acid. The separation is carried out with a 10^6 dalton molecular-weight cutoff polyaramid membrane with 0.35-m^2 active filtration area and 200-μm fiber lumens. Fiber walls are 20 μm thick. The cartridge is primed with methanol and flushed with pure water prior to operation. A challenge liquid is introduced at a flow rate of 1.2 L/min and the back pressure adjusted to give an initial permeate (including antibodies) flow rate of 0.4 L/in. Calculate the permeability of the hollow fibers. A second step can be performed to concentrate the antibody suspension. In this case the retentate is of interest. What size membrane pore diameter (in kilodaltons) do you recommend?

4.4-1. Mass transfer coefficient. If the value for the forced convective heat transfer coefficient has been found to be $3.7\,\mathrm{N\,m/(m^2\,sec\,{}^\circ C)}$, calculate the convective mass transfer coefficient for toluene moving into the surrounding air.

4.4-2. Lewis number. Calculate the Lewis number for water vapor in air at 30°C. Which is thicker, the heat transfer or the mass transfer boundary layer?

4.4-3. Microencapsulated flavors. Flavorings can be microencapsulated to remain secure from heat, moisture, oxidation, volatility, and adverse reactions to other food ingredients. The microcapsules can be easily handled, have no dust, and have no unwelcome odors. Many materials can be used for coating the microcapsules, including beeswax, starch, and gelatin. Assume a lemon flavor agent (density approximately 980 kg/m^3) 800 μm in diameter is coated with 22 μm of carboxymethyl cellulose. The coating has a mass diffusivity in water of about 3.7×10^{-7} m/sec. How long will it take the coating to disappear and the flavoring to be released?

4.4-4. Oxygen delivery to fish. An air block attached to an air line emits air bubbles at the bottom of a fish culture tank filled to 2 m with water. What is the rate of oxygen transfer from the bubbles into the water if the bubble diameters are in the range of 6 to 13 mm? The oxygen concentration in the water is at least 4 ppm in order for the fish to survive. The temperature of the water is 15°C.

4.4-5. Oxygen bubbles in water. A new type air block attached to an air line emits air bubbles at the bottom of a fish culture tank filled to 2.5 m with water and containing 100 kg of fish. How much oxygen will be transferred from the air bubble to the water transferred from the air bubble to the water if the bubble diameters are 1–2 mm? The oxygen concentration in the water is at least 4 ppm. If the biological oxygen demand of organic matter in the water is $100 \, kg \, O_2/day$, the oxygen required for nitrification is $22 \, kg \, O_2/day$, and the oxygen usage for the fish is $0.005 \, kg \, O_2/kg \, fish/day$, how many bubbles are required to supply the oxygen demands? Calculate the approximate bubble transit time using the drag coefficient and a force balance.

4.4-6. Microencapsulated islets. Porcine islets of Langerhans may be microencapsulated ($150-200 \, \mu m$ dia.) and injected into the portal circulation of patients with Type I diabetes. There they can produce insulin as needed to regulate blood glucose levels. Microcapsules attach to the walls of the portal vein (diameter approximately 0.25 cm) with blood flow of about $1200 \, mL/min$ in an adult. The mass diffusivity of insulin (molecular weight 6000) in blood is about $1.25 \times 10^{-9} \, m^2/sec$. Estimate the value of the mass transfer coefficient for insulin being released from the microcapsules.

4.4-7. Convective moisture loss in the trachea. Estimate the mass transfer coefficient value for moisture loss from the inside surface of the trachea (18 mm diameter, 120 mm length) with an airflow of $10 \, L/min$.

4.4-8. Moisture removal from plant leaf. Calculate the rate of moisture removal by convection from the underside of a plant leaf surface. Assume the water vapor pressure is saturated at $25°C$ ($3166 \, kN/m^2$) and the ambient vapor pressure is $1700 \, kN/m^2$. The wind is blowing across the narrow dimension of the leaf at $7.5 \, m/sec$. The leaf is $6 \times 4 \, cm$.

4.4-9. Phosphate in drainage water. Phosphate (PO_4^{2-}) is one of the leading causes of eutrophication in lakes and streams. It is therefore desired (and mandated in many cases) to eliminate phosphate runoff and leaching through soil. Phosphate is readily adsorbed on soil clay particles with an adsorption coefficient (concentration of PO_4^{2-} adsorbed divided by the concentration of PO_4^{2-} in solution) of about 0.90. The void fraction of a clay soil is about 0.47, and clay particle diameters are about $2 \, \mu m$. During a rainfall event, the concentration of phosphate at the soil surface is about $10 \, mg/L$, and the superficial velocity of water through unsaturated clay soil is $0.5-2 \, cm/h$. The mass diffusivity of phosphate in water is about $1.8 \times 10^{-9} \, m^2/sec$. How much phosphate will be found in drain tiles located 1 m beneath the surface of the soil, neglecting macropore flow through the soil?

4.4-10. Oxygen supply to biofilter. Biofilters are often used as chemical reactors. One common use is the transformation of ammonia to nitrate in waste water (Wheaton et al., 1991). A biofileter 1.8 m (length) $\times$ 1 m (width) $\times$ 2 m (depth) has particles 2.5 cm in diameter and a void fraction of 0.90. The flow rate of waste water through the filter is 7000 L/min, and the oxygen use of the filter is 24 kg O_2/day. If the oxygen concentration of the water entering the filter is 10 mg/L, what is the oxygen concentration of the water leaving the filter?

4.5-1. **Milk for lactose-intolerant consumers.** The lactose in milk cannot be digested by a significant minority of people. Lactase is sometimes added to milk to make the milk compatible with these drinkers. Lactase, also called β-galactosidase, transforms lactose (molecular weight 342.31 kg/kg mol) into galactose and glucose at a rate ($\dot{m}_{max}$) of 12,500 molecules lactose/sec for each molecule of lactase (molecular weight 43,000 from bovine liver source). Partially skim milk (2% butterfat) contains about 60 g of carbohydrate (mostly as lactose) per kg milk. How much lactase should be added to a thoroughly mixed 3800 L tank of 2% milk to convert all lactose into galactose and glucose in 30 min?

4.5-2. **Different levels of lactase.** If the maximum rate of lactose disappearance is 12,500 kg mol/sec in the presence of 1 kg mol of lactase (β-galactosidase), what is the rate of disappearance of lactose if the concentration of lactose is 20×10^{-5} kg mol/m^3? If the concentration of lactose is 100 kg mol/m^3? Assume the enzyme amount remains at 1 kg mol. What are your answers in terms of kg/sec, if the molecular weight of lactose is 342.31?

4.5-3. **Mucus lysozyme.** The value of the Michaelis–Menten constant for lysozyme, as found in Table 4.5.1, is a relatively small 6×10^{-6} kg mol/m^3 What advantage would this small value provide to the animal with lysozyme in its mucus?

4.5-4. **Clark electrode.** Diagram the two Clark electrode configurations, including the external power supply. Indicate what measurement would be made in each case.

4.5-5. **Chymotrypsin substrates.** As given in Table 4.5.1, the enzyme chymotrypsin operates on a number of substrates. Each substrate has a different Michaelis–Menten constant value. Explain how the digestion rates for the different protein substrates would differ.

4.5-6. **Nitrosomonas growth.** Wheaton et al. (1991) give the maximum specific growth rate of *Nitrosomonas* bacteria on a biofilter as $0.65\,d^{-1}$, and the half-saturation constant value as 1.0 mg ammonia per liter. Both of these values are for a temperature of 20°C. If the concentration of ammonia in the water running through the filter is 5 μg/L, what is the specific growth rate of the bacteria? What is the total bacterial growth rate if 10^9 bacteria are present?

4.6-1. **Cooling a solution.** If a saturated solution of ferrous chloride at 50°C is cooled to 10°C, what fraction of the salt will remain in solution?

4.6-2. **Mass capacitance of a tank.** A 1000-L tank is 1 m in diameter and 1.3 m high. What is its mass capacitance?

4.6-3. **Sucrose in solution.** If the 1000-L tank from the previous problem contains water at 30°C, what is the mass capacity of the tank for a sucrose solution?

4.6-4. **Osmotic pressure of salt solution.** Calculate the osmotic pressure produced by combining 1 kg NaCl with 1 kg H_2O at 30°C.

4.6-5. **Release of carbon dioxide.** A 1-liter tank contains CO_2 at a pressure of $50 \, kN/m^2$, and is connected to the atmosphere through a short 1-cm-diameter tube. At the end of the tube is a nonporous membrane with a resistance of $5 \times 10^{-4} \, sec/m^3$. How long will it take for 99% of the carbon dioxide to escape to the atmosphere? How long would it take if the pressure in the tank is doubled to $100 \, kN/m^2$?

4.6-6. **Gas diffusion sterlization.** Ethylene oxide (EtO) is commonly used to sterilize packaged medical devices, but stringent envrionmental regulations do not allow this to be performed haphazardly. The Sterijet system is designed for large-scale production of sterilized items such as surgical instruments or sutures (Andersen et al., 1997). In this system, the instruments are placed in a polyethylene (PE) bag that is sealed after the air is completely removed. This package is then placed inside a larger low-density polyethylene (LDPE) bag and a shot of EtO is injected between the bags before sealing. The packages are then transferred to a temperature-controlled room (35°C) with forced ventilation, where the EtO eventually diffuses through the outer LDPE bag and is removed. When the EtO level in the package falls below 250 ppm, the package may be removed and stored. The time for this to occur is 7–10 d.

A typical inner package has a volume of about 1 L, and the outer bag encloses an additional 1.5 L. The inner bag material is chosen purposely with a relatively low resistance to EtO diffusioin. A typical value is $1.2 \times 10^7 \, sec/m^3$. The outer bag has a typical EtO diffusion resistance of $9 \times 10^7 \, sec/m^7$. An initial injection of 1.25 g EtO in the package results in a concentration of 500 mg/L in the inner bag that is maintained for well over the 20 h necessary to achieve a 10^{-12} reduction in bacterial spore level.

The equivalent systems diagram appears below. Estimate:

1. The concentrations with time of EtO in the inner and outer bags.
2. The time that the EtO concentration in the inner bag remains above 250 mg/L.
3. The time necessary for the concentrations of EtO to fall to below 250 ppm in the package.

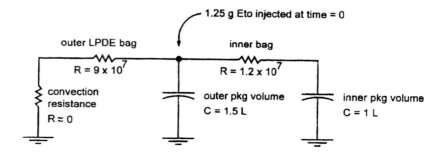

4.7-1. **Tracheae oxygen movement.** Insects, for the most part, do not breathe by mechanical ventilation, and none uses respiratory pigments such as hemoglobin. Instead, insects depend on a system of tubes, called tracheae, for their respiratory needs. The tracheae eventually branch out into fine tubules called tracheoles. Through this system, all the cells of the insect body are in close range of these tubes.

Model the tracheae, in an insect the size of a flea, as a cylinder with a diameter of 0.02 mm and a length of 0.5 mm. The approximate O_2 partial pressure difference between ends of the tracheae is about $2000\ \text{N/m}^2$. What is the resistance to mass flow in the tracheae?

4.7-2. **Desert golden moles.** Namib Desert golden moles are tiny blind burrowers about the size of large mice. After hunting termites on the surface, they dig about 0.3 m into the African sand and remain torpid there for 19 h or more as their bodies cool. The desert sand has uniform grains that permit the passage of air. Estimate the metabolic heat rate and the oxygen consumption of the mole. What is the concentration of oxygen at the location of the mole?

4.7-3. **Winter mulch.** Does mulch in the winter conserve soil moisture? Estimate the rate of moisture loss for the mulch in Example 4.7.2-1 during winter conditions.

4.7-4. **Atmospheric dispersion.** Make a list of biologically important cases where dispersion takes place. Classify each of these as likely being laminar or turbulent.

4.7-5. **Battery in a lake.** A lead–acid storage battery 15 cm (wide) × 25 cm (long × 18 cm (high) is mistakenly dropped in a lake where it settles on the bottom right-side up. The sulfuric acid contents begin to leak out. Assume the water in the lake is stagnant and the sulfuric acid is emitted as a pulse. Determine the time when the concentration of sulfuric acid at a point 1.5 m from the battery reaches its maximum, and determine what the concentration is at that time.

4.7-6. Péclet number in the respiratory bronchioles. Inspiration during quiet breathing occurs with a flow rate of about 12 L/min. The diameter of the respiratory bronchioles is about 0.8 mm, and there are about 8200 of them in parallel. Calculate the Péclet number for carbon dioxide.

4.7-7. Dispersion coefficient in the bronchioles. Calculate the dispersion coefficient for CO_2 in the bronchioles with the conditions given in the previous problem.

4.7-8. Intravenous drug. Three cubic centimeters (cc) of a drug with a concentration of active ingredient of 10^{-4} mg/cc are injected into a vein 0.5 cm in diameter and 6 cm long. The injection takes 6 sec. Blood flows through the vein at 1.2 L/min. What is the concentration of drug at the distal end of the vein 3 sec after the injection begins? Assume laminar flow in the vein.

4.7-9. Continuous bioreactor. During a pulse experiment in a small-scale continuous laboratory bioreactor, a pulse of 2.5 g of glycerol is injected into a pipe containing mostly water. The pipe is 1.5 cm in diameter and is 3 m long. The flow velocity is 0.09 m/sec. Determine the average glycerol concentration at the end of the pipe.

4.7-10. Dispersion in ecology. Discuss how dispersion models could be used to explain the spread of an introduced non-native species in the environment. What factors would make this situation more complicated than simple dispersion?

4.7-11. Ammonia from poultry house. The ammonia concentration in a 12 m (wide) × 170 m (long) × 3 m (high) poultry house containing 30,000 birds has been measured at 500 ppm (which is very high). Five ventilation fans 1.4 m above ground level cause a total of 21 m^3/sec air to be moved through the building. Wind speed in the area is about 4.5 m/sec. Will the ammonia be detectable at the location of the nearest dwelling 300 m downwind from the poultry house?

4.7-12. Pickle exhaust. Inside a small pickle processing plant the concentration of acetic acid is measured at 1.37×10^{-5} kg/m^3. A ventilation fan located 1.8 m above the ground removes 7.5 m^3/sec of air. Measurements of acetic acid concentration 100 m downwind from the building average 1×10^{-12} kg/m^3 on a bright, clear morning when the breeze is 2.8 m/sec. There is a filter in the ventilation duct that is supposed to remove 95% of the acetic acid vapor from the exhaust air. Is the filter working?

4.7-13. Gypsy moth pheromone. Gypsy moth sex attractant pheromone (glyplure, $C_{18}H_{35}O_3$) has a male threshold value of about 1.25×10^{-16} kg/m^3 (Wilson, 1974). Assume that: (1) the female gypsy moth releases 10^{-12} kg of gyplure over an area of 10^{-5} m^2, (2) the air is still, and (3) dispersion is one dimensional. If the effective distance of gyplure is 2500 m, what is the standard deviation of the substance during dispersion?

4.7-14. Chemical communication among plants. Tobacco plants emit a chemical vapor to give airborne warning to neighboring plants about attacking viruses. When attacked by mosaic virus, tobacco produces salicylic acid in defense. Some of that becomes methyl salicylate, which evaporates. Airborne methyl salicylate stimulates defense mechanisms of neighboring healthy plants. Assume a tobacco leaf 30 cm (long) × 20 cm (wide) × 4 mm (thick) that produces methyl salicylate uniformly throughout the leaf. The methyl salicylate then diffuses ($D \approx 1.4 \times 10^{-9}$ m^2/sec) through the leaf to the atmosphere. Air movement across the leaf is about 0.5 m/sec. Graph the fraction of the original concentration of methyl salicylate in the leaf against time.

4.7-15. Small fruit cakes. Cathy can also make small fruit cakes as gifts (yes, some people actually like them). These are sized 15 cm (long) × 10 cm (wide) × 8 cm (high). How long will these last before drying too much?

4.7-16. Orthopedic implants. Small pieces of bone are sometimes used as substrate splints to regenerate lost bone material in orthopoedic patients. These bone splits can be taken from cadavers and organ donors. Consider bone fragments 6 cm × 2 cm × 3 mm that are 9% moisture content (wet basis) when fresh. They are to be dried to 0.5% moisture content by suspending them in a blast of air moving at 25 m/sec and containing 0.004 kg H$_2$O/m^3 air. Assume the mass diffusivity of moisture through the bone has about the same value as water through a gel. Estimate the drying time.

4.7-17. Diffusion of water in bone. Describe an experiment to determine the diffusion coefficient for moisture in the orthopedic implant of the previous problem.

4.7-18. Drying of cookie dough. Dielectric heating can be used to reduce the moisture content in cookies, crackers, and other baked goods without burning their outside surfaces. Radio frequency energy is used to cause the evenly distributed dipolar water molecules to heat and evaporate inside the baked goods. If 56 N m/sec of RF energy is being delivered to each cracker 4 cm dia. × $\frac{1}{2}$ cm thickness during baking in a convection oven, how fast is moisture being removed from the crackers? Assume negligible end effects, negligible convective resistance to moisture removal at the cracker surface, original cookie dough moisture content of 55%, and an RF heating efficiency of 80%.

4.7-19. Placental oxygen transfer. Oxygen is exchanged by countercurrent flow in the human placenta. The fetal blood picks up oxygen and leaves through the fetal vein. Oxygen-poor maternal blood returns to the maternal heart.

Maternal blood enters the uterine artery at about $8.3 \times 10^{-6} \, m^3/sec$ and 98% oxygen saturation (corresponding to an O_2 partial pressure of $11.2 \, kN/m^2$) and leaves the uterine vein at 60% saturation $(4.7 \, kN/m^2)$. Fetal blood, rich in fetal hemoglobin that carries oxygen more readily than maternal hemoglobin, enters the umbilical artery at about $1.2 \times 10^{-6} \, m^3/sec$ and 60% saturation $(3.0 \, kN/m^2)$ and leaves the umbilical vein at 80% saturation $(4.1 \, kN/m^2)$. Use the ideal gas law to calculate all oxygen concentrations in kg/m^3. What is the oxygen consumption of the fetus? Is fetal or maternal blood flow the limiting fluid?

4.7-20. Extracorporeal membrane oxygenation (ECMO). ECMO is used to supplement oxygen exchange in lung-deficient patients. Hollow fibers made from silicone (to control clot formation) enclose the gas phase, while blood flows outside. If the blood enters with an oxygen partial pressure of $6000 \, N/m^2$ and leaves the ECMO device with an oxygen partial pressure of $11,200 \, N/m^2$, what is the oxygen concentration difference (kg/m^3) between blood entering and exiting the device? If the patient requires $400–600 \, mL/min$ of oxygen and his cardiac output is $5 \, L/min$, what portion of the cardiac output must be diverted through this device? If pure oxygen constitutes the gas phase, what is the mean partial pressure difference across the membrane? If 90% of the diffusional resistance to oxygen transfer occurs in the blood, what is the maximum membrane resistance that can be tolerated for adequate oxygen transfer? If the surface area of the hollow-fiber membrane is $4 \, m^2$, what is the maximum membrane resistance for a $1 \, m^2$ of membrane material?

4.7-21. Waste recovery. A large-scale biochemical process produces waste in solution in water at a rate of $440 \, L/min$. One of the constituents of the waste fluid is ferric chloride (molecular mass $= 162$) at a concentration of $1500 \, kg/m^3$. The other constituents all have molecular masses of $500 \, kg/kg \, mol$ or more.

It is desired to recover the ferric chloride by separating it from the other wastes using a membrane mass exchanger with pure water on the clean side. Suggest a membrane system that would do the job. Include in your recommendations: type of membrane, pore sizes, total area, flow rates of the two flow streams, and pressure difference between sides.

4.7-22. Drying of french fries. Potatoes that are cut into fries contain 93–100% moisture, but must be dried to a target 73% moisture in order to produce an acceptable product. Air is blown over the fries at $90°C$ and

1.3 m/sec. If care is taken to dry the fries slowly enough that a hard surface does not form to impede moisture movement from the potato surfaces, and if the fries are tumbled in order to expose all surfaces to the air, how long will it take the fries to dry?

4.7-23. **Your breadth is particularly appealing.** Biting midges (small flying biting insects) are sensitive to octanol, a volatile alcohol produced when plant material is digested, and are able to locate a potential victim by the octanol in the exhaled breath. Vegetarians produce more octanol than meat eaters. For any given octanol detection threshold possessed by a biting midge, how much farther away can a person be detected after eating a vegetarian meal compared to a juicy steak, assuming that the octanol concentration in the exhaled breath is thrice as high after the vegetarian meal? Assume atmospheric dispersion at night in mostly clear conditions with still air ($v = 1.5$ m/sec). Assume also that the breath of a meat eater can be detected 520 m away from the source at the same height as the breath is emitted. For simplicity, assume no reflection.

4.8-1. **Saturation vapor pressure.** Calculate the saturation vapor pressure of water at 25°C. Compare your result to the value listed in Table 4.8.2.

4.8-2. **Evaporation heat.** Calculate the latent heat of evaporation for water at 37°C. Compare this value to the value at 20°C.

4.8-3. **Respiratory evaporation heat loss.** If 6 L/min of perfectly dry air is inhaled and the exhaled air is saturated with water vapor, calculate the rate of heat lost from respiratory evaporation.

4.8-4. **Sum of partial pressures.** Show, by using data from Table 4.8.1 and Dalton's law, that the sum of atmospheric constituent partial pressures equals total atmospheric pressure.

4.8-5. **Minute volume measurement.** Respiratory minute volume (rate of air usage in L/min) is measured on a dry air basis at atmospheric pressure and 22°C as 8.6 L/min. Express this value in STPD and BTPS conditions.

4.8-6. **Humidity ratio in a refrigeration system.** Air in a refrigeration system used to cool produce usually contains a relatively high amount of water vapor. Air in a refrigeration system used to store biomedial supplies is often much drier. Calculate the humidity ratio of the air in a biomedical storage unit if the dry-bulb temperature is 0°C and the partial pressure of water vapor is 250 N/m². Calculate the humidity ratio of air in a produce storage unit if the dry-bulb temperature is 3°C and the partial pressure of water vapor is 600 N/m².

4.8-7. **Relative humidity in a refrigeration system.** Calculate the relative humidities for the conditions of the previous problem.

4.8-8. **WBGT in the work environment.** It is recommended (ACGIH, 1991) that, when the WBGT reaches 30.6°C, all nonacclimatized workers be allowed to rest and to cool off 15 min out of each hour. If the black globe temperature is measured as 46°C, and the natural wet-bulb temperature is measured as 25°C, what must the dry-bulb temperature value be to reach the WBGT limit?

4.8-9. **Wet-bulb temperature.** What is the difference in the way thermodynamic wet-bulb temperature and natural wet-bulb temperature are measured?

4.8-10. **Specific enthalpy difference.** Calculate the difference in specific enthalpy if the air has a humidity ratio of 0.012 kg H_2O/kg dry air and the air is heated by 7°C. If 250 kg of air are involved, what is the total amount of heat that had to be added to the air to raise it the 7°C?

4.8-11. **Psychrometric chart location.** A point on the psychrometric chart is given by a dry-bulb temperature of 30°C and wet-bulb temperature of 18°C. What are the humidity ratio, relative humidity, specific volume, and dew-point temperature values?

4.8-12. **More psychrometric chart location.** A point on the psychrometric chart is given by a relative humidity of 20% and a humidity ratio of 0.07. What are the dry-bulb temperature, wet-bulb temperature, dew-point temperature, and specific volume values?

4.8-13. **Sensible heating.** Heat is added to air at 10°C and 70% RH. The final temperature is 25°C. What is the final humidity ratio? Relative humidity? How much heat has been added per kg of dry air?

4.8-14. **Sensible cooling.** Air at 80°C and 7.7% RH is to be cooled to 30°C. What is the wet RH of the cooled air?

4.8-15. **Wood chip drying.** Air is to be used to dry wood chips. The air begins at 45°C and humidity ratio of 0.003. Each kg of dry air must absorb an additional 0.008 kg H_2O. What is the final temperature of the air?

4.8-16. **Drying of municipal solid waste (MSW).** MSW is to be dried by air at 35°C and 20% RH. What is the maximum amount of water that can be removed by each kg of air?

4.8-17. **Evaporative cooling for customers.** Many store fronts in Phoenix, AZ, have horizontal tubes above the doors and windows. From these tubes comes a wall of water mist that evaporates in the air and cools the air for customers. If the outside air in summer reaches 49°C with 10% RH, what is the potentially minimum temperature of the air that can be cooled in this way?

4.8-18. Evaporative cooling of chickens. Poultry houses can contain 50,000 birds. During extremely hot spells, ventilation systems in these houses cannot remove sufficient heat, and many chickens die. It has been suggested that evaporative cooling of the air in these poultry houses could economically reduce air temperature and help the birds survive. What is the difficulty with this scheme?

4.8-19. Ventilation system control. Modern ventilation systems are often controlled by mixing streams of warm air and cold air to produce the desired temperature. If one part of air at 5°C and 90% RH is mixed with two parts of air at 33°C and 0.029 kg H_2O/kg dry air, what is the final temperature of the mixture?

4.8-20. Hospital ventilation system. List as many heat and moisture sources in a hospital as you can. Estimate their magnitudes on a 100-bed basis. How would you design the ventilation system for this hospital? What special considerations need to be accounted for?

4.8-21. Ventilation system in an animal production facility. Breeding captive endangered species animals often requires careful environmental control. List as many heat and moisture sources as you can. What special considerations might there be for ventilation systems designed for different species?

4.8-22. Automobile assembly plant. List as many sources of heat and moisture in an automobile assembly plant as you can. What would be some design criteria for the ventilation system?

4.8-23. Greenhouse ventilation. Plants growing in a greenhouse need to be ventilated. List as many heat and moisture sources as you can. How would these change for closed-system plant production aboard a space vessel?

4.8-24. Storing gooseberries. Ten thousand kilograms of gooseberries are to be stored for one month at -1°C. Inlet air begins at -1°C and 50% RH. Outlet air is 0°C and moisture saturated. What is the minimum required ventilation rate, in m^3/sec, to remove all moisture loss from the gooseberries?

4.8-25. Overcrowded room. A room with a ventilation system designed to hold 100 seated humans holds 150 during a crowded meeting. Assume that the only heat load in the room comes from the occupants, and that air enters the room at 20°C and exits at 22°C for 100 occupants. Assume the humidity ratio of the air is constant at 0.008 kg H_2O/kg dry air. How hot will the exhaust air get for 150 seated occupants?

4.8-26. Deep-bed corn drying. Determine the required air flow rate, fan power, and heat for the corn in Example 4.8.2-1 if the dryer is to be 8 m in diameter. All other conditions remain the same.

4.8-27. Moisture content. Wood has a moisture content of 16%, dry basis. What is its moisture content on a wet basis?

4.8-28. Dried blood. If 1 kg of whole human blood were to be dried to 0.5% moisture content (wb), what would be its mass?

4.8-29. Moisture content of soil. Describe a procedure to determine the moisture content of a soil sample.

4.8-30. Dried peaches. Estimate the volume shrinkage of peaches dried to 5% moisture (wb).

4.8-31. Spoilage microorganisms. What class of microorganisms appearing in Table 4.8.4 would likely be the one to grow in a wet product? A dry product?

4.8-32. Drying of flaxseed. Flaxseed is to be dried to a final moisture content of 12% (wb). Air is to be supplied at 35°C dry-bulb temperature and humidity ratio of 0.005 kg H_2O/kg dry air. What is the value for θ_g?

4.8-33. Sweating heat loss. If a person sweats at a rate of 58 mg/sec and all the sweat evaporates from the skin surface, what is the rate of heat loss? If the air is at 20°C and the relative humidity is 65%, estimate the value for the heat transfer coefficient for evaporation.

4.8-34. Corn drying. Calculate the minimum fan power specification to dry shelled corn harvested from Maryland's Eastern Shore in early September from 35% moisture (wb) to 12% moisture (wb).

4.8-35. Mixing of air. Air at 0°C and 90% relative humidity is mixed with air at −30°C and 50% relative humidity in a 1:3 ratio. What is the final temperature of the mixture? The mixed air is then heated with 5400 N m/kg. What is the final temperature of the air?

4.9-1. Mushroom compost. Mushrooms require a nutrient and energy source in order to grow. This is supplied by compost. Mushroom compost is produced in two stages: The first stage accomplishes a general conditioning and heating function, while the second stage is meant to encourage thermophilic bacteria to decompose cellulose and hemicellulose into glucose. Stage two compost is brought to the composting site at 30°C, and the temperature is allowed to rise to 60°C to pasteurize the compost briefly. Thereafter, the temperature is reduced to 50°C and maintained there until composting is complete in about 120 h. During this time, the compost loses about 30% of its mass. Oxidation of the dry matter (converted to glucose) is expected to occur according to Eq. 1.7.2. Air is required to supply oxygen and remove carbon dioxide, excess water (both physical and metabolic), and excess heat. Answer the following questions for 1000 kg of compost that is 70% moisture (wb): How much dry matter is oxidized? How much water (both physical and metabolic) must be removed? If the air enters the compost pile at 20°C and 50% RH, and leaves saturated at 50°C, how much air is required to remove excess water? If the air leaving the compost should contain no more than 5% CO_2, how much air is required to remove excess CO_2? If the air leaving the compost should contain no less than 16% O_2, how much air is required to supply the compost oxygen needs? How much air is required to remove excess heat? Choose which amount of air you would recommend, and discuss the consequences of this decision.

4.9-2. **Orange juice concentrate.** Orange juice is concentrated by a factor of 4 by vacuum evaporation. Unfortunately, however, more than 90% of the volatile flavor compounds are also removed in the process.

About 0.15 kg of flavor essence is recovered from the distillate for each 100 kg of raw juice entering the concentration process. This recovered essence is added to the juice concentrate. Extra flavor compounds can be obtained from orange peel oil by squeezing. If the reconstituted juice (diluted concentrate) needs to contain at least 0.35 kg of flavor essence per 100 kg of juice, how much peel oil flavor essence needs to be added to the concentrate?

4.9-3. **Drying mustard seed.** Mustard seed must be dried from 10–12% moisture (wb) as newly harvested to 8% moisture (wb) to be stored indefinitely. Suggest a means to do this. Give specific recommendations for air flow rate and heating.

4.9-4. **Drying chopped onion.** Fresh onion contains 89% moisture (wb); it must be dried to 4–8% moisture to be stored. Suggest a means to do this. Give specific recommendations for air flow rate and heating.

4.9-5. **Don't cry.** Junior spills a 60 mL glass of water on the kitchen table. Two-thirds of the water spills on the floor. The table is 1.2 × 2.4 m. The wetted area of the floor is 4.5 m². If the temperature is 22°C and the relative humidity in the kitchen is 65%, estimate the time it would take for the water to dry on its own accord.

REFERENCES

ACGIH, 1991, Documentation of the Threshold Limit Values and Biological Exposure Indices, 6th ed., American Conference of Governmental Industrial Hygienists, Cincinnati, OH.

Amjad, Z. (Ed.), 1993, *Reverse Osmosis*, Van Nostrand Reinhold, New York.

Andersen, L., M. Delvers, and E. Hu, 1997, An Introduction to Gas-Diffusion Sterlization, *Med. Dev. Diag. Indust.* **19**(5): 137–50 (May).

Aronson, D. G., 1985, "The Role of Diffusion in Mathematical Biology: Skellam Revisited," in *Mathematics in Biology and Medicine*, V. Capasso, E. Grosso, and S. L. Paveri-Fontana (Eds.), Springer-Verlag, New York, NY, pp. 2–6.

ASAE, 1979, *Agricultural Engineers Yearbook*, American Society of Agricultural Engineers, St. Joseph, MI.

ASAE, 1989, *Standards*, R. H. Hahn and E. E. Rosentreter (Eds.), American Society of Agricultural Engineers, St. Joseph, MI.

ASHRAE, 1977, *Handbook and Product Directory: Fundamentals*, American Society of Heating, Refrigeration, and Air-Conditioning Engineers, Atlanta, GA.

ASHRAE, 1981, Thermal Environmental Conditions for Human Occupancy, ANSI/ASHRAE Standard 55-1981.

Ben Jebria, A., 1984, Convective Gas Mixing in the Airways of the Human Lung— Comparison of Laminar and Turbulent Dispersion, *IEEE Trans. Biomed. Eng.* **31**: 498–506.

Best, R. J., J. R. Booker, and C. Mackie, 1993, "Analysis of Contaminant Transport," in *Geotechnical Management of Waste Contamination*, R. Fell, T. Phillips, and C. Gerrard (Eds.), A. A. Balkema, Rotterdam, pp. 39–58.

Billings, C. E., 1973, Barometric Pressure in *Bioastronautics Data Book*, J. F. Parker, Jr., and V. R. West (Eds.), NASA, Washington, D.C. pp. 1–34.

Blum, L. J., and P. R. Couleet (Eds.), 1991, *Biosensor Principles and Applications*, Marcel Dekker, New York.

Brooker, D. B., F. W. Bakker-Arkema, and C. W. Hall, 1974, *Drying Cereal Grains*, AVI, Westport, CT.

Budavari, A. (Ed.), 1989, *The Merck Index*, Merck, Rahway, NJ.

Burgess, D. J., 1993, "Drug Delivery Aspects of Biotechnology Products," in *Biotechnology and Pharmacy*, J. H. Pezzuto, M. E. Johnson, and H. R. Manasse (Eds.) Chapman Hall, New York, NY, pp. 116–51.

Butchbaker, A. F., and M. D. Shanklin, 1964, Partitional Heat Losses of Newborn Pigs as Affected by Air Temperature, Absolute Humidity, Age, and Body Weight, *Trans. ASAE* **7**: 380–83, 387.

Carslaw, H. S., and J. C. Jaeger, 1959, *Conduction of Heat in Solids*, Oxford University Press, London.

Cass, A. E. G. (Ed.), 1990, *Biosensors: A Practical Approach*, Oxford University Press, Oxford.

Charm, S. E., 1971, *The Fundamentals of Food Engineering*, AVI, Westport, CT.

Cussler, E. L., 1984, *Diffusion: Mass Transfer in Fluid Systems*, Cambridge University Press, New York, NY.

Dawson, R. M. C., D. C. Elliott, W. H. Elliott, and K. M. Jones, 1986, *Data for Biochemical Research*, Clarendon Press, Oxford.

Diaz, D., 1996, Under Pressure, *Invention and Technology* **11**(4): 52–63.

Fasman, G. D., 1989, *Practical Handbook of Biochemistry and Molecular Biology*, CRC Press, Boca Raton, FL.

Fisher, J., D. Mahle, L. Bankston, R. Greene, and J. Gearhart, 1997, Lactational Transfer of Voltatile Chemicals in Breast Milk, *Amer. Indus. Hyg. Assoc. J.*, **58**, 425–31.

Fuller, E. N., P. D. Schettler, and J. C. Giddings, 1966, A New Method for Prediction of Binary Gas-Phase Diffusion Coefficients, *Ind. Eng. Chem.*, **58**(5): 19–27.

Galletti, P. M., C. K. Colton, and M. J. Lysaght, 1995, "Artificial Kidney," in *The Biomedical Engineering Handbook*, J. Bronzino (Ed.), CRC Press, Boca Raton, FL, pp. 1898–922.

Geankoplis, C. J., 1993, *Transport Processes and Unit Operations*, Prentice Hall, Englewood Cliffs, NJ.

Ghosh, T. K., and A. K. Banga, 1993a, Methods of Enhancement of Transdermal Drug Delivery: Part IIA, Chemical Permeation Enhancers, *Pharmaceutical Technology*, April, pp. 62–90.

Ghosh, T. K., and A. K. Banga, 1993b, Methods of Enhancement of Transdermal Drug Delivery: Part IIB, Chemical Permeation Enhancers, *Pharmaceutical Technology*, May, pp. 68–76.

Heldman, D. R., 1975, *Food Process Engineering*, AVI, Westport, CT.

Hellickson, M. A., A. F. Butchbaker, R. L. Witz, and R. L. Bryant, 1967, Performance of Growing Turkeys as Affected by Environmental Temperature, *Trans. ASAE* **10**: 793–95.

Hellman, T. M., and F. H. Small, 1974, Characterization of the Odor Properties of 101 Petrochemicals Using Sensory Methods, *J. Air. Pollu. Control Assoc.* **24**(10): 979–82.

Hertig, B. A., 1973, Thermal Standards and Measurement Techniques, in *The Industrial Environment—Its Evaluation and Control*, U.S. Government Printing Office, Washington, DC, pp. 413–30.

Johnson, A. T., 1991, *Biomechanics and Exercise Physiology*, John Wiley and Sons, New York, NY.

Johnson, A. T., and G. D. Kirk, 1981, Heat Transfer Study of the WBGT and Botsball Sensors, *Trans. ASAE* **24**: 410–17.

Jordan, R. C., and G. B. Priester, 1956, *Refrigeration and Air Conditioning*, Prentice-Hall, Englewood Cliffs, NJ.

Kemeny, D. M., 1991, *A Practical Guide to ELISA*, Pergamon Press, Oxford.

Langer, R., 1990, New Methods of Drug Delivery, *Science* **249**: 1527–33.

LaRue, J. H., and R. S. Johnson, 1989, *Peaches, Plums and Nectarines: Growing and Handling for Fresh Market*, University of California Division of Agriculture and Natural Resources Publication 3331, Oakland, CA.

Leonardos, G., D. Kendall, and N. Bernard, 1969, Odor Threshold Determinations of 53 Odorant Chemicals, *J. Air Pollu. Control Assoc.* **19**(2): 91–95.

Lide, D. R. (Ed.), 1995, *CRC Handbook of Chemistry and Physics*, CRC, Boca Raton, FL.

Lightfoot, E. N., 1995, The Roles of Mass Transfer in Tissue Function, in *The Biomedical Engineering Handbook*, J. D. Bronzino (Ed.), CRC Press, Boca Raton, FL, pp. 1656–70.

Longmore, D., 1968, *Spare Part Surgery: The Surgical Practice of the Future*, Doubleday, New York, pp. 53–59.

Maulé, C. P., and T. A. Fonstad, 1996, Ion Movement Immediately Beneath Earthen Hog Manure Storages; How Much, How Deep, and How Fast, Paper #962049, American Society of Agricultural Engineers, St. Joseph, MI, 49085–9659.

McGilvery, R. W., 1983, *Biochemistry: A Functional Approach*, W. B. Saunders, Philadelphia, pp. 309–10.

Middleman, A., 1972, *Transport Phenomena in the Cardiovascular System*, John Wiley and Sons, New York.

Murray, R. K., P. A. Mayes, D. K. Granner, and V. W. Rodwell, 1990, *Harper's Biochemistry*, Appleton & Lange, East Norwalk, CT.

Nagylaki, T., 1989, The Diffusion Model for Migration and Selection, in *Lectures on Mathematics in the Life Sciences*, Vol. 20, A. Hastings (Ed.), The American Mathematical Society, Providence, RI, pp. 55–75.

Okos, M. R., G. Narsimhan, R. K. Singh, and A. C. Weitnauer, 1992, Food Dehydration, in *Handbook of Food Engineering*, D. R. Heldman and D. B. Lund (Eds.), Marcel Dekker, New York, pp. 437–562.

Pearson, J. R. A., 1958, On Convection Cells Induced by Surface Tension, *J. Fluid Mech.* **4**: 489–500.

Rubinow, S. I., 1975, *Introduction to Mathematical Biology*, Wiley-Interscience, New York, NY.

Russell, E. W., 1961, *Soil Conditions and Plant Growth*, John Wiley and Sons, New York.

Schnoor, J. L., 1996, *Environmental Modeling: Fate and Transport of Pollutants in Water, Air, and Soil*, John Wiley and Sons, New York.

Schomburg, D., and M. Salzmann (Eds.), *Enzyme Handbook*, Springer-Verlag, Berlin.

Sirivicha, S., A. T. Johnson, L. W. Douglass, and A. Kramer, 1990, Diffusion of Carbon Monoxide, Carbon Dioxide, and Sulfur Dioxide in Potato Tissue, *Trans. ASAE* **33**: 189–98.

Spector, W. S., 1956, *Handbook of Biological Data* WADC Technical Report 56-273, Wright-Patterson Air Force Base, Ohio.

Stephen, H., and T. Stephen, 1963, *Solubilities of Inorganic and Organic Compounds*, Macmillan, New York.

Stryer, L., 1988, *Biochemistry*, W. H. Freeman, San Francisco, CA.

Tao, B., 1996, personal communication, Agricultural Engineering Department, Purdue University, West Lafayette, IN.

Turner, D. B., 1969, *Workbook of Atmospheric Dispersion Estimates*, Department of Health, Education, and Welfare, Washington, DC.

Ultman, J. S., 1981, Gas Mixing in the Pulmonary Airways, *Ann. Biomed. Engr.* **9**: 513–27.

Ursin, C., C. M. Hansen, J. W. Van Dyk, P. O. Jensen, I. J. Christensen, and J. Ebbehoj, 1995, Permeability of Commercial Solvents through Living Human Skin, *Amer. Indus. Hyg. Assoc. J.* **56**: 651–60.

USEPA, 1992, Dermal Exposure Assessment Principles and Applications: Interim Report, U.S. Environmental Protection Agency, Washington, D.C.

Vesiland, P. A., 1997, *Introduction to Environmental Engineering*, PWS Publishing, Boston.

Wagner, G., and G. G. Guilbaut, 1994, *Food Biosensor Analysis*, Marcel Dekker, New York.

Wardley, R. C., and J. R. Crowther (Eds.), 1982, *The ELISA: Enzyme-Linked*

Immunosorbent Assay in Veterinary Research and Diagnosis, Martinus Nijhoff, The Hague.

Wark, K., and C. F. Warner, 1981, *Air Pollution, Its Origin and Control*, Harper Collins, New York, NY.

Watt, B. K., and A. L. Merrill (Eds.), 1963, *Composition of Foods*, Washington, DC, U.S. Government Printing Office (Agriculture Handbook No. 8).

Wheaton, F., J. Hochheimer, and G. E. Kaiser, 1991, Fixed Film Nitrification Filters for Aquaculture, in *Advances in World Aquaculture*, Vol. 3, Aquaculture and Water Quality, D. E. Bruce and J. R. Tomasso (Eds.), World Aquaculture Society, Baton Rouge, LA, pp. 272–303.

Wheaton, F. W., J. N. Hochheimer, G. E. Kaiser, R. F. Malone, M. J. Krones, G. S. Libey, and C. C. Easter, 1994, Nitrification Filter Design Methods, in *Aquaculture Water Reuse Systems: Engineering Design and Management*, M. B. Timmons and T. M. Losordo (Eds.), Elsevier, Amsterdam, pp. 127–71.

Wheeler, E. F., L. D. Albright, L. P. Walker, R. M. Spanswick, and R. W. Lenghans, 1994a, Plant Growth and Nitrogen Uptake, Part I: Beyond the Michaelis–Menten Relationship. ASAE Paper 94-7506, ASAE—The Society for Engineering in Agriculture, Food, and Biological Systems, St. Joseph, MI.

Wheeler, E. F., L. D. Albright, L. P. Walker, R. M. Spanswick, and R. W. Langhans, 1994b, Plant Growth and Nitrogen Uptake, Part II: Effects of Light Level and Nitrate Nutrition. ASAE Paper 94-7505, ASAE—The Society for Engineering in Agriculture, Food, and Biological Systems, St. Joseph, MI.

White, A., P. Handler, E. L. Smith, and D. Stetten, Jr., 1959, *Princiles of Biochemistry*, McGraw-Hill, New York.

Williams, R. J., and E. M. Lansford, Jr., 1967, *The Encyclopedia of Biochemistry*, Reinhold, New York.

Wilson, E. O., 1974, Pheromones, in *Animal Engineering*, W. H. Freeman, San Francisco, pp. 94–104.

Wise, D. L., 1989, *Applied Biosensors*, Butterworths, Boston.

Wu, F., J., M. V. Peshwa, F. B. Cerra, and W.-S. Hu, 1995, Entrapment of Hepatocyte Spheroids in a Hollow Fiber Bioreactor as a Potential Bioartificial Liver, *Tissue Engineering* 1: 29–40.

5

LIFE SYSTEMS

"If of mortal goods thou art bereft,
and of thy slender store two leaves alone to thee are left, sell one,
and with the dole, buy hyacinths to feed thy soul."

Omar Khayam

I had the intention in this book to show you a very powerful concept for dealing with a huge variety of problems in a logical way. I have not yet found too many problems that cannot be dealt with by analyzing them in terms of effort and flow variables, and the relationships between these variables. So, by now, you should have some appreciation for this approach.

But life is more than logic and solving problems. Some problems can be left unsolved; some mysteries can remain unknown; some things need to be appreciated on a primal level without undue analysis.

Engineering students sometimes wonder why they are required to take courses in history, art appreciation, music, or other nonengineering subjects. It is precisely because they are not engineering or science courses that engineers should attend them, and attend them with gusto.

So, don't be afraid to be the best engineer that you can possibly be, but also listen to fine music, go to an art museum, stand on a street corner and watch people pass by, take off your shoes and walk in the mud, go to a fine restaurant and order exotic food, hug a young child, be active in your church, learn to dance, smile a lot in unexpected places, feel wonder when an airplane flies overhead, debate politics and religion with your friends, read *Les Miserables* and *Upon This Rock*, play sports just for the fun of it, watch a Road Runner cartoon followed by a Star Trek episode, carve a duck decoy, restore an old automobile, sing in the shower, volunteer to help nursing home patients, put a $20 bill in the Salvation Army pot, invite someone over, and grow flowers. After all, these are hyacinths to feed thy soul.

REFERENCES

Canfield, J., M. V. Hansen, M. Rogerson, M. Rutte, and T. Clauss, 1996, *Chicken Soup for the Soul at Work*, Health Communications, Deerfield Beach, FL.

Dewhurst, D. J., 1991, *On the Real Axis*, International Federation for Medical and Biological Engineering, Amsterdam.

Murphy, W. F., 1987, *Upon This Rock*, Macmillan, New York.

INDEX

Printed in the United States
69141LVS00001B/6